Environmental Technologies to Treat Selenium Pollution: Principles and Engineering

Environmental Technologies to Treat Selenium Pollution: Principles and Engineering

Edited by

Piet N. L. Lens and Kannan Pakshirajan

IWA PUBLISHING

Published by **IWA Publishing**
Unit 104–105, Export Building
1 Clove Crescent
London E14 2BA, UK
Telephone: +44 (0)20 7654 5500
Fax: +44 (0)20 7654 5555
Email: publications@iwap.co.uk
Web: www.iwapublishing.com

First published 2021

British Library Cataloguing in Publication Data
A CIP catalogue record for this book is available from the British Library

ISBN: 9781789061048 (paperback)
ISBN: 9781789061055 (eBook)

This eBook was made Open Access in August 2021.

Cover photographs

Selenium reducing biomass in anaerobic granular sludge or attached onto surfaces of various laboratory scale reactors fed with selenite or selenate. **Top right:** Red color formation showing the presence of elemental Se(0) due to reduction of Se-oxyanions, while dark grey colour showing the formation of PbSe due to the interaction of Pb(II) with selenide (HSe^-) after the bioreduction of Se-oxyanions. Photograph of data published in Mal *et al.* (2021) *Journal of Hazardous Waste* (In Press). **Central:** Architectural image of a bacterial biofilm exposed to 0.1 mM selenate captured using a Leica TCS-SP2 acousto-optical beam splitter confocal laser scanning microscope stained with SYTO 9 (green) and propidium iodide (red). Photograph of data published in Tan *et al.* (2018) *Journal of Chemical Technology and Biotechnology* **93**(8), 2380–2389. **Bottom left:** Cathodic selenium recovery in selenite fed dual-chambered bioelectrochemical system. Photograph of data published in Shanthi Sravan *et al.* (2020) *Journal of Hazardous Materials* **399**, 122843. **Bottom right:** Study of the selenium removal efficiency, biofilm structure and microbial community structure in biofilms growing in selenate, sulfate and nitrate rich wastewater in six-panel drip flow reactors, i.e. plug flow systems utilizing low-shear, laminar flow. Photograph of data published in Tan *et al.* (2018) *Journal of Chemical Technology and Biotechnology* **93**(8), 2380–2389.

Contents

Preface

Selenium is a naturally occurring yet scarce element with significant importance in health, electrical and manufacturing sectors. However, owing to its toxicity at a high concentration it is considered a double-edged sword. Selenium contamination in soil and water is a serious environmental concern, and it is attributed to growing industrial activities, in particular metal refining, nuclear and mining-related activities. Hence, with advancements in strict water quality requirements and pollution monitoring, the discharge limit of selenium in water is set as 5 µg/L. Several chemical and biological treatment strategies have been developed to treat selenium-contaminated water, and some of these techniques can even be applied to recover and reuse selenium for different applications.

Despite rigorous research in selenium removal/reduction being carried out at the laboratory scale as well as in the field, often understanding of the selenium cycle or of the mechanism of selenium reduction remains unclear for improving the treatment process efficiency. Hence, the first part of this book opens up by giving a clear account of the chemistry and biochemical mechanisms involved in selenium reduction. The second part of the book deals with various remediation technologies, including chemical and biological treatment techniques. Upcoming treatment techniques using plant systems are also presented in this book. Because of a strong history of selenium for its impact on the health sector, the book covers soil pollution and biofortification of agricultural crops with selenium. Finally, the synthesis of biologically-reduced selenium in the form of nanoparticles and quantum dots and its potential application is presented. Thus, the present book introduces its readers to the sources of selenium emissions up to the recovery of selenium based value-added products from wastewaters.

doi: 10.2166/9781789061055_xv

Each contributed chapter is presented on a stand-alone basis, so that the reader will find it helpful to consider only the theme of each chapter. There are nevertheless many connections between what may at first seem to be quite different topics. As in all the books of the *Integrated Environmental Technology* series, it was one of our purposes to draw out and emphasize these interdisciplinary linkages. For this reason, a comprehensive index is included to facilitate cross-referencing. We hope that the work described in this book will inspire those working in the field and will encourage those who are beginning to investigate this field.

We wish to thank all contributors to this book for their valuable contributions by sharing their expertise in the form of the various chapters included in the book. We also thank all past and present co-workers as well as all collaborators who joined in unraveling parts of the selenium cycle and its application in environmental technology as described in this book, especially those at Wageningen University, UNESCO-IHE, Indian Institute of Technology Guwahati and National University Ireland Galway. In addition, the national and international granting agencies who supported this work on the selenium cycle over the years are gratefully acknowledged, in particular the Science Foundation Ireland (SFI), who financially supported the open access publication of this book volume through the SFI Research Professorship Programme *Innovative Energy Technologies for Biofuels, Bioenergy and a Sustainable Irish Bioeconomy* (IETSBIO^3; grant number 15/RP/2763). We are also grateful to the editorial team of IWA Publishing, in particular Niall Cunniffe and Mark Hammond, for their help and editorial support in realizing this book.

Piet N. L. Lens
National University of Ireland Galway, Ireland

Kannan Pakshirajan
Indian Institute of Technology Guwahati, India

List of Contributors

Boltz Joshua P.
Biodesign Swette Center for
Environmental Biotechnology
Arizona State University
P.O. Box 85287-5701
Tempe, AZ 85287-5701
U.S.A.

Cappelletti Martina
Department of Pharmacy and
BioTechnology (FaBiT)
University of Bologna
Via Irnerio 42
40126 Bologna
Italy

Dinelli Enrico
Department of Biological, Geological and
Environmental Sciences (BiGeA)
University of Bologna
Piazza di Porta San Donato 1
40126 Bologna
Italy

Du Laing Gijs
Laboratory of Analytical Chemistry and
Applied Ecochemistry
Faculty of Bioscience Engineering
Ghent University
Coupure Links 653
9000 Gent
Belgium

Eiche Elisabeth
Laboratory for Environmental and Raw
Materials Analysis
Institute of Applied Geosciences
Karlsruhe Institute of Technology (KIT)
Adenauerring 20b
76131 Karlsruhe
Germany

Ivet Ferrer
GEMMA – Group of Environmental
Engineering and Microbiology
Department of Civil and Environmental
Engineering
Universitat Politecnica de
Catalunya·BarcelonaTech
c/ Jordi Girona 1-3,
Building D1
08034 Barcelona
Spain

Funari Valerio
Department of Earth System Sciences and Environmental Technologies
National Research Council,
ISMAR-CNR Bologna Research Area
Via P.Gobetti 101
40129 Bologna
Italy

Gasparotto Giorgio
Department of Biological, Geological and Environmental Sciences (BiGeA)
University of Bologna
Piazza di Porta San Donato 1
40126 Bologna
Italy

Gil-Díaz Teba
Institute of Geosciences
Friedrich-Schiller-Universität Jena
Burgweg 11
07749 Jena
Germany

Guan Xiaohong
State Key Laboratory of Pollution Control and Resources Reuse
College of Environmental Science and Engineering
Tongji University
Room 1801, Building 10,
1683# Zhoujiazui Street
Yangpu District
Shanghai 200092
China

Heberling Frank
Institute for Nuclear Waste Disposal
Karlsruhe Institute of Technology (KIT)
Hermann-von-Helmholtz-Platz 1
76344 Eggenstein-Leopoldshafen
Germany

Lampis Silvia
Department of Biotechnology
University of Verona
Strada le Grazie 15
37134 Verona
Italy

Lens Piet N. L.
Department of Microbiology
National University of Ireland Galway
University Road
H91 TK33 Galway
Ireland

Li Jinxiang
Key Laboratory of Chemical Pollution Control and Resources Reuse
School of Environmental and Biological Engineering
Nanjing University of Science and Technology
Room 316, Building 105B
200# Xiaolingwei Street
Xuanwu District
Nanjing 210094
China

Li Jun
Laboratory of Analytical Chemistry and Applied Ecochemistry
Faculty of Bioscience Engineering
Ghent University
Coupure Links 653
9000 Gent
Belgium

Lima Leonardo Warzea
Biology Department
Colorado State University
Fort Collins CO
U.S.A.

Pakshirajan Kannan
Department of Biosciences and Bioengineering
Indian Institute of Technology Guwahati
Guwahati 781039
Assam
India

Piacenza Elena
Department of Biological, Chemical, and Pharmaceutical Sciences and Technologies
University of Palermo
Via Giovanni Pascoli 6
90133 Palermo PA
Italy

Presentato Alessandro
Department of Biological, Chemical, and Pharmaceutical Sciences and Technologies
University of Palermo
Via Giovanni Pascoli 6
90133 Palermo PA
Italy

Raj Mohan Balakrishnan
Department of Chemical Engineering
National Institute of Technology, Karnataka
Surathkal 575025
India

Rittmann Bruce E.
Biodesign Swette Center for Environmental Biotechnology
Arizona State University
P.O. Box 85287-5701
Tempe, AZ 85287-5701
U.S.A.

Schiavon Michela
Department of Agricultural, Forestry and Food Sciences (DISAFA)
University of Turin
Largo Paolo Braccini 2 (già Via Leonardo da Vinci 44)
10095 Grugliasco (TO)
Italy

Simm Robert
Stantec
3133 West Frye Road, Suite 300
Chandler, Arizona 85226
U.S.A.

Sun Yuankui
School of Ecological and Environmental Sciences
East China Normal University
Room 206, Zihuan Building
500# Dongchuan Road
Shanghai, 200241
China

Turner Raymond J.
Microbial Biochemistry Laboratory
Department of Biological Sciences
Faculty of Science
University of Calgary
Calgary
Canada

Vallini Giovanni
Department of Biotechnology
University of Verona
Strada le Grazie 15
37134 Verona
Italy

Vrushali Vinayak Kadam
Department of Chemical Engineering
National Institute of Technology, Karnataka
Surathkal 575025
India

Wadgaonkar Shrutika L.
UNESCO-IHE Institute for Water Education
Westvest 7
2611 AX Delft
The Netherlands

Zannoni Davide
Department of Pharmacy and BioTechnology (FaBiT)
University of Bologna
Via Irnerio 42
40126 Bologna
Italy

Part I

The Selenium Cycle

Chapter 1

Selenium in the environment

Martina Cappelletti, Valerio Funari, Giorgio Gasparotto, Enrico Dinelli and Davide Zannoni

1.1 INTRODUCTION

Chapter 1 is intended to provide the essential background information to assist operators and researchers who use technologies for the treatment of selenium (Se) in environmental (soils, waters, sediments, air) and biological (foods) samples in addition to analysis of Se-molecular speciation in living organisms (plants, microorganisms, animals and humans). In this regard, it is important to underline that numerous reviews and reports have been published in the last decade on this and related topics (Hosnedlova *et al.*, 2018; Khurana *et al.*, 2019; Nancharaiah & Lens, 2015a, b; Preda *et al.*, 2015; Tan *et al.*, 2016; Winkel *et al.*, 2012). Therefore, all readers are kindly invited to refer to the recent literature on the subject – in addition to chapters 2 to 12 of this volume – for complete information on the role and importance of selenium in nature.

1.1.1 Historical background

Selenium ($^{34}{}_{79}Se$) is a chemical element of great importance for living beings, which was originally mistaken for tellurium (Te) until the Swedish chemist Jacob Berzelius (Berzelius, 1818) recognized it as a distinct element in 1817 (Trofast, 2011). Since Se was closely related to Te, which was named after

doi: 10.2166/9781789061055_0003

the Latin word for earth, *tellus* (Klaproth, 1798), Berzelius chose to name this new element selenium, in reference to *selene*, the Greek name for the moon goddess.

Selenium has a long history of being misunderstood since it has repeatedly been mistakenly considered as a strong toxicant, mainly due to inadequate analytical data. Indeed, the crucial role of selenium in the environment and for human beings became evident only in the 1930s, but it was exclusively associated with toxicity – selenosis – that is the result of excessive Se exposure. An example that illustrates the wrong reputation attributed to selenium as a notorious environmental toxin is the case of two livestock diseases, that is, alkali disease and blind staggers, which were initially fully attributed to Se. There are certain types of plants in western North America – commonly known as locoweed and/or crazyweed – which accumulate high amounts of Se that could have been the cause of severe neurological damage to the animals grazing on these lands. Locoweed is a general name for plants that produce swainsonine, a phytotoxin harmful to livestock mostly present in three plant genera of the Fabaceae's family: *Swainsona, Oxytropis* and *Astragalus* (see Section 1.3.3.4 of this chapter). Livestock poisoned by chronic ingestion of large amounts of swainsonine develop a medical condition known as locoism (also swainsonine disease) reported most often in cattle, sheep, and horses. Some locoweed species, including a few that produce swainsonine, accumulate selenium, and this has led to confusion between swainsonine poisoning and selenium poisoning. Even though Se did not cause as many poisoning episodes as were attributed to it, it is now well established that acute or chronically high Se exposure in food or water certainly causes debilitating symptoms and lethality. On the other hand, selenium's bad reputation is unfortunate because most of its roles in the environment are beneficial rather than harmful, and there are far more areas around the world where deficiency is a problem than regions where selenium excess is an issue (Combs, 2001; see also Section 1.1.3). The famous dictum of Philippus Theophrastus Aureolus Bombastus von Hohenheim, the German physician, universally known as Paracelsus, that "*...solely the dose determines what is or it is not a poison...*" seems particularly suitable for describing the role and effects of selenium on living organisms.

1.1.2 The rising of interest in selenium research

Research on selenium has undergone a surge in recent years following new data and discoveries on the importance of this metalloid in the fields of medical biotechnology and human health, but not only in these fields. In fact, selenium has a wide use in electronic applications, solar cells, the glass industry, photocopying, the cosmetic industry and as a dietary supplement (Bodnar *et al.*,

2016; Naumov, 2010) thus to be considered a high commercial value element. In agriculture, for example, selenium is applied as fertilizer in Se-deficient soils (Lavu *et al.*, 2012) and as an inhibitor of sulfate – mediated by toxic methyl-mercury production in paddy fields where the addition of selenium (in the form of selenite or selenate) removes free mercury in the soil generating stable Hg-Se complexes (Wang *et al.*, 2016b).

Estimates for world consumption of selenium are in the order of 40% in metallurgy, 25% in glass manufacturing, 10% in agriculture, 10% in chemicals and pigments, 10% in electronics, and 5% in other uses with its global annual production estimated at approximately 2800 tons in 2018 with China, Japan, Germany and Belgium being the major producers (U.S. Geological Survey, 2019) (see details in Section 1.2.3). Electrolytic manganese production is the main metallurgical end-use for selenium in China, where selenium dioxide is used in the electrolytic process to increase current efficiency and the metal deposition rate. Selenium consumption in China has increased in recent years and 51 electrolytic manganese producers were reported to have been operating and consuming selenium in 2017 (the latest information available), up from 47 reported in 2016 (U.S. Geological Survey, 2019). As a matter of fact, while recovery of selenium from anode slime formed as a by-product of metal refining may prove insufficient to meet the global demand of selenium in the future, there is need to reuse and recover selenium from soil/water habitats having an excess of selenium (Oldfield, 2002) using novel and hopefully non-invasive/environmentally compatible approaches to replenish the selenium resources (Nancharaiah & Lens, 2015a, b; Tan *et al.*, 2016) (see Section 1.2.3 of this chapter along with Chapters 6 and 7 of this book).

In addition to its use in industry and agriculture, selenium has a fundamental role in the metabolism of animals and humans (Fairweather-Tait *et al.*, 2010; Huawei, 2009; Papp *et al.*, 2007). To avoid both the lack and excessive intake of selenium, the World Health Organization (WHO, 2011) recommended a dietary allowance ranging from a minimum of 40 to a maximum of 400 μg of Se per day. Indeed, selenium deficiency and toxicity depend strictly on the location and environmental conditions. There are extreme conditions of Se concentration, that is, lack or excess of Se in the soil, only in specific areas of the planet (El-Ramady *et al.*, 2014a, b), such as China, India, the Middle-East and some European countries (e.g., Poland and western Russia), which are known for selenium deficiency in soils which was associated with heart disease and bone disorders (Lemly, 1999). One of the most striking examples is the so called Keshan's disease, a cardiomyopathy caused by a combination of dietary deficiency of selenium and the presence of a mutated strain of the Coxsackie-virus, named after Keshan County of Heilongjiang province in northeast China where symptoms were first noted. These symptoms were later found prevalent in a wide belt extending from north-east to south-west China, all due to selenium-deficient soil (Beck *et al.*, 2003; Ren *et al.*, 2004). Similarly, the disease known as Kashin-Beck, named after the two Russian military physicians N.I. Kashin and E.V. Beck who described it for the first time in regions of south-east Siberia bordering China, is an endemic osteo-arthropathy which is associated with the consumption of grains with a very low selenium content (Fordyce, 2005).

Although there are several million people in the world potentially subjected to selenium deficiency, it is equally true that many sanitary problems due to selenium excess have been recorded in various regions of the planet as well (El- Ramady *et al.*, 2014a). Indeed, selenium tends to bioaccumulate in the aquatic environment and becomes toxic to fish and birds that feed on fish containing elevated levels of this element (De La Riva *et al.*, 2014; Miller *et al.*, 2013; Zhang *et al.*, 2008). Further, its transfer to the food chain has caused additional toxic effects in humans and animals (selenosis) such as nausea, vomiting, dermal and neurological dysfunction or even paralysis and cardiovascular symptoms (De La Riva *et al.*, 2014; Duntas & Benvenga, 2014). Issues related to selenium toxicity are covered in more detail in Chapters 5, 6 and 7 of this book.

1.1.3 Microbial processing of selenium

As also reported in the following Section 1.2, selenium is found in four inorganic oxidation states (−2, 0, +4, +6). Elemental selenium Se(0) is highly insoluble, relatively non-toxic to humans and it occurs as a prevalent species under anoxic conditions (Barceloux, 1999), while selenide Se(−2) is both highly reactive and toxic although rapidly oxidized to Se(0) through inorganic and/or biochemical reactions (Turner *et al.*, 1998). In this respect, selenium bioprocessing in the environment is catalysed by Se-metabolizing bacteria which are able to either reduce or oxidize it, but also to generate methylated and volatile derivatives such as dimethylselenide (CH_3SeCH_3), dimethyl selenyl sulfide (CH_3SeSCH_3) and dimethyl diselenide ($CH_3SeSeCH_3$) (Chasteen & Bentley, 2003). The most oxidized selenium forms, selenite (SeO_3^{2-}) and selenate (SeO_4^{2-}), can be taken up by several species of microorganisms and reduce these oxyanions to less toxic elemental selenium, Se(0). Representative genera include *Wolinella*, *Pseudomonas*, *Sulfurospirillum*, *Enterobacter*, *Thaurea*, *Bacillus*, *Citrobacter*, *Azospirillum*, *Paenibacillus*, *Rhizobium*, *Stenotrophomonas* and *Rhodococcus* (Antonioli *et al.*, 2007; Hunter *et al.*, 2007; Presentato *et al.*, 2018; Sidique *et al.*, 2006; Tugarova *et al.*, 2014; Yao *et al.*, 2014; Zhang & Frankenberger, 2005). Deposition of Se(0) as particles of variable shape and size may occur in the extracellular milieu but also in the cytosolic and/or the periplasmic space in association with the cell wall or membrane (Gerrard *et al.*, 1974; Klonowska *et al.*, 2005; Losi &

Frankenberger, 1997; Oremland *et al.*, 2004; Presentato *et al.*, 2018) [see Chapters 3 and 10 for a complete discussion on this topic].

Several microorganisms have the ability to incorporate selenium derived from selenate or selenite into organo- Se compounds as selenocysteine and selenomethionine via the mechanism known as assimilatory Se-reduction (Lenz & Lens, 2009; Sarret *et al.*, 2005). Figure 1.1 shows how bacteria determine the distribution of the main Se oxidation- reduction forms [elemental selenium Se(0), selenite Se(+4) and selenate Se(+6)] in a contaminated habitat as a consequence of industrial and agricultural activities.

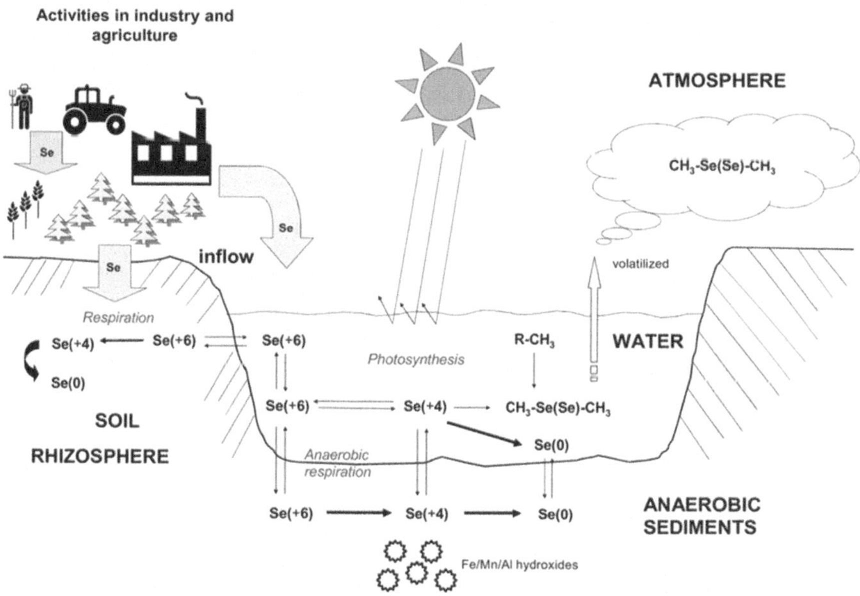

Figure 1.1 Schematic diagram showing the metabolic processes (*respiration, anaerobic respiration, photosynthesis*) by which bacteria determine the distribution of the main selenium oxidation-reduction forms: elemental selenium Se(0), selenite Se(+4), and selenate Se(+6) in a contaminated habitat as a consequence of activities in industry and agriculture. Natural processes such as weathering, soil leaching and volcanic emissions releasing selenium into the environment, are not shown. The scheme depicts four environmental phases: atmosphere, soil-rhizosphere, anaerobic soil-sediments and water. Symbols and abbreviations: Se, selenium element; R-CH_3, organic methyl groups; and the cloud, volatile CH_3-Se(Se)-CH_3. The thickness of the arrows indicates the prevalent bacterial activity; rough circles symbolize the binding of selenite, mainly Se(+4), to soil/sediment particles of iron (Fe), manganese (Mn) or aluminium (Al) hydroxides. (Modified from Sharma *et al.*, 2015 and Zannoni *et al.*, 2008; see text for further details).

1.2 SELENIUM CHEMISTRY

1.2.1 Chemical features of selenium

Selenium belongs to Group VIA of the Periodic Table of the elements and is located between sulfur and tellurium. Its chemical and physical properties are intermediate between those of a metal and a non-metal. Its atomic number is 34 and its atomic mass is 78.96. Selenium has six stable isotopes, namely: ^{74}Se [0.87%], ^{76}Se [9.02%], ^{77}Se [7.58%], ^{78}Se [23.52%], ^{80}Se [49.82%], and ^{82}Se [9.19%]; the relative abundances of these isotopes are indicated in brackets (Hoffman & King, 1997).

Selenium commonly exists in four valence states: −2, 0, +4, and +6, the −2 form being more common for organic compounds. Elemental selenium has a melting point of 217°C and a boiling point of 685°C. Between 74 and 217°C, exothermic modifications of crystalline forms are observed (Langner, 2000). Pure selenium, like sulfur, is allotropic and exists in three forms, including (1) a grey or 'metallic' thermodynamically stable hexagonal form displaying properties of a semiconductor, (2) a red monoclinic form, and (3) a vitreous form.

1.2.2 *Pourbaix* diagram of selenium in water

The behaviour of selenium in aqueous solution is dependent on redox state and pH, as described by the Eh-pH diagrams of Figure 1.2. Selenium is highly mobile under oxidizing, acidic, neutral and alkaline conditions mostly as selenate and selenite. In very oxidizing conditions the Se^{6+} is stable, and selenate [SeO_4^{2-}] is the dominant form in solution under a wide range of pHs, while under very acidic conditions biselenate [$HSeO_4^-$] is the common form. Under less oxidizing conditions, Se^{4+} is stable and can be present in solution as selenite [SeO_3^{2-}] at neutral to alkaline conditions or biselenite [$HSeO_3^-$] at pH ranging

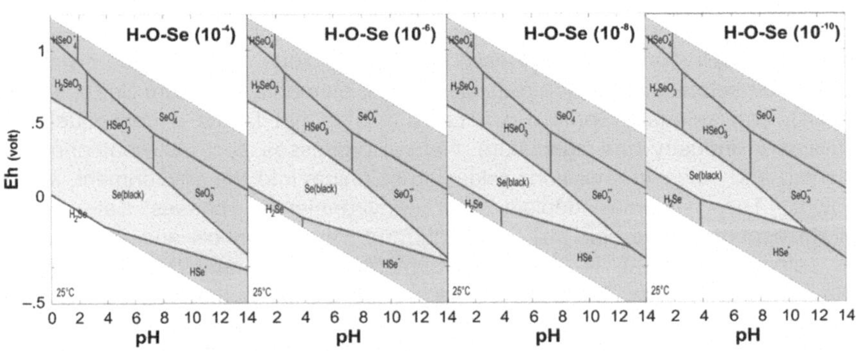

Figure 1.2 Eh-pH diagrams (*Pourbaix* diagrams) for the system Se-O-H at different selenium activities.

from neutral to acidic. At lower pH, selenous acid [H_2SO_3] is stable. Selenium mobility decreases in reducing conditions, at neutral to acidic pH, due to the formation of elemental Se whose stability field increases as dissolved selenium activities increase (Figure 1.2). Under reducing conditions, Se tends to form selenide ions that react with metals eventually present in solution, forming stable selenide minerals from aqueous solution even at low metal activities. As suggested by the Eh-pH diagrams, for example, involving Pb (lead) and Ag (silver), reducing conditions lead to the formation of clausthalite [PbSe] or naumannite [Ag_2Se] (Figure 1.3) and possibly other selenides in combination with Fe (iron), Co (cobalt), Ni (nickel) and Cu (copper). Under reducing conditions, Se also tends to replace sulfur in more common minerals such as pyrite, chalcopyrite, pyrrhotite and sphalerite.

The behaviour of selenium in solution is mostly as complexed oxyanion which is strongly influenced by the presence of natural particles of aluminium and iron oxy-hydroxides. These phases, in turn, with their high surface areas and point of zero charge (Chan *et al.*, 2009; Sparks, 2003) might adsorb dissolved selenium species, although it is reported that selenite is less prone to surface adsorption (Christophersen *et al.*, 2013; Meher *et al.*, 2020) due to its high tendency to remain in the aqueous phase. It must be stressed that selenite has been found to be more toxic than selenate and organo-selenium compounds due to its higher solubility and bioavailability (Torres *et al.*, 2011).

The redox chemistry of selenium is complex, given the importance of the biotic transformations involved in its cycling (Lenz & Lens, 2009) and with a range of reactions depending on the environmental conditions (assimilatory and dissimilatory reduction, alkylation, dealkylation and oxidation).

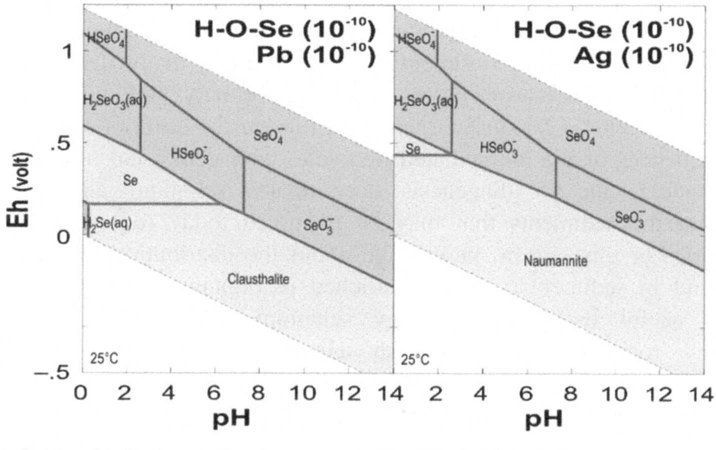

Figure 1.3 Eh-pH diagrams for the system Se-Pb-O-H and Se-Ag-O-H.

1.2.3 Global uses of selenium

The selenium content in the Earth's crust averages 0.05–0.09 mg/kg, approximating the typical concentration of cadmium and antimony and ranking above silver, mercury, and uranium. Higher concentrations of selenium are found in volcanic rocks, sandstones, uranium deposits and organic matter-rich rocks. However, the natural occurrence of these minerals can be rarely considered high-grade ore deposits. Although present in many minerals, selenium is generally recovered as a mining by-product from deposits mined for other elements, such as copper, lead, and silver (Langner, 2000). Therefore, selenium is mainly sourced from sulfidic ores in the context of primary ore mining, even if the mean Se/S ratio of magmatic sulfides is rather low ($\approx$0.1). Its principal supply comes from sulfide ores in Canada, the USA, Bolivia, and Russia.

Commercially exploited Se-ore deposits have been referred to four types of mineralization (Stillings, 2017), namely: magmatic, volcanic, hydrothermal, and exogenic. The latter is secondary mineral ore produced by chemical and physical weathering. Many of the secondary mineral sources are high in Se under peculiar conditions (e.g., volcanic soils, and marine shale) owing to erosion transport and deposition from seleniferous parent material or atmospheric selenium supplies. The occurrence of selenium depends upon the composition of the ore, the stratigraphic position of ore minerals, and tectonic settings (Figure 1.4).

Selenium enrichment has been observed in marine samples collected from black smokers and iron-copper massive sulfides associated with the East Pacific Rise. Here, Se-rich mineralization primarily occurs in the inner part of the deposit in equilibrium with hydrothermal fluids. On the other hand, clear examples of exogenic deposits are marine phosphorites that may contain around 300 mg/kg Se. Marine and sub-aerial highly organic shales contain as much as 1500 mg/kg Se because it may accumulate in detrital organic matter (Parnell *et al.*, 2016). High Se concentrations in sedimentary bedrock units likely derive from their depositional origin in marine basins with high primary biological productivity (Presser *et al.*, 2004). As such, organic detritus on the seafloor may experience the consequences of the biomediated Se cycle: with geological time, the loss of organic matter due to diagenesis accumulates selenium and other trace elements in the sediments that may be exhumed today (e.g., the Phosphoria Formation). For this reason, knowledge about the distribution of selenium-rich rocks found in sedimentary basins enriched in organic carbon (Presser *et al.*, 2004) is useful in identifying new selenium resources (Stillings, 2017). Carbon-rich rocks usually show high selenium concentrations owing to the possible selenium substitution for sulfur during diagenesis. Natural resources for potential Se cultivation are coal, marine phosphate deposits, and shales (Presser, 1999) (Figure 1.4).

When the host mineral is selenide, the ore deposits can be grouped into four types:

(a) telethermal selenide vein-type
(b) unconformity-related uranium deposits
(c) sandstone-hosted uranium deposits (roll-front type)
(d) gold-silver epithermal volcanic-hosted deposits

Examples of type (a) deposits are the Tilkerode-Zorge-Lerbach deposit in the Harz Mountains (Germany), the Pacajake and El Dragon deposits in Bolivia, Hope's Nose deposit in the United Kingdom, and the Sierra de Umango and Sierra de Cacheuta deposits in Argentina. In these deposits, selenides occur in veins hosted in sedimentary rocks, associated with carbonate minerals, hematite, and rare sulfides (e.g., those containing precious metals, gold and platinum group elements). Type (b) ores are found in veins of the Massif Central (France), the Bohemian Massif in the Czech Republic, and the Athabasca area of northern Saskatchewan in Canada, associated with sulfides, arsenides, hematite, carbonates, and uraninite minerals in Co-Cu-Ni ores. The occurrence of type (c) deposits is from uranium and vanadium sandstone ore deposits in the USA (Colorado Plateau, Oklahoma, Wyoming, and Texas) where Se spots locally at elevated concentrations (from thousands of mg/kg up to 3 weight per cent) were measured in uranium deposits of the Colorado Plateau (Coleman & Delevaux, 1957). Remarkable Se concentrations can be found in Fe-rich selenides (e.g., ferroselite), clausthalite (PbSe), and native selenium. Mineralization of economic importance occurred with other redox-active minerals in dynamic equilibrium between oxidizing and reducing environments. Hatten Howard (1977) hypothesized that continued cycling of oxidative dissolution (of reduced minerals in the unaltered rock) and redeposition (at the redox interface) could concentrate selenium, iron, and the uranium-ore minerals within the Fe-S-Se system. Selenium type (d) deposits occur at low-sulfidation epithermal gold-silver deposits (John, 2001) like those of Japan, Nevada (USA), New Zealand, Mexico, and Russia. In these ore deposits, significant Se concentrations are in silver selenides and, less commonly, in bismuth, copper, antimony, and lead selenides, all associated with sulfides, sulfosalts, gold and native elements. Primary geogenic Se ores have shown conditions that favour the formation of sulfides rather than selenides.

The average selenium content in magmatic and associated hydrothermal sulfides is 0.02%. The selenium estimated content is based on copper production figures: in porphyry copper ore with 0.5% copper, the selenium content is 2.5 mg/kg, or a copper-to-selenium ratio of 2000:1. For many years, selenium production was estimated based on copper production using a factor of 0.0215% or, for very productive ores such as for example the Sudbury basin of the Ontario region, 0.064% (George & Wagner, 2009). Selenium can be present in absolute concentrations of hundreds mg/kg in tuffaceous rocks, and several g per weight

colour scale
Se production by country
Mg/y
10
150
300
700
n.a.
10 Se-ore deposit
Petroleum basin (Presser et al., 2004)
Marine phosphate deposit (Presser et al., 2004)
Seafloor massive sulfides (Boschen et al., 2013)

Figure 1.4 Global map of active Se-ore deposits and predictive map of selenium sources associated with organic-rich depositional marine basins (modified from Presser *et al.*, 2004) and seafloor sulfide deposits (modified from Boschen *et al.*, 2013). Borders and country boundaries are from www.thematicmapping.org; Selenium deposits location are from www.mindat.org (Butterman & Brown, 2004; Kogel, 2006; Stillings, 2017). Se-ore deposits by geographic area – as labelled by numbers in figure – are the following: **Africa**: 33. Katanga deposit, Katanga province, Kinshasa, Democratic Republic of the Congo; 34. Shinkolobswe deposit, Democratic Republic of the Congo – **Eastern world (including the Middle East)**: 3. Agarak deposit, Armenia; 4. Dastakert deposit, Armenia; 5. Dzhindara deposit, Armenia; 6. Kadzharan deposit, Armenia; 7. Miskhana deposit, Armenia; 16. Yutangba Se deposit, Hubei, Central China, China; 21. Redzhang Lebong deposit, Sumatra, Indonesia; 22. Salida and Redzhang Lobong, Indonesia; 24. Hishikari, Japan; 25. Jordan Areva Resources, Amman Governorate, Jordan; 59. Murgul deposit, Turkey – **Europe**: 17. Bohemian Massif of the Czech Republic, Czechia; 18. Massif Central of France, France; 19. Eisenberg, Tilkerode-Zorge-Lerbach deposit in the Harz Mountains, near Korbach, Germany; 23. Lipari Islands, Italy; 53. Glava, Grusen, Moberg, Tinnsjå, Tjøstøltlaten, Sweden and Norway; 54. Skrikerum deposit, Småland, Sweden; 55. Idrija, Slovenia; 56. Almaden, Ciudad Real Province, Spain; 57. Providencia Mine, León Province, Spain; 58. Los Reales, Estepona, Málaga Province, Spain; 60. Traour ar Ru, Bahulien, Lannion, Côtes-d'Armor, Brittany Province, France – **Latin America**: 1. Cacheuta Mine, Cerro and Sierra de Cacheuta, Luján de Cuyo, Mendoza Province, Argentina; 2. Sierra de Umango, La Rioja Province, Argentina; 10. Pacajake Selenium Mine, Llallagua, Bolivia; 11. El Dragón mine, Antonio Quijarro Province, Potosí, Bolivia; 26. San Onofre mine, El Doctor, Queretaro, Mexico; 27. El Oro, Mexico; 28. Guanajuato, Mexico; 31. Cerro Negro volcano, León Department, Nicaragua; 32. La Oroya Complex, Peru – **North America**: 12. Sudbury basin, Ontario, Canada; 13. Lake Athabasca area of northern Saskatchewan, Canada; 14. Gaspé Copper mine, Murdochville, La Côte-de-Gaspé, Gaspésie-Îles-de-la-Madeleine, Québec, Canada; 15. Tombstone Mountains, Dawson mining district, Yukon, Canada; 20. Hawaiian Islands, Hawaii, USA; 61. Comstock and Goldfield, Nevada, USA; 62. Shoshone, Jasper, Riverton deposits, Wyoming, USA; 63. Colorado Plateau, Colorado, USA; 64. The western Great Basin, Nevada, USA; 65. Bristol Bay, Pebble deposit, Southern Brooks Range, Red Dog, USA and Alaska; 66. Duluth Complex, Bab-bitt deposit, Minnesota, USA; 67. Humboldt County, National District, Buckskin Mountain, Nevada, USA – **Oceania**: 8. Queen Victoria Springs Nature Reserve, Menzies Shire, Australia; 9. Milgun Variscite Mine (Mount Deverell), Milgun Station, Meekatharra Shire, Australia; 29. Native sulfur-selenium outcrops, New Zealand; 30. The Great Barrier Island, New Zealand – **Russia**: 35. Upper Seymchanskoye deposit, Verina River basin, Far Eastern; 36. Uchaley deposit, Urals; 37. Kirovskoye deposit, Krasnoyarsk, Siberia; 38. Bukuka deposit, Siberia; 39. Noril'sk, Siberia; 40. Kurile Islands, additional Kamchatka locations, Far Eastern; 41. Mendeleev Volcano, Kunashir Island, Far Eastern; 42. Mutnovskii Volcano, Kamchatka, Far Eastern; 43. Paramushir Island, Far Eastern; 44. Prasolovskoye, Kunashir Island, Far Eastern; 45. Karabash group, Tatarstan, Volga; 46. Mauk deposit, Urals; 47. Gay deposit, Urals; 48. III International deposit, Urals; 49. Levikha XIV deposit, Urals; 50. Sibay deposit, Urals; 51. Zyuzel'skoye deposit, Far Eastern; 52. Ust' Uyok deposit, Turan District, Tuva, Siberia.

cent (g/100 g) in sulfides that host mercury and antimony deposits. The volcanic deposits of andesite tuffs (e.g., Jasper, Riverton, and Shoshone deposits, USA) yield up to 120 mg/kg selenium while native sulfur deposits can display the highest selenium concentrations (>5% weight). In magmatic sulfide deposits like the Permian-to-Triassic-age Noril'sk group, located inside the Arctic Circle in Russia, and the Sudbury ores in Ontario, selenium may replace sulfur in sulfide veins. The estimated selenium content in such deposits varies between 0.002 and 0.01%, with measured concentrations as high as 200 mg/kg and as low as 3 mg/kg. At Noril'sk, the selenium of extractive interest is in sulfide minerals (e.g., chalcopyrite, millerite, pentlandite, pyrite and pyrrhotite), where it occurs as an isomorphous substitution for sulfur in concentrations of 2 to 74 mg/kg, and can reach up to about 150 mg/kg in massive vein deposits, and sometimes in magnetites. In sulfides of porphyry copper deposits, Se is generally sourced from hypogene fluids and subsequently remobilized during hydrothermal plumbing of the porphyry copper system.

From a mining perspective, no clear quantitative correlation has been found between selenium and any of the associated elements such as copper, lead, nickel, sulfur, or tellurium and precious metals. It is empirically evident that copper ores are a source for selenium while sulfides are prevalent ore minerals, so the close association of selenium with sulfur has a poor significance from the standpoint of absolute concentrations. Besides, selenium can be one of the most mobile elements during rock weathering, and so it can be enriched in a variety of environmental sinks. Fundamental steps in identifying new selenium sources would be to study the relationships between selenium and sulfur and their isotopes in other rock types, to define diagenetic processes of sulfide minerals better, and thereby to probe the interaction of Se-bearing minerals with organic matter and carbon. Despite efforts to standardize the estimation of selenium in sulfide ores, the wide variety of host minerals and, consequently, the highly variable Se concentrations found in ore minerals complicate the definition of Se grades and cut-off values in ore mining.

Selenium is sold in four commercial or refined grades, even though the industry has no specifications for the several classes and some producers provide customers with their specifications. A first, refined grade is 99.5% weight minimum selenium with Te, Fe, Pb and Cu as the main impurities. Materials containing >99.5% selenium are commercially available as powder, granules, and lump. Pigment grade selenium is typically used in colouration and has a purity of 99.7% weight. The high grade ranks as the third grade regarding purity with 99.999% weight minimum selenium. The ultra-high grade is claimed to contain from 99.999 to 99.9999% weight selenium (Hoffman & King, 1997), available as shot or powder. To reach higher purification somewhat inert contaminants (e.g., Na, Mg, Ca, Al, Si) and harmful impurities such as arsenic, iron, mercury, and tellurium must be removed. Generally, the concentration of sulfur, oxygen and halogens must be ensured to be as low as possible for high-purity selenium powder. When estimating elemental purity and grades, any analytical uncertainty linked to accurate selenium quantification should be carefully stated. The suitability of determination methods depends on several factors, such as (i) a proper conversion of selenium after distillation in the coordination compound wanted (that is, suitable for analytical measure), (ii) effective reduction/deposition to elemental selenium when a gravimetric determination is desired, (iii) the presence of possible interfering substances left behind in the sample preparative, (iv) selenium speciation and (v) type of environmental matrices (e.g., organic vs inorganic) in the test sample. With the development of organo-selenium chemistry, an increasing number of organo-selenium compounds have been prepared in the laboratory (mostly as organosulfur analogues): they include the simple alkyl selenide (CHSe) and carbon diselenide (CSe_2) as well as complex heterocyclic compounds and a variety of Se-based polymers thereof. An overview of the main organo-selenium compounds can be found elsewhere (Hoffman & King, 1997). Selenium compounds of strategic importance for the productive industry include selenium dioxide, selenium disulfide, cadmium selenide and sulfoselenide, sodium and other selenites, and some organic selenium compounds (e.g., phosphine selenides are widely used solvents).

The world Se production according to USGS (United States Geological Survey) was estimated around 2710 tons in 2017 (without the USA's quote, assumed around 150–750 tons per year), including China (930 tons per year), Japan (730 tons per year), Russia (150 tons per year), Germany and Belgium (500 tons per year) as main actors among all producers (Stillings, 2017) (Figure 1.4).

The glass industry uses approximately 25% of the total production of selenium. Selenium and its compounds with Na, Cu, Cd, and Te are required mainly to produce flat glass, pressed or blown glass and glassware for colouration and advanced optics in analytical devices (Langner, 2000). A significant percentage of the selenium production (25%) is employed in inorganic pigments (predominantly as cadmium sulfoselenide) used in plastics, paints, enamels, inks, rubber, and ceramics. Around 10–15% of selenium globally produced each year is used in other applications including accelerators, gaseous electric insulation (as selenium hexafluoride), solvents and lubricants, and in rubber production selenides are applied as vulcanizing agents. The remaining selenium globally produced is used in stainless steel and refractory metals, in the electrolytic production of manganese, in medical and pharmaceutical topical preparations (e.g., treatment of dandruff), as fungicide and insecticide in agricultural uses and, finally, it finds application as dietary supplements for humans and livestock. Selenium in the high-purity form is used in electronics due to its semiconductor and photoelectric characteristics, in thermoelements and in xerographic materials. However, applications of selenium in the chemical industry are considered largely dissipative, especially during distillation and recovery from low- concentration sources of primary supply and when required at a high grade.

Following the need to maintain the trade-off between supply and demand, selenium may be used to replace strategic metals (European Commission, 2014) in different applications with satisfactory performance: Se can enter in substitution of Sb in industrial applications for hardening of materials and alloys, it can be used in bimetallic catalysts to decrease rhenium's share of the existing catalyst market, or it can replace tellurium in many free-machining steels and rubber compounding. Some selenides show enhanced performance as high-temperature, high-vacuum lubricants instead of tellurides (e.g., selenides of niobium and tantalum in electrically-conducting solid lubricants). Copper indium diselenide is almost as efficient as cadmium telluride in photovoltaic power cells (Fiducia *et al.*, 2019). The substitutability of tellurium with selenium is particularly chased to lessen the criticality of supply for the much more expensive tellurium. However, this trend may revert soon from both a technical and economic standpoint since selenium is gaining increasing attention due to its critical importance in the production of solar cells (George & Wagner, 2009; Stillings, 2017). On the other hand, silicon is the major substitute for selenium in low- and medium-voltage rectifiers while new organic pigments substitute for cadmium sulfoselenide pigments. Sulfur dioxide can also be used as a replacement for selenium dioxide with limited efficiency. Bismuth and tellurium are rather efficient in place of Se due to their similar chemical properties, whereas lead and cerium oxide more frequently suit as substitutes, especially in the glass-making industry, rubber production and free-machining alloys. Amorphous silicon and cadmium telluride are considered the two principal competitors for replacing Se in thin-film photovoltaic solar cells (George & Wagner, 2009). However, market fluctuations control the supply trend of prime metal vs its substitute.

Refinery production of selenium is reported by 14 countries with more than 130 companies trading selenium, of which Mitsubishi Materials Corp., Nippon Mining and Metals Co. (Japan), Noranda Inc., Falconbridge, the Northwest Nonferrous International Investment Co. (Canada), Jiangxi Copper, Yunnan Copper, Jinduicheng Molybdenum Group (China), Norilsk Nickel, the Ural Mining-Metallurgical Company (Russia), Umicore S.A., Retorte Selenium Chemicals & Metals (Europe), Phelps Dodge Refining Corp. and Rio Tinto Zinc are the most famous. The majority of producers are electrolytic copper refineries and, according to USGS estimates, more than 90% of the USA selenium output and more than 8% of the world Se production is derived from the anode mud deposited during the electrolytic refining of copper. However, as only a fraction of the selenium is recovered from copper anode slimes that can contain between 0.5 and 280 kg of selenium per ton of copper effluent (Stillings, 2017), these slimes represent both a potential loss of Se and a contamination source when discharged as wastewater (Kilic *et al.*, 2013). Due to the close connection to major metal deposits, the production of minor metals like Se depends heavily on the production of the host metal (e.g., Cu), resulting in complex demand/supply and price patterns (Hageluken & Meskers, 2010). The dramatic price volatility of

Se is apparent due to its application for strategic energy technologies (Lenz & Lens, 2009). For example, the selenium price increased by 440% in the USA between 2003 and 2004, and it reached its all-time high in 2011 (146,000 US$ per ton). As such, innovative strategies like those coming from biotechnologies that allow for the recovery of minor metals are required because they can help to uncouple demand/supply and price patterns from major metals extraction. This, in turn, will open novel recycling possibilities, which may have been overlooked for merely technical reasons.

All commercial processes for the production of selenium involve a finely tuned combination of fundamental methods, such as chemical treatments, physical separation methods, thermal treatments, and electrodeposition. Electrowinning is the preferred option at the final stage of recovery, especially from Cu-Ni ores but also from fly dust and slag of metal foundries (Langner, 2000). Recent advances in membrane separation technologies are adding new perspectives for Se recovery from very fine precipitates or different types of physical and chemical phases. Microfiltration (MF), ultrafiltration (UF), nanofiltration (NF) and reverse osmosis (RO) can be applied, with RO and NF being the most effective for recovery of Se fine precipitates (CH2MHILL, 2010; Santos *et al.*, 2015). Soda ash roasting, roasting with sulfuric acid, wet chlorination, and alkaline autoclaving are primary recovery processes for selenium developed at the full scale (Hoffman & King, 1997).

In Europe, recovery of minor by-products like Se have minimal own production infrastructure, low efficiencies and represent a small contribution to revenues, typically less than 5% (Blengini *et al.*, 2017). Secondary selenium is processed by a small number of primary selenium refiners. No secondary selenium sources have in loco production. Most selenium consumed domestically is dissipated into the environment and assumed to be not recoverable. It is also a common belief that selenium from pigments, fertilizers, animal feeds, chemicals, and pharmaceuticals can be dispersed in soils and other environmental sinks without any possibility of avoiding loss of resource. The selenium from the glass and free-machining alloys industries is not accounted for recycling either, because it probably volatilizes during melting operations. However, mining from secondary and unconventional sources is of utmost importance to secure a stable supply and for diversifying sourcing. Superphosphates can contain about 20 mg/kg selenium, while soil Se concentrations can be higher than 20 mg/kg in the vicinity of copper and lead smelters and refineries. Coal generally contains between 0.5 and 12 mg/kg Se while soil samples associated with coal mine environments in Wyoming showed selenium concentrations up to 26 mg/kg (Stillings, 2017).

A high number of Se-bearing products may end up in landfills considering 1 mg/kg Se is the hazardous waste limit (Guo & Wu, 2017; USEPA [United States Environmental Protection Agency], 1992) or in waste incinerators and thermal treatment plants (Liu *et al.*, 2015), where Se likely volatilizes during combustion. In municipal incineration residues, named fly and bottom ashes, Se is in the range 0.05–10 mg/kg (Allegrini *et al.*, 2014). Subsequent studies have suggested a moderate enrichment factor for selenium calculated with respect to crustal abundance that is similar to the enrichments observed in fly ash from coal-fired power plants and refuse incinerators (Kogel, 2006). Although technically feasible, the recovery of selenium from coal fly ash and municipal solid waste incineration residues does not appear economical, and dedicated processes are thus far unidentified. As such, secondary sources for selenium recovery include coal ash, contaminated soils, municipal fly ash and bottom ash, wastewaters and sludge, landfill leachates and other anthropogenic residues. A most important source is inherently Waste Electrical and Electronic Equipment (hereafter: WEEE), such as factory scrap generated during the manufacture of selenium rectifiers, burned-out rectifiers, spent catalysts, used xerography-copying cylinders and heat-generators, outdated sensors, printed circuit boards (especially advanced microdevice graphics cards), relays, resistors (CdSe), and optical stores (George & Wagner, 2009; Schrauzer, 2004). Selenium can also be recovered from plastics and oil waste, treatment plants for phosphorus production, and abundant mining waste dumps experiencing significant selenium releases in some circumstances (Kogel, 2006).

The WEEE showing high selenium contents like scraps from document copiers and laser printers are the focus of urban mining. As an example, the selenium layer from for example, xerographic photocopier drums is either broken up mechanically, then cleaned and remelted; or dissolved in sodium sulfite or other solvents, then precipitated and eventually purified, generating no more than 50 tons of secondary selenium annually according to the USA report of mineral commodity in 2004 (Butterman & Brown, 2004). Other residual streams for practical urban mining are waste solutions from the production of trigonal selenium (used in photosensitive or photoconductive components) where selenium oxide may be in the form of sodium selenite and sodium selenate and for which a patented process may apply (Goodman *et al.*, 2000). Other methods currently under investigation rely on oxidizers/adsorbents such as Al, Fe, Ti, Si oxides (e.g., anatase), manganese oxide, binary metal oxides [Al(+3)/SiO_2, Fe(+3)/SiO_2, Fe(+3) and Mn(+3) hydrous oxides], layered double hydroxides (i.e., nanostructured anionic clays, mostly Mg-Al, Mg-Fe, and Zn-Al based), graphene and magnetic graphene oxide nanocomposites, and even peanut shell and rice husk. Membrane separation technologies are typically deployed for the final recovery stage (Figure 1.5a and b).

Industrial emitters are a compelling hot spot of selenium losses in the atmosphere compared to volcanic emissions and natural releases from plants, soils and animals which are minor sources. Selenium concentrations in air primarily depend on coal-burning power plants, copper-refining plants or selenium rectifier plants, zinc-cadmium smelters, fossil fuel combustion, lime industry, and refuse incineration. From the atmosphere, natural and anthropogenic selenium re-accumulates in environmental compartments (e.g., in lakes) or is taken up by

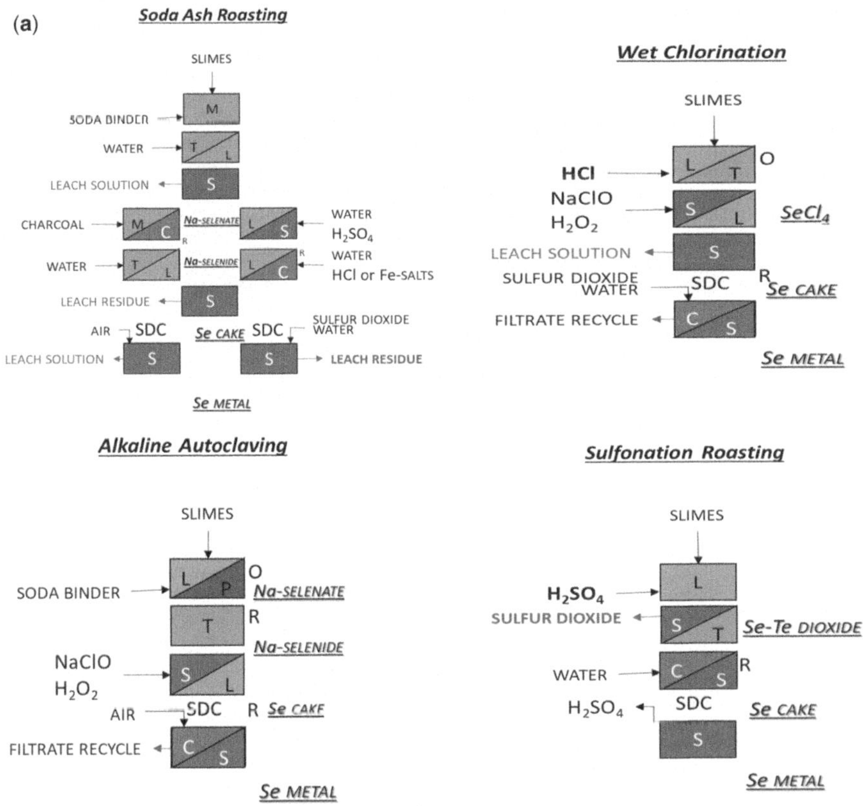

Figure 1.5 (a) Schematic flow designs of selenium recovery processes from primary and secondary ores. Commercial processes (a) and other methods (see Figure 1.5b). The blocks represent critical processing stages, where diverging blocks are possible alternate routes and letters indicate the key process acting: M = mixing, extruding, drying; T = severe thermal treatment (e.g., smelting, roasting); L = chemical leaching; S = physical separation methods; C = coagulation, sorption, flocculation, froth floatation; P = pressure; eC = electro-coagulation; B = bioleaching and biosorption; SDC = stripping, distillation, condensation (mainly) used at the final stage of recovery; MF, NF & RO specify the membrane separation technologies (see text for details); and R and O mean reducing and oxidizing conditions, respectively. Block halves represent co-participated processes. The bold format indicates high inflow/outflow, in red waste or by-products negatively impacting the process, and underlined the selenium fate during the process. *Sources*: Baldwin *et al.* (1985), Hoffman and King (1997), Goodman *et al.* (2000), Kashiwa *et al.* (2000), Overman (2000), Mollah *et al.* (2004), El-Shafey (2007a, b), Geoffroy and Demopoulos (2009, 2011), CH2MHILL (2010), Soda *et al.* (2011), Pickett and Sonstegard (2012), Fu *et al.* (2014), Santos *et al.* (2015), Mierzejewski *et al.* (2017), Stillings (2017).

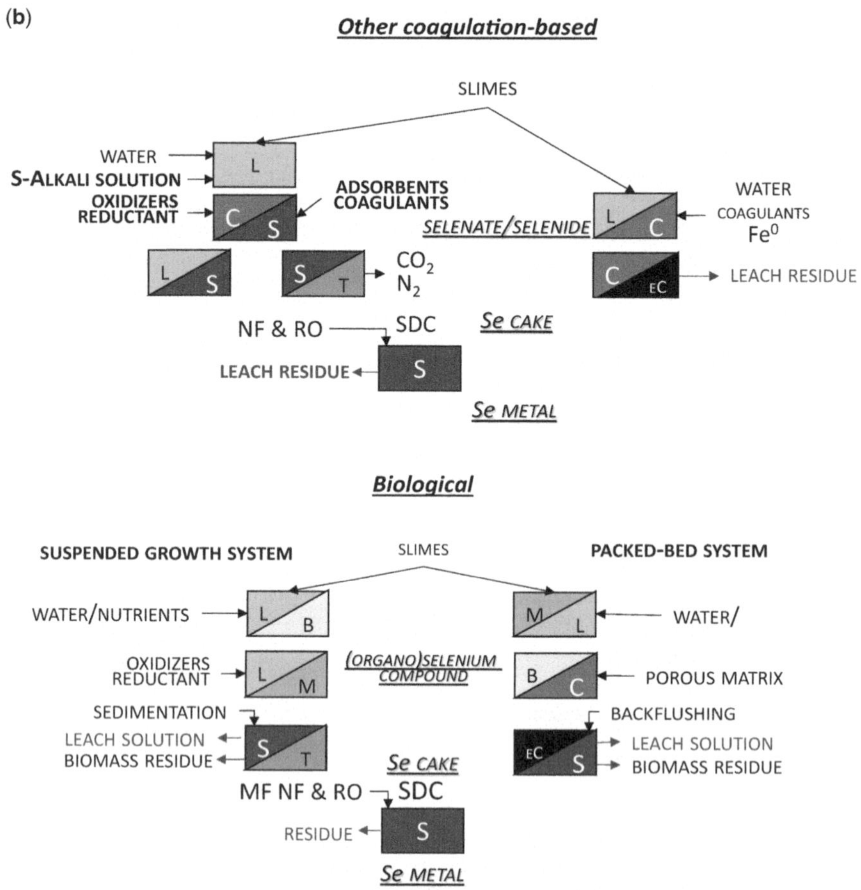

Figure 1.5 (*Continued*). **(b)** Schematic flow designs of other methods, in addition to those of Figure 1.5a, for selenium recovery processes from primary and secondary ores. The block-scheme flow, symbols, abbreviations and references are the same as those used in Figure 1.5a. See text for further details.

plants and converted into organic compounds such as selenomethionine, sometimes serving as a detoxifying mechanism. Microorganisms such as *Escherichia coli*, *Clostridium thermoaceticum*, and *Clostridium sticklandii* seem to crave selenium in their metabolic reactions. Accumulator plants can incorporate from 1000 to 10,000 mg/kg selenium (dry weight): they include *Astragalus* L., *Oonopsis* (Nutt.) Greene, *Oryzopsis* Michx., *Xylorhiza* Nutt., *Mentzelia* L., and *Stanleya* Nutt. (details in Section 1.3.3.4) (Presser, 1999). Significantly high selenium concentrations are also reported in some legumes, nuts, and mushrooms (see

Section 1.3.3.5). The presence of bioaccumulators and selenium-tolerant plants prompted the use of different microorganisms for biorecovery and bioremediation. Through bacterial leaching and phytovolatilization substantial amounts of selenium may be removed from Se-contaminated soil. Wetlands are a clear example of demonstrated biological passive treatment for selenium (CH2MHILL, 2010). Even though rapid recovery from biomass is virtually unfeasible, several technologies based on bacterial, fungal and plant metabolism have been tested in the laboratory. For example, species of the genus *Clostridium*, *Bacillus* and *Pseudomonas* (anaerobic and aerobic bacteria), *Aspergillus* spp., *Ganoderma lucidum* and *Phanerochaete chrysosporium* (fungi), *Saccharomyces cerevisiae* (baker's yeast), *Eichhornia crassipes* (water hyacinth), *Lemna minor* (lesser duckweed) and *Cladophora hutchinsiae* (green marine alga) have been tested for Se removal (Nancharaiah & Lens, 2015a, b; Santos *et al.*, 2015). Biological treatment relies on the enrichment and retention of microorganisms that convert soluble selenium oxyanions into the less-toxic elemental selenium. Furthermore, photobioreactors are emerging for the inclusion of microalgae as part of integrated algal-bacterial selenium removal systems, wherein selenium oxyanions are removed by biovolatalization and bioreduction. Selenate reduction has been investigated under methanogenic, sulfate-reducing, denitrifying, and hydrogenotrophic conditions (Nancharaiah & Lens, 2015a, b).

Different bioreactor configurations have been adopted (Figure 1.5a and b), such as up-flow anaerobic sludge blanket reactors, fluidized bed reactors, plug flow reactors, membrane biofilm reactors, and bioelectrochemical systems for retaining selenium-reducing microorganisms as flocs, granular sludge, or biofilms. For the latter purpose, soil, sand, cellulose, glass wool or glass beads can be used as a porous matrix. A significant fraction of the bioreduced selenium can still be present in the bioreactor effluent, necessitating a post-treatment step such as coagulation or electrocoagulation. It should be underlined that many of the suggested technologies are not demonstrated at full-scale for Se recovery and present the disadvantage of the high costs related to chemicals and solid waste disposal (Figure 1.5a and b). Biological selenium removal methods are more attractive than traditional metallurgical routes because of low operational costs and impacts on the environment. The performance of any technology should be evaluated on a case-specific basis, optimized and demonstrated at the pilot scale.

1.3 SELENIUM IN THE ENVIRONMENT

1.3.1 Selenium mineralogy

Since the first observation in 1817 of selenium in sulfides, there are today more than 100 known selenium minerals (123 approved by the International Mineralogical Association (IMA); Krivovichev *et al.*, 2019), classified into: native selenium (1), oxides (1), selenides (85), selenites (without H_2O, 14; with H_2O; 17), selenates (without H_2O, 1; with H_2O, 4) (Krivovichev *et al.*, 2019, Supplementary

Materials Table S1). This number is relatively high considering that most selenium is dispersed in other minerals as a substitute of sulfur. When taking into account the geochemical abundance of selenium, these minerals are rarely present as pure terms but rather as mixtures with the far more abundant sulfur-bearing minerals (sulfides and sulfates). The most up-to-date review on selenium minerals is the one of Krivovichev *et al.* (2019) with its Supplementary Materials Table S1. Wang *et al.* (2016a) adopt a slightly different classification into selenides, selenium sulfides, and oxygen-containing selenides. The paper by Stillings (2017) contains a careful listing of Se minerals with chemical formula, date of discovery or IMA approval and geological setting. This paper is an invaluable source of mineralogical and geochemical data for selenium mineralogy and occurrence. Another important source of data is the web site: www.mindat.org.

The crystal-chemistry of selenium minerals is complicated and was examined in depth in a paper by Krivorichev *et al.* (2019). Selenium, similarly to sulfur, can assume four oxidation states: −2, 0, +4, +6. The chemical formula resembles most common sulfides and due to the vicinity of the two elements, selenium minerals are very rarely pure terms, as summarized in Table 1.1, while the crystal structure of selenium minerals is variable and, in some cases, quite complex (see Figure 1.6).

A detailed account of the origin of Se minerals is described in Grundmann and Forster (2017) and Pažout *et al.* (2019). The most up-to-date study of the structure and crystal chemistry of Se minerals is the paper by Krivovichev *et al.* (2019) with the attempt to apply the concepts of *mineral ecology* (Hazen *et al.*, 2015). Many Se minerals are common in hydrothermal deposits (selenides), associated with fumaroles (selenites) or in the oxidation zone of sulfide and selenide deposits (secondary selenites and selenides). Selenium minerals rarely occur in high concentration; most often they are dispersed as small grains with other minerals. Most selenium minerals are composed of a narrow range of metals (Fe, Cu, Ni, Zn, Pb, Co, Ag, Pd, Pt, Hg) plus semimetals (As, Sb, Te, Bi) with scarce presence of other elements (K, Na, Al) in oxygen-containing selenides. As most of these minerals resemble most common sulfides, pure Se

Table 1.1 Selenium minerals.

Name	Ion	Example
Selenium	Se^0	Native selenium
Selenide	Se^{2-}	Clausthalite (PbSe), Berzelianite (Cu_2Se)
Diselenide	$[Se_2]^{2-}$	Ferroselite ($FeSe_2$), Dzharkenite ($FeSe_2$)
Selenite	$[Se^{4+}O_3]^{2-}$	Molybdomenite ($PbSeO_3$), Chalcomenite ($CuSeO_3.H_2O$)
Selenate	$[Se^{6+}O_4]^{2-}$	Olsacherite ($Pb_2(SeO_4)(SO_4)$), Ramaccioniite ($Cu_4(SeO_4)(OH)_6$)

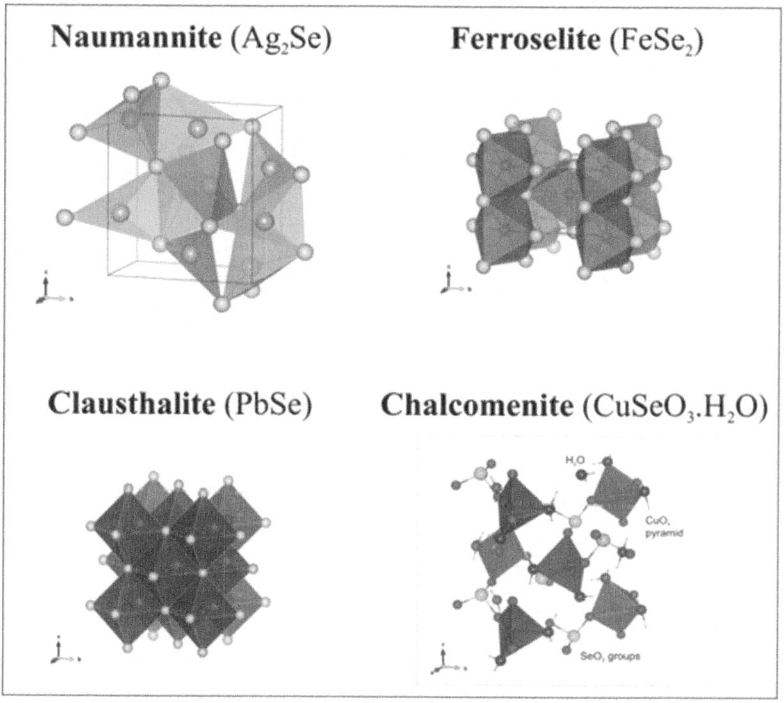

Figure 1.6 A few examples of crystal structures of selenium minerals. In all examples, the Se ion is presented in green. Naumannite (orthorhombic, a = 4.336, b = 7.067, c = 7.753, Z = 4, V = 237.57), Ferroselite (orthorhombic, a = 4.8001, b = 5.776, c = 3.585, Z = 2, V = 99.40), Clausthalite (isometric, cell dimensions a = 6.147, Z = 4, V = 232.37), Chalcomenite (orthorhombic, a = 6.671, b = 9.193, c = 7.384, Z = 4, V = 452.83). Drawings produced with VESTA (Momma & Izumi, 2011).

compounds are very rare. As previously stated, selenium mineralogy is complicated, and it is not possible to give a simple description of many Se minerals.

1.3.2 Selenium geochemistry

Selenium is a rare element, ranking 65th in average crustal abundance (0.09 mg/kg Se, Rudnick & Gao, 2003). Selenium concentrations in rocks, when reported, are generally very low (a summary of literature data is presented in Table 1.2). The highest concentrations, among the most common rock types are found in shales (average values however well below 1 mg/kg). In igneous rocks, the concentrations are generally low, likely higher in mafic than acidic material, for the occurrence of sulfides, although the impact of volcanic activity in selenium cycling is considered very high (Fordyce, 2013) with an estimate of a

Table 1.2 Selenium abundance in rocks.

Rock Type	Mean Value (Range) in mg/kg	Number of Observations	Reference
Igneous rocks	0.07		Koljonen (1992)
Ultramafic rocks			
Mafic intrusives (post Archean)	0.099	308	Gao *et al.* (1998)
Diorite (post Archean)	0.065	260	Gao *et al.* (1988)
Tonalite-Trondhjemite-Granodiorite (Archean)	0.056	553	Gao *et al.* (1988)
Tonalite-Trondhjemite-Granodiorite (post Archean)	0.052	641	Gao *et al.* (1988)
Granite (post Archean)	0.054	1226	Gao *et al.* (1988)
Granite (Archean)	0.036	402	Gao *et al.* (1988)
Granite	0.05		Bowen (1979)
Granite, Granodiorite	0.025		Koljonen (1992)
Volcanic rocks			
Mafic volcanics (post Archean)	0.111	632	Gao *et al.* (1988)
Felsic volcanics (post Archean)	0.066	972	Gao *et al.* (1988)
Basalt	0.05		Bowen (1979)
Basalt	0.12		Koljonen (1992)
Sedimentary rocks			
Sandstones (Archean)	0.13	121	Gao *et al.* (1988)
Sandstones (post Archean) Sandstone	0.269	2754	Gao *et al.* (1988)
Sandstone	<0.01		Bowen (1979)
	0.01		Koljonen (1992)
Pelites (Archean)	0.217	69	Gao *et al.* (1988)
Pelites (post Archean)	0.499	1341	Gao *et al.* (1988)

Shale	0.5		Bowen (1979)
Shale	0.3		Koljonen (1992)
North American shale composite	0.08		Morgan *et al.* (1978)
Marine pelagic clay	0.2		Li (1991)
Marine shales	0.6		Li (1991)
Carbonates (Archean)	0.042	50	Gao *et al.* (1988)
Carbonates (post Archean)	0.126	2038	Gao *et al.* (1988)
Limestone	0.03		Bowen (1979)
Limestone	0.025		Koljonen (1992)
Manganese nodules	0.6		Li (1991)
Black shales, UK and Ireland	18.7 (1.3–42)	44	Parnell *et al.* (2016)
Mediterranean sapropel	24 (0.2–44)		Nijenhuis *et al.* (1999)
Carbonaceous shale	0.77 (0.025–4.6)	12	Wen and Carignan (2011)
Average phosphorite	4.6 (0.5–13)	6	Altschuler (1980)
Phosphate rocks (Peru)	2.4 (0.5–6.0)	12	Bech *et al.* (2010a)
Phosphate rocks (other localities)	7.4 (0.11–44)	35	Bech *et al.* (2010b)
US, Coals	3.6 (0.02–75)	304	Coleman *et al.* (1993)
British, Coals	4.9 (0.3–61.9)	61	Bullock *et al.* (2018, 2019)
Coals	3		Koljonen (1992)
Metamorphic rocks			
Amphibolites (Archean)	0.235	189	Gao *et al.* (1988)
Mafic granulites (Archean)	0.254	128	Gao *et al.* (1988)
Intermediate granulites (Archean)	0.099	136	Gao *et al.* (1988)
Felsic granulites (Archean)	0.084	137	Gao *et al.* (1988)
Meta felsic volcanites (Archean)	0.037	41	Gao *et al.* (1988)

The table is a compilation of literature data regarding the abundance of selenium in rocks, when available the number of observations and the range are reported.

contribution of 0.1 g of selenium for every cm^2 of the Earth's surface. Given its volatility, during eruptions selenium is transferred to the atmosphere and not concentrated in rocks, perhaps except for some pyroclastic rocks (Fordyce, 2013). Floor and Román-Ross (2012) reported a range in Se concentrations in volcanic ash, ranging from <0.2 to 7 mg/kg Se, substantially higher than those in rocks, and with the peculiarity of being easily leachable compared to other sources. In metamorphic rocks, they could be related to the origin of the protolith. The highest concentrations reported (in the range of dozens mg/kg as maxima) in other geological material were observed in phosphatic rocks, eventually related to sulfides and organic matter (Altschuler, 1980), and are generally strongly enriched in black shales and coals, again in association with sulfides (Table 1.2).

1.3.3 Source of selenium in the environment

1.3.3.1 Selenium in soils

Soils represent a critical environmental matrix for the interaction with the biosphere, and selenium is usually present in trace amounts in soils, namely: (i) Fordyce (2013) reports a range of 0.01–2.0 mg/kg and an overall mean of 0.4 mg/kg Se; (ii) Kabata-Pendias (2011) reports a range of 0.05–1.5 mg/kg Se and a mean value of 0.44 mg/kg Se, while (iii) He *et al.* (2010) report a range of 0.1–2.0 mg/kg Se and a mean value of 0.2 mg/kg for agricultural soils and a wider variation (<0.1–4 mg/kg) and higher mean value (0.33 mg/kg) for worldwide surface soils. Although bedrock lithology is the primary factor influencing selenium in soils, additional inputs of selenium to soils occur following atmospheric deposition of selenium from natural (e.g., volcanoes and sea spray) or anthropogenic (e.g., fossil fuel combustion) sources or direct accumulation in soils due to sewage or fertilizer application.

The critical point for selenium occurrence in soils is that it can be transferred to the food chain and selenium can be concentrated in plants resulting in toxic concentrations for livestock (Fordyce, 2013), but it has also been realized that problems for livestock occur in Se-deficient areas, leading to a series of symptoms known as the white muscle disease (Levander, 1986). Low intake of selenium can also have effects in humans, in particular interfering with antioxidant systems, thyroid hormone metabolism, immune function and reproduction (Fordyce *et al.*, 2010).

A wider knowledge about selenium distribution in the environment has been obtained from several geochemical mapping projects carried out at different scales and densities around the world, as shown in Table 1.3. Given the important biological role played by selenium, the knowledge of the spatial distribution of this element is extremely important, not only for the identification of high concentrations areas, but also those where the Se concentration is low, eventually leading to Se deficiency in plants and livestock. As seen in Table 1.3, the average values are generally higher than those reported for rocks (Table 1.2) and in a few

Table 1.3 Selected data concerning selenium abundance in soils.

	Mean (Range) mg/kg	Median mg/kg	Number of Observations	Reference
Europe				
GEMAS Agricultural soils	(<0.05–3.8)	0.35	2108	Reimann *et al.* (2018a, b)
GEMAS Grazing land soils	(<0.05–6.8)	0.40	2024	Reimann *et al.* (2014)
Topsoil northern Europe	(0.12–9.02)		174	Reimann *et al.* (1997a, b)
Soil, C-horizon, Barents area	(<0.01–1.155)	0.046	605	Reimann *et al.* (1998)
Baltic soil survey				Reimann *et al.* (2003a)
Agricultural soil	(0.02–7.6)	0.14	752	Reimann *et al.* (2003b)
Bottom soil	(<0.01–6.7)	0.08	752	Reimann *et al.* (2003a)
Germany (West)	(0.016–0.65)		499	Hartfiel and Bahners (1988)
Netherlands	(0.05–2.0)		341	Mol *et al.* (2012)
Belgium	(0.14–0.70)		539	De Temmermann *et al.* (2014)
EIRE	(0.08–17.44)	0.74	1310	Fay *et al.* (2007)
Campania Region (Italy)	0.43 (0.065–2.4)	0.40	2333	Petrik *et al.* (2018)
Lombardia Region (Italy)	1.08 (0.5–2.4)		98	Beone *et al.* (2018)
Soil Murcia Region (Spain)	0.43 (0.003–2.7)	0.3	490	Perez-Sirvent *et al.* (2010)
Huelva municipality (Spain)	1.30 (0.08–68.6)	0.35	150	Guillén *et al.* (2011)
North Trøndelag forest soils				
O-horizon	(<0.15–5.27)	0.897	752	Reimann *et al.* (2015a, b)
C-horizon	(<0.5–3.54)	<0.5	752	Reimann *et al.* (2015a, b)
Southern Norway				
O-horizon	(0.4–6.3)	2	45	Reimann *et al.* (2009a)
C-horizon	(<0.1–2.7)	0.45	45	Reimann *et al.* (2009a)

(Continued)

Table 1.3 Selected data concerning selenium abundance in soils (*Continued*).

	Mean (Range) mg/kg	Median mg/kg	Number of Observations	Reference
Oslo area				
O-horizon	(0.3–3.4)	0.95	40	Reimann *et al.* (2007)
B-horizon	(<0.1–4)	0.5	40	Reimann *et al.* (2007)
C-horizon	(0.1–1.7)	0.4	40	Reimann *et al.* (2007)
England and Wales soils	0.71(0–16)	0.48	5670	Rawlins *et al.* (2012)
Cornwall (5–25 cm)	1(0.1–6.8)	0.9	1154	
Scotland	0.44 (0.115–0.877)	0.433	114	Fordyce *et al.* (2010)
London soil	0.67(<0.2–19.6)	0.6	6288	Scheib *et al.* (2011)
Belgian agricultural soil				De Temmermann *et al.* (2014)
Sandt loam	(0.14–0.4)	0.25	93	De Temmermann *et al.* (2014)
Sandy loam	(0.17–0.45)	0.26	145	De Temmermann *et al.* (2014)
Silt loam	(0.18–0.7)	0.34	297	De Temmermann *et al.* (2014)
Clay	(0.29–0.61)	0.46	4	De Temmermann *et al.* (2014)
Africa				
Malawy	0.19 (0.05–0.62)	0.16	73	Chilimba *et al.* (2011)
Egypt	0.79 (0.17–1.95)	0.73	15	Sabrynal *et al.* (2011)
Australia				
Topsoil	(<0.01–2.01)	0.06	1314	Reimann and de Caritat (2017)
Bottom soil	(<0.01–2.36)	0.06	1314	Reimann and de Caritat (2017)

Asia				
Floodplain sediments, China				
top samples (5–25 cm)	0.213 (0.012–128)	0.5	846	Xie and Cheng (2001)
deep samples (80–120 cm)	0.267 (0.044–11.2)	0.17	468	Xie and Cheng (2001)
Average content in Chinese soils	0.29		2904	Chen *et al.* (1991)
Soils in Chinese cities	0.23 (0.03–10.8)	0.23	3799	Cheng *et al.* (2014)
Japan agricultural soil		0.42	180	Yanai *et al.* (2012)
Soil United States				
Topsoil (0–5 cm)	(<0.2–6.9)	0.2	4841	Smith *et al.* (2019)
A-horizon	(<0.2–8.3)	0.2	4841	Smith *et al.* (2019)
C-horizon	(<0.2–7.5)	<0.2	4841	Smith *et al.* (2019)
South America				
Sao Paulo soils	0.12 (0.01–0.7)		30	Rodrigues Nogueira *et al.* (2018)

The data reported include mean (range) and median as well as the number of samples considered in each study. Preference has been given to large-scale geochemical studies.

cases very high concentrations were observed (128 mg/kg Se in soils from China) (Xie & Cheng, 2001). Conversely, in large-scale geochemical surveys maxima are below 10 mg/kg Se (e.g., in Europe 6.8 mg/kg Se for grazing land and 3.8 mg/kg Se for agricultural soil; 6.9 mg/kg Se in topsoil from the United States; 2.01 mg/kg Se in topsoil from Australia). When different topsoil textures were considered (e.g., De Temmerman *et al.* (2014) for Belgian soils, Table 1.3) it appears that the mean and median values increase with finer grain sizes. In the cases where different soil layers were sampled, it is frequent to observe higher median and maxima in the upper layers or topsoil (in Table 1.3, results from the Baltic Soil Survey and floodplain sediments from China). Where sampled, the organic rich layers of soils display the highest median among different soil horizons (Table 1.3, North Trøndelag soils and Southern Norway; Oslo area, Norway), confirming a great affinity of Se for organic matter and, in particular, an origin possibly related to sea spray contributions. In soils developed under a warmer and more humid climate, Se could more likely be associated with colloids or Al-compounds (Dhillon *et al.*, 2019; Malisa, 2001).

Results from surveys in urban areas (Table 1.3, Greater London; Chinese cities; Huelva Municipality) reach relatively high concentrations, whose origin has been associated with localized and historical industrial activity. Especially, coal combustion leads to increased Se atmospheric emissions causing wide dispersion of Se, and its deposition negatively affects wide areas in the environment and public health around the combustion plants (Cheng *et al.*, 2014).

The results presented in Table 1.3 are intended to provide a broad picture of natural selenium concentrations and its spatial variability. Several seleniferous areas are recognized in different parts of the world, and in those cases, much higher concentrations in soils can be observed. Selenium-rich soils in the world have been reported in the USA, Canada, Mexico, Colombia, Ireland, Australia and China (Dhillon & Dhillon, 2003; Dhillon *et al.*, 2019) and other places (Oldfield, 2002). In the USA, selenium-rich soils are common in arid and semi-arid regions where selenium-rich shales form the bedrock. Concentrations reaching values of between 20 and 95 mg/kg Se have been reported for several seleniferous areas in the central USA (Dhillon *et al.*, 2019). Seleniferous areas in India are known in Punjab (Dhillon & Dhillon, 1991, 2014) where soils reach a maximum of 4.52 mg/kg Se, with a mean of 2.1 mg/kg Se compared to a mean of 0.4 mg/kg Se for non-seleniferous areas (Dhillon *et al.*, 2019). In Ireland, England and Wales several selenium-rich areas were recognized in central-southern Ireland, with soils reaching concentrations as high as 324 mg/kg Se in County Limerick and 132 mg/kg Se in County Tipperary (Walsh & Fleming, 1952), whereas in England Webb *et al.* (1966) reported maxima of 7 mg/kg Se for North Staffordshire and Derbyshire soils, 4 mg/kg Se for Devon soils and 5 mg/kg Se for Caernarvon soils; further confirmed by Davies and Houghton (1983) who reported a maximum amount of 7.5 mg/kg Se and a median of 2.2 mg/kg Se for some other seleniferous areas in Wales, also

Table 1.4 Classification of selenium impact on human health based on soil selenium content.

Soil Selenium Classification	Selenium Content (mg/kg)
Deficient	<0.125
Marginal	0.125–0.175
Moderate	0.175–0.40
High	0.40–1.00
Toxic	1.00–3.00
Excessive	>3.00

associated with mining activities. In China, several areas are reported to be enriched in selenium (Dhillon *et al.*, 2019), for example Chang *et al.* (2019) reported, based on cited references, concentrations up to 7007 mg/kg for a seleniferous area in central China and a geometric mean of 133 mg/kg for soils developed on black shales. Zhu *et al.* (2008) reported values as high as 2018 mg/kg Se for soils, eventually affected by discarded coal spoils in Yutangba (Hubei Province, Central China), whereas the maximum concentration reported for non-polluted soils was 42.3 mg/kg Se, which is far above the normal values reported worldwide.

Based on studies relating soil selenium with human health (Tan *et al.*, 2009; Zhang *et al.*, 2008), a classification of soil types based on selenium concentration has been proposed (Table 1.4), although other authors consider different concentrations as toxic (Bailey, 2017).

Apart from total soil concentrations, the bioavailable forms of selenium demonstrate a critical effect in soils, which is basically controlled by the combination of pH and redox conditions that, in turn, control Se speciation (see Section 1.2.2), and the soil texture and mineralogy. The selenate form represents the most common form in neutral to alkaline soils: it is mobile (Figure 1.2) and readily available for plant uptake (Neal, 1995). Actually, the most widely accepted indicator for plant bioavailability is the water-soluble selenium concentration (Fordyce *et al.*, 2000; Jacobs, 1989) although speciation studies are increasing (Natasha Shahid *et al.*, 2018 and references therein) to further define selenium behaviour in soil systems.

1.3.3.2 Selenium in waters

The estimated representative selenium concentration in surface water (Lemly, 2004; Schrautzer, 2004) is approximately 0.2 μg/L Se in river water and 0.09 μg/L Se in seawater. Selenium is a very mobile element (Gaillardet *et al.*, 2005) which tends to be more concentrated in groundwater as compared to surface water, likely due to water-rock interactions (Hem, 1985).

Table 1.5 summarizes data available on Se concentration in waters. The data confirm that slightly higher average values are seen in groundwater as compared to surface water although there is a large variation that tends to decrease in small-scale studies such as for example those for surface waters in Norway (Reimann *et al.*, 2009a, b, 2018a, b) and Canada (Hu *et al.*, 2009). Notably, the values reported in US groundwaters and aquifers from France and Denmark are particularly high, up to 969 and 247 μg/L, respectively (De Simone *et al.*, 2014; Shand & Edmunds, 2008). Oxidizing conditions and higher pH values are likely to determine the high dissolved Se concentration in the Denver basin aquifer (Bauch *et al.*, 2014), the US area where the highest dissolved Se concentration has been observed (De Simone *et al.*, 2014). High levels of Se were also observed in a shallow alluvial aquifer, whereas deeper wells, generally used for drinking water supply, presented much lower concentrations (<0.4 μg/L Se) (Musgrove *et al.*, 2014). European surface waters (Salminen *et al.*, 2005) are characterized by a large-scale variability, with high values observed in southern Italy and in several areas of Spain and the Netherlands, possibly due to the presence of sulfide mineralization and the effect of seawater intrusion in coastal sand aquifers (Salminen *et al.*, 2005). Groundwaters in volcanic areas might record high values, such as in the case of Mount Etna, Volcano Island (Table 1.5) and other volcanic areas (Floor & Román-Ross, 2002).

European bottled drinking water indicated a low average value (0.054 μg/L Se). Results spread over four orders of magnitude with spikes on wells that could be affected by sulfide oxidation, while the highest value observed was recorded in water from the Czech Republic (Reimann & Birke, 2010). Limited effect is related to bottle leaching since results indicate only 0.018 μg/L Se contribution from glass leaching and 0.012 μg/L Se from soft plastic (polyethylene terephthalate, PET) leaching (Reimann & Birke, 2010). Tap water from Europe and Italy indicated lower median values and generally fits with the drinking water standards of 10 μg/L Se required in many countries, although the World Health Organization (WHO) sets a higher threshold value (40 μg/L). These are not the highest concentrations reported for selenium in tap water, since in many seleniferous areas around the world larger dissolved concentrations (from 1300 up to 2000 μg/L) have been observed, namely: (i) in the Coast Range alluvial aquifer of the San Joaquin Valley in California, (ii) in springs from the Soan-Sakesar valley in Pakistan, or (iii) in shallow wells of the upper reaches of the Colorado River in Utah (Afzal *et al.*, 2000; Deverel *et al.*, 1994; Engberg, 1999; Plant *et al.*, 2014).

1.3.3.3 Selenium in air

Concentrations of Se in the atmosphere are highly variable due to different sources of Se emissions. These can be of natural (e.g. crustal weathering, volcanic eruptions, sea salt and continental and marine biosphere) or of anthropogenic (e.g. industrial processes that involve combustion of coal, oil biomass; nonferrous metal smelting, manufacturing and utilization of agricultural products) origin

Table 1.5 Selected data concerning Se abundance in surface water, groundwater, bottled water and drinking water.

	Mean (Range) µg/L	Median µg/L	Number of Observations	Reference
Surface water Europe	(0.01–15)	0.340	307	Salminen *et al.* (2005)
Manitoba, Canada	(0.34–11.6)	0.34	25	Hu *et al.* (2009)
Oppdal area, Norway	(0.01–0.15)	0.04	200	Reimann *et al.* (2018a, b)
Oslo area, Norway	(0.5–1.4)	<0.5	39	Reimann *et al.* (2009b)
River water				
World average	0.07 (0.0.1–0.3)			Gaillardet *et al.* (2005)
Amazon river, Brazil/Colombia	0.051 (0.032–0.050)			Gaillardet *et al.* (2005)
Orinoco, Colombia/Venezuela				Gaillardet *et al.* (2005)
Lena, Siberia	(0.022–0.23)			Gaillardet *et al.* (2005)
Groundwater				
Europe	1.83 (<0.015–247)	0.50	577	Shand and Edmunds (2008)
Norway bedrock	(<0.01–21)	0.2	476	Frengstad *et al.* (2000)
Mt. Etna area, Italy	2.9 (0.6–66.8)		53	Aiuppa *et al.* (2000a)
Volcano, Italy	24.9 (0.8–237)	11	24	Aiuppa *et al.* (2000b)
Groundwater, USA	2.43 (0.02–969)	1	5335	De Temmermann *et al.* (2014)
Drinking water				
European bottled water	(<0.02–371)	0.054	884	Reimann and Birke (2010)
European bottled water	(<0.01–49.3)	0.635	66	Misund *et al.* (1999)
European tap water	(<0.02–15)	0.12	579	Reimann and Birke (2010)
Italian bottled water	(<0.01–2.67)	0.193	152	Dinelli *et al.* (2012)
Italian tap water	(<0.01–2.03)	0.164	169	Dinelli *et al.* (2012)
Ethiopian drinking water	(0.015–7.58)	0.615	138	Reimann *et al.* (2003a, b)

The data reported include mean, range and median as well as the number of samples under analysis in each study. Preference has been given to studies including a large number of samples to explore natural variability.

(Wen & Carignan, 2011). Among the natural sources, the marine and continental biosphere represents the most important source of atmospheric selenium (Mosher & Duce, 1987; Nriagu, 1989; Nriagu & Pacyna, 1988), followed by volcanic emissions and contribution from sea spray. Among the anthropogenic sources, coal combustion is dominant followed by nonferrous metal production (Mosher & Duce, 1987; Nriagu, 1989; Nriagu & Pacyna, 1988). The Se concentration in air above the South Pole is 0.06 ng/m^3 and the average value for worldwide air from remote regions is 0.2 ng/m^3 whereas the median for polluted areas is 4.0 ng/m^3 (Reimann & de Caritat, 1998). There is evidence that the ocean is a significant source of Se to coastal areas. Significant enrichment of Se in marine aerosols results from the formation of volatile organoselenium compounds, mainly dimethyl selenide $(CH_3)_2Se$. Increased Se levels in mosses (>1 mg/kg) and peat (>2 mg/kg) in the marine regions clearly indicate the impact of Se volatilization from seawater aerosols (Steinnes, 2003). Selenium is released into the air as hydrogen selenide generated by plants, and as elemental selenium, selenites and selenates in particulate form. The level of selenium in most urban air ranges from 0.1 to 10 ng/m^3, but higher levels may be found in certain areas as well, such as in the vicinity of copper smelters (WHO, 2011). The USEPA has established inhalation exposure limits as follows (in $\mu g/m^3$): 12,700 for hydrogen selenide, 400 for Se-hexafluoride, and 200 for other Se-compounds (Fordyce, 2005). According to the guidelines presented by ATSDR (2002), the Se concentration in air may vary from 160 to 1000 $\mu g/m^3$. The recommended threshold limit value for Se in a workplace is 200 $\mu g/m^3$, whereas in Germany, the MAK (maximum permissible concentration) value at the workplace has been established at 50 $\mu g/m^3$, and in Russia at 100 $\mu g/m^3$ (Schrautzer, 2004). Selenium released during fossil fuel combustion in Europe was 420 tons in 1979 (Schrautzer, 2004), this source provides for more than 6 kt/y, either as small particles and volatile compounds, accounting for about 40% of the total aerial Se abundance (Kabata-Pendias & Mukherjee, 2007).

1.3.3.4 Selenium in plants

The Se content of plants depends on the amounts of element available in soils (Combs, 2001) also taking into account the fact that there is little evidence that selenium is required for plant growth (Fordyce, 2013; Terry *et al.*, 2000). Indeed, there seems to be a direct relation between the amount of soluble selenium in the soil environment and the one in plants (Kabata-Pendias & Pendias, 2001), with the Se amount in the plant affected by soil pH (Winkel *et al.*, 2015), temperature and rainfall. In plants, similarly to other living organisms, selenium mimics sulfur biochemical properties and it can replace sulfur in amino acids and other biological processes.

Assuming that grasses and other forage plants are the main routes for grazing animals, the concentrations of selenium are in the ng/g dry weight range

(Table 1.6). In moderately low Se-soils, *Medicago sativa* grass accumulates more Se than many other forage plants, probably because of the different rooting depth and different selenium translocation to new buds (Mayland, 1994). Interestingly, a few plants are able to accumulate or hyper-accumulate Se in their tissues and are mostly found on seleniferous soils (White, 2016). They can be classified on the basis of the amount of selenium they contain (Plant *et al.*, 2014; Rosenfeld & Beath, 1964), namely:

- Selenium accumulator plants having more than 1000 mg/kg Se
- Secondary selenium absorbers with 50–100 mg/kg Se
- Plants with a concentration <50 mg/kg Se

Among the hyper-accumulators, some plants are operationally defined as 'obligate' because they require selenium for their growth. These include species from the genera *Astragalus*, *Conopsis*, *Xylorhiza*, and *Stanleya* (Ellis & Salt, 2003). In this respect, the species *Astragalus bisulcatus* (Hook) A. Gray is able to accumulate selenium up to 15,000 mg/kg Se dry weight with a seasonal distribution (Galeas *et al.*, 2007). This plant species is common on seleniferous soils and in the arid climate of the Western USA (see Section 1.1.1). There are also plants that are known as 'facultative' accumulators since they do not require selenium for growth but are able to bind selenium in organic form and concentrate it in their tissues. This category includes plants belonging to the genera *Acacia*, *Artemisia*, *Aster*, *Atriplex*, *Castilleja*, *Penstemon*, and *Grindelia* (Plant *et al.*, 2014).

As stated in section 1.3.3.1, and further detailed in Chapter 8, the uptake of selenium in plants occurs in the prevailing form of selenate (SeO_4^{2-}) via sulfate transporters; then, it is first reduced and subsequently incorporated into selenoamino acids by the sulfur (S) assimilation pathway (Sors *et al.*, 2005; Terry *et al.*, 2000; Valdez Barillas *et al.*, 2012). Selenate is reduced to selenite (SeO_3^{2-}) and then to selenide (Se^{2-}), which is incorporated into selenocysteine (SeCys). The non-specific incorporation of SeCys into proteins is presumed to be toxic (Stadtman, 1990). Notably, selenium hyper-accumulators avoid Se toxicity by methylating SeCys to methylselenocysteine (MeSeCys) through the use of a unique enzyme, SeCys methyltransferase, effectively circumventing the misincorporation of SeCys into protein (Neuhierl & Böck, 1996).

1.3.3.5 Selenium in food and feed

Since selenium is necessary for a balanced metabolism, animals and humans acquire Se from their daily diet. In animal tissues selenium is mostly found bound in proteins, as is the case in plants (see Section 1.3.3.4). Due to this, the most important food sources of selenium are cereals (0.1–10 mg/kg) as well as meat and seafood (0.3–0.5 mg/kg), because of their high protein contents. In contrast, foods with relatively low protein levels, such as vegetables and fruits, have low

Table 1.6 Representative data concerning Se abundance in plants and grass (ng/g dry weight).

	Mean (Range) ng/g DW	Number of Observations	Reference
Clover/*M. sativa*, Sweden	18–40		Fergusson (1990)
Clover/*M.sativa*, Germany	90 (50–130)		Kabata-Pendias and Pendias (2001)
Clover/*M.sativa*, France	38 (36–39)		Kabata-Pendias and Pendias (2001)
Clover/*M.sativa*, Canada	15 (5–31)		Kabata-Pendias and Pendias (2001)
Grass, Canada	13 (5–23)		Kabata-Pendias and Pendias (2001) Fergusson (1990)
Grass, France	47 (19–134)		Fergusson (1990)
Grass, USA	32 (10–40)		Fergusson (1990)
Grass, Sweden	30 (11–64)		Kabata-Pendias and Pendias (2001)
Grass, Germany	110 (30–210)		Kabata-Pendias and Pendias (2001)
Grass, Finland	11 (1–54)		Kabata-Pendias and Pendias (2001)
Birch leaves, Norway	0.1* (<0.1–0.3)	45	Reimann *et al.* (2015a)
Willow leaves, Norway	0.1* (<0.1–0.4)	45	Reimann *et al.* (2015b)
Juniper leaves, Norway	<0.1* (<0.1–0.2)	46	Reimann *et al.* (2015a)
Heather leaves, Norway	0.2* (<0.1–0.5)	46	Reimann *et al.* (2015b)

*Median values and concentrations as mg/kg.

selenium contents (<0.01 mg/kg) while still applying the general principle that the selenium content of a food reflects the available Se-content of the soils used to produce it (and the feedstuffs used to produce livestock) (Fairweather-Tait *et al.*, 2010; Chapter 9).

The main food groups providing selenium in the diet of a western country such as the UK are: bread and cereals (26%), meat (26%), milk/dairy products (21%), fish (10%), vegetables and fruits (7%) and eggs (4%). As shown in Table 1.7, the selenium content can vary widely from ~ 0.03 up to 30 mg/kg in bread and cereals, respectively (Barclay *et al.*, 1995; Rayman *et al.*, 2008).

Table 1.7 Selected data concerning Se abundance in foods (mg/kg) fresh weight.

	Mean (Range) mg/kg	Number of Observations	Reference
Cereals, USA	(0.06–0.66)		Combs (2001)
Cereals, Germany	(0.03–0.88)		Combs (2001)
Cereals, Venezuela	(0.123–0.51)		Combs (2001)
Wheat grain, Italy	0.071 (0.007–0.245)		Spadoni *et al.* (2007)
Breakfast cereals, France	0.025		Leblanc *et al.* (2005)
Rice Grains, Hainan Island, China	0.08 (0.01–0.43)		Xu *et al.* (2020)
Bread	(0.03–0.09)	$\leq 17 \leq 27$	Barclay *et al.* (1995)
Bread, UK	0.044		Ysart *et al.* (2000)
Bread, France	0.016		Leblanc *et al.* (2005)
Vegetables, USA	(0.005–0.14)		Combs (2001)
Vegetables, England	(0.01–0.09)		Combs (2001)
Brazil nuts	(0.03–500)		Combs (2001) Rayman *et al.* (2008)
Potatoes, UK	0.003		Ysart *et al.* (2000)
Brazil nuts	(0.03–500)		Combs (2001), Rayman *et al.* (2008)
Nuts, UK	0.25		Ysart *et al.* (2000)
Fish, cod	1.5		Fairweather-Tait *et al.* (2010)
Fish, tuna	5.6		Fairweather-Tait *et al.* (2010)
Fish, shark	2.0		Fairweather-Tait *et al.* (2010)
Fish, France	0.17		Leblanc *et al.* (2005)
Chicken, USA	0.2		Fairweather-Tait *et al.* (2010)
Poultry, UK	0.19		Ysart *et al.* (2000)
Beef, USA	(0.25–0.3)		Rayman *et al.* (2008)
Offal, kidney	4.5		British Nutrition Foundation (2001)

(Continued)

Table 1.7 Selected data concerning Se abundance in foods (mg/kg) fresh weight (*Continued*).

	Mean (Range) mg/kg	Number of Observations	Reference
Offal, liver	0.93		British Nutrition Foundation (2001)
Offal, heart	0.55		British Nutrition Foundation (2001)
Offal, muscle	0.2		British Nutrition Foundation (2001)
Hen's egg	(0.34–0.58)		Lipiec *et al.* (2010)
Eggs, UK	0.19		Ysart *et al.* (2000)

The predominant species of selenium in wheat and bread are selenomethionine (usually ~55–85%), selenocysteine (~4–12%), and selenate/selenite (~12–19%) (Whanger, 2002).

The selenium content of meat depends on many factors, for example offal contains relatively high levels of selenium, in particular liver and kidneys (Table 1.7). In the US, the average selenium content of chicken meat is ~0.2 mg/kg and beef ~0.25–0.3 mg/kg (Fairweather-Tait *et al.*, 2011). Meat represents a large fraction of the selenium intake in the USA and UK providing one-quarter of the total estimated Se intake. The predominant species of selenium in meat may be selenomethionine (~50–60% of total extractable selenium species) and selenocysteine (20–31 and ~50% of the total extractable selenium species in chicken and lamb, respectively).

The selenium content in fish is relatively high, being between 0.1 and ~5.0 mg/kg (Fairweather-Tait *et al.*, 2010; Reyes *et al.*, 2009). The main selenium species in fish are selenomethionine (29–70%) and selenite/selenate (12–45%) with the species profile differing between fish species and the total selenium content. Hens' eggs contain from ~3 to ~25 mg selenium per whole egg (Lipiec *et al.*, 2010). Selenium supplementation of the hen's diet may increase the selenium content of eggs to 0.34–0.58 mg/kg as Se-enriched eggs are widely produced around the world (Fisinin *et al.*, 2009). The main selenium species in eggs are selenocysteine, selenomethionine, and possibly selenite. Selenomethionine and selenocysteine are the predominant species (>50%) in egg white and egg yolk, respectively (Lipiec *et al.*, 2010). The selenium content of milk and dairy products varies widely; in the UK, milk and dairy products contain ~0.01–0.03 mg/kg selenium. The predominant selenium species in cows' milk are selenocysteine and selenite. Supplementation of dairy cows with Se-enriched yeast alters the species profile in the milk and the major species after supplementation are selenocysteine, selenomethionine, and selenite (Muniz-Naveiro *et al.*, 2007).

Fruit and vegetables typically contain relatively small amounts of selenium in the form of selenate in onions or selenomethionine (53%), γ-glutamyl-Se-methylselenocysteine (31%), Se-methylselenocysteine (12%), and selenate (4%) in garlic with a natural selenium content of <0.5 mg/kg. However, vegetables such as onions, garlic, and broccoli when grown on Se-rich soil can accumulate selenium, resulting in selenium enrichment from <0.5 mg/kg up to 140–300 mg/kg. The main selenium species in Se-enriched vegetables are: γ-glutamyl-Se-methyl-selenocysteine (accounting for ~45–73% of the species), selenate (~4–20%) and selenomethionine (5–13%) plus other species at lower levels (Finley *et al.*, 2001; Hurst *et al.*, 2010). Notably, these forms of selenium in vegetables have received attention due to purported protection against cancer in animal models when compared with other forms of selenium (Fairweather-Tait *et al.*, 2011). Therefore, it can be concluded that the amount of selenium in the diet largely depends on where crops are grown and cultivated, the soil/fodder to which animals are exposed, and the actual foods consumed.

1.3.3.6 Selenium in animals and humans

Selenium is an essential trace element for humans and animals, although even low concentrations of selenium, in the order of a few mg/kg, can provoke health disturbances (Vinceti *et al.*, 2001). Also in humans and animals, Se-Cys is incorporated into a very specific location in the amino acid sequence of selenoproteins, and the uncontrolled S substitution of Se may cause toxic effects in humans and animals. At least 11 selenoproteins have been characterized in animal systems. The first characterized was the glutathione peroxidases; more precisely, four selenium-containing glutathione peroxidases (GPx) have been identified: cellular or classical GPx, plasma or extracellular GPx, phospholipid hydroperoxide GPx, and gastrointestinal GPx (Holben & Smith, 1999). These enzymes reduce damaging reactive oxygen species (ROS) oxidizing glutathione. A well-characterized example of this enzyme family is phospholipid hydroperoxide GPx, an antioxidant enzyme protecting cells from oxidative damages. Se-Cys is present also in the active site of the thioredoxin reductase that, in conjunction with the compound thioredoxin, is involved in the regeneration of several antioxidant systems in animal cells, possibly including vitamin C (Mustacich & Powis, 2000). Moreover, maintenance of thioredoxin in a reduced form by thioredoxin reductase is important for regulating cell growth and viability (Di Gregorio, 2008).

Selenium occurs in mammalian tissues in the range from 0.7 in heart tissue to 2.5 mg/kg in muscles. The average Se content in human soft tissues is estimated as 0.11 mg/kg (Li, 2000). Concentrations of Se in kidneys of European people were reported by Zduôska *et al.* (1994) as follows (in mg/kg of fresh weight):

Bulgaria, 2.5; Germany 0.7; and Italy, 1.9. In human fluids, mean Se concentrations are (in μg/L): blood, 107; serum, 80; and urine, 22 (Li, 2000). In Finland, the Se level in serum was between 0.63 and 0.76 μmol/L, and after the Se supplementation of fertilizers, it increased to the range of 1.2–1.4 μmol/L (Hartikainen, 2005).

Deficiency symptoms of improper Se supply to humans can be summarized as follows: muscle weakness and pain, inflammation of muscles, fragile red blood cells, degeneration of pancreas, abnormal skin colouration, heart muscle dysfunction, prolonged illness condition, susceptibility to cancer, Keshan disease (KD) and Kashin-Beck disease (KBD). Conversely, the toxicity symptoms are: liver and kidney damage, blood clotting, necrosis of heart and liver, skin lesions, hair and nail loss, nausea and vomiting (Kabata-Pendias & Mukherjee, 2007). On the other hand, selenium seems to be required to stimulate the immune response in humans (Roy *et al.*, 1994), playing a role in regulating the expression of cell signalling molecules called cytokines, which orchestrate the immune response (Baum *et al.*, 2000). Many studies suggest that selenium supplementation at high levels reduces the incidence of cancer in animals and that the methylated forms of selenium are the active species against tumours (Combs & Gray, 1998). In this respect, several mechanisms have been proposed for the cancer prevention effects of selenium: (1) maximizing the activity of antioxidant selenoenzymes and improving antioxidant status, (2) improving immune system function, (3) affecting the metabolism of carcinogens, and (4) increasing the levels of selenium metabolites that inhibit tumour cell growth (Di Gregorio, 2008). Some aspects related to the therapeutic effects of selenium are reported in Section 1.4.

1.4 EFFECTS AND BIOAVAILABILITY OF NANO-SELENIUM (SeNPs)

Various methods have been reported for the synthesis of selenium nanoparticles (hereafter: SeNPs), namely: biological and/or chemical methods (see Chapters 10, 11 and 12). Nanoparticles (NPs) are, in fact, a promising alternative in drug delivery (Ensign *et al.*, 2012) as well as nutritional supplements (Zhang *et al.*, 2011). Numerous studies have dealt with the effects of bioactive compound supplements in nano-particulate preparations (Anal & Singh, 2007; Hadrup *et al.*, 2016; Hu *et al.*, 2012; Zhai *et al.*, 2017; Zhang *et al.*, 2004) and some applications of NPs in nutrition and medicine have also been approved for clinical use (Wacker, 2014). Inclusion of nanotechniques in human nutrition is justified by the possibility of increasing the drug solubility, its protection against oxidation and enzymatic degradation, prolongation of residence time, and effective passage through the gastrointestinal tract enhancing the bioavailability of supplemented substances (Zhang *et al.*, 2004). Figure 1.7 shows some advantages of using NPs. SeNPs have quite promising and interesting results as a unique approach to either prevent or cure a variety of human diseases.

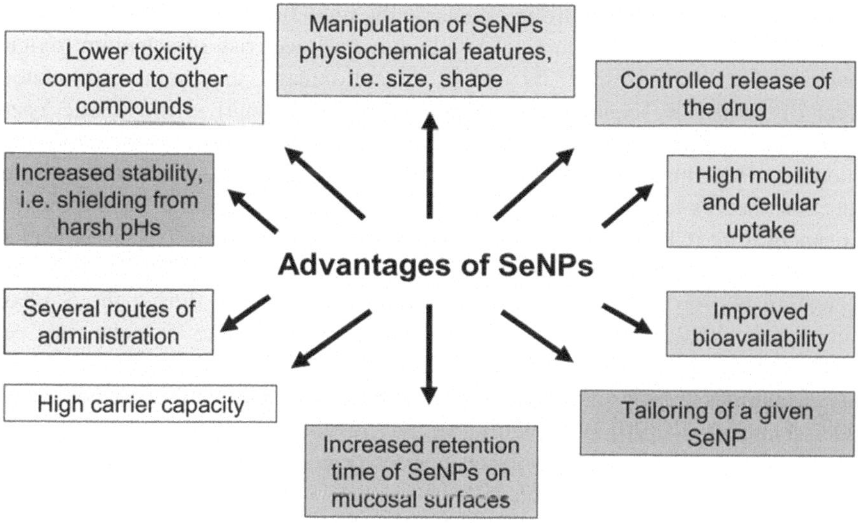

Figure 1.7 Diagram depicting the main advantages of using Se-nanoparticles (SeNPs) in nanomedicine applications (see text and references for details). Modified from Hosnedlova *et al.* (2018) and Khurana *et al.* (2019).

Therapeutic applications of SeNPs were recently reviewed by Khurana *et al.* (2019) and Hosnedlova *et al.* (2018). Besides, many studies have shown the antimicrobial effect (Piacenza *et al.*, 2017) and antifungal activity of nano-Se (Kheradmand *et al.*, 2014). In addition, its protective effects against metal intoxication have been well documented (Hassanin *et al.*, 2013; Prasad & Selvaraj, 2014).

The nano-form of selenium attracts even more attention, thanks to its higher bioavailability and lower toxicity than inorganic (selenite and selenate) and organic (selenomethionine, selenocysteine, and methylselenocysteine) forms (Fajt *et al.*, 2009; Zhang *et al.*, 2008). The advantage of nano-selenium (nano-Se) is the possibility of using selenium in zero oxidation state [Se(0)], which has low toxicity and excellent bioavailability as compared to other oxidation states [Se(+4) and Se(+6)] (Torres *et al.*, 2012; Wang *et al.*, 2007). On the other hand, Se(0) is very unstable and easily transformed into an inactive form although, its stabilization can be achieved by encapsulation into suitable nano-vehicles, for example, chitosan (CS) (Zhai *et al.*, 2017). As expected from the chemical-physical features of the nanoparticles, the biological properties of SeNPs depend on their size: smaller particles have a greater activity (Torres *et al.*, 2012). Indeed, the particle size affects the cellular intake of NPs; for example, *in vitro* absorption of 0.1 μm particles was found to be 2.5 and 6 times higher compared to 1 and 10 μm particles, respectively (Desai *et al.*, 1997). For this reason, in the preparation of dietary supplements, appropriate particle size, morphology, and encapsulation material should be chosen (De Jong & Borm, 2008).

Nanoscale selenium has strong effects on the reduction of oxidative stress and it has also been shown that spherical SeNPs have a lower risk of selenium toxicity (Gao *et al.*, 2002). Besides its use as an antioxidant, the immunostimulatory effect of nanoscale selenium has been confirmed (Kojouri *et al.*, 2012; Yazdi *et al.*, 2013) along with its beneficial effects on a number of physiological functions (Hegedüs *et al.*, 2012; Yin *et al.*, 2017). Even more interesting is the fact that SeNPs can be helpful in cancer chemo-prevention as a potential anticancer drug (Chen *et al.*, 2008; Sonkusre *et al.*, 2014) as well as an anticancer drug delivery carrier (Estevez *et al.*, 2014; Liu *et al.*, 2012). SeNPs have interesting anti-proliferation activity and inhibit HeLa cells during the S phase (Luo *et al.*, 2012; Ramya *et al.*, 2015). In the case of A375 melanoma cell lines, SeNPs were decorated with *Spirulina* and/or *Undaria pinnatifida* polysaccharides to improve the NPs biocompatibility and stability (Chen *et al.*, 2008; Yang *et al.*, 2012). Notably, similar results were obtained by capping SeNPs with water soluble polysaccharides-protein complexes from various species of fungi (Wu *et al.*, 2012, 2013).

In addition to biocompatibility and stability, SeNPs need to be selective in their capacity to enter the cells. In this respect, Yu *et al.* (2012) reported that SeNPs decorated with folate-chitosan were found to selectively endocytose inside cancer cells. This latter result is important because it shows the possibility to construct SeNPs for active targeting. Indeed, Zhang *et al.* (2013) synthesized adenosine triphosphate (ATP) surface-functionalized SeNPs which were able to specifically bind to purino-receptors in tumour tissues, thus causing a significant cancer cell apostosis. A further approach, such as the conjugation of SeNPs with other drugs, has also been used to elicit NPs cellular internalization and overcome multi-drug resistance. For example, the combination of adriamycin and SeNPs exhibited synergistic anticancer activity at low concentration compared to the drug used alone in Bel7402 hepatic cancer cell lines (Tan *et al.*, 2009), while bimetallic Se-AgNPs showed antitumour activity against Dalton's lymphoma cells (Kumar *et al.*, 2015). Another aspect related to multidrug resistance is that it may occur due to the mutation in the drug efflux pumps as seen in the case of the overexpression of ATP binding cassettes including P-glycoproteins (reviewed in Khurana *et al.*, 2019). In this respect, SeNPs are interesting carriers for drugs such as doxorubicin and curcumin to be used against *in vitro* lung-cancer cells (Zhao *et al.*, 2017) and Ehrlich's ascites carcinoma mouse model, respectively (Kumari *et al.*, 2017). Similarly, since small interfering RNAs (siRNA) have been used in treating many types of cancer, Xia *et al.* (2017) reported successful targeted delivery of siRNA using functionalized SeNPs against liver carcinoma.

Based on *in vitro* and *in vivo* studies, biogenic SeNPs can be considered as a novel therapeutic agent for the treatment of localized lesions typical of cutaneous leishmaniasis caused by *Leishmania major* (Beheshti *et al.*, 2013). The anti-leishmanial activities of SeNPs against *Leishmania infantum* were also described. SeNPs have more growth-inhibitory effect on promastigotes than selenium dioxide (SeO_2), while the IC_{50} (half minimal [50%] inhibitory concentration) was determined to be 25 and 50 ng/L, respectively (Soflaei *et al.*, 2014).

REFERENCES

Afzal S., Younas M. and Ali K. (2000). Selenium speciation studies from Soan-Sakesar valley, Salt Range, Pakistan. *Water International*, **25**, 425–436.

Aiuppa A., Allard P., D'Alessandro W., Michel A., Parello F., Treuil M. and Valenza M. (2000a). Mobility and fluxes of major, minor and trace metals during basalt weathering and groundwater transport at Mt. Etna volcano (Sicily). *Geochimica et Cosmochimica Acta*, **64**(11), 1827–1841.

Aiuppa A., Dongarrà G., Capasso G. and Allard P. (2000b). Trace elements in the thermal groundwaters of Vulcano Island (Sicily). *Journal of Volcanology and Geothermal Research*, **98**(1–4), 189–207.

Allegrini E., Maresca A., Olsson M. E., Holtze M. S., Boldrin A. and Astrup T. F. (2014). Quantification of the resource recovery potential of municipal solid waste incineration bottom ashes. *Waste Management*, **34**, 1627–1636.

Altschuler Z. S. (1980). The geochemistry of trace elements in marine phosphorites. Part I. Characteristic, abundances and enrichment. *Society of Economic Paleontologists and Mineralogists*, **29**, 19–30.

Anal A. K. and Singh H. (2007). Recent advances in microencapsulation of probiotics for industrial applications and targeted delivery. *Trends Food Science and Technology*, **18**(5), 240–251.

Antonioli P., Lampis S., Chesini I., Vallini G., Rinalducci S., Zolla L. and Righetti P. G. (2007). Stenotrophomonas maltophilia SeITE02, a new bacterial strain suitable for bioremediation of selenite-contaminated environmental matrices. *Applied Environmental Microbiology*, **73**, 6854–6863.

ATSDR (2002). Draft Toxicological Profile for Several Trace Elements. U.S. Dept. Health & Human Services. Agency for Toxic Substances and Disease Registry, Atlanta GA, USA.

Bailey R. T. (2017). Review: Selenium contamination, fate, and reactive transport in groundwater in relation to human health. *Hydrogeological Journal*, **25**, 1191–1217.

Baldwin R. A., Stauter J. C., Kauffman J. W. and Laughlin W. C. (1985). U.S. Patent No. 4 519 913. U.S. Patent and Trademark Office, Washington, DC.

Barceloux D. G. (1999). Selenium. *Journal of Toxicology and Clinical Toxicology*, **37**, 145–172.

Barclay M. N. I., MacPherson A. and Dixon J. (1995). Selenium content of a range of UK foods. *Journal of Food Composition and Analysis*, **8**(4), 307–318.

Bauch N. J., Musgrove M., Mahler B. J. and Paschke S. S. (2014). The quality of our Nation's waters - Water quality in the Denver Basin aquifer system, Colorado. *U.S. Geological Survey Circular*, **1357**, 100.

Baum M. K., Miguez-Burbano M. J., Campa A. and Shor-Posner G. (2000). Selenium and interleukins in persons infected with human immunodeficiency virus type 1. *Journal Infectious Disease*, **182**(1), S69–S73.

Beheshti N., Soflaei S., Shakibaie M., Yazdi M. H., Ghaffarifar F., Dalimi A. and Shahverdi A. R. (2013). Efficacy of biogenic selenium nanoparticles against *Leishmania major*: in vitro and in vivo studies. *Journal Trace Elements in Medicine and Biology*, **27**(3), 203–207.

Beck M. A., Levander O. A. and Handy J. (2003). Selenium deficiency and viral infection. *The Journal of Nutrition*, **133**(5), 1463S–1467S.

Bech J., Suarez M., Reverter F., Tume P., Sánchez P., Bech J. and Lansac A. (2010a). Selenium and other trace elements in phosphate rock of Bayovar–Sechura (Peru). *Journal of Geochemical Exploration*, **107**(2), 136–145.

Bech J., Suarez M., Reverter F., Tume P., Sánchez P., Roca N. and Lansac A. (2010b). Selenium and other trace element in phosphorites: a comparison between those of the Bayovar-Sechura and other provenances. *Journal of Geochemical Exploration*, **107**(2), 146–160.

Beone G. M., Carini F., Guidotti L., Rossi R., Gatti M., Fontanella M. C. and Cenci R. M. (2018). Potentially toxic elements in agricultural soils from the Lombardia region of northern Italy. *Journal of Geochemical Exploration*, **190**, 436–452.

Berzelius J. (1818). Letters to C.L. Berthollet. *Annales de Chimie et de Physique*, **9**, 160–180, 225–267, 337–365.

Blengini G. A. A., Blagoeva D., Dewulf J., Torres de Matos C., Nita V., Vidal-Legaz B., Latunussa C., Kayam Y., Talens, Peirò L., Baranzelli C. E. L., Manfredi S., Mancini L., Nuss P., Marmier A., Alves-Dias P., Pavel C., Tzimas E., Mathieux F., Pennington D. and Ciupagea C. (2017). JRC Technical reports – Assessment of the Methodology for Establishing the EU List of Critical Raw Materials – Annexes. https://doi.org/10.2760/73303.

Bodnar M., Szcyglowska M., Konieczka P. and Namiesnik J. (2016). Methods of selenium supplementation: bioavailability and determination of selenium compounds. *Critical Reviews in Food Science and Nutrition*, **56**(1), 36–55.

Boschen R. E., Rowden A. A., Clark M. R. and Gardner J. P. A. (2013). Mining of deep-sea seafloor massive sulfides: a review of the deposits, their benthic communities, impacts from mining, regulatory frameworks and management strategies. *Ocean & Coastal Management*, **84**, 54–67.

Bowen H. J. M. (1979). Environmental Chemistry of the Trace Elements. Academic Press, London.

British Nutrition Foundation. (2001). Briefing paper: Selenium and Health. www.nutrition.org.uk, **145**, 1–36.

Bullock L. A., Parnell J., Perez M., Armstrong J. G., Feldmann J. and Boyce A. J. (2018). High selenium in the Carboniferous Coal Measures of Northumberland, North East England. *International Journal of Coal Geology*, **195**, 61–74.

Bullock L. A., Parnell J., Feldmann J., Armstrong J. G., Henn A. S., Mesko M. F., Mello P. A. and Flores E. M. M. (2019). Selenium and tellurium concentrations of Carboniferous British coals. *Geological Journal*, **54**(3), 1401–1412.

Butterman W. C. and Brown R. D., Jr. (2004). Mineral Commodity Profiles-Selenium. U.S. Geological Survey Open-File Report **03–018**. Retrieved from https://pubs.usgs.gov/of/2003/of03-018/of03-018.pdf

CH2MHILL (2010). Review of available technologies for the removal of selenium from water – final report. Prepared for North American Metals Council by CH2M HILL Companies, Ltd, Englewood, Colorado. Retrieved from https://www.namc.org/docs/FINAL-NAMC-SWG_Removal_Technologies.pdf

Chan Y. T., Kuan W. H., Chen T. Y. and Wang M. K. (2009). Adsorption mechanism of selenate and selenite on the binary oxide systems. *Water Resources*, **43**, 4412–4420.

Chang C., Yin R., Zhang H. and Yao L. (2019). Bioaccumulation and Health Risk Assessment of Heavy Metals in the Soil–Rice System in a Typical Seleniferous Area in Central China. *Environmental Toxicology and Chemistry*, **38**(7), 1577–1584.

Chasteen T. G. and Bentley R. (2003). Biomethylation of selenium and tellurium: microorganisms and plants. *Chemical Reviews*, **103**, 1–25.

Chen J., Wei F., Zheng C., Wu Y. and Adriano D. C. (1991). Background concentrations of elements in soils of China. *Water, Air, and Soil Pollution*, 57–58, 699–712.

Chen T., Wong Y. S., Zheng W., Bai Y. and Huang L. (2008). Selenium nanoparticles fabricated in *Undaria pinnatifida* polysaccharide solutions induce mitochondria-mediated apoptosis in A375 human melanoma cells. *Colloids and Surface B: Biointerfaces*, **67**(1), 26–31.

Cheng H., Li M., Zhao C., Li K., Peng M., Qin A. and Cheng X. (2014). Overview of trace metals in the urban soil of 31 metropolises in China. *Journal of Geochemical Exploration*, **139**, 31–52.

Chilimba A. D. C., Young S. D., Black C. R., Rogerson K. B., Ander E. L., Watts M. J., Lamme J. and Broadley M. R. (2011). Maize grain and soil surveys reveal suboptimal dietary selenium intake is widespread in Malawi. *Scientific Reports*, **1**(72), 1–9.

Christophersen O. A., Lyons G., Haug A. and Steinnes E. (2013). Chapter 16 – Selenium. In: Heavy Metals in Soils: Trace Metals and Metalloids in Soils and their Bioavailability, B. Alloway (ed.), Environmental Pollution Series, 22, Springer Science+Business Media, Dordrecht, pp. 429–463.

Coleman R. G. and Delevaux M. H. (1957). Occurrence of selenium in sulfides from some sedimentary rocks of the Western United States. *Economic Geology*, **52**, 499–527.

Coleman L., Bragg L. J. and Finkelman R. B. (1993). Distribution and mode of occurrence of selenium in US coals. *Environmental Geochemistry and Health*, **15** (4), 215–227.

Combs G. A. (2001). Selenium in global food systems. *British Journal of Nutrition*, **85**, 517–547.

Combs G. F. and Gray W. P. (1998). Chemopreventive agents: selenium. *Pharmacology and Therapeutics*, **79**, 179–192.

Davies B. E. and Houghton N. J. (1983). The selenium content of Welsh soils with special reference to bedrock and contamination from sulphide ores. *Minerals and the Environment*, **5**, 67–70.

De La Riva D. G., Vindiola B. G., Castañeda T. N., Parker D. R. and Trumble J. T. (2014). Impact of selenium on mortality, bioaccumulation and feeding deterrence in the invasive Argentine ant, *Linepithema humile* (*Hymenoptera: Formicidae*). *Science of the Total Environment*, **481**, 446–452.

Desai M. P., Labhasetwar V., Walter E., Levy R. J. and Amidon G. L. (1997). The mechanism of uptake of biodegradable microparticles in Caco-2 cells is size dependent. *Pharmaceutical Research*, **14**(11), 1568–1573.

Deverel S. J., Fio J. L. and Dubrovsky N. M. (1994). Distribution and Mobility of Selenium in Groundwater in the Western San Joaquin Valley of California. In: Selenium in the Environment, W. T. Frankenberger and S. Benson (eds.), Dekker, New York, pp. 157–183.

De Jong W. H. and Borm P. J. (2008). Drug delivery and nanoparticles: applications and hazards. *International Journal of Nanomedicine*, **3**(2), 133–149.

De Simone L. A., McMahon P. B. and Rosen M. R. (2014). The quality of our Nation's waters – Water quality in Principal Aquifers of the United States, 1991–2010. *U.S. Geological Survey Circular*, **1360**, 1–15.

De Temmerman L., Waegeneers N., Thiry C., Du Laing G., Tack F. and Ruttens A. (2014). Selenium content of Belgian cultivated soils and its uptake by field crops and vegetables. *Science of The Total Environment*, **468–469**, 77–82.

Dhillon K. S. and Dhillon S. K. (1991). Selenium toxicity in soils, plants and animals in some parts of Punjab, India. *International Journal of Environmental Studies*, **37**, 15–24.

Dhillon K. S. and Dhillon S. K. (2003). Distribution and management of seleniferous soils. *Advances in Agronomy*, **79**, 119–184.

Dhillon K. S. and Dhillon S. K. (2014). Development and mapping of seleniferous soils in north-western India. *Chemosphere*, **99**, 56–63.

Dhillon K. S., Dhillon S. K. and Bijay-Singh (2019). Genesis of seleniferous soils and associated animal and human health problems. *Advances in Agronomy*, **154**, 1–80.

Di Gregorio S. (2008). Selenium: a versatile trace element in life and environment. In: Trace Elements in Human Health and Disease, A. S. Prasad (ed.), Vol 2. Academic Press, New York, USA, pp. 593–622.

Dinelli E., Lima A., Albanese S., Birke M., Cicchella D., Giaccio L., Valera P. and De Vivo B. (2012). Comparative study between bottled mineral and tap water in Italy. *Journal of Geochemical Exploration*, **112**, 368–389.

Duntas L. H. and Benvenga S. (2014). Selenium: an element for life. *Endocrine*, **48**, 756–775.

Ellis D. R. and Salt D. E. (2003). Plants, selenium and human health. *Current Opinion in Plant Biology*, **6**(3), 273–279.

El-Ramady H., Abdalla N. A., Alshaal T., Domokos-Szabolcsy E., Elhawat N., Prokisch J., Sztrik A., Fári M, El-Marsafawy S. and Shams M. S. (2014a). Selenium in soils under climate change, implication for human health. *Environmental Chemistry Letters*, **13**(1), 1–19.

El-Ramady H. R., Domokos-Szabolcsy E., Abdalla N. A., Alshaal T. A., Shalaby T. A., Sztrik A., Prokisch J. and Fári M. (2014b). Selenium and nano-selenium in agroecosystems. *Environmental Chemistry Letters*, **12**(4), 495–510.

El-Shafey E. I. (2007a). Removal of Se(IV) from aqueous solution using sulphuric acid treated peanut shell. *Journal of Environmental Management*, **84**, 620–627.

El-Shafey E. I. (2007b). Sorption of Cd(II) and Se(IV) from aqueous solution using modified rice husk. *Journal of Hazardous Material*, **147**, 546–555.

Engberg R. A. (1999). Selenium budgets for Lake Powell and the upper Colorado River Basin. *Journal of the American Water Resources Association*, **35**, 771–786.

Ensign L. M., Cone R. and Hanes J. (2012) Oral drug delivery with polymeric nanoparticles: the gastrointestinal mucus barriers. Advanced Drug Delivery *Reviews*, **64**(6), 557–570.

Estevez H., Garcia-Lidon J. C., Luque-Garcia J. L. and Camara C. (2014). Effects of chitosan-stabilized selenium nanoparticles on cell proliferation, apoptosis and cell cycle pattern in HepG2 cells: comparison with other selenospecies. *Colloids and Surfaces B: Biointerfaces*, **122**, 184–193.

European Commission (2014) Report on Critical Raw Materials for the Eu. *Eucom*, **39**, 1–41. https://doi.org/Ref.Ares(2015)1819595 (accessed 29 April 2015).

Fairweather-Tait S. J., Collings R. and Hurst R. (2010). Selenium bioavailability: current knowledge and future research requirements. *The American Journal of Clinical Nutrition*, **91**(5), 1484S–1491S.

Fairweather-Tait S. J., Bao Y., Broadley M. R., Collings R., Ford D., Hesketh J. E. and Hurst R. (2011). Selenium in human health and disease. *Antioxidants and Redox Signaling*, **14** (7), 1337–1383.

Fajt Z., Drabek J., Steinhauser L. and Svobodova Z. (2009). The significance of pork as a source of dietary selenium – an evaluation of the situation in the Czech Republic. *Neuro Endocrinology Letters*, **30**(1), 17–21.

Fay D., Kramers G., Zhang C., McGrath D. and Grennan E. (2007). Soil Geochemical Atlas of Ireland.

Fergusson J. E. (1990). The Heavy Elements: Chemistry, Environmental Impact and Health Effects. Pergamon, London.

Fiducia T. A. M., Mendis B. G., Li K., Grovenor C. R. M., Munshi A. H., Barth K., Sampath W. S., Wright L. D., Abbas A., Bowers J. W. and Walls J. M. (2019) Understanding the role of selenium in defect passivation for highly efficient selenium-alloyed cadmium telluride solar cells. *Nature Energy*, **4**, 504–511.

Finley J. W., Ip C., Lisk D. J., Davis C. D., Hintze K. J. and Whanger P. D. (2001). Cancer-protective properties of high- selenium broccoli. *Journal of Agricultural and Food Chemistry*, **49**, 2679–2683.

Fisinin V. I., Papazyan T. T. and Surai P. F. (2009). Producing selenium-enriched eggs and meat to improve the selenium status of the general population. *Critical Reviews in Biotechnology*, **29**(1), 18–28.

Floor G. H. and Román-Ross G. (2012). Selenium in volcanic environments: a review. *Applied Geochemistry*, **27**(3), 517–531.

Fordyce F. M. (2013). Selenium deficiency and toxicity in the environment. In: Essentials of Medical Geology, O. Selinus (ed.), Springer, Dordrecht, The Netherlands, pp. 375–416.

Fordyce F. M. (2005). Selenium deficiency and toxicity in the environment. In: Essentials of Medical Geology, O. Selinus, B. Alloway, J. A. Centeno, R. B. Finkelman, R. Fuge, U. Lindh and P. Smedley (eds.), Elsevier, London, UK, pp. 373–415.

Fordyce F. M., Zhang G., Green K. and Liu X. (2000). Soil, grain and water chemistry and human selenium imbalances in Enshi District, Hubei Province, China. *Applied Geochemistry*, **15**(1), 117–132.

Fordyce F. M., Brereton N., Hughes J., Luo W. and Lewis J. (2010). An initial study to assess the use of geological parent materials to predict the Se concentration in overlying soils and in five staple foodstuffs produced on them in Scotland. *Science of the Total Environment*, **408**(22), 5295–5305.

Frengstad B., Midtgård Skrede A. K., Banks D., Krog J. R. and Siewers U. (2000). The chemistry of Norwegian groundwaters: III. The distribution of trace elements in 476 crystalline bedrock groundwaters, as analysed by ICP- MS techniques. *Science of the Total Environment*, **246**(1), 21–40.

Fu Y., Wang J., Liu Q. and Zeng H. (2014). Water-dispersible magnetic nanoparticle–graphene oxide composites for selenium removal. *Carbon*, **77**, 710–721.

Gaillardet J., Viers J. and Dupré B. (2005). Trace element in river waters. In: Treatise on Geochemistry, volume 5. Surface and Ground Water Weathering and Soils, J. J. Drever, H. D. Holland and K. K. Turekian (eds.), Elsevier, NY, pp. 225–272.

Galeas M. L., Zhang L. H., Freeman J. L., Wegner M. and Pilon-Smits E. A. H. (2007). Seasonal fluctuations of selenium and sulfur accumulation in selenium hyperaccumulators and related nonaccumulators. *New Phytologist*, **173**, 517–525.

Gao S., Luo T.-C., Zhang B.-R., Zhang H.-F., Han Y.-W., Zhao Z.-D. and Hu Y.-K. (1998). Chemical composition of the continental crust as revealed by studies in east China. *Geochimica et Cosmochimica Acta*, **62**(11), 1959–1975.

Gao X., Zhang J. and Zhang L. (2002). Hollow sphere selenium nanoparticles: their *in vitro* anti hydroxyl radical effect. Advanced *Materials*, **14**(4), 290–293.

Geoffroy N. and Demopoulos G. P. (2009). Reductive precipitation of elemental selenium from selenious acidic solutions using sodium dithionite. *Industrial and Engineering Chemistry Research*, **48**, 10240–10246.

Geoffroy N. and Demopoulos G. P. (2011). The elimination of selenium(IV) from aqueous solution by precipitation with sodium sulfide. *Journal of Hazardous Material*, **185**, 148–154.

George M. W. and Wagner L. A. (2009). Selenium recycling in the United States in 2004. *U. S. Geological Survey Circular*, **1196-T**, 1–7.

Gerrard T. L., Telford J. N. and Williams H. H. (1974). Detection of selenium deposits in *Escherichia coli* by electron microscopy. *Journal of Bacteriology*, **173**, 3463–3467.

Goodman D. J., Kerollis R. E., Mindler A. B. and Bruns B. (2000). U.S. Patent No. 6 090 290. U.S. Patent and Trademark Office, Washington, DC.

Guillén M. T., Delgado J., Albanese S., Nieto J. M., Lima A. and De Vivo B. (2011). Environmental geochemical mapping of Huelva municipality soils (SW Spain) as a tool to determine background and baseline values. *Journal of Geochemical Exploration*, **109**(1–3), 59–69.

Guo L. and Wu D.-Q. (2017). Study of recycling Singapore solid waste as land reclamation filling material. *Sustainable Environment Research*, **27**, 1–6.

Grundmann G. and Förster H.-J. (2017). Origin of the El Dragón Selenium Mineralization, Quijarro Province, Potosí, Bolivia. *Minerals*, **7**(5), 68.

Hadrup N., Loeschner K., Skov K., Ravn-Haren G., Larsen E. H., Mortensen A., Lam H. R. and Frandsen H. L. (2016). Effects of 14-days oral low-dose selenium nanoparticles and selenite in rat – as determined by metabolite pattern determination. *Peer J – Life and Environment*, **4**, e2601.

Hageluken C. and Meskers C. (2010). Complex life cycles of precious and special metals. In: Linkages of Sustainability, T. E. Graedel and E. van der Voet (eds.), The MIT Press, Cambridge, pp. 163–198.

Hartfiel W. and Bahners N. (1988). Selenium Deficiency in the Federal Republic of Germany. In: Selenium, G. N. Schrauzer (ed.), Humana Press.

Hartikainen H. (2005). Biogeochemistry of selenium and its impact on food chain quality and human health. *Journal of Trace Elements in Medicine and Biology*, **18**, 309–318.

Hassanin K. M., Abd El-Kawi S. H. and Hashem K. S. (2013). The prospective protective effect of selenium nanoparticles against chromium-induced oxidative and cellular damage in rat thyroid. *International Journal of Nanomedicine*, **8**, 1713–1720.

Hatten Howard J. H., III. (1977). Geochemistry of selenium—Formation of ferroselite and selenium behavior in the vicinity of oxidizing sulfide and uranium deposits. *Geochimica et Cosmochimica Acta*, **41**(11), 1665–1678.

Hazen R. M., Grew E. S., Downs R. T., Golden J. and Hystad G. (2015). Mineral ecology: Chance and necessity in the mineral diversity of terrestrial planets. *Canadian Mineralogist*, **53**, 295–324.

Hegedüs V., Prokisch J., Fébel H., Kleiner D., Ditrói K., Szijártó A. and Blázovics A. (2012). Nanoselenium treatment in fatty liver. *Zeitschrift fur Gastroenterolologie*, **50**(5), A29.

He Z. L., Xiao J. S. and Yang E. (2010). Manganese and Selenium. In: Trace Elements in Soils, P. S. Hooda (ed.), Blackwell Publishing Co, Chichester, UK, pp. 481–495.

Hem J. D. (1985). Study and Interpretation of the Chemical Characteristics of Natural Water. *USGS Professional Paper*, **2254**, 1–264.

Hoffman J. E. and King M. G. (1997). Selenium and selenium compounds. In: Encyclopedia of Chemical Technology, J. I. Kroschwitz and M. H. Howe-Grant (eds.), John Wiley & Sons, New York, USA, pp. 686–719.

Holben D. H. and Smith A. M. (1999). The diverse role of selenium within selenoproteins: a review. *Journal of the American Dietetic Association*, **99**, 836–843.

Hosnedlova B., Kepinska M., Skalickova S., Fernandez C., Ruttkay-Nedecky B., Peng Q., Baron M., Melcova M., Opatrilova R., Zidkova J., Bjørklund G., Sochor J. and Kizek R. (2018). Nano-selenium and its nanomedicine applications: a critical review. *International Journal of Nanomedicine*, **13**, 2107–2128.

Hu C. H., Li Y. L., Xiong L., Zhang H. M., Song J. and Xia M. S. (2012). Comparative effects of nano-elemental selenium and sodium selenite on selenium retention in broiler chickens. *Animal Feed Science and Technology*, **177**(3), 204–210.

Hu X., Wang F., Mark L. and Hanson M. L. (2009). Selenium concentration, speciation and behavior in surface waters of the Canadian prairies. *Science of the Total Environment*, **407**(22), 5869–5876.

Huawei Z. (2009). Selenium as an essential micronutrient: roles in cell cycle and apoptosis. *Molecules*, **14**(3), 1263–1278.

Hunter W. J., Kuykendall L. D. and Manter D. K. (2007). Rhizobium selenireducens sp. nov., a selenite-reducing α-Proteobacterium isolated from a bioreactor. *Current Microbiology*, **55**, 455–460.

Hurst R., Armah C. N., Dainty J. R., Hart D. J., Teucher B., Goldson A. J., Broadley M. R., Motley A. K. and Fairweather- Tait S. J. (2010). Establishing optimal selenium status: results of a randomized, double blind, placebo-controlled trial. *American Journal of Clinical Nutrition*, **91**, 923–931.

Jacobs L. W. (ed.) (1989). Selenium in Agriculture and the Environment. Soil Science Society of America Special Publication, Madison, 23.

John D. A. (2001). Miocene and early Pliocene epithermal gold-silver deposits in the northern Great Basin, Western United States—Characteristics, distribution, and relationship to magmatism. *Economic Geology*, **96**, 1827–1853.

Kabata-Pendias A. (2011). Trace Elements in Soils and Plants, 4th edn. CRC Press, Boca Raton.

Kabata-Pendias E. and Pendias H. (2001). Trace Elements in Soils and Plants, 3rd edn. LLC, CRC Press/Taylor & Francis Group, Boca Raton.

Kabata-Pendias A. and Mukherjee A. B. (2007). Trace Elements from Soil to Human. Springer Science + Business Media, Berlin, Germany.

Kashiwa M., Nishimoto S., Takahashi K., Ike M. and Fujita M. (2000). Factors affecting soluble selenium removal by a selenate-reducing bacterium *Bacillus* sp SF-1. *Journal of Biosciences and Bioengineering*, **89**, 528–533.

Kheradmand E., Rafii F., Yazdi M. H., Sepahi A. A., Shahverdi A. R. and Oveisi M. R. (2014). The antimicrobial effects of selenium nanoparticle-enriched probiotics and their fermented broth against *Candida albicans*. *DARU Journal of Pharmacological Sciences*, **22**(48), 1–6.

Khurana A., Tekula S., Saifi M. A., Venkatesh P. and Godugu C. (2019). Therapeutic applications of selenium nanoparticles. *Biomedicine and Pharmacotherapy*, **111**, 802–812.

Kilic Y., Kartal G. and Timur S. (2013). An investigation of copper and selenium recovery from copper anode slimes. *International Journal of Mineral Processing*, **124**, 75–82.

Klaproth M. H. (1798). On the Transylvanian gold ore, and the new metal contained in it. Chemical Annals for the Friends of Science, Medicine, Economics, and Manufacturing, **1**, 91–104.

Klonowska A., Heulin T. and Vermeglio A. (2005). Selenite and tellurite reduction by *Shewanella oneidensis*. *Applied and Environmental Microbiology*, **71**, 5607–5609.

Kogel J. E. (2006). Industrial Minerals and Rocks: Commodities, Markets, and Uses. Society for Mining Metallurgy and Exploration, Littleton, Colorado, USA.

Kojouri G. A., Sadeghian S., Mohebbi A. and Dezfouli M. R. M. (2012). The effects of oral consumption of selenium nanoparticles on chemotactic and respiratory burst activities of neutrophils in comparison with sodium selenite in sheep. *Biological Trace Elements Research*, **146**(2), 160–166.

Koljionen T. (ed.) (1992). The Geochemical Atlas of Finland, Part II: Till. Geological Survey of Finland, Espoo, Finland.

Krivovichev V., Krivovichev S. and Charykova M. (2019). Selenium minerals: structural and chemical diversity and complexity. *Minerals*, **9**(455), 1–20.

Kumar S., Tomar M. R. and Acharya A. (2015). Carboxylic group-induced synthesis and characterization of selenium nanoparticles and its antitumor potential on Dalton's lymphoma cells. *Colloids and Surfaces B: Biointerfaces*, **126**, 546–552.

Kumari M., Ray L., Purohit M., Patnaik S., Pant A., Shukla Y., Kumar P. and Gupta K. (2017). Curcumin loading potentiates the chemotherapeutic efficacy of selenium nanoparticles in HCT116 cells and Ehrlich's ascites carcinoma bearing mice. *European Journal of Pharmaceutics and Biopharmaceutics*, **117**, 346–362.

Langner B. E. (2000). Selenium and Selenium Compounds. In: Ullmann's Encyclopedia of Industrial Chemistry, M. Bohnet (ed.), 6th edn. Vol 32. Wiley-VCH Verlag GmbH & Co. KGaA, Weinheim, pp. 203–215.

Lavu R. V. S., Du Laing G., Van De Wiele T., Pratti V. L., Willekens K., Vandecasteele B. and Tack F. (2012). Fertilizing soil with selenium fertilizers: impact on concentration, speciation, and bioaccessibility of selenium in leek (*Allium ampeloprasum*). *Journal of Agricultural and Food Chemistry*, **60**(44), 10930–10935.

Leblanc J.-C., Guérin T., Noël L., Calamassi-Tran G., Volatier J.-L. and Verger P. (2005). Dietary exposure estimates of 18 elements from the 1st French Total Diet Study. *Food Additives & Contaminants*, **22**(7), 624–641.

Lemly A. D. (1999). Selenium transport and bioaccumulation in aquatic ecosystems: a proposal for water quality criteria based on hydrological units. *Ecotoxicology and Environmental Safety*, **42**(2), 150–156.

Lemly A. D. (2004). Aquatic selenium pollution is a global environmental safety issue. *Ecotoxicology and Environmental Safety*, **59**(1), 44–56.

Lenz M. and Lens P. N. L. (2009). The essential toxin: the changing perception of selenium in environmental sciences. *The Science of the Total Environment*, **407**, 3620–3633.

Levander O. A. (1986). Selenium. In: Trace Elements in Human and Animal Nutrition, W. Mertz (ed.), Academic Press, London, pp. 139–197.

Li Y.-H. (1991). Distribution patterns of the elements in the ocean: a synthesis. *Geochimica et Cosmochimica Acta*, **55**(11), 3223–3240.

Li Y.-H. (2000). A Compendium of Geochemistry: from Solar Nebula to the Human Brain. Princeton University Press, Princeton.

Lipiec E., Siara G., Bierla K., Ouerdane L. and Szpunar J. (2010). Determination of selenomethionine, selenocysteine, and inorganic selenium in eggs by HPLC-inductively coupled plasma mass spectrometry. *Analytical and Bioanalytical Chemistry*, **397**, 731–741.

Liu W., Li X., Wong Y. S., Zheng W., Zhang Y, Cao W. and Chen T. (2012). Selenium nanoparticles as a carrier of 5- fluorouracil to achieve anticancer synergism. *ACS Nano*, **6**(8), 6578–6591.

Liu A., Ren F., Lin W. Y. and Wang J. Y. (2015). A review of municipal solid waste environmental standards with a focus on incinerator residues. *International Journal of Sustainable Built Environment*, **4**, 165–188.

Losi M. E. and Frankenberger W. T. (1997). Reduction of selenium oxyanions by *Enterobacter cloacae* SLD1a-1: isolation and growth of the bacterium and its expulsion of selenium particles. *Applied and Environmental Microbiology*, **63**, 3079–3084.

Luo H., Wang F., Bai Y., Chen T. and Zheng W. (2012). Selenium nanoparticles inhibit the growth of HeLa and MDA- MB-231 cells through induction of S phase arrest. *Colloids and Surfaces B: Biointerfaces*, **94**, 304–308.

Malisa E. P. (2001). The behaviour of selenium in geological processes. *Environmental Geochemistry and Health*, **23**, 137–158.

Mayland H. F. (1994). Selenium in plant and animal nutrition. In: Selenium in the Environment, W. T. Frankenberger, Jr. and S. Benson (eds.), Marcel Dekker Inc, New York, USA, pp. 29–45.

Meher A. K., Jadhav A., Labhsetwar N. and Bansiwal A. (2020). Simultaneous removal of selenite and selenate from drinking water using mesoporous activated alumina. *Applied Water Science*, **10**, 1–12.

Mierzejewski M., Wismer M. W. and Sathrugnan K. (2017). U.S. Patent No. 9 567 249. U.S. Patent and Trademark Office, Washington, DC.

Miller L. L., Rasmussen J. B., Palace V. P., Sterling G. and Hontela A. (2013). Selenium bioaccumulation in stocked fish as an indicator of fishery potential in pit lakes on reclaimed coal mines in Alberta, Canada. *Environmental Management*, **52**(1), 72–84.

Misund A., Frengstad B., Siewers U. and Reimann C. (1999). Variation of 66 elements in European bottled mineral waters. *Science of the Total Environment*, **243–244**, 21–41.

Mol G., Spijker J., van Gaans P. and Römkens P. (2012). Geochemische bodematlas van Nederland. Wageningen Academic Publishers, Wageningen (in Dutch).

Mollah M. Y. A., Morkovsky P., Gomes J. A. G., Kesmez M., Parga J. and Cocke D. L. (2004). Fundamentals, present and future perspectives of electrocoagulation. *Journal of Hazardous Material*, **114**, 199–210.

Momma K. and Izumi F. (2011). VESTA 3 for three-dimensional visualization of crystal, volumetric and morphology data. *Journal of Applied Crystallography*, **44**, 1272–1276.

Morgan J. W., Higuchi H., Takahashi H. and Hertogen J. (1978). Chondritic Eucrite Parent Body – Inference from Trace- Elements. *Geochimica et Cosmochimica Acta*, **42**(1), 27–38.

Mosher B. W. and Duce R. A. (1987). A global atmospheric selenium budget. *Journal of Geophysical Research*, **92**, 13289–13298.

Musgrove M., Beck J. A., Paschke S. S., Bauch N. J. and Mashburn S. L. (2014). Quality of groundwater in the Denver Basin aquifer system, Colorado. U.S. Geological Survey Scientific Investigations Report **2014–5051**, 1–107.

Muniz-Naveiro O., Dominguez-Gonzalez R., Bermejo-Barrera A., Bermejo-Barrera P., Cocho J. A. and Fraga J. M. (2007). Selenium speciation in cow milk obtained after supplementation with different selenium forms to the cow feed using liquid chromatography coupled with hydride generation atomic fluorescence spectrometry. *Talanta*, **71**, 1587–1593.

Mustacich D. and Powis G. (2000). Thioredoxin reductase. *Biochemical Journal*, **346**, 1–8.

Nancharaiah Y. V. and Lens P. N. L. (2015a). The ecology and biotechnology of selenium respiring bacteria. *Microbiology and Molecular Biology Reviews*, **79**(1), 61–80.

Nancharaiah Y. V. and Lens P. N. L. (2015b). Selenium biomineralization for biotechnological applications. *Trends in Biotechnology*, **33**(6), 323–330.

Natasha Shahid M., Niazi N. K., Khalid S., Murtaza B., Bibi I. and Rashid M. I. (2018). A critical review of selenium biogeochemical behavior in soil-plant system with an inference to human health. *Environmental Pollution*, **234**, 915–934.

Naumov A. V. (2010). Selenium and tellurium: state of the markets, the crisis, and its consequences. *Metallurgist*, **54**, 197–200.

Neal R. H. (1995). Selenium. In: Heavy Metals in Soils, B. J. Alloway (ed.), Blackie Academic & Professional, London, pp. 260–283.

Neuhierl B. and Böck A. (1996). On the mechanism of selenium tolerance in selenium-accumulating plants: Purification and characterization of a specific selenocysteine methyltransferase from cultured cells of *Astragalus bisulcatus*. *European Journal of Biochemistry*, **239**, 235–238.

Nijenhuis I. A., Bosch H. J., Sinninghe Damsté J. S., Brumsack H. J. and De Lange G. J. (1999). Organic Matter and Trace Element Rich Sapropels and Black Shales: A Geochemical Comparison. *Earth and Planetary Science Letters*, **169**(3–4), 277–290.

Nriagu J. O. (1989). A global assessment of natural sources of atmospheric trace metals. *Nature*, **338**, 47–49.

Nriagu J. O. and Pacyna J. M. (1988). Quantitative assessment of worldwide contamination of air, water and soils by trace metals. *Nature*, **333**, 134–139.

Oldfield J. E. (2002). Selenium World atlas. Selenium-Tellurium Development Association (STDA). Grimbergen, Belgium.

Oremland R. S., Herbel M. J., Blum J. S., Langley S., Beveridge T. J., Ajayan P. M., Sutto T., Ellis A. V. and Curran S. (2004). Structural and spectral features of selenium nanospheres produced by Se-respiring bacteria. *Applied Environmental Microbiology*, **70**, 52–60.

Overman S. D. (2000). U.S. Patent No. 6 156 191. U.S. Patent and Trademark Office, Washington, DC.

Papp L. V., Lu J., Holmgren A. and Khanna K. K. (2007). From selenium to selenoproteins: synthesis, identity, and their role in human health. *Antioxidants and Redox Signalling*, **9**(7), 775–806.

Parnell J., Brolly C., Spinks S. and Bowden S. (2016). Selenium enrichment in Carboniferous Shales, Britain and Ireland: Problem or opportunity for shale gas extraction? *Applied Geochemistry*, **66**, 82–87.

Pažout R., Sejkora J. and Šrein V. (2019). Ag-Pb-Sb Sulfosalts and Se-rich Mineralization of Anthony of Padua Mine near Poličany–Model Example of the Mineralization of Silver Lodes in the Historic Kutná Hora Ag-Pb Ore District, Czech Republic. *Minerals*, **9**, 430, 1–47.

Pérez-Sirvent C., Martínez-Sánchez M. J., García-Lorenzo M. L., Molina J., Tudela M. L., Mantilla W. and Bech J. (2010). Selenium content in soils from Murcia Region (SE, Spain). *Journal of Geochemical Exploration*, **107**(2), 100–109.

Petrik A., Thiombane M., Lima A., Albanese S., Buscher J. T. and De Vivo B. (2018). Soil contamination compositional index: a new approach to quantify contamination demonstrated by assessing compositional source patterns of potentially toxic elements in the Campania Region (Italy). *Applied Geochemistry*, **96**, 264–276.

Piacenza E., Presentato A., Zonaro E., Lemire J. A., Demeter M., Vallini G., Turner R. J. and Lampis S. (2017). Antimicrobial activity of biogenically produced spherical Se-nanomaterials embedded in organic material against *Pseudomonas aeruginosa* and *Staphylococcus aureus* strains on hydroxyapatite-coated surfaces. *Microbial Biotechnology*, **10**(4), 804–818.

Pickett T. M. and Sonstegard J. N. (2012). 'Selenium Separation and Recovery from Bioreactor Sludge.' U.S. Patent Application No. 12/971,585. U.S. Patent and Trademark Office, Washington, DC.

Plant J. A., Bone J., Voulvoulis N., Kinniburgh D. G., Smedley P. L., Fordyce F. M. and Klinck B. (2014). Chapter 11.2: arsenic and selenium. In: Environmental Geochemistry, B. Sherwood Lollar (ed.), H.D. Holland and K.K. Turekian (executive editors). Treatise on Geochemistry, 2nd edn, volume 11. Elsevier, Oxford, pp. 13–57.

Prasad K. S. and Selvaraj K. (2014). Biogenic synthesis of selenium nanoparticles and their effect on As(III)-induced toxicity on human lymphocytes. *Biology of Trace Elements Research*, **157**(3), 275–283.

Preda C., Vasiliu I., Bredetean O., Ciobanu D. G., Ungureanu M.-C., Leustean E. L., Grigorovici A., Oprisa C. and Vulpoi C. (2015). Selenium in the environment: essential or toxic to human health? *Environmental and Engineering Management Journal*, **15**, 913–921.

Presentato A., Piacenza E., Anikovskiy M., Cappelletti M., Zannoni D. and Turner R. J. (2018). Biosynthesis of selenium-nanoparticles and -nanorods as a product of selenite bioconversion by the aerobic bacterium *Rhodococcus aetherivorans* BCP1. *New Biotechnology*, **41**, 1–8.

Presser T. S. (1999). Selenium pollution. In: Encyclopedia of Environmental Science, P. E. Alexander and R. W. Fairbridge (eds.), Springer-Verlag, Berlin, Germany, pp. 554–556.

Presser T. S., Piper D. Z., Bird K. J., Skorupa J. P., Hamilton S. J., Detwiler S. J. and Huebner M. A. (2004). The Phosphoria Formation – a model for forecasting global selenium sources to the environment. In: Life Cycle of the Phosphoria Formation – From Deposition to the Post-mining Environment, J. R. Hein (ed.), Elsevier, Amsterdam, *Handbook of Exploration and Environmental Geochemistry*, **8**, 299–319.

Rayman M. P., Infante H. G. and Sargent M. (2008). Food-chain selenium and human health: spotlight on speciation. *British Journal of Nutrition*, **100**, 238–253.

Ramya S., Shanmugasundaram T. and Balagurunathan R. (2015). Biomedical potential of antibacterially synthesized selenium nanoparticles with special reference to anti-biofilm, anti-oxidant, wound healing, cytotoxic and anti-viral activities. *Journal of Trace Elements in Medicine and Biology*, **32**, 30–39.

Rawlins B. G., McGrath S. P., Scheib A. J., Breward N., Cave M., Lister T. R., Ingham M., Gowing C. and Carter S. (2012). The Advanced Soil Geochemical Atlas of England and Wales. British Geological Survey, Keyworth. www.bgs.ac.uk/gbase/advsoilatlasEW.html.

Reimann C. and Birke M. (eds.) (2010). Geochemistry of European Bottled Water. Borntraeger Science Publishers, Stuttgart.

Reimann C. and de Caritat P. (1998). Chemical Elements in the Environment. Springer, Berlin.

Reimann C. and de Caritat P. (2017). Establishing geochemical background variation and threshold values for 59 elements in Australian surface soil. *Science of the Total Environment*, **578**, 633–648.

Reimann C., Ayras M., Chekushin V. A., Bogatyrev I., Boyd R., de Caritat P., Dutter R., Finne T. E., Halleraker J. H., Jæger Ø., Kashulina G., Lehto O., Niskavaara H., Pavlov V. A. and Raisanen L. (1997a). Environmental geochemical atlas of the central Barents region. Norges geologiske undersôkelser (NGU) Geological Survey of Norway.

Reimann C., Boyd R., de Caritat P., Halleraker J. H., Kashulina G., Niskavaara H. and Bogatyrev I. (1997b). Topsoil (0–5 cm) composition in eight arctic catchments in northern Europe (Finland, Norway and Russia). *Environmental Pollution*, **95**(1), 45–56.

Reimann C., Äyräs M., Chekushin V., Bogatyrev I., Boyd R., Caritat P., Dutter R., Finne T. E., Halleraker J. H., Jæger Ø., Kashulina G., Lehto O., Niskavaara H., Pavlov V. K., Räisänen M. L., Strand T. and Volden T. (1998). Geochemical analysis of soil in the central Barents Region. *PANGAEA*, https://doi.org/10.1594/PANGAEA.686641

Reimann C., Bjorvatn K., Frengstad B., Melaku Z., Tekle-Haimanot R. and Siewers U. (2003a). Drinking water quality in the Ethiopian section of the East African Rift Valley I – Data and health aspects. *Science of the Total Environment*, **311**(1–3), 65–80.

Reimann C., Siewers U., Tarvainen T., Bityukova L., Eriksson J., Giucis A., Gregorauskiene V., Lukashev V. K., Matinian N. N. and Pasieczna A. (2003b). Agricultural Soils in Northern Europe. A geochemical atlas. Geologisches Jahrbuch Sonderhefte Reihe D, Heft 5, E. Schweizerbart'sche Verlangsbuchhandlung, Stuttgart.

Reimann C., Arnoldussen A., Englmaier P., Filzmoser P., Finne T. E., Garrett R. G., Koller F. and Nordgulen Ø. (2007). Element concentrations and variations along a 120-km transect in southern Norway – Anthropogenic vs. geogenic vs. biogenic element sources and cycles. *Applied Geochemistry*, **22**(4), 851–871.

Reimann C., Englmaier P., Flem B., Gough L., Lamothe P., Nordgulen Ø. and Smith D. (2009a). Geochemical gradients in soil O-horizon samples from southern Norway: Natural or anthropogenic? *Applied Geochemistry*, **24**(1), 62–76.

Reimann C., Finne T. E., Nordgulen Ø., Saether O. M., Arnoldussen A. and Banks D. (2009b). The influence of geology and land-use on inorganic stream water quality in the Oslo region, Norway. *Applied Geochemistry*, **24**, 1862–1874.

Reimann C., Birke M., Demetriades A., Filzmoser P. and O'Connor P. (eds.) (2014). Chemistry of Europe's Agricultural Soils, Part A. Geologisches Jahrbuch, Band B 102, E. Schweizerbart'sche Verlangsbuchhandlung, Stuttgart.

Reimann C., Englmaier P., Fabian K., Gough L., Lamothe P. and Smith D. (2015a). Biogeochemical plant-soil interaction: Variable element composition in leaves of four plant species collected along a south–north transect at the southern tip of Norway. *Science of the Total Environment*, **506–507**, 480–495.

Reimann C., Fabian K., Schilling J., Roberts D. and Englmaier P. (2015b). A strong enrichment of potentially toxic elements (PTEs) in Nord-Trøndelag (central Norway) forest soil. *Science of the Total Environment*, **536**, 130–141.

Reimann C., Birke M., Flem B., Eggen O. A. and Gasser D. (2018a). Background values of gold, potentially toxic elements and emerging high-tech critical elements in surface water collected in a remote northern European environment. *Geochemistry: Exploration, Environment, Analysis*, **18**(3), 185–195.

Reimann C., Fabian K., Birke M., Filzmoser P., Demetriades A., Négrel P., Oorts K., Matschullat J and de Caritat P., The GEMAS Project Team. (2018b). GEMAS: Establishing geochemical background and threshold for 53 chemical elements in European agricultural soil. *Applied Geochemistry*, **88**(Part B), 302–318.

Ren L. Q., Li X. J., Li G. S., Zhao Z. T., Sun B. and Sun F. (2004). Coxsackievirus B3 infection and its mutation in Keshan disease. *World Journal of Gastroenterology*, **10**(22), 3299–3302.

Reyes L. H., Mar J. L., Rahman G. M., Seybert B., Fahrenholz T. and Kingston H. M. (2009). Simultaneous determination of arsenic and selenium species in fish tissues using microwave assisted enzymatic extraction and ion chromatography inductively coupled plasma mass spectrometry. *Talanta*, **78**, 983–990.

Rodrigues Nogueira T. A., Abreu-Junior C. H., Ferracciú Alleoni L. R., He Z., Soares M. R., dos Santos Vieira C., Frediani Lessa L. G. and Capra G. F. (2018). Background concentrations and quality reference values for some potentially toxic elements in soils of São Paulo State, Brazil. *Journal of Environmental Management*, **221**, 10–19.

Rosenfeld I. and Beath O. A. (1964). Selenium: Geobotany, Biochemistry, Toxicity and Nutrition. Academic Press, New York, USA.

Roy M., Kiremidjian-Schumacher L., Wishe H. I., Cohen M. W. and Stotzky G. (1994). Supplementation with selenium and human immune cell functions. I. Effect on lymphocyte proliferation and interleukin 2 receptor expression. *Biology of Trace Elements Research*, **41**, 103–114.

Rudnick R. L. and Gao S. (2003). 3.01 – Composition of the Continental Crust. In: Treatise on Geochemistry, H. D. Holland and K. K. Turekian (eds.), Pergamon, pp. 1–64.

Sabrynal H. H. A., Yahiya N. S. M., Abdel-Rahman E. A. M. and Radwan S. A. (2011). Selenium status in some Egyptian soils. *Environmental Sciences*, **6**, 123–131.

Salminen R., Batista M. J., Bidovec M., Demetriades A., De Vivo B., De Vos W., Duris M., Gilucis A., Gregorauskiene V., Halamic J., Heitzmann P., Lima A., Jordan G., Klaver G., Klein P., Lis J., Locutura J., Marsina K., Mazreku A., O'Connor P. J., Olsson S. Å., Ottesen R.-T., Petersell V., Plan J. A., Reeder S., Salpeteur I., Sandström H., Siewers U., Steenfelt A. and Tarvainen T. (eds.). (2005). Geochemical Atlas of Europe. Part 1 – Background Information, Methodology and Maps. Geological Survey of Finland, Espoo, Finland.

Santos S., Ungureanu G., Boaventura R. and Botelho C. (2015). Selenium contaminated waters: An overview of analytical methods, treatment options and recent advances in sorption methods. *Science of the Total Environment*, **521–522**, 246–260.

Sarret G., Avoscan L., Carriere M., Collins R., Geoffroy N., Carrot F., Coves J. and Gouget B. (2005). Chemical forms of selenium in the metal-resistant bactyerium *Ralstonia metallidurans* CH34 exposed to selenite and selenate. *Applied and Environmental Microbiology*, **71**, 2331–2337.

Scheib C., Flight D., Lister T. R., Scheib A. and Breward N. (2011). Selenium in topsoil. Summary interpretation. Downloaded from https://www.bgs.ac.uk/gbase/londonearth.html

Schrautzer G. N. (2004). Selenium. In: Elements and Their Compounds in the Environment, 2nd edn, E. Merian, M. Anke, M. Ihnat and M. Stoeppler (eds.), Wiley-VCH Verlag GmbH & Co, Weinheim, pp. 1365–1406.

Shand P. and Edmunds W. M. (2008). The Baseline Inorganic Chemistry of European Groundwaters. In: Natural Groundwater Quality, W. M. Edmunds and P. Shand (ed.), Blackwell Publishing Co., pp. 22–58.

Sharma V. K., McDonald T. J., Sohn M., Anquandah G. A., Pettine M. and Zboril R. (2015). Biogeochemistry of selenium: a review. *Environmetal Chemistry Letters*, **13**(1), 49–58.

Sidique T., Zhang Y., Oleke B. C. and Frankenberger W. T. (2006). Characterization of sediment bacteria involved in selenium reduction. *Bioresource Technology*, **97**, 1041–1049.

Smith D. B., Solano F., Woodruff L. G., Cannon W. F. and Ellefsen K. J. (2019). Geochemical and mineralogical maps, with interpretation, for soils of the conterminous United States. U.S. Geological Survey Scientific Investigations Report, **2017–5118**.

Soda S., Kashiwa M., Kagami T., Kuroda M., Yamashita M. and Ike M. (2011). Laboratory scale bioreactors for soluble selenium removal from selenium refinery wastewater using anaerobic sludge. *Desalination*, **279**, 433–438.

Soflaei S., Dalimi A., Abdoli A., Kamali M., Nasiri V., Shakibaie M. and Tat M. (2014). Anti-leishmanial activities of selenium nanoparticles and selenium dioxide on *Leishmania infantum*. *Comparative Clinical Pathology*, **23**(1), 15– 20.

Sonkusre P., Nanduri R., Gupta P. and Cameotra S. S. (2014). Improved extraction of intracellular biogenic selenium nanoparticles and their specificity for cancer chemoprevention. *Journal of Nanomedicine and Nanotechnology*, **5**(194), 1–9.

Sors T. G., Ellis D. R. and Salt D. E. (2005). Selenium uptake, translocation, assimilation and metabolic fate in plants. *Photosynthesis Research*, **86**, 373–389.

Spadoni M., Voltaggio M., Carcea M., Coni E., Raggi A. and Cubadda F. (2007). Bioaccessible selenium in Italian agricultural soils: comparison of the biogeochemical approach with a regression model based on geochemical and pedoclimatic variables. *Science of the Total Environment*, **376**(1–3), 160–177.

Sparks D. L. (2003). Environmental Soil Chemistry, 2nd edn. Academic Press, Amsterdam, The Netherlands.

Stadtman T. C. (1990). Selenium biochemistry. *Annual Review of Biochemistry*, **59**, 111–127.

Steinnes E. (2003). Biogeochemical cycling of iodine and selenium and potential geomedical relevance. In: Geology and Health: Closing Gap, H. C. Skinner and A. R. Berger (eds.), Oxford University Press, Oxford, pp. 57–60.

Stillings L. L. (2017). Selenium. In: Critical Mineral Resources of the United States—Economic and Environmental Geology and Prospects for Future Supply, K. J. Schulz, J. H. DeYoung, Jr., R. R. Seal, II and D. C. Bradley (eds.), *U.S. Geological Survey Professional Paper* **1802-Q**, U.S. Geological Survey, Reston, VA U.S.A., 1–55.

Torres J., Pintos V., Gonzatto L., Domínguez S., Kremer C. and Kremer E. (2011). Selenium chemical speciation in natural waters: protonation and complexation behavior of selenite and selenate in the presence of environmentally relevant cations. *Chemical Geology*, **288**, 32–38.

Trofast J. (2011). Berzelius' Discovery of Selenium. *Chemistry International*, **33**(5). http://publications.iupac.org/ci/2011/3305/5_trofast.html

Tugarova A. V., Vetchinkina E. P., Loshchinina E., Burov A. M., Nikitina V. E. and Kamnev A. (2014). Reduction of selenite by *Azospirillum brasilense* with the formation of selenium nanoparticles. *Microbial Ecology*, **68**, 495–503.

Tan L. C., Nancharaiah Y. V., van Hullebusch E. D. and Lens P. N. L. (2016). Selenium: environmental significance, pollution, and biological treatment technologies. *Biotechnology Advances*, **34**, 886–907.

Tan L. C., Jing X., Zhang Y., Tang H., Yao S. and Xie Q. (2009). In vitro study on the individual and synergistic cytotoxicity of adriamycin and selenium nanoparticles against Bel7402 cells with a quartz crystal microbalance. *Biosensors and Bioelectronics*, **24**, 2268–2272.

Terry N., Zayed A. M., de Souza M. P. and Tarun A. S. (2000). Selenium in higher plants. *Annual Review in Plant Physiology and Plant Molecular Biology*, **51**, 401–432.

Torres S. K., Campos V. L., León C. G., Rodríguez-Llamazares S. M., Rojas S. M., González M., Smith C. and Mondaca M. A. (2012). Biosynthesis of selenium nanoparticles by *Pantoea agglomerans* and their antioxidant activity. *Journal of Nanoparticle Research*, **14**(1236), 1–9.

Turner R. J., Weiner J. H. and Taylor D. E. (1998). Tellurite-mediated thiol oxidation in *Escherichia coli. Microbiology*, **145**, 33–40.

U.S. Geological Survey (2019). Selenium and Tellurium statistics and information. https://www.usgs.gov/centers/nmic/, pp. 146–147.

USEPA (1992). Method 1311. Toxicity Characteristic Leaching Procedure. *The United States Environmental Protection Agenc*, pp. 1–27.

Valdez Barillas J. R., Quinn C. F., Freeman J. L., Lindblom S. D., Fakra S. C., Marcus M. A., Gilligan T. M., Alford É. R., Wangeline A. L. and Pilon-Smits E. A. (2012). Selenium distribution and speciation in the hyperaccumulator *Astragalus bisulcatus* and associated ecological partners. *Plant Physiology*, **159**(4), 1834–1844.

Vinceti M., Wei E. T., Malagoli C., Bergomi M. and Vivoli G. (2001). Adverse health effects of selenium in humans. *Reviews on Environmental Health*, **16**, 233–251.

Wacker M. G. (2014). Nanotherapeutics – product development along the 'nanomaterial' discussion. *Journal of Pharmaceutical Sciences*, **103**(3), 777–784.

Walsh T. and Fleming G. A. (1952). Selenium levels in rocks, soils and herbage from a high selenium locality in Ireland. *Transaction of the 5th International Congress on Soil Sciences*, **2**(A), 178–183.

Wang H., Zhang J. and Yu H. (2007). Elemental selenium at nano size possesses lower toxicity without compromising the fundamental effect on selenoenzymes: comparison with selenomethionine in mice. *Free Radicals in Biology and Medicine*, **42**(10), 1524–1533.

Wang C., Li S., Wang H. and Fu J. (2016a). Selenium minerals and the recovery of selenium from copper refinery anode slimes. *Journal of the Southern African Institute of Mining and Metallurgy*, **116**(6), 593–600.

Wang Y., Dang F., Zhao J. and Zhong H. (2016b). Selenium inhibits sulphate-mediated methylmercury production in rice paddy soil. *Environmental Pollution*, **213**, 232–239.

Webb J. S., Thornton I. and Fletcher K. (1966). Seleniferous soils in parts of England and Wales. *Nature*, **211**(5046), 327–330.

Wen H. and Carignan J. (2011). Selenium isotopes trace the source and redox processes in the black shale-hosted Se- rich deposits in China. *Geochimica et Cosmochimica Acta*, **75**(6), 1411–1427.

Whanger P. D. (2002). Selenocompounds in plants and animals and their biological significance. *Journal of the American College of Nutrition*, **21**, 223–232.

White P. J. (2016). Selenium accumulation by plants. *Annals of Botany*, **117**(2), 217–235.

Winkel L. H., Johnson C. A., Lenz M., Grundl T., Leupin O. M., Amini M. and Charlet L. (2012). Environmental selenium research: from microscopic processes to global understanding. *Environmental Science Technology*, **46**(2), 571–579.

Winkel L. H. E., Vriens B., Jones G. D., Schneider L. S., Pilon-Smits E. and Bañuelos G. S. (2015). Selenium cycling across soil-plant-atmosphere interfaces: a critical review. *Nutrients*, **7**(6), 4199–4239.

World Health Organization (2011). Selenium in drinking water. *Guidelines for Drinking-water Quality*. https://www.who.int/water_sanitation_health/dwq/chemicals/selenium.pdf, pp. 1–22.

Wu H., Zhu H., Li X., Liu Z., Zhen W., Chen T., Yu B. and Wong K.-H. (2013). Induction of apoptosis and cell cycle arrest in A549 human lung adenocarcinoma cells by surface-capping selenium nanoparticles: an effect enhanced by polysaccharide-protein complexes from *Polyporus rhinocerus*. *Journal of Agricultural and Food Chemistry*, **61**, 9859–9866.

Wu H., Li X., Liu W., Chen T., Li Y., Zhen W., Man C. W.-Y., Wong M.-K. and Wong K.-H. (2012). Surface decoration of selenium naoparticles by mushroom polysaccharide-protein complexes to achieve enhanced cellular uptake and anti-proliferative activity. *Journal of Materials Chemistry*, **22**, 9602–9610.

Xia Y., Lin Z., Li Y., Zhao M., Wang C., Guo M., Zhang B. and Zhu B. (2017). Targeted delivery of siRNA using RGDfC-conjugated functionalized selenium nanoparticles for anticancer therapy. *Journal of Materials Chemistry*, **B5**, 6941–6952.

Xie X. and Cheng H. (2001). Global geochemical mapping and its implementation in the Asia–Pacific region. *Applied Geochemistry*, **16**(11–12), 1309–1321.

Xu Y., Hao Z., Li Y., Li H., Wang L., Zang Z., Liao X. and Zhang R. (2020). Distribution of selenium and zinc in soil-crop system and their relationship with environmental factors. *Chemosphere*, **242**, 125289.

Yao R., Wang R., Wang D., Su J., Zheng S. and Wang G. (2014). Paenibacillus selenitireducens sp. nov., a selenite reducing bacterium isolated from a selenium mineral soil. *International Journal of System and Evolutionary Microbiology*, **64**, 805–811.

Yanai J., Okada T. and Yamada H. (2012) Elemental composition of agricultural soils in Japan in relation to soil type, land use and region. *Soil Science and Plant Nutrition*, **58**(1), 1–10.

Yang F., Tang Q., Zhong X., Bai Y., Chen T., Zhang Y., Li Y. and Zheng W. (2012). Surface decoration by Spirulina polysaccharide enhances the cellular uptake and anticancer efficacy of selenium nanoparticles. *International Journal of Nanomedicine*, **7**, 835–844.

Yazdi M. H., Mahdavi M., Setayesh N., Esfandyar M. and Shahverdi A. R. (2013). Selenium nanoparticle-enriched *Lactobacillus brevis* causes more efficient immune responses in vivo and reduces the liver metastasis in metastatic form of mouse breast cancer. *DARU. Journal of Pharmaceutical Sciences*, **21**(1), 33.

Yin J., Hou Y., Yin Y. and Song X. (2017). Selenium-coated nanostructured lipid carriers used for oral delivery of berberine to accomplish a synergic hypoglycemic effect. *International Jornal of Nanomedicine*, **12**, 8671–8680.

Ysart G., Miller P., Croasdale M., Crews H., Robb P., Baxter M., De L'Argy C. and Harrison N. (2000). 1997 UK Total Diet Study dietary exposures to aluminium, arsenic, cadmium, chromium, copper, lead, mercury, nickel, selenium, tin and zinc. *Food Additives & Contaminants*, **17**(9), 775–786.

Yu B., Zhang Y., Zhen W., Fan C. and Chen T. (2012). Positive charge enhances selective cellular uotake and anticanver efficacy of selenium nanoparticles. *Inorganic Chemistry*, **51**, 8956–8963.

Zannoni D., Borsetti F., Harrison J. J. and Turner J. R. (2008). The bacterial response to the chalcogen metalloids Se and Te. *Advances in Microbial Physiology*, **53**, 1–71.

Zduôska H., Bloch-BogusÝawska E., Przygoôska J., Þliwka K. and Zachara B. A. (1994). Selenium concentration and glutathione peroxidase activity in the organs of people died a violent death. In: Arsenic and Selenium in the Environment. Ecologic and Analytic Problem, A. Kabata-Pendias and B. Szteke (eds.), Polish Academy of Sciences, pp. 190–193 (in Polish).

Zhai X., Zhang C., Zhao G., Stoll S., Ren F. and Leng X. (2017). Antioxidant capacities of the selenium nanoparticles stabilized by chitosan. *Journal of Nanobiotechnology*, **15**(1), 4.

Zhang Y. and Frankenberger W. T., Jr. (2005). Removal of selenium from river water by microbial community enhanced with *Enterobacter taylorae* in organic carbon coated sand columns. *Science of the Total Environment*, **346**, 280–285.

Zhang J., Wang H., Bao Y. and Zhang L. (2004). Nano red elemental selenium has no size effect in the induction of seleno-enzymes in both cultured cells and mice. *Life Sciences*, **75**(2), 237–244.

Zhang J., Wang X. and Xu T. (2008). Elemental selenium at nano size (Nano-Se) as a potential chemopreventive agent with reduced risk of selenium toxicity: comparison with Se-methylselenocysteine in mice. *Toxicological Sciences*, **101**(1), 22–31.

Zhang S., Luo Y., Zeng H., Wang Q., Tian F., Song J. and Wen-Hsing Cheng W.-H. (2011). Encapsulation of selenium in chitosan nanoparticles improves selenium availability and protects cells from selenium-induced DNA damage response. *Journal of Nutritional Biochemistry*, **22**(12), 1137–1142.

Zhang Y., Li X., Huang Z., Zhen W., Fan C. and Chen T. (2013). Enhancement of cell permeabilization apostosis- inducing activity of selenium nanoparticles by ATP surface decoration. *Nanomedicine: Nanotechnology in Biology and Medicine*, **7**, 74–84

Zhao S., Yu Q., Pan J., Zhou Y., Cao C., Ouyang J.-M. and Liu J. (2017). Redox-responsive mesoporus selenium delivery of doxorubicin targets MCF-7 cells and synergistically enhances its anti-tumor activity. *Acta Biomaterials*, **54**, 294–306.

Zhu J., Wang N., Li S., Li L., Su H. and Liu C. (2008). Distribution and transport of selenium in Yutangba, China: impact of human activities. *Science of the Total Environment*, **392**(2–3), 252–261.

Chapter 2

Radioactive selenium: origin and environmental dispersion scenarios

Teba Gil-Díaz, Frank Heberling and Elisabeth Eiche

2.1 INTRODUCTION

Radionuclides can be present in the environment from both natural and anthropogenic origins, showing characteristic biogeochemical behaviours according to the specific properties of the element. The environmental mobility and bioavailability of selenium (Se) strongly depends on the chemical species which, in turn, depends on aspects like redox state and microbiology. Among the most common oxidation states, species of Se(IV) and Se(VI) are considered relatively mobile and bioavailable. Once incorporated within an organism, Se shows a narrow band between dietary deficiency (e.g., used as a co-factor in functional proteins and RNA) and toxicity (e.g., selenosis, dependent on the concentrations and the chemical species involved; Jeffery *et al.*, 2002). The recommended daily intake for adult humans is limited to 1 μg kg^{-1} of body weight, with a maximum allowable concentration in drinking water of 10 μg L^{-1} (WHO [World Health Organization], 2011). In addition, Se has no essential metabolism in plants but it is still readily taken up and accumulated due to its structural similarity with other oxyanion forms of bio-essential elements like sulfate and phosphate (Pilon-Smits *et al.*, 2017).

doi: 10.2166/9781789061055_0061

Potential incorporation and accumulation due to the use of Se in the metabolism gives rise to the radiotoxicity of radioactive Se isotopes. Of all the radioactive Se isotopes, the long-lived fission product ^{79}Se shows the biggest concern with respect to the long-term dispersion from disposal sites for high-level nuclear waste and its potential radiotoxicity from beta emissions. Environmental risk assessment of radioactive Se from anthropogenic sources is still a topic of ongoing research aiming to better determine the specific dispersion paths and potential environmental impacts. This chapter will present (i) basic knowledge on radioactive Se, (ii) examples on environmental analytical techniques, (iii) current understanding of its environmental reactivity, dispersion and fate from deep geological repositories to different compartments of the geosphere, and (iv) considerations on distribution and effective doses in the biosphere.

2.2 CHARACTERISTICS OF RADIOACTIVE SELENIUM

2.2.1 Environmental persistence: half-lives and decay modes

Selenium has six stable isotopes: ^{74}Se (0.89%), ^{76}Se (9.37%), ^{77}Se (7.63%), ^{78}Se (23.77%), ^{80}Se (49.61%) and ^{82}Se (8.73%). Despite the fact that ^{82}Se is strictly an unstable nucleus, decaying into stable, gaseous krypton (^{82}Kr), it is mainly seen as a stable isotope when considering environmental processes given its extremely long half-life ($t_{1/2}$ in the order of 10^{19} y; Audi *et al.*, 2003).

From over 21 radioactive Se-isotopes, only two show half-lives long enough to render them persistent in a system ($t_{1/2} > 1$ h): ^{75}Se ($t_{1/2} = 199.78$ d; Audi *et al.*, 2003) decaying into stable ^{75}As by electron capture, and ^{79}Se ($t_{1/2} = 3.27.10^5$ y; Jörg *et al.*, 2010) decaying into stable ^{79}Br by emitting beta radiation (β^-, negatively charged high-energy electrons). From these two, only ^{75}Se releases gamma radiation (γ, photon emissions).

The production of other Se radioisotopes from parent radionuclides produced during nuclear fission (e.g., germanium, Ge) is considered to have a small contribution to the global budget of Se radioisotopes. An example would be ^{81}Se, produced via ^{81}Ge decay ($^{81}Ge \rightarrow {}^{81}As \rightarrow [{}^{81m}Se \rightarrow]\ {}^{81}Se \rightarrow {}^{81}Br$; Gray *et al.*, 2007), presenting short half-lives for both ^{81}Se ($t_{1/2} = 18.45$ min) and its metastable form (^{81m}Se, $t_{1/2} = 57.28$ min). Similarly, radionuclides of ^{72}Se ($t_{1/2} = 8.40$ d) and ^{73}Se ($t_{1/2} = 7.1$ h) show relatively longer half-lives but are only produced by the radioactive decay of Br (Br → Se → As → Ge), resulting from nuclear fusion reactions in research accelerators (e.g., bombardment of ^{58}Ni with ^{16}O to produce ^{72}Br; Mihailescu & Cata-Danil, 2010). Therefore, these Se radioisotopes will not be discussed in this chapter.

2.2.2 Sources and applications

2.2.2.1 Natural geogenic ^{82}Se

The only naturally occurring radionuclide of Se (^{82}Se) originates from both volcanic emissions and the Earth's crust. The average concentrations of ^{82}Se in the Earth's crust are very low with 4–8 µg kg^{-1} (i.e., total Se content of 50–90 µg kg^{-1}; Salminen *et al.*, 2005). Its characteristic radioactive decay ($2\beta^-$) is extremely rare among radioactive emissions, also present in ^{76}Ge, ^{96}Zr, ^{100}Mo, ^{116}Cd and ^{136}Xe. Nevertheless, this radioactive decay has been extensively used in several research programmes (e.g., NEMO-3, SuperNEMO, LUCIFER/CUPID-0; Artusa *et al.*, 2017; Kaizer *et al.*, 2018) with the aim of evidencing neutrinoless double beta decay patterns, which could provide insights into the matter-antimatter asymmetry in the universe. For example, the SuperNEMO programme designs experiments improving the detection limits of double beta decay emissions for several isotopes, tracking the trajectories of the emitted electrons and determining their individual energies (Arnold *et al.*, 2010). Otherwise, specific uses of ^{82}Se include isotopically-labelled experiments in both dissolved and organic chemical forms, given its relatively low natural abundance compared to stable ^{77}Se and ^{78}Se. These studies comprise a wide range of experimental approaches, mostly in biological/ecotoxicological studies for understanding metabolic routes of Se (e.g., Ogra *et al.*, 2018; Shiobara *et al.*, 2000; Suzuki *et al.*, 2006), detoxification mechanisms of Se-bearing proteins concerning the uptake of mercury (Spiller, 2018; Yoneda & Suzuki, 1997) and isotopic fractionation during uptake (Banning *et al.*, 2013; Schilling *et al.*, 2015, 2020; Xu *et al.*, 2020).

2.2.2.2 Anthropogenic radiotracer ^{75}Se

Due to its relatively short half-life ($t_{1/2} = 199.78$ d), ^{75}Se is not naturally present nor relevant for environmental studies. In fact, it can only be produced artificially from proton bombardment of natural arsenous oxide or neutron activation from stable ^{74}Se. It may be present in the cooling pond for UO_2 fuel from light water reactors (e.g., Neeb, 1997), but it is mostly known for its paramount application related to isotopically-labelled experiments. Experimental studies widely use ^{75}Se as a radiotracer in all kinds of chemical forms, ranging from labelled organic ^{75}Se–selenomethionine to inorganic ^{75}Se–selenite, and ^{75}Se–selenate species. The reason for its extensive use in different studies is related to several points such as: (i) its ease of production and simple commercialization, (ii) its half-life, rendering it sufficiently persistent during the time period of usual laboratory experiments, (iii) its decay into stable ^{75}As in a relatively short time, facilitating the management of the radioactive waste, and (iv) its electron capture decay producing characteristic gamma lines detectable by radiometric techniques, allowing it to reach environmentally relevant concentration-levels due to the achievable detection limits in the order of

$<10^{-15}$ mol Se, equivalent to ~75 fg, and simplifying sample preparation (*c.f. Section 2.3.2*).

The use of ^{75}Se as a radiotracer serves the general interest in expanding the knowledge on the biogeochemical behaviour of Se, from both biological and environmental viewpoints. Original works initially focused on improving the understanding of sample loss during pre-treatment and the production of satisfactory figures of merit for analytical techniques quantifying Se species (e.g., hydride generation and atomic absorption spectroscopy; Campbell, 1992; Dočekal *et al.*, 1997; Krivan, 1982; Reamer *et al.*, 1981). Biological studies include a wide range of applications. For instance, many ecotoxicology studies aim at developing a fundamental understanding of Se distribution, bioaccumulation and biomagnification (e.g., in marine organisms, terrestrial plants and mammals; Araie *et al.*, 2008; Beilstein & Whanger, 1988; Einoder *et al.*, 2018; Fowler *et al.*, 1999; Reinfelder & Fisher, 1994). Other works use this knowledge on the metabolism of Se for radioactive diagnostic materials in medical science (Irons *et al.*, 2006; Yang *et al.*, 2017) and the development of new biotechnological solutions (e.g., alternatives to fertilizers for Se-deficient crops; Galinha *et al.*, 2012; Riaz *et al.*, 2018). Additionally, environmentally-oriented research works mainly focus on the sorption of Se to different solid phases in various liquid matrices in order to determine solid/liquid partitioning in pure and natural sediments for purposes of environmental risk assessment, for example, related to radionuclide mobility or plant uptake in Se-deficient areas. Some studies include scenarios of potential radionuclide dispersion from deep geological repositories (e.g., underground research laboratories such as Grimsel in Switzerland; Alexander *et al.*, 2009; Eikenberg *et al.*, 1997). The fate of Se after radionuclide accidental releases and the potential dispersion in specific environments was also studied using ^{75}Se labels (i.e., Gil-Díaz *et al.*, 2020a; Hesslein, 1987; Lee *et al.*, 2012). Other applications focus on remediation purposes concerning the decontamination of Se, among other trace elements, using environmental matrices (e.g., Santschi *et al.*, 1980; Suzuki *et al.*, 2014; Tuğrul *et al.*, 2015).

2.2.2.3 *Anthropogenic ^{79}Se from nuclear fission*

Many radionuclides are products of nuclear fission and may be released to the environment during nuclear power plant (NPP) accidental events (e.g., Chernobyl, Ukraine 26th April 1986, and Fukushima Dai-ichi, Japan 11th March 2011), from fuel reprocessing plants or from nuclear waste repositories. Some of these radionuclides show low fission yields (i.e., low probability of being formed during nuclear fission) but long half-lives (high persistence), and thus accumulate in the fuel and become of relevant concern over time. Out of all the potential fission products of ^{235}U and ^{239}Pu related to Se radionuclides (from ^{75}Se to ^{95}Se), only ^{79}Se fits within these criteria and is by far the most ubiquitous radioisotope of Se in nuclear waste, contributing to the estimated total cumulative radioactive dose (ANDRA, 2005; Ikonen *et al.*, 2016). Such conditions are also the reason why ^{79}Se is scarcely registered nor followed after NPP accidental releases or from admissible daily discharges of nuclear installations (i.e., compared to other known fission products such as ^{137}Cs, ^{131}I, ^{132}Te and ^{125}Sb). In fact, ^{79}Se is rather included in the lists of radionuclides of relevance or of 'potential criticality' during decommissioning of NPPs and from potential radioactive leaks from nuclear waste disposal areas (i.e., from geological repositories; Tanaka *et al.*, 2014). In addition, ^{79}Se is also a main activation product in vitrified waste, spent fuel, hulls and endpieces surrounding the nuclear waste (NIROND, 2001). Upon potential contact of the spent fuel with groundwater, ^{79}Se is expected to be released relatively quickly. Therefore, the interest in ^{79}Se is related to its estimated, future potential environmental mobility and radiation doses from nuclear waste disposal/storage sites to the geosphere and biosphere, essential for the long-term safety assessment of geological repositories (Asai *et al.*, 2011; Hoving, 2018; Ikonen, 2017).

At a minor scale, low and intermediate level wastes are also expected to contain ^{79}Se, not only originating from the nuclear industry, but also as activation products from nuclear applications, research laboratories, and hospitals (Aguerre & Frechou, 2006). Concerning environmental evidence, only one study has reported activities of ^{79}Se in trees at the Fukushima Daiichi NPP site one year after the accident. This study classified ^{79}Se within the 'highly volatile fission products or neutron activated nuclides' together with ^{90}Sr, ^{99}Tc and ^{129}I (Tanaka *et al.*, 2014). They measured values between the detection limit of <0.05 Bq g^{-1} and maximum activities of 0.22 Bq g^{-1} (equivalent to 0.10–0.43 µg kg^{-1}). For comparison, the German radio-protection-regulations (StrlSchV, 2018) specify a maximum limit of 0.1 Bq g^{-1} of ^{79}Se for the unrestricted use of solid and liquid materials. Other studies related to risk assessment of ^{79}Se are based on ^{75}Se radiotracer behaviour in hypothetical conditions with environmentally relevant solid and liquid phases (*c.f. Section* 2.2.2.2) or on studies with stable Se-isotopes.

2.3 SAMPLE COLLECTION AND QUANTIFICATION TECHNIQUES

2.3.1 Environmental sampling

Environmental biogeochemical processes hardly distinguish between stable or radioactive isotopes. For this reason, it is generally assumed that the biogeochemical behaviour of radionuclides follows that of their analogue stable isotopes (e.g., IAEA TRS [International Atomic Energy Agency Technical Reports Series] 422). This also implies that sampling for Se radionuclides in natural matrices follows the same procedures and precautions as for studies on stable environmental Se. A general remark can be made that for Se, no ashing procedures or strong heating should be applied as some Se species may be volatile at >70°C.

The major difference between sampling procedures for radioactive and stable isotopes would be related to the amount of sample collected. In general, sampling of radioactive elements requires larger volumes of water (e.g., >5 L, particularly in seawater) and solid masses than for the analogue stable isotopes, due to the generally low activities found in the environment. For instance, Tanaka *et al.* (2014) point out that plant material should weigh more than 30 g dry weight for radiochemical analysis of ^{79}Se. Another reason for higher sample amounts is the necessity for extensive sample pre-treatment and separation steps. In environmental samples, the radionuclide of interest often has to be pre-concentrated. Separation steps help to avoid interference with other radionuclides, enabling an accurate quantification due to, for example, the uncharacteristic beta emissions from ^{79}Se decay. In contrast, ^{75}Se can be quantified with only minor or no pre-treatment steps due to its characteristic signal from gamma radiation. Therefore, there are no established standard protocols regarding the minimum mass required for optimum quantification of Se radionuclides for specific matrices and sites. General sampling procedures and sample storage strategies for standard radiological monitoring in specific matrices are published in IAEA guidelines, for example, for water, foodstuffs, soil size fractionation and vegetation (Barnekow *et al.*, 2019; Guidebook, 1989; Joint, 2016; Martinčič, 1999).

2.3.2 Analytical methods

The main analytical methods used to quantify both ^{75}Se (gamma-emitter) and ^{79}Se (beta-emitter) are based on the respective radioactive decay signals, summarized for specific cases in Table 2.1. The most commonly applied method for ^{75}Se is gamma spectroscopy (e.g., based on a high purity Germanium HPGe or a NaI (Tl) crystal), related to its characteristic gamma energy lines (e.g., 264.7 and 136 keV energy lines). When applicable, ^{79}Se may be quantified by liquid scintillation counting (LSC) through the excitation of a mixture of reagents, sensitive to beta-radiation. Speciation of Se can be determined by using X-ray spectroscopic techniques, though few studies report measurements on radioactive Se (Table 2.1).

Noteworthy, the quantification of ^{75}Se by gamma spectroscopy is much less affected by matrix effects compared to classical spectrometric techniques (mass-MS, atomic absorption-AAS, or optical emission-OES), and can be directly applied to salt brines, seawater, or digestion solutions. These characteristics present big advantages in experimental approaches using radiotracers of ^{75}Se over spikes of Se stable isotopes. Until recently, analytical challenges in the

Table 2.1 Examples of quantification methods used in different matrices and experimental conditions concerning both ^{75}Se and ^{79}Se.

Context	Radionuclide and Matrix	Pre-treatment and/or Separation Method	Quantification Method	Reference
Spikes of ^{75}***Se***				
Selenoproteins in coccolithophores	Biological samples	Sonication then several purification and size fractionation steps	NaI(Tl) gamma-detector	Araie *et al.* (2008)
Se transfer from zooplankton to fish	Biological samples	Centrifugation and biochemical fractionation steps	NaI(Tl) gamma-detector (264 keV)	Reinfelder and Fisher (1994)
Se enrichment in seeds	Water and plant seeds	None specified	HPGe (136 keV)	Galinha *et al.* (2011)
Se solid fractionation in forest floors	Lake water	Soil extractions with KH_2PO_4 and NaOH; Water spikes fractionated in HPLC columns; Centrifugation	'Well-type scintillation counter'	Gustafsson and Johnsson (1994)
Se solid/liquid partitioning in estuarine sediment suspensions	River and estuarine water	Centrifugation	HPGe (264.7 keV) (DL: <0.03 Bq mL^{-1} in 10 mL)	Gil-Díaz *et al.* (2020a)
Se retention by mangrove sediment columns	Tidal water	None specified	HPGe	Suzuki *et al.* (2014)
Batch experiments with modified bentonites	Ultra-pure water	None specified	'High resolution gamma spectrometry'	Tuğrul *et al.* (2015)
Batch experiments with bentonite and quartz	Synthetic ground and freshwater	Centrifugation	NaI(Tl) gamma-detector (500–700 keV)	Lee *et al.* (2012)

(*Continued*)

Table 2.1 Examples of quantification methods used in different matrices and experimental conditions concerning both ^{75}Se and ^{79}Se (*Continued*).

Context	Radionuclide and Matrix	Pre-treatment and/or Separation Method	Quantification Method	Reference
Injected radioactive tracer at Grimsel Test Site	Synthetic solution	None specified	HPGe (DL: <1 Bq m^{-1} and <0.01 Bq mL^{-1} in 1 L)	Eikenberg *et al.* (1997)
Radiochemical protocol to quantify ^{79}Se	Simulated intermediate level wastes (solids)	Microwave digestion including HF	n-type Be detector (264.7 keV)	Aguerre and Frechou (2006)
Direct measurements of ^{79}Se				
Radioactive contamination after the Fukushima NPP accident	Natural tree samples (30 g)	Acid extraction with 100 mL 10M HNO_3; Elution from TEVA® Resin with HBr; Reduction by hydroxylamine; Precipitate re-dissolved in HNO_3	LSC (DL: <0.05 Bq g^{-1})	Tanaka *et al.* (2014)
Quantification in high level waste (HLW)	Spent nuclear fuel (5 g)	Digestion with 50 mL 4M HNO_3; cation exchange resin (BIO-RAD AG), eluted with 1M HNO_3	ICP-QMS	Asai *et al.* (2011)
In-vivo tracers in humans	Biological samples	None specified	AMS	Kutschera (1998)
Speciation of radioactive Se in nuclear fuel from a BWR	Spent nuclear fuel micro-particles	Preparation of sample pellets	Micro XANES	Curti *et al.* (2015)
Speciation of Se vitrified waste	Vitrified nuclear waste	Glass fragment in sample container	XAFS	Dardenne *et al.* (2015)

Abbreviations: HPGe, high purity germanium detector; DL, detection limits; HPLC, high-performance liquid chromatography; LSC, liquid scintillation counting; ICP-QMS, inductively coupled plasma quadrupole mass spectrometry; AMS, accelerator mass spectrometry; BWR, boiling water reactor; XANES, X-ray absorption near edge spectroscopy; XAFS, X-ray absorption fluorescence spectroscopy.

quantification of Se stable isotopes included important matrix effects and polyatomic/isobaric interferences of bromine hydrides (HBr), doubly-charged rare earths (REE, e.g., such as $^{164}Dy^{++}$) and/or from the argon gas (e.g., ^{82}Kr or $^{38}Ar^{40}Ar^{+}$; ^{80}Se is usually not measured due to the high abundance of $^{40}Ar^{40}Ar^{+}$). For this reason, the detection limit of all classical methods is relatively high so that it is nearly impossible to measure natural concentrations in many samples, especially when matrix dilution is required. Existing techniques to measure low levels of Se in the past included HR-ICP-MS (high resolution inductively coupled plasma mass spectrometry) or sector-field ICP-MS. But these techniques are very expensive and thus not widely distributed. Nevertheless, the current advent of new generation ICP-MS apparatus (e.g., Triple Quadrupole generation, TQ-ICP-MS) allows such interferences to be successfully overcome. This advancement provides better detection limits for quantifying trace levels of stable Se in natural samples and allows future experimental studies to use more environmentally relevant concentrations; though not yet reaching the levels achieved in radiotracer studies.

Several optical and electrochemical techniques have been suggested to show strong potential applications for on-site, field-deployable devices to monitor Se directly in freshwater and seawater (Devi *et al.*, 2017). Actual portable devices may have been developed for on-site monitoring of Se species, though with sporadic uses. An example is the portable device measuring the speciation of Se for a month in wastewaters of thermal power plants by hydride generation coupled to a chemiluminescence detector, showing relatively high detection limits (DL) of ~8 μg L^{-1} (Ezoe *et al.*, 2016). Other examples include electrochemical probes using gold nanoparticles showing DL of 0.12–0.27 μg L^{-1} for potential applications in seawater and agricultural food but no proven use in environmental monitoring (Segura *et al.*, 2015; Tan *et al.*, 2020). The use of portable XRF (X-ray fluorescence) analysis for on-site screening of Se in solid samples has been reported for nail clippings, soils, soil suspensions and animal tissues, showing DL of 0.83–1.00 mg kg^{-1} for direct analyses (Beaudette *et al.*, 2009; Fleming *et al.*, 2015). However, as the sample-instrument geometry and sample matrix strongly influence the XRF measurements, accuracy and precision, the DL can strongly vary when using portable XRF instruments. Specific pre-extraction methods applied on digested sample solutions in combination with total reflection XRF are reported to bring down the DL to 0.05 mg kg^{-1} (Marguí *et al.*, 2010). Nevertheless, such pre-treatment steps may not be easy to implement on site on a routine basis. Thus, to the best of our knowledge, there are no current on-site/field-based analytical techniques or monitoring probes being used systematically to quantify radioactive or stable Se species. Monitoring programmes until now have generally relied on classical field sample collection of environmental matrices and laboratory analyses (e.g., 30-year monitoring of the Kesterson case in San Francisco; Ohlendorf *et al.*, 2020).

2.4 PRODUCTION AND MOBILITY OF ^{79}Se IN NUCLEAR WASTE REPOSITORIES

2.4.1 Estimated activities in the nuclear waste

It is important to bear in mind that there are different types of nuclear fuel and according to its specific composition, burn-up and delay until its reprocessing or final disposal point, it may develop and present different radionuclides in variable amounts. Several simulations exist, estimating the potential radioactive levels of ^{79}Se present in spent fuel for specific conditions (i.e., nuclear reactors, fuel types, and burn-ups). For instance, a 10-y aged high-level waste from a light water reactor (LWR) may produce 3.7×10^{10} Bq y^{-1} of ^{79}Se (i.e., equivalent to ~72 g y^{-1}), which is in the same order of magnitude as estimations for ^{129}I and lower than those for ^{135}Cs, ^{137}Cs, ^{126}Sn, ^{99}Tc, ^{90}Sr and several actinides like ^{239}Pu, ^{241}Am and ^{244}Cm (WHO, 1982). Spent fuel with 4.2 wt.% enrichment in ^{235}U may contain 3.2–8.75 g ^{79}Se t^{-1} after 6 years of cooling (Magill *et al.*, 2003); 3.2 wt.% enriched spent nuclear fuel with a burn-up of 44.9 GWd t^{-1} from a pressurized water reactor (PWR) will produce 5.2 ($\pm$1.5) g ^{79}Se t^{-1} U (Asai *et al.*, 2011). The composition of a high level waste solution recovered after reprocessing a 3.5–4.9 wt.% enriched uranium oxide (UOX) spent fuel from a PWR after 3 y may contain even higher quantities of 77.04–136 g SeO_2 t^{-1} U (i.e., higher than those estimated for In_2O_3, SnO_2, Sb_2O_3, PuO_2, CmO_2, Tb_2O_3, and Dy_2O_3; Caurant *et al.*, 2009). These estimations are still part of ongoing research as they depend on the accuracy of nuclear reaction cross sections. Nevertheless, they are fundamental for understanding the amount of Se expected to be formed within the nuclear fuel cycle and to justify its relevance for research in nuclear waste disposal.

2.4.2 Underground reactivity and dispersion

2.4.2.1 The multi-barrier system: from the fuel to the host rock

Radionuclides in the spent fuel accumulate within the fuel matrix, at the grain boundaries, in the cladding structure or in the gap in between. In general, ^{79}Se is assumed to diffuse out of the fuel grains and migrate to the periphery of the pellets, like other volatile elements such as I and Cs. A considerable part of ^{79}Se in the nuclear fuel accumulates in the gap between the UO_2 pellets and the Zircaloy cladding, or along grain boundaries in the fuel pellets (Johnson & McGinnes, 2002). For instance, a spent fuel from Belgium with an average burn up of 47.3 GWd tHM^{-1} and 4.05% ^{235}U enrichment shows an average distribution of ^{79}Se in Bq per primary waste package of 6.82×10^4 in the structural metals, 6.35×10^6 in the claddings and 9.82×10^8 in the fuel matrix (Yu & Weetjens, 2016).

Nuclear waste is intended to be isolated and contained within a multi-barrier system upon disposal. This multi-barrier consists of both engineered/artificial retention structures (e.g., waste forms, canisters, buffers, backfills, seals and plugs) and geological/natural materials (e.g., the host rock). Over time and depending on the host rock type, these barriers may become saturated with groundwater and may lose their ability to provide complete containment, thus releasing radionuclides into the environment (Olszewska *et al.*, 2015). In fact, upon contact with groundwater, up to 20% of the ^{79}Se content in the nuclear waste may be released within the 'instant release fraction (IRF)' after canister failure together with ^{36}Cl and ^{129}I (Johnson & McGinnes, 2002; Yu & Weetjens, 2016). The fate of the remaining ^{79}Se will be linked to the solubility of the fuel components over time. This means that long-term ^{79}Se releases will be determined by the dissolution rate of the fuel/cladding matrix (not straightforward to estimate; Yu & Weetjens, 2016). It also implies that a long lifetime of the matrix will cause a long-time spread of ^{79}Se releases at low rates and doses. Short lifetimes of the matrix would favour fast release rates and higher doses.

The transport processes of released radionuclides will follow radionuclide-dependent conditions, initially related to the solubility limits of each elemental oxidation state. Direct measurements in the matrix of spent nuclear fuel samples suggest that Se may be bound to U atoms in selenide (Se(-II)) forms (e.g., case in Switzerland; Curti *et al.*, 2015). Additionally, the intrinsic method of constructing deep geological waste repositories implies an initial entrapment of atmospheric (oxic) conditions. Over time, the oxygen may be consumed by microbial activities (De Cannière *et al.*, 2010) or due to corrosion of the canister material, which may take $\sim 10^4$ y. Such effects eventually result in strongly reducing conditions in the near-field to the nuclear waste. These conditions would favour the presence of reduced species of Se like elemental (Se(0)) and selenide (Se (-II)), which generally have a low solubility. As an example, a simple calculation with 1 μM Se concentration in interstitial water (as defined by Gimmi *et al.*, 2014) at pH 7.28 simulating groundwater conditions in Opalinus clay at 25°C, shows precipitation of Se(0) in reducing conditions (i.e., a decrease in the concentration of the soluble species and in parallel the precipitation of solid species for the case study without Fe, Figure 2.1). The presence of other elements such as S, Fe, Mn, Ca or Sr will also influence the solubility of Se, particularly in highly reducing conditions. Continuing with the previous example, in the presence of 1 mM aqueous Fe, Se will show lower solubility than in the absence of Fe for the same chemical system, favouring the precipitation of solid selenides like ferrosilite (e.g., hashed area in the lower panel, and solid species in the upper panel, in Figure 2.1).

Nevertheless, in the direct vicinity of nuclear waste, radiolytic effects may play a role, that is, water molecules may be split by the emitted radiation from the fuel and produce oxidizing radicals like $OH^{\cdot}$ or peroxo-species like HOOH. These molecules together with other oxidizing species present in the water (e.g., nitrate salts from the nuclear waste; Bleyen *et al.*, 2018) will favour oxidative dissolution from the host phases, thus mobilizing radionuclides. The formation of oxidized species of Se

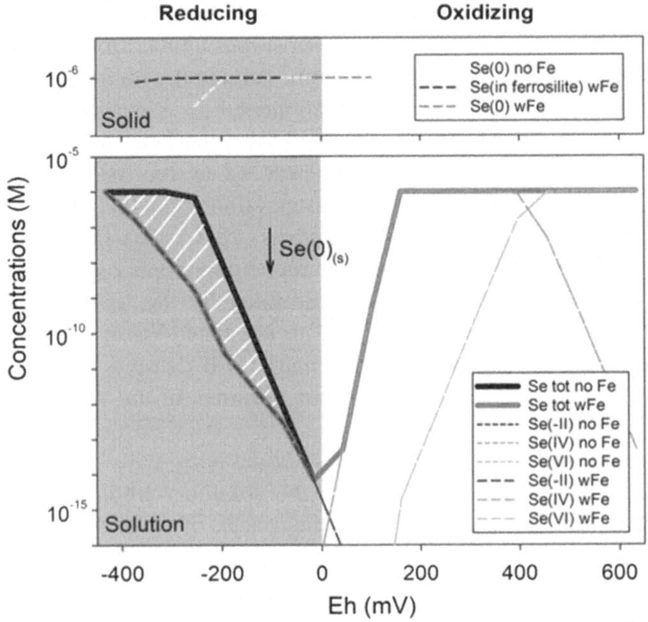

Figure 2.1 Example of the solubility of different Se oxidation states at pH 7.28 for 1 μM total Se content. A comparison between results in the presence (wFe) and without (no Fe) 1 mM aqueous iron is highlighted with a hashed area. Solid species (top panel) and species in solution (bottom panel) are presented. Precipitation of elemental Se is marked with an arrow (bottom panel). Calculations were done with PhreeqC3 (Parkhurst & Appelo, 2013) using the sit.dat (thermochimie) database (Giffaut *et al.*, 2014). The water composition corresponds to that used by Gimmi *et al.* (2014) to simulate groundwater conditions in Opalinus clay at 25°C.

could imply a higher mobility and dispersion than the reduced species, given the high solubility limits of Se(IV) and Se(VI) (e.g., increased concentration of Se in solution, lower panel in Figure 2.1). Nevertheless, oxidizing conditions might not necessarily imply mobilization of Se. For instance, Se(IV) and Se(VI) may also be incorporated into Fe-bearing mineral phases in oxic conditions (e.g., max. 15% retention of Se(VI) and complete uptake of 10^{-3} M of Se(IV) by both adsorption and irreversible co-precipitation during aging of 9.0 g L^{-1} of ferrihydrite into hematite at pH 7.5, or immobilization of 10^{-3} M Se(IV) and Se(VI) at pH 9.2 by co-precipitation with Fe during magnetite formation via Fe(II) hydroxide and green rust; Börsig *et al.*, 2017, 2018).

In any case, a general, inherent gradient between the source and the surrounding environment develops, becoming less reducing from the near-field towards the biosphere. Cementitious systems (e.g., mostly AFm phases, i.e., a family of hydrated calcium aluminates) as well as bentonite clays (presenting microbially-mediated processes) used within the multi-barrier system at the interface between the nuclear waste and the host rock may retain Se species (e.g., formation of solid solutions mostly for Se(IV) and Se(VI) species; Duro *et al.*, 2020; Ruiz-Fresneda *et al.*, 2018). More detailed information about the multi-barrier system and the interfacial solution chemistry of Se can be found in Abdelouas and Grambow (2012).

2.4.2.2 Reactivity within the host rock: mobility and dispersion of Se species

The extent of the radionuclide transport by groundwater, through advection and/or diffusion, depends on the porosity and permeability of the host rock (i.e., rock salt, clay rock or crystalline rock), the mineralogy of the host rock and on the on-site physico-chemical conditions (e.g., reducing environment with heterogeneous assemblage of minerals and organic phases in the host rock; Altmann, 2008). Case study areas include Boom Clay at Mol (Belgium, e.g., NIROND, 2001), Opalinus clay at Mont Terri and granitic rock at the Grimsel Fels Labor (Switzerland, e.g., Alexander *et al.*, 2009; Bleyen *et al.*, 2018; Eikenberg *et al.*, 1997; Ikonen, 2017), the Asse salt mine (Germany, e.g., Kienzler *et al.*, 2016; Rabung *et al.*, 2018), granitic rock at the Äspö laboratory (Sweden, e.g., TR-99-06), and the Callovo-Oxfordian clay at Bure (France, e.g., Descostes *et al.*, 2008). More information on the underground research infrastructures can be found on the internet (last accessed on the 30th October 2020): https://www.sckcen.be/en/expertises/environment/waste-and-disposal (Belgium), https://www.mont-terri.ch/ and https://www.nagra.ch/de/felslaborgrimsel.htm (Switzerland), https://www.bge.de/de/asse/ (Germany), https://www.skb.com/research-and-technology/laboratories/the-aspo-hard-rock-laboratory/ (Sweden), and https://meusehautemarne.andra.fr/landra-en-meusehaute-marne/installations/le-laboratoire-souterrain (France).

Nevertheless, there is actually little certainty on the true composition of pore waters of host rocks (e.g., Gaucher *et al.*, 2009; Gimmi *et al.*, 2014; Kienzler *et al.*, 2016), or, even more concerning, on the speciation of Se in the pore water of the host rock (e.g., De Cannière *et al.*, 2010; Hoving, 2018). Uncertainties within developed transport models include: (i) specific solubility limits (e.g., from the formation of ferroselite, $FeSe_2$ to the precipitation of Se(0) in strongly reducing conditions; Iida *et al.*, 2009), (ii) the assumption of shared oxidation states due to slow transformation kinetics (e.g., reduction of Se(VI) to lower oxidation states; Grambow, 2008), (iii) the aqueous speciation of Se (e.g., formation of $CaSeO_3$ complexes in solution for Se(IV) in hyperalkaline solutions in contact with cementitious materials or bentonite from the retention structures; Alhajji, 2007; Mace, 2006) or Se adsorption/complexation to other phases (e.g., in the presence of organic matter or mineral surfaces), and (iv) the presence or absence of equilibrium conditions (Altmann, 2008).

For instance, existing models assume releases of Se as Se(IV) and Se(VI) oxyanions. However, thermodynamic calculations suggest that Se should be in lower oxidation states (-II and 0) due to interactions with the iron from the canister materials or due to developed reducing conditions within the host rock (e.g., diffusion through clays; Basu *et al.*, 2007; Beauwens *et al.*, 2005; Descostes *et al.*, 2008; Gautschi, 2017). In fact, solid samples containing Se(0) and Se(IV) next to each other originating from reducing environments in Boom Clay have been reported (e.g., Breynaert *et al.*, 2010; Hoving, 2018). Under these conditions, 14–92% of 4–400 mg L^{-1} of Se(IV), respectively, was adsorbed suggesting variable retention mechanisms depending on the Se(IV) concentration in the pore water. Microbially-mediated reactions influencing Se oxidation states within the host rock are also a potential scenario (De Cannière *et al.*, 2010; Ruiz-Fresneda *et al.*, 2018), the cause of which is most likely the high organic matter content mainly present in claystone.

Mobile species migrating through the host rock will be subjected to electrostatic interactions, potentially influencing their diffusion rates. Diffusion may be very low due to anion exclusion in the small, few nanometres pore sizes of claystone where the electric double layers overlap, hindering transport of Se species through the pores (Grambow, 2008). Low anion sorption, on the other hand, is expected in argillaceous rocks due to the electrostatic repulsion developed on the surface, and may favour transport. Conservative approaches usually assume un-retarded transport for all Se species, resulting in a high mobility/dispersion. Clay-Se interactions remain a field of research under development for which electrostatic models can greatly contribute to migration scenarios, particularly within the context of nuclear waste barrier efficiency (e.g., Charlet *et al.*, 2007; Ervanne *et al.*, 2016).

In repositories composed of granite, groundwater will disperse radionuclides relatively rapidly through rock fractures unless retained by diffusion into the rock pores or sorption onto mineral surfaces (Ikonen, 2017; Ikonen *et al.*, 2016). For instance, geological units containing apatite, pyrite and iron oxides may readily immobilize selenite (ANDRA, 2005; Börsig *et al.*, 2017; Hoving, 2018). Effective diffusion coefficients of $2.5\ (\pm 1.5) \times 10^{-12}\ m^2\ s^{-1}$ and $7\ (\pm 2) \times 10^{-13}\ m^2\ s^{-1}$ were obtained for Se in Grimsel granodiorite and Kuru grey granite, respectively (Ikonen, 2017). Sorption interactions between dissolved species and mineral surfaces can be approximated by a solid/liquid partition coefficient (*Kd*). Examples of experimental studies aiming at obtaining representative *Kd* values for nuclear waste repositories are summarized in Table 2.2. Noteworthy, it has been suggested that many *Kd* values used in transport models for underground dispersion of radionuclides may be overestimated if originally based on batch experiments using crushed materials (i.e., potentially not representing proper solid/liquid partitioning along rock cracks, especially in the case of Se; Altmann, 2008; Ikonen, 2017). Enhanced

Table 2.2 Examples of solid/liquid partitioning (*Kd*) values used for radioactive risk assessment of radionuclide mobility in underground and surface environments.

Materials/Conditions	*Kd* (L kg^{-1})	Reference
Applied in nuclear waste near-field conditions		
Crushed quartz in synthetic ground/seawater	<1.0	Lee *et al.* (2012)
Crushed-bentonite in synthetic ground/seawater	10–22	Lee *et al.* (2012)
Mudrock in synthetic ground and seawater	40–80	Lee *et al.* (2008)
Block scale diffusion through granitic rock	0.1–1.5	Ikonen *et al.* (2016)
Crushed granitic rock	6.2–7.0	Ikonen *et al.* (2016)
Solid claystones in reducing environment	~11	Grambow (2008)
Crushed rocks from overburden of Asse salt mine[a]	7–137	Rabung *et al.* (2018)
Applied in surface environmental compartments		
All soils	4.0–2100 (200, $N = 172$)[b]	IAEA TRS 422
Sandy soils	56[b]; 4.0–1600 (56, $N = 15$)[b]	Gil-García *et al.* (2009) IAEA TRS 422
Clay soils	240[b]	Gil-García *et al.* (2009)
Loam + clay soils	12–2100 (220, $N = 134$)[b]	IAEA TRS 422
Organic soils	230–1800 (1000, $N = 2$)[b]	IAEA TRS 422
Average upper crust and ocean water	~100	Grambow (2008)
0.3 ng ^{75}Se L^{-1} added to estuarine SPM[c]	9.0–84.1	Gil-Díaz *et al.* (2020a)
100 μg ^{77}Se L^{-1} added to estuarine SPM[c]	315–630	Gil-Díaz *et al.* (2020b)
SPM of 100 mg L^{-1} in San Francisco Estuary	~100–31 600	Benoit *et al.* (2010)
19 Japanese coastal regions	~400–7900	Takata *et al.* (2016)
Open ocean	1000	IAEA TRS 422
Ocean margin	3000	IAEA TRS 422

[a]Se(IV) in a cocktail solution of radionuclides; [b]Geometric means; [c]SPM concentrations of 10, 100 and 1000 mg L^{-1}.
Abbreviations: SPM, suspended particulate matter.

solubility in natural solids may also be related to the presence of organic complexing agents in the interstitial water (Grambow, 2008). Further understanding of the processes behind sorption of Se on mineral surfaces and *Kd* values can be achieved when the surface site properties and the pore water chemistry are well known and explained by mechanistic models such as surface complexation models (e.g., Davis & Leckie, 1980; Grambow, 2008; Nie *et al.*, 2017). In any case, all *Kd* values are relatively low, suggesting a high mobility of Se and, thus, eventual dispersion into the geologic overburden.

Another process potentially retaining traces of Se species within the host rock is related to its structural incorporation into minerals. For instance, some studies suggest that ^{79}Se may become immobilized when interacting with uranyl phases present as low concentration impurities in the nuclear fuel waste due to substitution of Si by Se in solid structures of, for example, α-uranophane – ($Ca(UO_2)_2(SiO_3OH)_2 \cdot 5H_2O$) and boltwoodite – $HK(UO_2)(SiO_4) \cdot 1.5(H_2O)$ (Chen *et al.*, 1999). Others suggest favoured structural incorporation of Se(IV) into calcite, particularly on the surface and potentially in deeper layers upon precipitation at highly supersaturated conditions (Heberling *et al.*, 2014; Polly *et al.*, 2017). The most effective process regarding mineral inclusion could be Se incorporation to Fe-bearing minerals, present as container corrosion products or constituent of host rocks and in bentonite backfills. In fact, studies in this field have shown >98% incorporation of Se(-II) and Se(IV) from 10^{-3} M total Se into pyrite (FeS_2) and mackinawite (FeS) for highly supersaturated solutions under acidic and anoxic conditions (Diener *et al.*, 2012; Finck *et al.*, 2012). Additionally, aging processes such as the transformation of ferrihydrite into hematite and/or Fe(II) hydroxides or green rust into magnetite may efficiently incorporate Se(IV), and in lower quantities Se(VI) (Börsig *et al.*, 2017, 2018).

2.4.2.3 Simulated environmental releases

Radionuclide dispersion scenarios are developed to foresee the long-term consequences of storing nuclear waste underground and to protect future generations from potential environmental releases. These scenarios are based on long-term radioactive risk, as radionuclide releases to the environment are expected between ten thousand and several hundred thousand years. This time frame is related to expected durability of the waste containers and the inherent properties of the long-lived ^{79}Se. It is estimated from conservative approaches as applied in worst case scenarios, that is, independent of the estimated timescales of all the retention mechanisms mentioned in the previous sections: transport subjected to retardation factors as determined by solid/liquid partitioning (Eikenberg *et al.*, 1997; Hampel *et al.*, 2013).

Estimated concentrations in the near-field of nuclear waste repositories could be in the order of 10^{-11} M of ^{79}Se (i.e., <1 ng L^{-1}, with 10^{-9} M of total Se corresponding to ~80 ng L^{-1}; Grambow, 2008). These levels are actually below saturated conditions (*c.f. Section 2.4.2.1*) and will decrease in the far-field from the source. This decrease is also reflected by estimated outgoing fluxes from the deep geological units towards surface environments. For instance, fluxes of ~72 g y^{-1} ^{79}Se from a 10-y aged high-level waste may be expected in the near-field (*c.f. Section 2.4.1*). Preliminary simulations of far-field conditions suggest a maximum flux peak of 2×10^7 Bq y^{-1} (equivalent to ~39 mg y^{-1}; NIROND, 2001) after 150 000–200 000 y into the interface between Boom Clay and a Neogene Aquifer in Belgium. Maximum releases of ~10^{-6} mol ^{79}Se y^{-1} (equivalent to ~79 μg y^{-1}; Altmann, 2008) at around 100 000 y are expected from a model reference claystone geological barrier in the geologic overburden.

These releases lead to effective dose levels which are in general low, compared to environmental activities, as will be discussed in *Section 2.6*. The reported concentrations do not contribute significantly to the environmental budget of Se, or to the local chemotoxicity of Se. These concentrations are below average Se environmental levels (e.g., median 340 ng L^{-1} in European rivers; Salminen *et al.*, 2005) and below the limits of drinking water quality for both total Se (1.3×10^{-7} M, equivalent to ~10 µg L^{-1}) and ^{79}Se (47 Bq L^{-1}, equivalent to ~92 ng L^{-1}; Grambow, 2008). Likewise, reported flux estimates are also very low compared to currently estimated annual fluxes of stable Se from rivers/estuaries subjected to anthropogenic activities into coastal areas. For example, 390 tons y^{-1} have been estimated for the Yangtze River (~3.5 tons y^{-1} ^{74}Se, the least abundant stable isotope of Se; Yao *et al.*, 2007) and ~62000 tons y^{-1} for the Lingdingyang Estuary (~550 tons y^{-1} ^{74}Se; Yao *et al.*, 2006). Nevertheless, better estimations of both far-field ^{79}Se releases and current gross/net environmental fluxes are required to confirm such statements.

2.5 ENVIRONMENTAL DISPERSION SCENARIOS

2.5.1 Conceptual model and assumptions

The main concern in the radioecological risk assessment of ^{79}Se is its transport/remobilization from deep strata and underground waters to surface aquatic and terrestrial systems. A visual diagram is present in Figure 2.2 as an example of a geological setting containing a nuclear waste repository and the environmental compartments involved in the potential transport, dispersion and fate of ^{79}Se from the source to the biosphere and within the biosphere. The presence of ultra-trace levels (or below) of ^{79}Se in natural samples is expected to be subjected to a complex mix of processes (e.g., various oxidation states, precipitation/dissolution and sorption/desorption from various phases). These processes are site-specific, thus, cannot be generalized or expected to be comparable. Nevertheless, most studies agree on the general bioavailability of ^{79}Se, which, however, still depends on the chemical species, as well as the specific organisms and environmental compartments involved (*c.f. Section 2.6*).

Furthermore, differences between the geochemical behaviour of radionuclides and their homologue stable isotopes may exist in specific occasions where recoil effects, mass differences, and radiolytic effects play a role in the system. However, recoil effects are of limited relevance for beta-emitters and radiolytic effects play a role directly next to the waste, not in surface environments where low activities are expected. Isotopic mass effects have been reported for stable Se, as will be discussed below, however, as ^{79}Se is between the most abundant isotopes ^{78}Se and ^{80}Se, mass effects will be small, and likely within the quantification uncertainties of, for example, *Kd* determinations. Therefore, as a general rule, works aiming at understanding the risk assessment of nuclear discharges in surface environments of the geosphere and the biosphere often

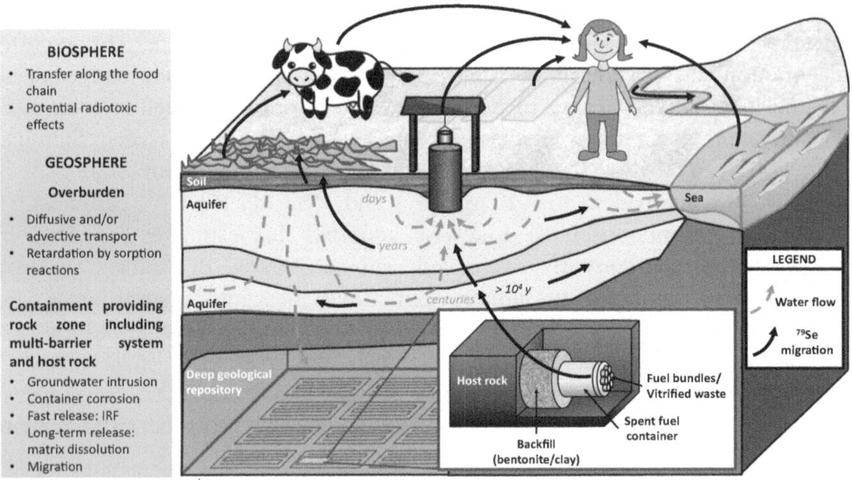

Figure 2.2 Visual scheme to represent the interaction pathways in aquatic and terrestrial systems between potential releases of ^{79}Se from deep geological repositories of spent nuclear fuel (*c.f. Section 2.4*) and accumulation in the biosphere, including ingestion pathways for human uptake (*c.f. Section 2.6*). Examples of the processes involved in each environmental compartment are summarized in the left panel. The geosphere is considered as the geologic underground, while the biosphere includes the pedosphere, the hydrosphere, and (to a limited extent) the atmosphere. Similar schemes should be adapted for site-specific scenarios. Abbreviation: instant release fraction (IRF).

apply the known environmental behaviour of stable Se isotopes as analogues of their radioactive counterparts.

2.5.2 Biogeochemical behaviour in aquatic systems

Once the groundwater has been in contact with the waste repository and is contaminated with ^{79}Se, it may eventually transport ^{79}Se to surface aquatic systems such as rivers and estuaries (e.g., Figure 2.2). In these systems, Se is considered to be relatively mobile, given reported *Kd* values (e.g., Table 2.2). Nevertheless, attention should be paid when extracting conclusions and applying field-based *Kd* values to dispersion and transport models for nuclear risk assessment. In fact, it has been recently suggested that both solid fractionation and *Kd* of added elements to the environment (e.g., anthropogenic releases) may differ from those registered from inherited (geologically-derived) analogue counterparts. The potential added/inherited duality may also be the reason why field- and laboratory-based *Kd* values vary for certain elements in aquatic systems (e.g., Ciceri *et al.*, 1988). Field-based *Kd* values integrate inherited (refractory) trace elements and laboratory-based *Kd* values address solid/liquid partitioning of spiked elements (better simulating anthropogenic releases). Therefore, reporting proper *Kd* values is of great relevance, particularly from added elements at environmentally representative conditions. Nevertheless, the recommended *Kd* values by the IAEA (IAEA TRS 422) for radionuclide transport models in surface environments are still based indistinctly on experimental or field *Kd* values (e.g., Table 2.2). This may lead to scenarios suggesting a higher particle affinity of added Se compared to expected reactivity in underground conditions.

It is difficult to ascertain if this difference between added and inherent behaviour also applies to Se. This is particularly the case in solid fractionation studies aiming at determining Se solid carrier phases. Indeed, it is difficult to draw sound conclusions from studies where the added chemicals might change the original conditions of the Se species and/or the minerals, inducing a biased behaviour of Se independent of the targeted, operationally-defined carrier phases (e.g., enhanced mobility due to reagents containing organic compounds or a change in the oxidation state of Se; Gil-Díaz *et al.*, 2020b; Gruebel *et al.*, 1988; Lenz *et al.*, 2008). It is also extremely challenging to measure Se in complicated matrices of most of the extraction solutions. Furthermore, isotopically-labelled studies using ^{75}Se radiotracer and/or stable Se do not focus on determining both inherited and added Se, nuancing a more conclusive statement on the matter. One study suggests that, under oxic and abiotic conditions with natural water/sediment matrices, added Se may be more easily exchangeable and mobile during early diagenetic processes than inherited forms (Gil-Díaz *et al.*, 2020b). A further point of interest is to understand what happens to added elements over time. The added Se might be more mobile initially, but over time the question remains whether it could get immobilized more strongly. For environmental scenarios this knowledge and especially the kinetics of such transformations are very important.

In any case, the transport of ^{79}Se in aquatic systems is expected to be mostly in the dissolved phase, thus, subjected to the specific, local biogeochemistry in its soluble forms, which are considered highly to moderately bioavailable (Altmann, 2008). Speciation of Se in natural waters would be expected to involve oxyanion species of Se(IV) and Se(VI) such as $HSeO_3^-$, SeO_3^{2-}, and SeO_4^{2-} in oxidizing conditions at pH >5 (Torres *et al.*, 2011). Other redox states such as Se(-II) and Se(0) may also occur in stagnant waters or in sites dominated by biological activity under reducing conditions (e.g., Wang *et al.*, 2001). Depending on the environmental sites, elemental concentrations, and pH and Eh conditions, interactions with metal ions may occur resulting in cation species such as $[Cu(HSeO_3)]^+$, $[Ca(HSeO_3)]^+$ and $[Mg(HSeO_3)]^+$ at pH >4 (Torres *et al.*, 2011). In any case, this transport in soluble forms would imply a rate of dispersion in surface aquatic systems most likely linked to inherent timescales of hydrological processes (e.g., river discharges, seasonal patterns, and tidal influence). ^{79}Se may also show a more or less important reactivity along the salinity and turbidity gradients of estuarine systems (i.e., conservative vs non-conservative biogeochemical behaviour) according to the specific continent-ocean transition system involved (e.g., Chang *et al.*, 2016; Cutter, 1989; Measures & Burton, 1978; Seyler & Martin, 1991; van den Berg *et al.*, 1991; van der Sloot *et al.*, 1985; Yao *et al.*, 2006). Given the long half-life of the expected Se radioactive releases, ^{79}Se may eventually reach the ocean and, further, cycle several times within the biogeochemical cycle of Se in accordance with its nutrient-like behaviour (Cutter & Cutter, 1995). Radioactive Se may also transfer to the atmosphere when methylation processes are favoured within estuaries or in the ocean (e.g., Amouroux & Donard, 1997; Feinberg *et al.*, 2020).

2.5.3 Biogeochemical behaviour in terrestrial systems

Contaminated groundwater in terrestrial systems may involve biological uptake by humans from drinking water and/or from food intake due to the direct contact of soils with contaminated groundwater or the use of such groundwater for irrigation (e.g., Figure 2.2). The main factors controlling the speciation and behaviour/mobility of ^{79}Se in soils are redox potential (Eh) and pH soil levels (i.e., classical Eh-pH diagrams), sorption/complexation processes onto inorganic/organic phases (e.g., competition of Se sorption with other anions such as phosphate, introduced by fertilizers; Nothstein *et al.*, 2019), and active biological fixation/transformations (Dhillon & Dhillon, 1991, 2019; Koch-Steindl & Pröhl, 2001; Natasha *et al.*, 2018). Site-specific conditions will be related to the intrinsic characteristics of the soil, the affinity of plants for Se uptake, the climate of the region and soil management, which in turn will determine the daily intake by animals and humans.

Soils may act as long-term secondary sources of radiation when solid adsorption/complexation of ^{79}Se as selenite is favoured and/or in waterlogged systems where precipitation of elemental selenium may occur. For example, reduction reactions may be induced abiotically from heterogeneous reactions with Fe(II)-containing minerals (Février & Martin-Garin, 2005). Soils may also present microenvironments where, for example, reduced species can be formed in soil aggregates even if the soils are considered as totally oxic (Eiche *et al.*, 2015). In oxic conditions, a certain reactivity may be inferred in specific soils (e.g., as seen in *Kd* values, Table 2.2), though a general mobility is also assumed for terrestrial environments, particularly in the presence of other anions in solution such as phosphate (Dhillon & Dhillon, 2019; Gustafsson & Johnsson, 1994).

Nevertheless, soil systems are highly influenced by biological activities, and *Kd* transport models alone may not be sufficient to represent migration in soil horizons, particularly in the presence of microorganisms (Février & Martin-Garin, 2005; Natasha *et al.*, 2018). In fact, an active biotic role such as microbial reduction may significantly enhance immobilization of Se compared to abiotic reduction processes (e.g., >85% of environmentally-relevant spikes of ^{75}Se were incorporated into the low molecular weight organic fraction of forest soils in

<64 h; Darcheville *et al.*, 2008; Gustafsson & Johnsson, 1994). This role is in accordance with solid fractionation studies suggesting a high association of Se with the organic fraction. Furthermore, depending on the location of the deep geological site, released ^{79}Se may also be subjected to complex speciation changes and uptake by plants, characteristic but not exclusively for seleniferous areas (e.g., area of Punjab in India; Eiche *et al.*, 2019). Microorganisms can also produce nano-elemental Se from Se(IV) under relatively oxic conditions (Bajaj *et al.*, 2012). Potential biomethylation processes in soils due to microbial communities would also favour a transfer of radioactive, methylated Se species to the atmosphere (Feinberg *et al.*, 2020), with the consequent dispersion of the corresponding radioactivity (expected to be very low, *Section 2.4.2.3*).

Interestingly, when mobile Se is introduced into the soil through irrigation, it can be easily taken up by plants, where Se may be reduced mainly to organic species (Eiche *et al.*, 2019; Schilling *et al.*, 2015). Depending on the Se concentration (toxicity) or the plant species, some Se might be volatilized. Most importantly, as soon as the plant dies, Se remains in some parts as organic Se within the soil producing a long-term pool. Furthermore, some selenate might be reduced within the soil to selenite which is adsorbed or further reduced into Se(0), which would also lead to an on-site long-term enrichment (Eiche *et al.*, 2019). Consequently, once Se has intruded into a soil-plant system, it can remain there for a long time, especially if the plants are not fully removed during harvest. The behaviour of organic-Se species also differs from that of inorganic forms. Radionuclide dispersion scenarios should consider these implications of the soil-plant system.

In the terrestrial environments, where there is a significant role of the (micro) biosphere, a question arises concerning the effect of potential isotopic fractionation in radionuclide scenarios. In fact, a compromise between released concentrations and molecular weight of ^{79}Se could be expected to interplay with the local, stable isotopes of Se (e.g., ^{79}Se being lighter, though potentially less abundant, than ^{80}Se and ^{82}Se). Evidence in wetland environments suggests that microbial reduction can be subjected to slight isotopic fractionation, favouring transformations of the lighter isotopes (e.g., maximum fractionation of $-5.7‰$ for ^{80}Se/^{76}Se in a wide range of both added and inherited Se(IV) and Se(VI) concentrations, from 22 µg L^{-1} to 8 mg L^{-1}; Ellis *et al.*, 2003). Microbial reduction of Se(VI) may even show greater isotope fractionation (^{82}Se/^{76}Se from $-9.2‰$ to $-11.8‰$) than that of Se(IV) (max. $-7.8‰$), suggesting different metabolic pathways (Schilling *et al.*, 2020). Surprisingly, though potentially less relevant, this process could also take place during abiotic reduction (e.g., favoured adsorption of the lighter isotopes of Se(IV) onto Fe and Mn oxides, ~1‰ for ^{82}Se/^{76}Se; Xu *et al.*, 2020). Plants may also selectively uptake Se isotopes, for example, inducing high fractionation of ^{82}Se/^{76}Se of up to $+3.5‰$ of Se(VI) and $+1.9‰$ of Se(IV), suggesting an enrichment in heavy isotopes within the plant (Banning *et al.*, 2013, 2018), particularly in irrigated systems (e.g., Schilling *et al.*, 2015). These studies indicate that ^{79}Se may be subject to isotope fractionation. However, delta values are expected to be closer to zero compared to the extreme isotope pairs $^{82}Se/^{76}Se$ or $^{80}Se/^{76}Se$.

Furthermore, dispersion scenarios in terrestrial systems are based on the current knowledge of site-specific soil conditions and known physico-chemical/biological processes. However, ideally future changes in the organic matter content and quality of soils (happening within decades to centuries) and changes in weathering and pedogenesis (occurring within millennia; Dhillon & Dhillon, 2019; Koch-Steindl & Pröhl, 2001) should be included in risk assessment scenarios as they fall within the timescales of the half-life of ^{79}Se. Additionally, potentially forthcoming changing conditions and their consequences (e.g., climate change and future soil management) should also be taken into account. Important aspects also include irrigation, introducing Se but also potentially leading to salinization; industry bringing in anthropogenic molecules, which might be very persistent; and heavy rain, washing out Se or leading to water-logged intervals. A summary of contrasting examples of climatic scenarios and potential impacts in soil processes concerning Se are described in Koch-Steindl and Pröhl (2001) and a more recent case scenario predicting site-specific trends of future Se in soils worldwide can be found in Jones *et al.* (2017). For instance, an increase in temperature, precipitation and positive water balance could accelerate all soil processes and favour soil leaching, low pH and higher turnover of organic matter, all of which could lead to an increased mobilization of Se. This would enhance the bioavailability of ^{79}Se for the local biosphere (both microorganisms as well as plant uptake) and/or its transport to underground water systems. In arid environments, high temperatures may slow down organic matter decomposition and mineral weathering, decreasing the mobilization of radionuclides attached/sorbed to such particles. Increased erosion by wind and water in both future arid and humid conditions may favour the spatial distribution of radionuclides (higher dispersion, less concentrated on-site).

2.6 IMPACT OF RADIOACTIVE Se ON THE BIOSPHERE: INSIGHTS FROM ECOLOGICAL MODELS

2.6.1 Bioaccumulation factors in aquatic and terrestrial systems

There are several mathematical approaches used to simulate the biological uptake and acute/chronic accumulation of trace elements in different environmental compartments. These are essential tools for predicting worst-case scenarios of potential intake of radionuclides into the biosphere in order to develop management strategies to protect the environment and humans, especially concerning potential external and internal exposure doses. This is particularly important for radionuclide risk assessment where direct measurements are not always available due to the lack of environmental evidence (e.g., case of ^{79}Se).

These models quantify the uptake of radionuclides by organisms through concentration factors or bioaccumulation factors, and include the dynamic transfer/retention within the organisms by taking into account the biological half-lives of the elements (Beresford *et al.*, 2015). Concentration factors account for the relative concentration/activity of the radionuclide in the organism compared to that present in the water (direct pathway) and/or the food (trophic pathway).

Existing wildlife projects/models for risk assessment and management of ionizing radiation as well as α/β-emitters (e.g., the ERICA/FREDERICA Tool, the Ecopath-with-Ecosim model, the ENVIRHOM Program; Beresford *et al.*, 2008; Booth *et al.*, 2020; Henner, 2008; Hosseini *et al.*, 2008) mostly rely on databases composed of known concentration factors from both field and experimental conditions. Current databases show elevated concentration factors for Se in many trophic levels of both aquatic and terrestrial organisms (Table 2.3). Indeed, other radionuclides of radioecological relevance show lower intakes. Such is the case for the 'geochemical pair' of Sc, Tc (e.g., <700 L kg^{-1} in aquatic and <1 L kg^{-1} in terrestrial systems), or the commonly followed ^{137}Cs and ^{131}I (e.g., $\sim 10^2$–10^3 L kg^{-1} in the aquatic and $\sim 10^{-2}$ L kg^{-1} in the terrestrial system; Gil-Díaz, 2019). As expected, biomagnification patterns for Se can also be discerned from these databases, particularly in marine environments (Table 2.3). This means that, despite the expected, low environmental activities of ^{79}Se reaching the aquatic and terrestrial systems, organisms may accumulate ^{79}Se and transfer it along the trophic chain. Nevertheless, these databases still require further research regarding the assessment of ^{79}Se as, for instance, there are few studied groups in freshwater and brackish systems, occasionally equal concentration factors are applied to different groups, and there are generally high standard deviations, when specified (Table 2.3).

2.6.2 Human radiotoxicity: exposure pathways and estimated doses

Within the aim of providing dose rates in realistic scenarios, important advancements in radioecological studies have been achieved. Particularly after the Fukushima Dai-ichi NPP accidental event, improvements in post-accidental assessment were made in marine ecosystems for commonly followed radionuclides, which have not yet been applied for ^{79}Se (Beresford *et al.*, 2015; Booth *et al.*, 2020; Vives i Batlle *et al.*, 2018). Nevertheless, terrestrial models including a comprehensive view of both element-dependent parameters (e.g., transfer factors and soil-plant distribution coefficients) and element-independent parameters (e.g., irrigation, agricultural practices and human consumption rates) have been developed and applied for ^{79}Se. For instance, the BIOGEM model was used to identify the main exposure doses for humans at five national geological repositories (e.g., Kowe *et al.*, 2005). Results suggested that, in general, the

Table 2.3 Recommended concentration factors of Se used for radionuclide risk assessment in aquatic and terrestrial organisms.

Organisms	Conditions	Concentration factor (L kg^{-1})	Reference
Freshwater			
Zooplankton	*In situ*	6600 ± 3900 (f.w.) ($N = 3$)	IAEA TRS 479
Macroalgae	*In situ*	3100 ± 1300 (f.w.) ($N = 3$)	IAEA TRS 479
Vascular plants	*Ex situ*	1000 (f.w.) ($N = 1$)	Hosseini *et al.* (2008)
	In situ	220 ± 57 (f.w.) ($N = 3$)	IAEA TRS 479
Molluscs (Gastropods)	*In situ*	3200 ± 2900 (f.w.) ($N = 3$)	IAEA TRS 479
Fish	*In situ*	4800 ± 3300 (f.w.) ($N = 127$)	IAEA TRS 479
Benthic feeding	*In situ*	6200 ± 3700 (f.w.) ($N = 51$)	IAEA TRS 479
Piscivorous	*In situ*	4200 ± 2700 (f.w.) ($N = 70$)	IAEA TRS 479
Insect larvae	*In situ*	2400 ± 1900 (f.w.) ($N = 9$)	IAEA TRS 479
Amphibians	—	63.2 (f.w.) ($N = 1$)	Beresford *et al.* (2008)
Reptiles	*In situ*	2700 ± 2500 (f.w.) ($N = 11$)	IAEA TRS 479
Brackish			
Vascular plants	*In situ*	4200 ± 1000 (f.w.) ($N = 3$)	IAEA TRS 479
Fish	*In situ*	2300 ± 2900 (f.w.) ($N = 8$)	IAEA TRS 479

Marine			
Phytoplankton	*In situ*	30000	IAEA TRS 422
	Ex situ	3600 ± 13000 (f.w.) ($N = 94$)	Hosseini *et al.* (2008)
Zooplankton	*In situ*	6000 (f.w.)	IAEA TRS 422
Macroalgae	*In situ*	1000 (f.w.)	IAEA TRS 422
	In situ	430 ± 790 (f.w.) ($N = 36$)	IAEA TRS 479
	In situ	310 ± 300 (f.w.) ($N = 35$)	Hosseini *et al.*, (2008)
Annelids (worms)	*Ex situ*	4500 (f.w.) ($N = 1$)	Hosseini *et al.* (2008)
Cnidaria (anemones)	*Ex situ*	10 (f.w.) ($N = 1$)	Hosseini *et al.* (2008)
Bivalves	*Ex situ*	5000 ± 3700 (f.w.) ($N = 3$)	Hosseini *et al.* (2008)
Gastropods	*In situ*	34.7 ± 31.2 (f.w.) ($N = 7$)	Beresford *et al.* (2008)
Molluscs *(general)*	*In situ*	9000 (d.w.)	IAEA TRS 422
	Ex situ/In situ	6700 ± 4600 (f.w.) ($N = 4$)	IAEA TRS 479
Crustaceans	*In situ*	10000 (f.w.)	IAEA TRS 422
	Ex situ/In situ	7100 ± 4800 (f.w.) ($N = 4$)	Hosseini *et al.*, (2008)
Fish	*In situ*	10000 (f.w.)	IAEA TRS 422
	Ex situ/In situ	9300 ± 4600 (f.w.) ($N = 3$)	Hosseini *et al.* (2008)
Mammals (general)	*In situ*	8300 ± 2700 (f.w.) ($N = 720$)	Hosseini *et al.* (2008)
Pinnipeds (seals and sea lions)	*In situ*	10000 muscle 700000 liver	IAEA TRS 422
Polar bears	*In situ*	8000 muscle 100000 liver	IAEA TRS 422
Cetaceans (whales, dolphins and porpoises)	*In situ*	80000 muscle 400000 liver	IAEA TRS 422

(Continued)

Table 2.3 Recommended concentration factors of Se used for radionuclide risk assessment in aquatic and terrestrial organisms (*Continued*).

Organisms	Conditions	Concentration factor (L kg^{-1})	Reference
Terrestrial*			
Grasses and herbs	*In situ*	1.0 ± 2.1 (*N* = 364)	IAEA TRS 479
Grasses	*In situ*	1.8 ± 1.6 (*N* = 48)	IAEA TRS 479
Herbs	*In situ*	1.4 ± 2.2 (*N* = 132)	IAEA TRS 479
Lichen and Bryophytes	*Ex situ*	20 (*N* = 1)	Beresford *et al.* (2008)
		0.36 ± 0.20 (*N* = 18)	IAEA TRS 479
Shrubs	*In situ*	1.81 ± 1.40 (*N* = 73)	Beresford *et al.* (2008)
	In situ	1.5 ± 1.4 (*N* = 94)	IAEA TRS 479
Trees	*In situ*	1.81 ± 1.40 (*N* = 73)	Beresford *et al.* (2008)
Detritivorous invertebrates	–	1.48 (*N* = 1)	Beresford *et al.* (2008)
Soil invertebrates	–	1.48	Beresford *et al.* (2008)
Annelids	–	1.50 (*N* = 1)	IAEA TRS 479
Molluscs (Gastropods)	*In situ*	0.035 ± 0.031 (*N* = 7)	IAEA TRS 479
Reptiles and mammals	*In situ*	0.0632 ± 0.381 (*N* = 12)	Beresford *et al.* (2008)

These values are based on experimental setups (*ex situ*) and/or from environmental sites (*in situ*, from exposure to stable Se). Arithmetic averages and standard deviations (SD), number of samples (*N*) and fresh weight (f.w.) vs dry weight (d.w.) values are specified when provided by the original reference.

*Resulting units are kg of dry soil per kg of fresh weight of organism.

ingestion pathway is the dominating exposure route, particularly from drinking water and food intake, related to national consumption habits.

It is difficult to precisely estimate the exposure doses from direct intake of drinking water due to the site-specific conditions of each geological repository and the nearby land use. For instance, estimated dose rates for a potential contaminated well 20 km away from a given repository could start showing detectable exposures to ^{79}Se after ~35000 y and reach maximum values of 0.1 mSv y^{-1} after 100000 y (Magill *et al.*, 2003). This dose rate is actually equal to the total admissible effective dose allowed to be discharged from NPPs in liquid and gas releases. It is also of the same order of magnitude as the world average dose of 0.3 mSv y^{-1} for ingestion of natural radionuclides, and below the admissible average doses of ~1 mSv y^{-1} for exposure of the population to background radiation. These comparisons imply a significant, but still acceptable exposure from drinking water for this specific example, though not generalizable and highly dependent on the specific case study. The impact of such estimations could reach doses of concern depending on the activities released from the waste repository and the annual intake of water. More refined predictions are still required as the potential doses of mobile nuclides such as Se are generally largely overestimated due to all the conservative assumptions related to lack of knowledge (Grambow, 2008).

Concerning food consumption, examples of known transfer factors of Se for human consumption of poultry, milk and eggs can be found in IAEA TRS 472. Examples of simulations calculating the radiation effect of ^{79}Se in humans from contaminated seafood suggest non-negligible dose coefficients, particularly from fish (1.7×10^{-7} Sv m^3 kg^{-1} Bq^{-1}), crustaceans (1.4×10^{-7} Sv m^3 kg^{-1} Bq^{-1}), molluscs (1.7×10^{-7} Sv m^3 kg^{-1} Bq^{-1}) and algae (2.8×10^{-8} Sv m^3 kg^{-1} Bq^{-1}). These dose coefficients are comparable to estimations for ^{60}Co, ^{119m}Sn, ^{121m}Sn, ^{129}I, ^{131}I, ^{226}Ra, ^{227}Ac, ^{232}Th, ^{242}Pu and ^{242m}Am (Ganul *et al.*, 2006). These values are higher than the recommended values for ingestion pathways from the International Commission on Radiation Protection (ICRP) using human internal dose conversion factors for ^{79}Se of 2.0×10^{-9}–2.3×10^{-9} Sv Bq^{-1} (Palattao *et al.*, 1997). They are also higher than the effective dose allowed for public ingestion of ^{79}Se, ranging between 1.9×10^{-8}–4.9×10^{-8} Sv Bq^{-1} (Decree 783/2001). In these simulations, intakes of ^{79}Se could reach levels of significant concern in specific conditions. In specific scenarios for solubility limited releases of radionuclides from a high-level nuclear waste repository in Belgian Boom Clay, dose rates related to ^{79}Se were $<10^{-5}$ Sv y^{-1} (e.g., from vitrified waste and spent fuel after >150000 y; Hoving, 2018; NIROND, 2001; Yu & Weetjens, 2016). Estimated internal doses for CANDU fuel in a conceptual clay-based backfill and a plutonic host rock are much lower, falling in the range of 10^{-11}–10^{-7} Sv y^{-1} (e.g. Johnson *et al.*, 1996). Therefore, further research should aim at reducing conservative assumptions in order to proceed towards more realistic descriptions of the Se distribution in the environment.

This knowledge should serve to locate optimal disposal sites and waste management strategies.

Concerning external exposure routes due to proximity to radioactive releases, simulated maximum annual doses for ^{79}Se vary depending on the fuel type, the host rock, and the environmental scenario considered. For granite rock repositories in Sweden, the annual dose of ^{79}Se to the geosphere has been shown to be four orders of magnitude below the local annual limit of 0.15 mSv (TR-99-06). In general, external exposure routes of ^{79}Se are not expected to be critical, though the chemistry of Se in these simulations still has to be refined.

2.7 CONCLUSIONS

Selenium is known to show a mobility and bioavailability in the environment which is dependent on the chemical species. The only radioactive isotope of concern for the environment (^{79}Se) is expected to be released in the long-term ($>10^4$ y) from deep geological nuclear waste repository sites in relatively low concentrations. These radioactive releases may be subjected to more or less complex transformations, showing a characteristic, site-specific behaviour according to the multiple chemical forms in which Se may exist in each environmental compartment. Dispersion scenarios for nuclear risk assessment are still developing and require the complementary information provided by environmental studies on stable Se and experimental studies using stable Se or benefiting from use of the radiotracer, ^{75}Se. Existing analytical challenges such as polyatomic and isobaric interferences encountered when quantifying stable isotopes of Se in environmental samples may be overcome more easily nowadays with new generation ICP-MS. Biological studies suggest a potentially relevant intake of Se by organisms, particularly in aquatic systems. In any case, the incorporation of ^{79}Se into the biogeochemical cycle of Se implies that, in the far future the highest radioactive doses to humans are expected to be linked to internal exposure pathways, whereas the external exposure pathway will contribute with very low doses of ^{79}Se, eventually becoming part of the environmental radioactive background.

REFERENCES

Abdelouas A. and Grambow B. (2012). Aquatic chemistry of long-lived mobile fission and activation products in the context of deep geological disposal. In: Radionuclide Behaviour in the Natural Environment, C. Poinssot and H. Geckeis (eds.), Woodhead Publishing, pp. 70–102.

Aguerre S. and Frechou C. (2006). Development of a radiochemical separation for selenium with the aim of measuring its isotope 79 in low and intermediate nuclear wastes by ICP-MS. *Talanta*, **69**(3), 565–571.

Alexander W. R., Frieg B., Ota K. and Bedrock Geosciences A. (2009). Grimsel test site investigation Phase IV. The Nagra-JAEA in situ study of safety relevant radionuclide retardation in fractured crystalline rock. III: The RRP project final report (No. NTB--00-07). Bedrock Geosciences.

Alhajji E. (2007). Etude des propriétés de sorption de la bentonite mx-80 vis-à-vis de se (iv), ni (ii) et cs (i). du système disperse au système compacte. Doctoral dissertation, Paris, 11.

Altmann S. (2008). 'Geo'chemical research: a key building block for nuclear waste disposal safety cases. *Journal of Contaminant Hydrology*, **102**(3–4), 174–179.

Amouroux D. and Donard O. F. (1997). Evasion of selenium to the atmosphere via biomethylation processes in the Gironde estuary, France. *Marine Chemistry*, **58**(1–2), 173–188.

ANDRA (2005). Dossier 2005 Argile. Evaluation de sureté du stockage géologique. Agence National pour la gestion des Déchets Radioactifs, Paris.

Araie H., Suzuki I. and Shiraiwa Y. (2008). Identification and characterization of a selenoprotein, thioredoxin reductase, in a unicellular marine haptophyte alga, *Emiliania huxleyi. Journal of Biological Chemistry*, **283**(51), 35329–35336.

Arnold R., Augier C., Baker J., Barabash A. S., Basharina-Freshville A., Bongrand M., Brudanin V., Caffrey A. J., Cebrián S., Chapon A. and Chauveau E. (2010). Probing new physics models of neutrinoless double beta decay with SuperNEMO. *The European Physical Journal C*, **70**(4), 927–943.

Artusa D. R., Balzoni A., Beeman J. W., Bellini F., Biassoni M., Brofferio C., Camacho A., Capelli S., Cardani L., Carniti P. and Casali N. (2017). The LUCIFER/CUPID-0 demonstrator: searching for the neutrinoless double-beta decay with Zn 82 Se scintillating bolometers. *Journal of Physics: Conference Series*, **888**(1), 012077.

Asai S., Hanzawa Y., Okumura K., Shinohara N., Inagawa J., Hotoku S., Suzuki K. and Kaneko S. (2011). Determination of ^{79}Se and ^{135}Cs in spent nuclear fuel for inventory estimation of high-level radioactive wastes. *Journal of Nuclear Science and Technology*, **48**(5), 851–854.

Audi G., Bersillon O., Blachot J. and Wapstra A. H. (2003). The NUBASE evaluation of nuclear and decay properties. *Nuclear Physics A*, **624**, 1–124.

Bajaj M., Schmidt S. and Winter J. (2012). Formation of Se (0) Nanoparticles by Duganella sp. and *Agrobacterium* sp. isolated from Se-laden soil of North-East Punjab, India. *Microbial Cell Factories*, **11**(1), 64.

Banning H. M., Stelling M., Eiche E., Nothstein A. K., Neumann T., Schoenberg R., Riemann M. and Nick P. (2013). Use of stable isotope signatures in plants as a tool to explore the selenium cycle in the 'critical zone'. *Selenium in the Environment and Human Health*, **22**, 22–23.

Banning H., Stelling M., König S., Schoenberg R. and Neumann T. (2018). Preparation and purification of organic samples for selenium isotope studies. *PloS One*, **13**(3), e0193826.

Barnekow U., Fesenko S., Kashparov V., Kis-Benedek G., Matisoff G., Onda Y., Sanzharova N., Tarjan S., Tyler A. and Varga B. (2019). Guidelines on Soil and Vegetation Sampling for Radiological Monitoring. International Atomic Energy Agency: Vienna, Austria.

Basu R., Haque S. E., Tang J., Ji J. and Johannesson K. H. (2007). Evolution of selenium concentrations and speciation in groundwater flow systems: Upper Floridan (Florida) and Carrizo Sand (Texas) aquifers. *Chemical Geology*, **246**(3–4), 147–169.

Beaudette D. E., Stupi L. K., Swarowsky A., O'Geen A. T., Chang J. F. and Gallagher B. (2009). Watershed-scale geochemical inventory of soils by portable x-ray fluorescence. *AGUFM*, **2009**, B11A-0467.

Beauwens T., De Cannière P., Moors H., Wang L. and Maes N. (2005). Studying the migration behaviour of selenate in Boom Clay by electromigration. *Engineering Geology*, **77**(3–4), 285–293.

Beilstein M. A. and Whanger P. D. (1988). Glutathione peroxidase activity and chemical forms of selenium in tissues of rats given selenite or selenomethionine. *Journal of Inorganic Biochemistry*, **33**(1), 31–46.

Benoit M. D., Kudela R. M. and Flegal A. R. (2010). Modeled trace element concentrations and partitioning in the San Francisco estuary, based on suspended solids concentration. *Environmental Science and Technology*, **44**(15), 5956–5963.

Beresford N. A., Barnett C. L., Howard B. J., Scott W. A., Brown J. E. and Copplestone D. (2008). Derivation of transfer parameters for use within the ERICA Tool and the default concentration ratios for terrestrial biota. *Journal of Environmental Radioactivity*, **99**(9), 1393–1407.

Beresford N. A., Beaugelin-Seiller K., Burgos J., Cujic M., Fesenko S., Kryshev A., Pachal N., Real A., Su B. S., Tagami K. and Vives i Batlle J. (2015). Radionuclide biological half-life values for terrestrial and aquatic wildlife. *Journal of Environmental Radioactivity*, **150**, 270–276.

Bleyen N., Smets S., Small J., Moors H., Leys N., Albrecht A., De Cannière P., Schwyn B., Wittebroodt C. and Valcke E. (2018). Impact of the electron donor on in situ microbial nitrate reduction in Opalinus Clay: results from the Mont Terri rock laboratory (Switzerland). In: Mont Terri Rock Laboratory, 20 Years. Swiss Journal of Geosciences Supplement, P. Bossart and A. Milnes (eds.), Birkhäuser, Cham, vol. **5**.

Booth S., Walters W. J., Steenbeek J., Christensen V. and Charmasson S. (2020). An Ecopath with Ecosim model for the Pacific coast of eastern Japan: describing the marine environment and its fisheries prior to the Great East Japan earthquake. *Ecological Modelling*, **428**, 109087.

Börsig N., Scheinost A. C., Shaw S., Schild D. and Neumann T. (2017). Uptake mechanisms of selenium oxyanions during the ferrihydrite-hematite recrystallization. *Geochimica et Cosmochimica Acta*, **206**, 236–253.

Börsig N., Scheinost A. C., Shaw S., Schild D. and Neumann T. (2018). Retention and multiphase transformation of selenium oxyanions during the formation of magnetite via iron (II) hydroxide and green rust. *Dalton Transactions*, **47**(32), 11002–11015.

Breynaert E., Scheinost A. C., Dom D., Rossberg A., Vancluysen J., Gobechiya E., Kirschhock C. E. and Maes A. (2010). Reduction of Se (IV) in boom clay: XAS solid phase speciation. *Environmental Science and Technology*, **44**(17), 6649–6655.

Campbell A. D. (1992). A critical survey of hydride generation techniques in atomic spectroscopy (Technical Report). *Pure and Applied Chemistry*, **64**(2), 227–244.

Caurant D., Loiseau P., Majerus O., Aubin-Chevaldonnet V., Bardez I. and Quintas A. (2009). Glasses, Glass-Ceramics and Ceramics for Immobilization of Highly Radioactive Nuclear Wastes. Nova Science, Hauppauge.

Chang Y., Zhang J., Qu J., Zhang G., Zhang A. and Zhang R. (2016). The behavior of dissolved inorganic selenium in the Changjiang Estuary. *Journal of Marine Systems*, **154**, 110–121.

Charlet L., Scheinost A. C., Tournassat C., Greneche J. M., Géhin A., Fernández-Martı A., Coudert S., Tisserand D. and Brendle J. (2007). Electron transfer at the mineral/water interface: selenium reduction by ferrous iron sorbed on clay. *Geochimica et Cosmochimica Acta*, **71**(23), 5731–5749.

Chen F., Burns P. C. and Ewing R. C. (1999). 79Se: geochemical and crystallo-chemical retardation mechanisms. *Journal of Nuclear Materials*, **275**(1), 81–94.

Ciceri G., Traversi A. L., Martinotti W. and Queirazza G. (1988). Radionuclide partitioning between water and suspended matter: comparison of different methodologies. In: Studies in Environmental Science, L. Pawlowski, E. Mentasti, W. J. Lacy and C. Sarzanini (eds.), Elsevier, Vol. **34**, pp. 353–375.

Curti E., Puranen A., Grolimund D., Jädernas D., Sheptyakov D. and Mesbah A. (2015). Characterization of selenium in UO_2 spent nuclear fuel by micro X-ray absorption spectroscopy and its thermodynamic stability. *Environmental Science: Processes and Impacts*, **17**(10), 1760–1768.

Cutter G. A. (1989). The estuarine behaviour of selenium in San Francisco Bay. *Estuarine, Coastal and Shelf Science*, **28**(1), 13–34.

Cutter G. A. and Cutter L. S. (1995). Behavior of dissolved antimony, arsenic, and selenium in the Atlantic Ocean. *Marine Chemistry*, **49**(4), 295–306.

Darcheville O., Février L., Haichar F. Z., Berge O., Martin-Garin A. and Renault P. (2008). Aqueous, solid and gaseous partitioning of selenium in an oxic sandy soil under different microbiological states. *Journal of Environmental Radioactivity*, **99**(6), 981–992.

Dardenne K., González-Robles E., Rothe J., Müller N., Christill G., Lemmer D., Praetorius R., Kienzler B., Metz V., Roth G. and Geckeis H. (2015). XAS and XRF investigation of an actual HAWC glass fragment obtained from the Karlsruhe vitrification plant (VEK). *Journal of Nuclear Materials*, **460**, 209–215.

Davis J. A. and Leckie J. O. (1980). Surface ionization and complexation at the oxide/water interface. 3. Adsorption of anions. *Journal of Colloid and Interface Science*, **74**(1), 32–43.

De Cannière P., Maes A., Williams S., Bruggeman C., Beauwens T., Maes N. and Cowper M. (2010). Behaviour of selenium in Boom Clay. External Report, SCK• CEN-ER-120.

Decree R. (2001). Real Decreto 783/2001, de 6 de julio por el que se aprueba el Reglamento sobre Protección Sanitaria contra Radiaciones Ionizantes.

Descostes M., Blin V., Bazer-Bachi F., Meier P., Grenut B., Radwan J., Schlegel M. L., Buschaert S., Coelho D. and Tevissen E. (2008). Diffusion of anionic species in Callovo-Oxfordian argillites and Oxfordian limestones (Meuse/Haute–Marne, France). *Applied Geochemistry*, **23**(4), 655–677.

Devi P., Jain R., Thakur A., Kumar M., Labhsetwar N. K., Nayak M. and Kumar P. (2017). A systematic review and meta-analysis of voltammetric and optical techniques for inorganic selenium determination in water. *TrAC Trends in Analytical Chemistry*, **95**, 69–85.

Dhillon K. S. and Dhillon S. K. (1991). Selenium toxicity in soils, plants and animals in some parts of Punjab, India. *International Journal of Environmental Studies*, **37**(1–2), 15–24.

Dhillon K. S. and Dhillon S. K. (2019). Genesis of seleniferous soils and associated animal and human health problems. In: Advances in Agronomy, D. L. Sparks (ed.), Academic Press, Vol. **154**, pp. 1–80.

Diener A., Neumann T., Kramar U. and Schild D. (2012). Structure of selenium incorporated in pyrite and mackinawite as determined by XAFS analyses. *Journal of Contaminant Hydrology*, **133**, 30–39.

Dočekal B., Dědina J. and Krivan V. (1997). Radiotracer investigation of hydride trapping efficiency within a graphite furnace. *Spectrochimica Acta Part B: Atomic Spectroscopy*, **52**(6), 787–794.

Duro L., Altmaier M., Holt E., Mäder U., Claret F., Grambow B., Idiart A., Valls A. and Montoya V. (2020). Contribution of the results of the CEBAMA project to decrease uncertainties in the Safety Case and Performance Assessment of radioactive waste repositories. *Applied Geochemistry*, **112**, 104479.

Eiche E., Bardelli F., Nothstein A. K., Charlet L., Göttlicher J., Steininger R., Dhillon K. S. and Sadana U. S. (2015). Selenium distribution and speciation in plant parts of wheat (*Triticum aestivum*) and Indian mustard (*Brassica juncea*) from a seleniferous area of Punjab, India. *Science of the Total Environment*, **505**, 952–961.

Eiche E., Nothstein A. K., Göttlicher J., Steininger R., Dhillon K. S. and Neumann T. (2019). The behaviour of irrigation induced Se in the groundwater-soil-plant system in Punjab, *India. Environmental Science: Processes & Impacts*, **21**(6), 957–969.

Eikenberg J., Ruethi M., Alexander W. R., Frieg B. and Fierz T. (1997). The excavation project in the grimsel test site: in situ high-resolution gamma and alpha spectrometry of ^{60}Co, ^{75}Se, ^{113}Sn, ^{152}Eu, ^{235}U, and ^{237}Np/^{233} Pa. MRS Online Proceedings Library Archive, 506.

Einoder L. D., MacLeod C. K. and Coughanowr C. (2018). Metal and isotope analysis of bird feathers in a contaminated estuary reveals bioaccumulation, biomagnification, and potential toxic effects. *Archives of Environmental Contamination and Toxicology*, **75** (1), 96–110.

Ellis A. S., Johnson T. M., Herbel M. J. and Bullen T. D. (2003). Stable isotope fractionation of selenium by natural microbial consortia. *Chemical Geology*, **195**(1–4), 119–129.

Ervanne H., Hakanen M. and Lehto J. (2016). Selenium sorption on clays in synthetic groundwaters representing crystalline bedrock conditions. *Journal of Radioanalytical and Nuclear Chemistry*, **307**(2), 1365–1373.

Ezoe K., Ohyama S., Hashem M. A., Ohira S. I. and Toda K. (2016). Automated determinations of selenium in thermal power plant wastewater by sequential hydride generation and chemiluminescence detection. *Talanta*, **148**, 609–616.

Feinberg A., Maliki M., Stenke A., Sudret B., Peter T. and Winkel L. H. (2020). Mapping the drivers of uncertainty in atmospheric selenium deposition with global sensitivity analysis. *Atmospheric Chemistry and Physics*, **20**(3), 1363–1390.

Février L. and Martin-Garin A. (2005). Biogeochemical behaviour of anionic radionuclides in soil: Evidence for biotic interactions. *Radioprotection*, **40**(1), S79.

Finck N., Dardenne K., Bosbach D. and Geckeis H. (2012). Selenide retention by mackinawite. *Environmental Science and Technology*, **46**(18), 10004–10011.

Fleming D. E., Groves J. W., Gherase M. R., George G. N., Pickering I. J., Ponomarenko O., Langan G., Spallholz J. E., Alauddin M., Ahsan H. and Ahmed S. (2015). Soft tissue measurement of arsenic and selenium in an animal model using portable X-ray fluorescence. *Radiation Physics and Chemistry*, **116**, 241–247.

Fowler S. W., Teyssie J. L., Acosta A., Gattuso J. P. and Jaubert J. (1999). Radiotracer studies on radionuclide and trace element cycling in corals (No. IAEA-TECDOC--1094).

Fresneda M. A. R., Martín J. D., Bolívar J. G., Cantos M. V. F., Bosch-Estévez G., Moreno M. F. M. and Merroun M. L. (2018). Green synthesis and biotransformation of amorphous Se nanospheres to trigonal 1D Se nanostructures: impact on Se mobility within the concept of radioactive waste disposal. *Environmental Science: Nano*, **5**(9), 2103–2116.

Galinha C., Freitas M. C., Pacheco A. M. G., Coutinho J., Maçãs B. and Almeida A. S. (2012). Radiotracing selenium in bread-wheat seeds for a Se-biofortification program: an optimization study in seed enrichment. *Journal of Radioanalytical and Nuclear Chemistry*, **291**(1), 193–195.

Ganul M. N., Kuchin N. L. and Sergeev I. V. (2006). Prediction of the radiation effect resulting from the contamination of water in marine fisheries. *Atomic Energy*, **100**(2), 79–82.

Gaucher E. C., Tournassat C., Pearson F. J., Blanc P., Crouzet C., Lerouge C. and Altmann S. (2009). A robust model for pore-water chemistry of clayrock. *Geochimica et Cosmochimica Acta*, **73**(21), 6470–6487.

Gautschi A. (2017). Safety-relevant hydrogeological properties of the claystone barrier of a Swiss radioactive waste repository: an evaluation using multiple lines of evidence. *Grundwasser*, **22**(3), 221–233.

Giffaut E., Grivé M., Blanc P., Vieillard P., Colàs E., Gailhanou H., Gaboreau S., Marty N., Made B. and Duro L. (2014). Andra thermodynamic database for performance assessment: ThermoChimie. *Applied Geochemistry*, **49**, 225–236.

Gil-Díaz T. (2019). Tellurium radionuclides produced by major accidental events in nuclear power plants. *Environmental Chemistry*, **16**(4), 296–302.

Gil-Díaz T., Heberling F., Keller V., Fuss M., Böttle M., Eiche E. and Schäfer J. (2020a). Tin-113 and Selenium-75 radiotracer adsorption and desorption kinetics in contrasting estuarine salinity and turbidity conditions. *Journal of Environmental Radioactivity*, **213**, 106133.

Gil-Díaz T., Schäfer J., Keller V., Eiche E., Dutruch L., Mößner C., Lenz M. and Eyrolle F. (2020b). Tellurium and selenium sorption kinetics and solid fractionation under contrasting estuarine salinity and turbidity conditions. *Chemical Geology*, **532**, 119370.

Gil-García C., Tagami K., Uchida S., Rigol A. and Vidal M. (2009). New best estimates for radionuclide solid–liquid distribution coefficients in soils. Part 3: miscellany of radionuclides (Cd, Co, Ni, Zn, I, Se, Sb, Pu, Am, and others). *Journal of Environmental Radioactivity*, **100**(9), 704–715.

Gimmi T., Leupin O. X., Eikenberg J., Glaus M. A., Van Loon L. R., Waber H. N., Wersin P., Wang H. A., Grolimund D., Borca C. N. and Dewonck S. (2014). Anisotropic diffusion at the field scale in a 4-year multi-tracer diffusion and retention experiment–I: insights from the experimental data. *Geochimica et Cosmochimica Acta*, **125**, 373–393.

Grambow B. (2008). Mobile fission and activation products in nuclear waste disposal. *Journal of Contaminant Hydrology*, **102**(3–4), 180–186.

Gray T., Mann N. and Whitby M. (2007). Periodic Table of Isotopes. Accessed 10th March 2015 through http://periodictable.com/Isotopes/051.123/index.p.full.html

Gruebel K. A., Davis J. A. and Leckie J. O. (1988). The feasibility of using sequential extraction techniques for arsenic and selenium in soils and sediments. *Soil Science Society of America Journal*, **52**(2), 390–397.

Guidebook A. (1989). Measurement of Radionuclides in Food and the Environment. International Atomic Energy Agency, Vienna. Retrieved from https://www.IAEA.org/publications/1398/measurement-of-radionuclides-in-food-and-the-environment Accessed 9th September 2020.

Gustafsson J. P. and Johnsson L. (1994). The association between selenium and humic substances in forested ecosystems—laboratory evidence. *Applied Organometallic Chemistry*, **8**(2), 141–147.

Hampel A., Hansen F. D., Salzer K., Minkley W., Lux K. H., Herchen K., Pudewills A., Yildirim S., Staudtmeister K., Rokahr R. and Zapf D. (2013). Benchmark Calculations of the Thermo-Mechanical Behavior of Rock Salt—Results from a US-German Joint Project. 47th US Rock Mechanics/Geomechanics Symposium, American Rock Mechanics Association, San Francisco, California.

Heberling F., Vinograd V. L., Polly R., Gale J. D., Heck S., Rothe J., Bosbach D., Geckeis H. and Winkler B. (2014). A thermodynamic adsorption/entrapment model for selenium (IV) coprecipitation with calcite. *Geochimica et Cosmochimica Acta*, **134**, 16–38.

Henner P. (2008). Bioaccumulation of Radionuclides and Induced Biological Effects in Situations of Chronic Exposure of Ecosystems a Uranium Case Study. Loads and Fate of Fertilizer Derived Uranium. Backhuys Publishers Leiden, The Netherlands.

Hesslein R. H. (1987). Whole-lake metal radiotracer movement in fertilized lake basins. *Canadian Journal of Fisheries and Aquatic Sciences*, **44**(S1), s74–s82.

Hosseini A., Thørring H., Brown J. E., Saxén R. and Ilus E. (2008). Transfer of radionuclides in aquatic ecosystems–default concentration ratios for aquatic biota in the Erica Tool. *Journal of Environmental Radioactivity*, **99**(9), 1408–1429.

Hoving A. L. (2018). Redox Properties of Boom Clay and their Effect on Reactions with Selenium and Uranium. Doctoral dissertation, UU Dept. of Earth Sciences.

International Atomic Energy Agency (2004). Sediment Distribution Coefficients and Concentration Factors for Biota in the Marine Environment, Technical Reports Series No. 422, IAEA, Vienna.

International Atomic Energy Agency (2010). Handbook of Parameter Values for the Prediction of Radionuclide Transfer in Terrestrial and Freshwater Environments, Technical Reports Series No. 472, IAEA, Vienna.

International Atomic Energy Agency (2014). Handbook of Parameter Values for the Prediction of Radionuclide Transfer to Wildlife, Technical Reports Series No. 479, IAEA, Vienna.

Iida Y., Yamaguchi T., Tanaka T., Kitamura A. and Nakayama S. (2009). Determination of the solubility limiting solid of selenium in the presence of iron under anoxic conditions. Proc. Mobile Fission and Activation Products in Nuclear Waste Disposal, La Baule, pp. 135–145.

Ikonen J. (2017). Sorption and Matrix Diffusion in Crystalline Rocks–In-situ and Laboratory Approach. Doctoral thesis, University of Helsinki, Finland.

Ikonen J., Voutilainen M., Söderlund M., Jokelainen L., Siitari-Kauppi M. and Martin A. (2016). Sorption and diffusion of selenium oxyanions in granitic rock. *Journal of Contaminant Hydrology*, **192**, 203–211.

Irons R., Carlson B. A., Hatfield D. L. and Davis C. D. (2006). Both selenoproteins and low molecular weight selenocompounds reduce colon cancer risk in mice with genetically impaired selenoprotein expression. *The Journal of Nutrition*, **136**(5), 1311–1317.

Jeffery E. H., Wallig M. A. and Tumbleson M. E. (2002). Nutritional toxicologic pathology. In: Handbook of Toxicologic Pathology, 2nd edn, W. M. Haschek, C. G. Rousseaux and M. A. Wallig (eds.), Academic Press, Amsterdam, Boston, Heidelberg, London, New York, Oxford, Paris, San Diego, San Francisco, Singapore, Sydney, Tokyo, pp. 595–629.

Johnson L. H. and McGinnes D. F. (2002). Partitioning of radionuclides in Swiss power reactor fuels (No. NTB--02-07). National Cooperative for the Disposal of Radioactive Waste (NAGRA).

Johnson L. H., LeNeveu D. M., King F., Shoesmith D. W., Kolar M., Oscarson D. W., Sunder S., Onofrei C. and Crosthwaite J. L. (1996). The disposal of Canada's nuclear fuel waste: a study of postclosure safety of in-room emplacement of used CANDU fuel in copper containers in permeable plutonic rock. Volume 2: Vault model (No. AECL--11494-2). Atomic Energy of Canada Ltd.

Joint F. A. O. (2016). Supporting Sampling and Sample Preparation Tools for Isotope and Nuclear Analysis (No. IAEA-TECDOC--1783). Joint FAO/IAEA Division of Nuclear Techniques in Food and Agriculture.

Jones G. D., Droz B., Greve P., Gottschalk P., Poffet D., McGrath S. P., Seneviratne S. I., Smith P. and Winkel L. H. (2017). Selenium deficiency risk predicted to increase under future climate change. *Proceedings of the National Academy of Sciences*, **114** (11), 2848–2853.

Jorg G., Bühnemann R., Hollas S., Kivel N., Kossert K., Van Winckel S. and Gostomski C. L. V. (2010). Preparation of radiochemically pure 79Se and highly precise determination of its half-life. *Applied Radiation and Isotopes*, **68**(12), 2339–2351.

Kaizer J., Dulanská S., Horváthová B., Ješkovský M. and Povinec P. P. (2018). Development of separation procedures for determination of uranium and thorium in the 82 Se source of the SuperNEMO experiment: first steps. *Journal of Radioanalytical and Nuclear Chemistry*, **318**(3), 2321–2327.

Kienzler B., Schlieker M., Rabung T., Borkel C., Geyer F., Hilpp S. and Walschburger C. (2016). Schachtanlage Asse II, Sorptionsuntersuchungen an Gesteinsproben aus der Bohrung Remlingen 15 Teil 1a: Grundwasseranalysen. Auftrag des Bundesamts für Strahlenforschung (BfS), Vol. KIT-INE 002/2016.

Koch-Steindl H. and Pröhl G. (2001). Considerations on the behaviour of long-lived radionuclides in the soil. *Radiation and Environmental Biophysics*, **40**(2), 93–104.

Kowe R., Mobbs S., Proehl G., Bergstrom U., Olyslaegers G., Zeevaert T. and Simon I. (2005). Application of biosphere models in the BIOMOSA project: a comparative assessment of five European radioactive waste disposal sites. *Radioprotection*, **40**(1), S701.

Krivan V. (1982). Role of radiotracers in the development of trace element analysis. *Talanta*, **29**(11), 1041–1050.

Kutschera W. (1998). Dating and environment using long-lived radionuclides. *Il Nuovo Cimento A (1971–1996)*, **111**(8–9), 1019–1031.

Lee C. P., Tsai S. C., Wei Y. Y., Teng S. P. and Hsu C. N. (2008). Comparative study on sorption of individual and coupling cesium and selenium on mudrock. *Journal of Radioanalytical and Nuclear Chemistry*, **275**(1), 115–119.

Lee C. P., Wu M. C., Tsai T. L., Wei H. J., Men L. C. and Hsu C. N. (2012). Comparative study on sorption additivity of individual and coexisting cesium and selenium on bentonite/quartz sand mixtures. *International Journal of Nuclear Energy Science and Engineering*, **2**, 23–27.

Lenz M., Hullebusch E. D. V., Farges F., Nikitenko S., Borca C. N., Grolimund D. and Lens P. N. (2008). Selenium speciation assessed by X-ray absorption spectroscopy of sequentially extracted anaerobic biofilms. *Environmental Science and Technology*, **42** (20), 7587–7593.

Mace N. (2006). Influence de la température sur la rétention du sélénite par une pâte de ciment altérée et par ses phases pures constitutives. Doctoral dissertation.

Magill J., Berthou V., Haas D., Galy J., Schenkel R., Wiese J. W., Heusener G., Tommasi J. and Youinou G. (2003). Impact limits of partitioning and transmutation scenarios on the radiotoxicity of actinides in radioactive waste. *Nuclear Energy-London*, **42**(5), 263–278.

Marguí E., Floor G. H., Hidalgo M., Kregsamer P., Román-Ross G., Streli C. and Queralt I. (2010). Analytical possibilities of total reflection X-ray spectrometry (TXRF) for trace selenium determination in soils. *Analytical Chemistry*, **82**(18), 7744–7751.

Martinčič R. (ed.). (1999). Generic Procedures for Monitoring in a Nuclear or Radiological Emergency. IAEA, Vienna.

Measures C. I. and Burton J. D. (1978). Behaviour and speciation of dissolved selenium in estuarine waters. *Nature*, **273**(5660), 293–295.

Mihailescu A. and Cata-Danil G. H. (2010). Gamma-ray production in the 58 Ni (16 O, X) nuclear reaction. *Romanian Journal of Physics*, **55**(7–8), 712–723.

Natasha T., Shahid M., Niazi N. K., Khalid S., Murtaza B., Bibi I. and Rashid M. I. (2018). A critical review of selenium biogeochemical behavior in soil-plant system with an inference to human health. *Environmental Pollution*, **234**, 915–934.

Neeb K. H. (1997). The Radiochemistry of Nuclear Power Plants with Light Water Reactors. Walter de Gruyter, Berlin and New York.

Nie Z., Finck N., Heberling F., Pruessmann T., Liu C. and Lützenkirchen J. (2017). Adsorption of selenium and strontium on goethite: EXAFS study and surface complexation modeling of the ternary systems. *Environmental Science and Technology*, **51**(7), 3751–3758.

Nothstein A. K., Eiche E., Riemann M., Nick P., Maier P., Tenspolde A. and Neumann T. (2019). Coupling Langmuir with Michaelis-Menten—A practical alternative to estimate Se content in rice?. *PloS One*, **14**(4), e0214219.

Ogra Y., Shimizu M., Takahashi K. and Anan Y. (2018). Biotransformation of organic selenium compounds in budding yeast, Saccharomyces cerevisiae. *Metallomics*, **10** (9), 1257–1263.

Ohlendorf H. M., Santolo G. M., Byron E. R. and Eisert M. A. (2020). Kesterson reservoir: 30 years of selenium risk assessment and management. *Integrated Environmental Assessment and Management*, **16**(2), 257–268.

Olszewska W., Miśkiewicz A., Zakrzewska-Kołtuniewicz G., Lankof L. and Pająk L. (2015). Multibarrier system preventing migration of radionuclides from radioactive waste repository. *Nukleonika*, **60**(3), 557–563.

Ondraf/Niras (NIROND) (2001). Technical Overview of the SAFIR 2 Report, Safety Assessment and Feasibility Interim Report 2. Belgian Agency for Radioactive Waste and Enriched Fissile Materials.

Palattao M. V. B., Hajas W. C. and Goodwin B. W. (1997). Evaluation of the use of ICRP 60 dose conversion factors in a postclosure assessment of a deep geological disposal system (No. AECL--11734). Atomic Energy of Canada Limited.

Parkhurst D. L. and Appelo C. (2013). Description of input and examples for PHREEQC version 3—a computer program for speciation, batch-reaction, one-dimensional transport, and inverse geochemical calculations. *US Geological Survey Techniques and Methods*, **6**, 497.

Pilon-Smits E. A., Winkel L. H. and Lin Z. Q. (eds.). (2017). Selenium in Plants: Molecular, Physiological, Ecological and Evolutionary Aspects. Springer, Cham, Switzerland, Vol. **11**.

Polly R., Heberling F., Schimmelpfennig B. and Geckeis H. (2017). Quantum chemical investigation of the selenite incorporation into the calcite (1014) surface. *The Journal of Physical Chemistry C*, **121**(37), 20217–20228.

Post-closure safety of a deep repository for spent nuclear fuel, Technical Report TR-99-06, Swedish Nuclear Fuel and Waste Management Co. (1999).

Rabung T., Kienzler B., Skerencak-Frech A., Schlieker M., Fellhauer D., Plaschke M., Geyer F., Walschburger C., Kisely T., Böttle M., Fuss M., Altmaier M., Metz V. and Geckeis H. (2018). Schachtanlage Asse II, Sorptionsuntersuchungen an den Gesteinsproben aus der Bohrung Remlingen 15 (Teil 3a). Auftrag des Bundesamts für Strahlenforschung (BfS).

Reamer D. C., Veillon C. and Tokousbalides P. T. (1981). Radiotracer techniques for evaluation of selenium hydride generation systems. *Analytical Chemistry*, **53**(2), 245–248.

Reinfelder J. R. and Fisher N. S. (1994). Retention of elements absorbed by juvenile fish (Menidia menidia, Menidia beryllina) from zooplankton prey. *Limnology and Oceanography*, **39**(8), 1783–1789.

Riaz A., Abbas M. A., Mubeen H., Ibrahim N. and Raza S. (2018). Methods to Enhance Selenium in Wheat through Biofortification: a review. *Journal of Biotechnology and Biomaterials*, **8**(2), 1000282.

Salminen R., Batista M. J., Bidovec M., Demetriades A., De Vivo B., De Vos W., Duris M., Gilucis A., Gregorauskiene V., Halamic J., Heitzmann P., Lima A., Jordan G., Klaver G., Klein P., Lis J., Locutura J., Marsina K., Mazreku A., O'Connor P.J., Olsson S. A., Ottesen R.T., Petersell V., Plant J.A., Reeder S., Salpeteur I., Sandström H., Siewers U., Steenfelt A. and Tarvainen T. (2005). Geochemical Atlas of Europe. Part 1 – Background Information, Methodology and Maps.

Santschi P. H., Li Y. H. and Carson S. R. (1980). The fate of trace metals in Narragansett Bay, Rhode Island: Radiotracer experiments in microcosms. *Estuarine and Coastal Marine Science*, **10**(6), 635–654.

Schilling K., Johnson T. M., Dhillon K. S. and Mason P. R. (2015). Fate of selenium in soils at a seleniferous site recorded by high precision Se isotope measurements. *Environmental Science and Technology*, **49**(16), 9690–9698.

Schilling K., Basu A., Wanner C., Sanford R. A., Pallud C., Johnson T. M. and Mason P. R. (2020). Mass-dependent selenium isotopic fractionation during microbial reduction of seleno-oxyanions by phylogenetically diverse bacteria. *Geochimica et Cosmochimica Acta*, **276**, 274–288.

Segura R., Pizarro J., Díaz K., Placencio A., Godoy F., Pino E. and Recio F. (2015). Development of electrochemical sensors for the determination of selenium using gold nanoparticles modified electrodes. *Sensors and Actuators B: Chemical*, **220**, 263–269.

Seyler P. and Martin J. M. (1991). Arsenic and selenium in a pristine river-estuarine system: the Krka (Yugoslavia). *Marine Chemistry*, **34**(1–2), 137–151.

Shiobara Y., Ogra Y. and Suzuki K. T. (2000). Exchange of endogenous selenium for dietary selenium as 82Se-enriched selenite in brain, liver, kidneys and testes. *Life Sciences*, **67** (25), 3041–3049.

Spiller H. A. (2018). Rethinking mercury: the role of selenium in the pathophysiology of mercury toxicity. *Clinical Toxicology*, **56**(5), 313–326.

Suzuki K. T., Doi C. and Suzuki N. (2006). Metabolism of 76Se-methylselenocysteine compared with that of 77Se-selenomethionine and 82Se-selenite. *Toxicology and Applied Pharmacology*, **217**(2), 185–195.

Suzuki K. N., Machado E. C., Machado W., Bellido A. V. B., Bellido L. F., Osso J. A. and Lopes R. T. (2014). Removal efficiency of ^{75}Se, ^{51}Cr and ^{60}Co from tidal water by mangrove sediments from Sepetiba Bay (SE Brazil). *Journal of Radioanalytical and Nuclear Chemistry*, **299**(1), 357–361.

StrlSchV (2018). Verordnung zum Schutz vor der schädlichen Wirkung ionisierender Strahlung. Accessed 1st October 2020 through https://www.gesetze-im-internet.de/strlschv_2018/

Tan Z., Wu W., Yin N., Jia M., Chen X., Bai Y., Wu H., Zhang Z. and Li P. (2020). Determination of selenium in food and environmental samples using a gold nanocages/fluorinated graphene nanocomposite modified electrode. *Journal of Food Composition and Analysis*, **94**, 103628.

Tanaka K., Shimada A., Hoshi A., Yasuda M., Ozawa M. and Kameo Y. (2014). Radiochemical analysis of rubble and trees collected from Fukushima Daiichi Nuclear Power Station. *Journal of Nuclear Science and Technology*, **51**(7–8), 1032–1043.

Takata H., Aono T., Tagami K. and Uchida S. (2016). A new approach to evaluate factors controlling elemental sediment–seawater distribution coefficients (Kd) in coastal regions, Japan. *Science of the Total Environment*, **543**, 315–325.

Torres J., Pintos V., Gonzatto L., Domínguez S., Kremer C. and Kremer E. (2011). Selenium chemical speciation in natural waters: Protonation and complexation behavior of selenite and selenate in the presence of environmentally relevant cations. *Chemical Geology*, **288**(1–2), 32–38.

Tuğrul B., Erentürk S., Haciyakupoğlu S., Karatepe N., Altinsoy N., Baydoğan N., Baytaş F., Büyük B., Demir E. and Gedik S. (2015). Kinetic and thermodynamic behavior of selenium on modified bentonite and activated carbon using radiotracer technique. *Acta Physica Polonica A*, **128**(2), 180–181.

van den Berg C. M. G., Khan S. H., Daly P. J., Riley J. P. and Turner D. R. (1991). An electrochemical study of Ni, Sb, Se, Sn, U and V in the estuary of the Tamar. *Estuarine, Coastal and Shelf Science*, **33**(3), 309–322.

van der Sloot H. A., Hoede D., Wijkstra J., Duinker J. C. and Nolting R. F. (1985). Anionic species of V, As, Se, Mo, Sb, Te and W in the Scheldt and Rhine estuaries and the Southern Bight (North Sea). *Estuarine, Coastal and Shelf Science*, **21**(5), 633–651.

Vives i Batlle J., Aoyama M., Bradshaw C., Brown J., Buesseler K. O., Casacuberta N., Christl M., Duffa C., Impens N. R. E. N., Iosjpe M. and Masqué P. (2018). Marine radioecology after the Fukushima Dai-ichi nuclear accident: are we better positioned to understand the impact of radionuclides in marine ecosystems? *Science of the Total Environment*, **618**, 80–92.

Wang Z., Gao Y. X. and Belzile N. (2001). Microwave digestion of environmental and natural waters for selenium speciation. *Analytical Chemistry*, **73**(19), 4711–4716.

World Health Organization (1982). Nuclear power: management of high-level radioactive waste: report on a Working group, Bruges, 2–6 June 1980. World Health Organization. Regional Office for Europe.

World Health Organization (2011). Selenium in drinking-water: background document for development of WHO guidelines for drinking-water quality (No. WHO/HSE/WSH/ 10.01/14). World Health Organization.

Xu W., Zhu J. M., Johnson T. M., Wang X., Lin Z. Q., Tan D. and Qin H. (2020). Selenium isotope fractionation during adsorption by Fe, Mn and Al oxides. *Geochimica et Cosmochimica Acta*, **272**, 121–136.

Yang Y., Xie Q., Zhao Z., He L., Chan L., Liu Y., Chen Y., Bai M., Pan T., Qu Y. and Ling L. (2017). Functionalized selenium nanosystem as radiation sensitizer of ^{125}I seeds for precise cancer therapy. *ACS Applied Materials and Interfaces*, **9**(31), 25857–25869.

Yao Q. Z., Zhang J., Qin X. G., Xiong H. and Dong L. X. (2006). The behavior of selenium and arsenic in the Zhujiang (Pearl River) estuary, South China Sea. *Estuarine, Coastal and Shelf Science*, **67**(1–2), 170–180.

Yao Q. Z., Zhang J., Wu Y. and Xiong H. (2007). Hydrochemical processes controlling arsenic and selenium in the Changjiang River (Yangtze River) system. *Science of the Total Environment*, **377**(1), 93–104.

Yoneda S. and Suzuki K. T. (1997). Detoxification of mercury by selenium by binding of equimolar Hg–Se complex to a specific plasma protein. *Toxicology and Applied Pharmacology*, **143**(2), 274–280.

Yu L. and Weetjens E. (2016). Application of solubility and transport limited radionuclide release in performance assessment of a spent fuel repository in Boom Clay.

Chapter 3

Microbial reduction of selenium oxyanions: energy-yielding and detoxification reactions

Silvia Lampis and Giovanni Vallini

3.1 INTRODUCTION

Selenium (Se), a semi-metallic chemical element in the oxygen group (group 16 [VIa]) of the periodic table, can be beneficial – even essential in some instances – for microbes and animals, including humans, when present at a suitable concentration, whereas no essential Se requirement has been shown for higher plants (Lenz & Lens, 2009; Winkel *et al.*, 2015). Se is an essential trace element required for the biosynthesis of seleno-amino acids such as selenocysteine (Se-Cys) (Bock *et al.*, 1991; Gromer *et al.*, 2005) and selenomethionine (Se-Met, the major dietary form) (Schrauzer, 2000). These are potent antioxidants as well as a source of Se for the synthesis of Se-dependent antioxidant and repair proteins such as glutathione peroxidases, thioredoxin reductases, and methionine sulfoxide reductases (Flohe *et al.*, 1973; Kim & Gladyshev, 2007; Mustacich & Powis, 2000; Zoidis *et al.*, 2018). Multiple selenoproteins have been identified in eukaryotes, ranging from yeasts (Tastet *et al.*, 2008) to humans (Papp *et al.*, 2018), but Se is also found in prokaryotic proteins such as formate dehydrogenase from *Methanococcus jannaschii* (Jones *et al.*, 1983) formylmethanofuran dehydrogenase from *Methanopyrus kandleri* (Vorholt *et al.*, 1997), and thiol/disulfide oxidoreductase from *Geobacter sulfurreducens* (Kryukov & Gladyshev, 2004).

doi: 10.2166/9781789061055_0101

Although Se is an essential mineral element, there is a very narrow window between necessary and toxic concentrations. The oxyanions selenite (SeO_3^{2-}) and selenate (SeO_4^{2-}) are the primary Se species found in oxic environments. They are highly soluble and bioavailable, which makes them toxic at low concentrations in the parts per million range (Lenz & Lens, 2009; Nancharaiah & Lens, 2015). Selenite is the more toxic of the two species (Frankenberger & Engberg, 1998). This makes it imperative to understand how the distribution of Se in the environment is controlled. Such an understanding will lead to efficient strategies for the detoxification of environmental matrices contaminated with Se, including soil, sediment, surface water, groundwater and wastewater.

Selenium behaves chemically in a similar manner to sulfur (S). Both Se and S can exist in the 2^-, 0, 4^+ and 6^+ oxidation states and can therefore form structurally analogous compounds, although those containing Se are more toxic because Se has a lower electronegativity than S and forms weaker bonds (Whitham, 1995). The concentration, speciation and association of Se in a given habitat depends on the pH and redox conditions, the solubility of Se salts, the complexing ability of soluble and solid ligands, biological interactions, and reaction kinetics. Redox transformations that occur in natural systems can increase or decrease the mobility and bioavailability of Se, and can involve both chemical and biotic mechanisms (Myneni *et al.*, 1997; Zhang *et al.*, 2004b).

Prokaryotic and eukaryotic microbes play a prominent role in the biogeochemical cycle of Se by performing both oxidation and reduction reactions (Nancharaiah & Lens, 2015; Ojeda *et al.*, 2020). Se metabolic transformations have been reported in all domains of life – *Bacteria* (Böck, 2001), *Archaea* (Bini, 2010), and *Eukarya* (Rayman, 2012) – as well as in viruses (Shisler *et al.*, 1998). Microbes can transform Se oxyanions via assimilatory or dissimilatory reduction mechanisms as well as alkylation/dealkylation and oxidation reactions (Ojeda *et al.*, 2020). The main bacterial mechanisms of selenate and selenite reduction are shown in Figure 3.1.

Assimilatory reduction refers to the incorporation of Se into seleno-amino acids, whereas dissimilatory reduction primarily involves the anaerobic respiration of selenite or selenate for energy production. If the reduction of Se oxyanions occurs in the absence of energy generation, it is interpreted as a detoxification mechanism. Prokaryotes can use selenite and selenate as terminal electron acceptors, resulting in the formation of insoluble, nanostructured elemental Se^0 (Stolz & Oremland, 1999). Some bacteria and fungi can also reduce Se oxyanions to Se^0 under aerobic or microaerophilic conditions as a detoxification reaction (Tejo Prakash *et al.*, 2009) or, in phototrophic bacteria, as a redox homeostasis mechanism (Kessi *et al.*, 1999).

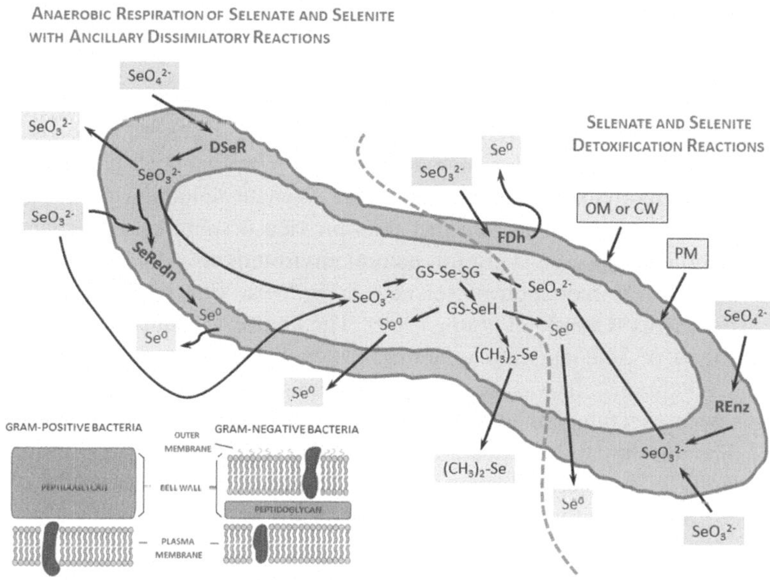

Figure 3.1 The dissimilatory reduction of selenate and selenite by anaerobic respiration (or other energy uncoupled reactions) and selenium detoxification mechanisms in bacteria. Abbreviations: CW cell wall; $(CH_3)_2$-Se dimethylselenide; DSeR dissimilatory selenate reductase; FDh fumarate reductase; GS-Se-SG selenodiglutathione; GS-SeH selenoglutathione; OM outer membrane; PM plasma membrane; REnz reductase enzyme; seRedn selenite reductase.

3.2 SELENIUM OXYANIONS AS FINAL ELECTRON ACCEPTORS IN BACTERIAL ENERGY METABOLISM

Bacterial strains that support their growth by using selenate and/or selenite for respiration under anaerobic/anoxic conditions are known as selenium-respiring bacteria (Nancharaiah & Lens, 2015). The reduction potentials, under standard conditions, for the couples SeO_4^{2-}/SeO_3^{2-} and SeO_3^{2-}/Se^0 are +0.48 and +0.21 V, respectively. This is lower than the potential required for nitrate reduction (NO_3^-/N_2) but much higher than that required for sulfate reduction SO_4^{2-}/SO_3^{2-}. The free energies for the reduction of selenate and selenite coupled to H_2 oxidation are (Newman *et al.*, 1998; Stolz & Oremland, 1999):

$$1/2\ SeO_4^{2-} + 1/2\ H^+ + 1/2\ H_2 \rightarrow 1/2\ HSeO_3^- + 1/2\ H_2O$$
$$\Delta G' - 15.53\ \text{kcal/mol}\ e^- \qquad (3.1)$$

$$1/4\ HSeO_3^- + 1/4\ H^+ + 1/2\ H_2 \rightarrow 1/4\ Se^0 + 3/4\ H_2O$$
$$\Delta G' - 8.93\,\text{kcal/mol}\,e^- \quad (3.2)$$

The reduction of selenate, therefore, occurs at a slightly lower redox potential than that required for nitrate reduction but at a higher redox potential than sulfate reduction. Thermodynamic values show that selenate reduction is energetically favorable for microorganisms and can thus provide a significant mechanism for certain bacteria to conserve energy in natural environments.

The dissimilative reduction of selenate to Se^0 was first reported in sediment slurry experiments (Oremland *et al.*, 1989). The complete reduction of millimolar concentrations of selenate into quantitative levels of Se^0 was observed under anaerobic conditions. It was accelerated in the presence of H_2 or acetate but inhibited by autoclaving or in the presence of O_2, NO_3^- or MnO_2, but not by sulfate. The dissimilative reduction of selenate was clearly shown to occur independently of sulfate, indicating that the Se and S biogeochemical cycles are not related. Subsequently, enriched cultures from agricultural drainage waters in the San Joaquin Valley (California, USA) yielded two bacterial strains that metabolized Se in axenic cultures (Macy *et al.*, 1989). One was a strain named *Pseudomonas* sp. AX that reduced selenate to selenite by respiration, and the other was a strictly anaerobic, Gram-positive, rod-shaped bacterium (strain E) that reduced selenite to Se^0 under anaerobic growth conditions.

Further studies have identified other prokaryotes that can use selenate and/or selenite as final electron acceptors for dissimilative reduction, including phylogenetically diverse groups of Gram-positive and Gram-negative *bacteria* as well as *archaea* (Oremland & Stolz, 2000; Stolz *et al.*, 2002). This indicates that Se respiration is a widespread phenomenon.

3.2.1 Bacterial selenate respiration

Bacterial selenate respiration requires the sequential reduction of the Se oxyanions selenate and selenite, resulting in the precipitation of red-colored Se^0 and the formation of Se nanostructures. The overall process for the biological reduction of selenate to Se^0 is described by two sequential reactions (Debieux *et al.*, 2011; Nancharaiah & Lens, 2015):

$$SeO_4^{2-} + 2e^- + 2H^+ \rightarrow SeO_3^{2-} + H_2O \quad (3.3)$$

$$SeO_3^{2-} + 4e^- + 6H^+ \rightarrow Se^0 + 3H_2O \quad (3.4)$$

The respiration of Se oxyanions to Se^0 is followed by the formation of Se nanostructures. Selenate respiration is driven by metalloproteins that utilize cofactors containing iron and molybdenum, and this has been well described in a few model bacterial strains (Butler, 2012). Analysis of the processes concerning the formation of Se nanostructures is widely discussed elsewhere (see Chapter 10).

Multiple bacterial strains are capable of respiring selenate under anaerobic conditions. These have mainly been sourced from aquifers with high levels of Se pollution and were isolated in selenate respiration enrichment cultures. They mainly represent the phyla *Proteobacteria*, *Firmicutes* and *Chrysiogenetes* (Table 3.1). Different organic substrates can serve as electron donors in the respiratory conversion of selenate to Se^0 (Narasingarao & Häggblom, 2007a; Stolz & Oremland, 1999). Both H_2 and methane can drive the reduction of selenate under anaerobic conditions (Haroon *et al.*, 2013; Orphan *et al.*, 2001; Shi *et al.*, 2020; Stolz & Oremland, 1999). *Gammaproteobacteria* isolated via enrichment cultures from sediments of Arthur Kill and the Kesterson Reservoir (San Joaquin Valley, California, USA) were also found to be capable of selenate respiration while using aromatic compounds such as benzoate, 3-hydroxybenzoate or 4-hydroxybenzoate as a carbon source (Knight *et al.*, 2002).

Thauera selenatis AX^T was isolated from the Se-contaminated drainage waters of the San Joaquin Valley and was initially classified as the first member of a new species of *β-Proteobacteria*: *T. selenatis* sp. (Macy *et al.*, 1993). *T. selenatis* AX^T was found to grow under anaerobic conditions using selenate as the sole electron acceptor. Moreover, it could simultaneously reduce both selenate and nitrate, thus indicating that distinct reductases were acting on each substrate (Rech & Macy, 1992). However, it was unable to use selenite as a final electron acceptor. The *T. selenatis* AX^T selenate reductase (Ser) was purified and characterized. It is a soluble metalloenzyme, located in the periplasmic space, that possesses both iron and molybdenum redox centers (Shröder *et al.*, 1997). Ser is a trimeric complex of the proteins SerA, SerB and SerC. A fourth component (SerD) is a cytoplasmic chaperone probably involved in the insertion of the molybdenum cofactor into the catalytic subunit SerA. SerB is an iron-sulfur protein which transfers electrons from SerC to SerA, whereas SerC is a cytochrome *b*. Electron paramagnetic resonance (EPR) spectroscopy revealed that SerA and SerB contain cysteine-rich motifs (Dridge *et al.*, 2007). The cysteine-rich motif in SerA is proposed to bind a [4Fe–4S] iron–sulfur cluster, whereas the multiple cysteine-rich motifs in SerB are predicted to bind one [3Fe–4S] and three [4Fe–4S] iron–sulfur clusters. The reduction of selenate to selenite requires the transfer of two electrons from quinole (QH_2) and the presence of a cytochrome *c*4 protein (Lowe *et al.*, 2010). The mechanism of electron transfer to periplasmic cytochrome *c*4 is proposed to involve both quinol cytochrome *c* oxidoreductase (QCR) and quinol dehydrogenase (QDH). A periplasmic nitrite reductase may be involved in the reduction of selenite into Se^0 under anaerobic conditions (DeMoll-Decker & Macy, 1993). However, once selenite has passed the cytoplasmic membrane it can be reduced to Se^0 via a thiol-mediated mechanism that probably involves glutathione (GSH) (Debieux *et al.*, 2011). The protein SefA, isolated from the spent medium of *T. selenatis* AX^T grown under aerobic conditions, was shown to be involved in the stabilization of the Se^0

Table 3.1 Bacterial strains capable of selenate and/or selenite respiration to sustain cell growth (n.d. not determined).

Microbial Strain	Taxonomical Classification	Source	Electron Donors	Electron Acceptors	Formation/ Location of Se Nanostructures	Reference
Thauera selenatis AX^T (ATCC 55363^T)	*β-Proteobacteria*	Drainage waters of the San Joaquin Valley, CA, USA	Acetate, H_2, O_2	SeO_4^{2-}; NO_3^-	Yes intracellular and extracellular	Debieux *et al.*, 2011; Macy *et al.*, 1993
Enterobacter cloacae SLD1a–1 (ATCC 700258)	*γ-Proteobacteria*	Freshwater samples from the San Luis Drain, CA, USA	Glucose, Complex medium (TSB)	SeO_4^{2-}; NO_3^- O_2	Yes extracellular	Losi and Frankenberger, 1997; Ridley *et al.*, 2006; Watts *et al.*, 2003
Citrobacter freundii RLS1	*γ-Proteobacteria*	Se-contaminated sediment	Glucose, Acetate, Lactate	SeO_4^{2-}; NO_3^- O_2	Yes	Zhang *et al.*, 2004a
Ferrimonas futtsuensis FUT3661^T	*γ-Proteobacteria*	Sediment from Tokyo Bay	Lactate, Pyruvate, Tryptone	SeO_4^{2-}; S_2O^{3-} Fe(III) AsO_4^{2-} MnO_2; S^{2-}; O_2	Yes	Nakagawa *et al.*, 2006
Ferrimonas kyonanensis Asr22–7^T	*γ-Proteobacteria*	Alimentary tract of littleneck clams from Tokyo Bay	Lactate, Pyruvate, Tryptone	SeO_4^{2-}; Fe(III) AsO_4^{2-} MnO_2; S^{2-}; O_2	Yes	Nakagawa *et al.*, 2006
Sedimenticola selenatireducens AK4OH1	*γ-Proteobacteria*	Estuarine sediments		SeO_4^{2-};	Yes	Narasingarao and Häggblom, 2006
Sulfurospirillum barnesii SES–3	*δ-Proteobacterium*	Se-contaminated marsh freshwater	Lactate, Pyruvate, Formate, Fumarate	SeO_4^{2-}; NO_3^- SO_4^{2-}; Fe(III) AsO_4^{2-}; O_2	Yes	Oremland *et al.*, 1994; Stolz *et al.*, 1999

Pelobacter seleniigenes KM[T]	*δ-Proteobacterium*	Sediment from a wetland system	Acetate, Lactate, Pyruvate	SeO_4^{2-}; NO_3^- Fe(III)E S^2	Yes extracellular and cell-associated	Narasingarao and Häggblom, 2007b
Bacillus arsenicoselenatis E1H	*Firmicutes*, *Bacilli*	Anoxic muds of Mono Lake, CA, USA	Lactate	SeO_4^{2-}; AsO_4^{2-}	Yes	Switzer-Blum *et al.*, 1998
Bacillus selenatarsenatis SF–1	*Firmicutes*, *Bacilli*	Wastewater sediment from a glass manufacturing plant	Lactate	SeO_4^{2-}; NO_3^- AsO_4^{2-}	Yes intracellular and extracellular	Fujita *et al.*, 1997; Kuroda *et al.*, 2011
Bacillus selenitireducens MLS10	*Firmicutes*, *Bacilli*	Mono Lake, CA, USA	Lactate; Pyruvate	SeO_3^{2-}; NO_3^-; NO_2^- AsO_4^{2-}; fumarate	Yes periplasmic and extracellular	Switzer-Blum *et al.*, 1998; Wells *et al.*, 2019
Bacillus beveridgei MLTeJB	*Firmicutes*, *Bacilli*	Sediment from Mono Lake, CA, USA	Lactate	SeO_4^{2-}; SeO_3^{2-}; NO_3^-; NO_2^- AsO_4^{2-}; TeO_4^{2-}; TeO_3^-	Yes	Baesman *et al.*, 2009
Desulfuribacillus stibiiarsenatis MLFW–2[T]	*Firmicutes*, *Bacilli*	Sediment from the drainage area of a geothermal spring near Mono Lake, CA, USA	Lactate, Pyruvate, Formate, H_2	SeO_4^{2-}; SeO_3^{2-}; NO_3^-; NO_2^- DMSO	n.d.	Abin and Hollibaugh, 2017
Selenihalanaerobacter shriftii DSSe–1 ATCC BAA–73	*Firmicutes*, *Clostridia*	Dead Sea	Glucose, Glycerol	SeO_4^{2-}; NO_3^{2-}; NO_2^- AsO_4^{2-}; S^2; fumarate	Yes	Blum *et al.*, 2001
Desulfurispirillum indicum S5	*Chrysiogenetes*	Sediment from an estuarine canal	Acetate, Lactate, Pyruvate	SeO_4^{2-}; SeO_3^{2-} NO_3^-; AsO_4^{2-}	Yes intracellular	Rauschenbach *et al.*, 2011

nanostructures by preventing aggregation and may also help in the secretion of these structures from the cell (Butler, 2012; Debieux *et al.*, 2011).

Enterobacter cloacae SLD1a-1 was isolated from freshwater samples from the San Luis Drain (California, USA). It is a facultative anaerobe that can use nitrate and selenate as terminal electron acceptors during anaerobic growth (Losi & Frankenberger, 1997). Selenate reduction is catalyzed by a membrane-associated reductase (Ser), and is followed by rapid expulsion of the Se^0 particles. Moreover, the strain was shown to reduce selenate and selenite under either anaerobic or aerobic conditions with no induction or conditioning process necessary, and the concomitant reduction of nitrate, selenate and selenite was also confirmed. The membrane-associated Ser protein is a molybdenum-dependent selenate reductase expressed under both aerobic and anaerobic conditions, with the catalytic site facing the periplasmic compartment (Watts *et al.*, 2003). The protein is selective for selenate and cannot reduce nitrate. Unlike *T. selenatis* AX^T Ser, *E. cloacae* SLD1a-1 Ser is unable to support bacterial growth, and may instead be used for detoxification, in accordance with the location of the active site, enabling the bacterium to protect itself against elevated levels of both selenate and the reduction product selenite (Watts *et al.*, 2003). Isolation and characterization of the enzyme revealed that it is a heterotrimeric complex ($\alpha\beta\gamma$) with an apparent molecular mass of ~600 kDa containing molybdenum, heme, and non-heme iron as prosthetic constituents and cytochrome *b* in the active complex (Ridley *et al.*, 2006).

Citrobacter freundii RLS1 (class *γ-Proteobacteria*, family *Enterobacteriaceae*) is similarly proposed to achieve the dissimilative reduction of selenate to Se^0 via the formation of selenite. This strain, isolated from a Se-contaminated sediment from the New River (California, USA), can grow on selenate as a final electron acceptor under microaerophilic conditions and starts to reduce selenate once O_2 is depleted in the medium, resulting in the formation of amorphous red selenium (Zhang *et al.*, 2004a, b). Genome sequencing and functional annotation suggested that selenate reduction is mediated by a membrane-bound metalloenzyme with a molybdenum cofactor, regulated by the FNR protein (fumarate nitrate reduction) that monitors the availability of oxygen in the cytoplasm. FNR also regulates selenate reductase in *E. cloacae* SLD1a-1 (Yee *et al.*, 2007). Genes for selenate reductase in the *C. freundii* genome are organized in the *ynfEGHdmsD* operon, which is induced by the FNR protein under anaerobic conditions. Once bound to the molybdenum cofactor, the selenate reductase protein folds into the appropriate conformation and anchors onto the inner membrane with the active site of the YnfE subunit facing the periplasmic compartment. Electrons are transferred from the menaquinone pool to YnfE via the iron–sulfur protein YnfG, allowing the reduction of selenate to selenite (Theisen & Yee, 2014).

The Gram-negative, strictly anaerobic bacterial strain *Sulfurospirillum barnesii* SES-3 was isolated from a freshwater marsh in the Stillwater Wildlife Management Area of western Nevada (USA), using a selenate respiration enrichment culture (Oremland *et al.*, 1994). This strain was shown to convert selenate to selenite by dissimilative reduction under anaerobic conditions coupled with the oxidation of lactate to acetate and CO_2. No growth was observed on selenite, but cell suspensions readily reduced selenite to Se^0 with a final conversion of ~5% of the initial selenate. *S. barnesii* SES-3 can therefore achieve the complete reduction of selenate to Se^0 and can utilize selenate as final electron acceptor to support growth (Stolz *et al.*, 1999).

Four different species of *Proteobacteria* capable of dissimilatory selenate reduction were isolated by selenate respiration enrichment cultures from a variety of sources, including sediments from three different water bodies in Chennai (India) and a tidal estuary in New Jersey (USA) (Narasingarao & Häggblom, 2007a). *Pelobacter seleniigenes* KM^T is a strict anaerobe isolated from Kearny Marsh, a New Jersey wetland system. It was classified as a novel species of the class *δ-Proteobacteria* due to unique physiological and taxonomic characteristics (Narasingarao & Häggblom, 2007b). The stoichiometric respiration of selenate to Se^0 involves the use of acetate as the electron donor and carbon source. Selenate reduction is followed by a transient accumulation of selenite, which is eventually reduced to Se^0, resulting in the production of a bright red precipitate. The utilization of 5.4 mM acetate was accompanied by the reduction of 7.3 mM selenate to 4.6 mM selenite and 2.7 mM Se^0. Interestingly, this strain can also ferment short-chain organic acids, such as pyruvate, citrate and lactate. Selenate-respiring cultures produce abundant Se^0 granules which are closely associated with the cells (Narasingarao & Häggblom, 2007b). *Ferrimonas futtsuensis* $FUT3661^T$ and *Ferrimonas kyonanensis* Asr22-7^T are facultative anaerobic, selenate-reducing chemo-organotrophs of the class *γ-Proteobacteria*, isolated from sediments and littleneck clams (*Ruditapes philippinarum*), respectively, collected from the coast of Tokyo Bay (Japan). They can both reduce selenate by using it as a final electron acceptor coupled with lactate oxidation, resulting in the appearance of orange-to-red precipitates representing insoluble Se^0 (Nakagawa *et al.*, 2006). *Sedimenticola selenatireducens* AK4OH1 is a strict anaerobe isolated from estuarine sediments. The reduction of selenate is followed by the stoichiometric accumulation of selenite and the formation of red colonies due to the precipitation of Se^0. This strain has the unique ability to oxidize aromatic acids such as benzoate, 4-hydroxybenzoate and 3-hydroxybenzoate coupled to selenate respiration (Narasingarao & Häggblom, 2006).

Bacillus selenatarsenatis SF-1 is a Gram-positive species that was isolated from Se-rich sediment collected in an effluent drain that had received contaminated discharge from a glass-manufacturing plant (Fujita *et al.*, 1997). Also this strain uses selenate as a terminal electron acceptor coupled with the oxidation of lactate (Kuroda *et al.*, 2011; Yamamura *et al.*, 2007). Selenate is reduced to Se^0 through the intermediate selenite (Fujita *et al.*, 1997; Kashiwa *et al.*, 2000). The strain can reduce selenate at concentrations up to 20 mM, but the reduction of selenate to selenite is faster than the subsequent reduction of selenite to Se^0 (Kashiwa *et al.*, 2000). This strain can reduce nitrate to nitrite and arsenate to arsenite under anaerobic conditions as well. In a laboratory-scale continuous bioreactor, the strain removed selenate from synthetic wastewater (41.8 mg Se/L) in the presence of excess lactate as the carbon and energy source. The bioreactor was operated with a cell retention time between 2.2 and 95.2 h. At short cell retention times selenate was removed, but accumulation of selenite was observed. At long cell retention times soluble selenium was reduced to Se^0 (Fujita *et al.*, 2002).

The selenate reductase SrdBCA from *Bacillus selanatarsenatis* SF-1 has been isolated and characterized as a membrane-bound, trimeric, molybdopterin-containing oxidoreductase (Kuroda *et al.*, 2011). The mechanism of selenate reduction to selenite involves the shunting of an electron pool from the quinole QH_2 to selenate via the SrdC, SrdB and SrdA subunits. QH_2 is oxidized to quinone by SrdC, releasing two protons outside the cell membrane and transferring two electrons to SrdB. These electrons pass through the [4Fe–4S] clusters of SrdB and transfer to the [4Fe–4S] cluster of SrdA, and selenate then receives the electrons via the molybdenum cofactor.

Bacillus arsenicoselenatis strain E1H is a strictly anaerobic, spore-forming bacterium isolated from the anoxic muds of the alkaline, hypersaline, and arsenic-rich Mono Lake in California (USA). It converts arsenate to arsenite and selenate to selenite with the concomitant oxidation of lactate to acetate and CO_2 (Switzer-Blum *et al.*, 1998). Similarly, *Selenihalanaerobacter shriftii* DSSe-1 is an obligate anaerobic halophilic bacterium from the Dead Sea that can grow via the respiration of selenate to selenite plus Se^0 and the concomitant oxidation of glycerol or glucose to acetate and CO_2 (Blum *et al.*, 2001). *Desulfuribacillus stibiiarsenatis* MLFW-2^T, isolated from anoxic sediments collected from the drainage area of a geothermal spring near Mono Lake, can grow anaerobically by using both selenate and selenite as final electron acceptors and using lactate, pyruvate, formate or H_2 as electron donors (Abin & Hollibaugh, 2017).

3.2.2 Bacterial selenite respiration

Although selenite respiration plays an important role in the Se biogeochemical cycle, it has received much less attention than selenate respiration. Three bacterial strains have been shown to reduce selenite to Se^0 under anaerobic conditions, using selenite as the final electron acceptor for respiration (Table 3.1): *Bacillus selenitireducens* MLS10 (Switzer-Blum *et al.*, 1998), *Bacillus beveridgei* MLTeJB (Baesman *et al.*, 2009) and *Desulfurispirillum indicum* S5 (Rauschenbach *et al.*, 2011). *Bacillus selenitireducens* MLS10 is a Gram-positive haloalkaliphilic bacillus isolated from Mono Lake, and was the first bacterial strain shown to grow rapidly via the respiration of selenite (Switzer-Blum *et al.*, 1998). A putative respiratory selenite reductase (Srr) was characterized by combined biochemical, genomic and proteomic analysis, identifying the protein as a member of the complex iron–sulfur molybdoenzyme (CISM) family (Wells *et al.*, 2019). The 80 kDa catalytic subunit (SrrA) is a periplasmic complex with one putative 4Fe–4S binding site and specificity for selenite as an electron acceptor, with no activity in the presence of arsenate, selenate, or thiosulfate. The K_m for selenite was 145 ($\pm$ 53) μM, much higher than the values reported for other selenium oxidoreductases. Genomic analysis revealed the presence of an operon containing six genes (*srrE*, *srrA*, *srrB*, *srrC*, *srrD*, and *srrF*). In addition to the catalytic subunit SrrA, the operon coded for a small electron transfer subunit (SrrB, 17.7 kDa) with four putative 4Fe–4S binding sites, an anchoring subunit (SrrC, 43 kDa), a chaperone (SrrD, 24 kDa) and two rhodanese domain-containing proteins (SrrE, 38 kDa; and SrrF, 45.6 kDa).

Bacillus beveridgei MLTeJB is a facultative anaerobe isolated from Mono Lake sediment by a tellurite enriched culture. This strain can use multiple electron acceptors to grow under anaerobic conditions, including tellurite, tellurate, selenite, selenate, nitrite, nitrate and arsenate. The strain achieves the complete reduction of selenite to Se^0 plus selenide over an incubation period of 25 days, coupling selenite reduction with lactate oxidation to acetate and formate (Baesman *et al.*, 2009).

Desulfurispirillum indicum $S5^T$ is a strictly anaerobic strain isolated from the sediment of an estuarine canal in Chennai (India), based on its ability to reduce selenate and selenite to Se^0. In addition to Se oxyanions, this strain also achieves the dissimilatory reduction of arsenate and nitrate coupled with the oxidation of pyruvate, lactate and acetate (Rauschenbach *et al.*, 2011).

3.3 STRATEGIES FOR THE DETOXIFICATION OF SELENIUM OXYANIONS IN *BACTERIA*

Reduction chemistry may provide a basic level of resistance to particular elements and compounds in the environment (Avazeri *et al.*, 1997). Indeed, the most widespread mechanism of selenate/selenite detoxification is its reduction to insoluble Se^0 and subsequent precipitation, producing extracellular or intracellular particles that may be deposited in the cytoplasm or on the cell wall or membrane to form Se nanostructures (Zannoni *et al.*, 2008). This process occurs under both aerobic and anaerobic conditions and is generally more rapid for selenite than selenate.

The reduction of selenate/selenite by bacteria may involve (1) specific or non-specific enzymes, mainly periplasmic or cytosolic oxidoreductases that function under anaerobic or aerobic growth conditions; (2) reaction with thiols via the so-called Painter-type reaction; and (3) interactions with siderophores such as pyridine-2,6-bisthiocarboxylic acid (PDTC).

3.3.1 Enzymatic detoxification

Various enzymatic systems may be involved in the reduction of selenate/selenite to Se^0, including sulfate, sulfite, nitrate, nitrite and fumarate reductases. For example, a nitrite reductase may be responsible for selenite reduction under anaerobic conditions in *T. selenatis* AX^T, because mutants lacking this activity are unable to reduce either nitrite or selenite (DeMoll-Decker & Macy, 1993). Similarly, *Rhizobium sullae* HCNT1 was shown to reduce millimolar concentrations of selenite using a copper-containing nitrite reductase (Nir) (Basaglia *et al.,* 2007). The periplasmic nitrate reductase NapA or the cytoplasmic nitrate reductases NarGHIJ and NarZUWV may be responsible for the reduction of selenate to selenite in *Escherichia coli* (Avazeri *et al.*, 1997).

Shewanella oneidiensis MR-1 possesses multiple respiration reductases and can therefore reduce a wide range of electron acceptors. By exploiting its broad respiratory ability, MR-1 can survive in different environments and shows high resistance to different toxic compounds. *S. oneidiensis* MR-1 mutants lacking different reductase activities were tested for their ability to reduce selenite. This revealed that fumarate reductase (FccA) is involved in selenite reduction under anaerobic conditions. The authors tested mutants lacking periplasmic terminal reductases, namely nitrate reductase (NapA), nitrite reductase (NrfA) or fumarate reductase (FccA), as well as mutants deficient for individual genes encoding Mtr proteins, the periplasmic mediators of anaerobic extracellular respiration (Li *et al.*, 2014). The analysis of *ΔnapB* and *ΔnrfA* mutants revealed that neither nitrate reductase nor nitrite reductase contributed to selenite reduction in *S. oneidensis* MR-1, nor did the extracellular respiratory proteins. Conversely, selenite reduction by mutant *ΔfccA* was severely suppressed, with a 60% decrease in activity within the initial 12 h of selenite exposure. The cytoplasmic membrane-anchored cytochrome CymA was shown to bridge electron transfer from the quinone pool to several respiratory reductases during anaerobic respiration, including the fumarate reductase FccA. The authors therefore proposed a mechanism in which FccA catalyzes selenite reduction and CymA shunts electrons from the quinol pool to FccA (Li *et al.*, 2014). It seems that anaerobic selenite reduction in *S. oneidiensis* MR-1 does not support growth, but is instead a mechanism to detoxify the bacterial environment to prevent selenite uptake (Li *et al.*, 2014; Nancharaiah & Lens, 2015). The reduction of selenite under anaerobic conditions not coupled to growth has also been reported for *Veilonella atypica*. In this case, the formation of Se^0 and selenide was observed in bacterial cultures exposed to selenite and supplied with the extracellular electron shuttle anthraquinone-2,6-disulphonate (AQDS), suggesting that selenite reduction was achieved via a hydrogenase-coupled reduction, mediated by ferredoxin (Pearce *et al.,* 2009).

In the case of *Alishewanella* sp. WH16-1, a facultative anaerobe from mining soil, a novel aerobic selenite reductase was identified, named chromate selenite reductase flavoenzyme (CrsF). This flavoprotein, with 37% identity to *E. coli* chromate reductase (ChrR), is able to reduce selenite or chromate *in vitro*, as well as sulfate and ferric anions, using NADPH as a cofactor (Xia *et al.*, 2018). An interesting study of selenate and selenite reduction under aerobic conditions in the obligate aerobic bacterium *Comamonas testotsteroni* S44 revealed evidence that selenate is reduced via the sulfate reduction pathway (Tan *et al.*, 2018). The strain was first isolated from a metal-polluted site and was shown to transform both selenite and selenate to Se^0, resulting in the formation of intracellular Se nanostructures (Zheng *et al.*, 2014). Transposon mutagenesis experiments revealed that the Cys regulon transcriptional activator CysB (involved in sulfur metabolism and responsible for regulating cysteine anabolism) is involved in the reduction of selenate to Se^0 (Tan *et al.*, 2018). The authors also identified an aerobic selenite reductase (SerT), a member of the molybdenum-containing sulfite oxidase family localized in the periplasmic space. This concurred with electron microscopy results showing the presence of Se nanoparticles in the cytoplasm when *C. testosteroni* was exposed to selenate, but in the periplasm when it was exposed to selenite (Tan *et al.*, 2018).

Under anaerobic conditions, several bacterial strains have been shown to reduce both selenate and selenite in the cytoplasm or periplasm without using them as final electron acceptors for respiration, including *Desulfovibrio desulfuricans* DSM1924 (Tomei *et al.*, 1995) and *Wolinella succinogenes* (Tomei *et al.*, 1992). *D. desulfuricans* DSM1924 can grow in the presence of 10 mM selenate or 0.1 mM selenite by using sulfate or fumarate as an electron acceptor and formate as an electron donor. It can reduce both selenate and selenite to Se^0, resulting in the precipitation of Se deposits in the periplasm and cytoplasm, respectively. *Wolinella succinogenes* can grow in the presence of 10 mM selenate or 0.1 mM selenite by using fumarate as an electron acceptor and formate as an electron donor, but only after an adaptation step with sublethal doses of Se oxyanions.

3.3.2 Thiol driven reactions

3.3.2.1 Reaction mechanisms

The reduction of selenite to Se^0 is mediated by reaction with thiols. Painter (1941) studied the chemical reaction of selenite with thiol groups and was the first to demonstrate the formation of selenium disulfides (RS-Se-SR) through the following reaction (Equation. 3.5):

$$4RSH + H_2SeO_3 \rightarrow RS{-}Se{-}SR + RSSR + 3H_2O \tag{3.5}$$

Ganther (1971) investigated the reaction of selenite with the thiol-containing molecule glutathione (GSH), the most abundant thiol found in eukaryotic cells, cyanobacteria, and the bacterial phylum Proteobacteria. The reaction of selenite with GSH was shown to form an intermediate (Se-diglutathione) that provides an excellent substrate for glutathione reductase, with K_m and V_{max} values comparable to those of GSH itself. The reaction of Se-diglutathione with NAD(P)H glutathione reductase leads to the formation of a highly unstable product, the selenium persulfide anion GS-Se$^-$ (Equation 3.6). Ganther (1971) also suggested that the persulfide anion reverted quickly to elemental Se^0 as shown in (Equation 3.7):

$$GS{-}Se{-}SG + NAD(P)H \rightarrow GSH + GS{-}Se^- + NAD(P)^+ \quad (3.6)$$

$$GS{-}Se^- + H^+ \rightarrow Se^0 + GSH \quad (3.7)$$

Kessi and Hanselmann (2004) proposed a modification of (Equation 3.5) which leads to the formation of a superoxide anion (Equation 3.8). This was based on the comparison of the abiotic (chemical) reduction of selenite by glutathione and a reaction mediated by *Rhodospirillum rubrum* and *E. coli* when exposed to selenite.

$$6GSH + 3SeO_3^{2-} + 4H^+ \rightarrow 3GSSeSG + 2O_2^- + 5H_2O \quad (3.8)$$

Superoxide anions are removed from biological systems by superoxide dismutase (SOD) and catalase to protect cells from oxidative stress, explaining the induction of two types of SOD in *E. coli* grown in the presence of selenite (Bébien *et al.*, 2002). Thiol-containing biomolecules other than glutathione may also be involved, such as the reduction of selenite and Se-diglutathione by the thioredoxin system of *E. coli* (Björnstedt *et al.*, 1992). The authors proposed that Se-diglutathione is reduced by thioredoxin that is in turn reduced by NADPH-dependent thioredoxin reductase to regenerate reduced thioredoxin as shown below (Equations 3.9 and 3.10) (Björnstedt *et al.*, 1992; Kumar *et al.*, 1992):

$$Trx(SH)_2 + GS{-}Se{-}SG \rightarrow Trx{-}S_2 + GSH + GS{-}Se^- + H^+ \quad (3.9)$$

$$Trx{-}S_2 + NADPH + H^+ \rightarrow Trx(SH)_2 + NADP^+ \quad (3.10)$$

The selenite reaction mechanism proposed by Kessi and Hanselman (2004) may therefore involve three steps: an abiotic step in which selenite is transformed into Se-diglutathione with the concomitant release of reactive oxygen species (ROS) as reported in (Equation 3.8), an enzymatic step in which NAD(P)H-dependent reductases such as glutathione or thioredoxin reductases lead to the production of the unstable persulfide anion via (Equation 3.6), and finally the abiotic formation of Se^0 via (Equation 3.7). The initial reducing power is thereby restored by regenerating six molecules of GSH.

3.3.2.2 *Microbial strategies for thiol based Se detoxification*

3.3.2.2.1 Gram negative bacteria

Several bacterial strains have been identified and characterized for their ability to reduce selenite into Se^0 under either anaerobic or aerobic conditions using thiols. A well-documented case is *Stenotrophomonas maltophilia* SeITE02, a bacterial isolate obtained from the rhizosphere of the Se-hyperaccumulating legume *Astragalus bisulcatus* grown on a seleniferous soil. This strain can survive exposure to 50 mM selenite and can reduce up to 2.0 mM selenite in 48 hours after a conditioning step, with the concomitant formation of Se^0 nanostructures (Di Gregorio *et al.*, 2005). These structures were initially localized in the cytoplasmic fraction, but later also accumulated in the extracellular space (Di Gregorio *et al.*, 2005; Lampis *et al.*, 2017).

It remains largely unclear how Se^0 nanostructures reach the extracellular space. One potential release mechanism is cell lysis, because Se^0 nanostructures have been detected by electron microscopy in the spent medium close to cell ghosts (Di Gregorio *et al.*, 2005; Lampis *et al.*, 2017). This hypothesis was also proposed for *Desulfovibrio desulfuricans* exposed to selenate and selenite (Tomei *et al.*, 1995). More recently, *S. maltophilia* SeITE02 cells grown in defined medium supplied with glucose or pyruvate as carbon sources and exposed to selenite were found to generate membrane vesicles that surrounded the extracellular Se^0 nanostructures, indicating that such vesicles might play a key role in the excretion of Se^0 nanostructures into the extracellular environment (Piacenza *et al.*, 2018) (Figure 3.2). The export of Se^0 nanostructures has already been proposed for other bacterial strains, including *R. rubrum* and *T. selenatis* AX^T (Debieux *et al.*, 2011; Kessi *et al.*, 1999).

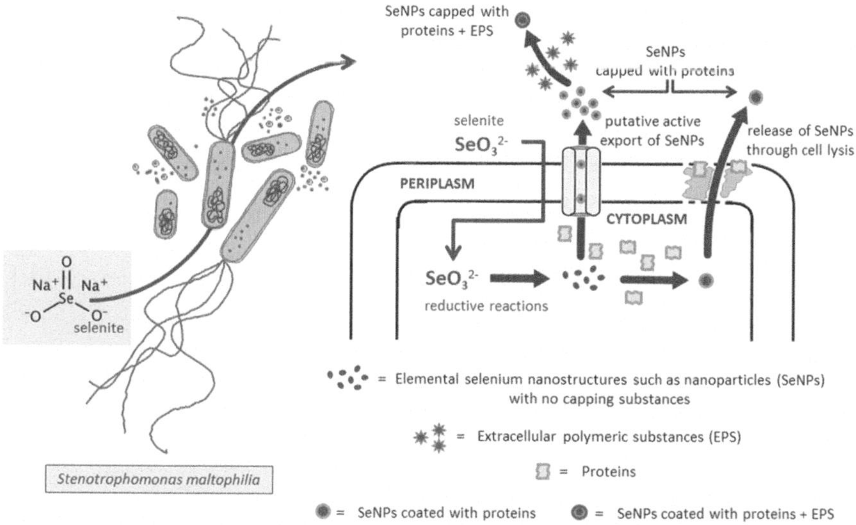

Figure 3.2 Selenite transformation reactions in *Stenotrophomonas maltophilia* SeITE02. Once inside the cell, selenite is reduced in the cytoplasmic compartment to form Se^0 nanostructures such as nanoparticles (SeNPs). These are exported via a release mechanism based on cell lysis or by an unknown active process. Se^0 nanostructures are surrounded by an organic coating composed mostly of proteins and/or extracellular polymeric substances (EPS).

Combined biochemical, physiological and proteomic analysis revealed that *S. maltophilia* SeITE02 does not use nitrite reductase for the reduction of selenite, but there is evidence these two oxyanions may share the same transport system (Antonioli *et al.*, 2007). Moreover, *in vitro* enzymatic assays demonstrated that selenite reduction occurs in the cytoplasmic fraction and the enzyme is NAD(P)H-dependent. Afterwards, growth tests in the presence of the GSH synthesis inhibitor *S-n*-butyl homocysteine sulfoximine suggested that GSH is involved in the first-stage response of the cells to selenite exposure. Proteomic analysis revealed that the main functional classes of proteins upregulated by exposure to either selenite or nitrite are those related to damaged-protein catabolism, DNA metabolism and cell division, as well as oxidative stress responses. Interestingly, selenite seems to strongly induce the expression of at least two different GSH-related enzymes (glutamate-cysteine ligase and GSH synthetase) that are responsible for the biosynthesis of GSH in prokaryotes. Selenite reduction via glutathione was also proposed for *Ochrobactrum* sp. MPV1, isolated from an arsenopyrites dump. This strain can survive exposure to 80 mM selenite and accumulates intracellular Se^0 nanostructures mainly in the cytoplasm (Zonaro *et al.*, 2017).

The mechanism of selenite reduction has also been investigated in methane-oxidizing bacteria such as *Methylococcus capsulatus* (Bath), which are capable of the methane-driven conversion of selenite to Se^0 nanoparticles and methylated selenium species (Eswayah *et al.*, 2017). Chromatographic and spectroscopic analyses of *M. capsulatus* (Bath) cells exposed to selenite in the presence of methane to allow the formation of Se^0 nanoparticles, indicated that methylselenol (CH_3SeH) is the key intermediate in the reduction of selenite into methylated Se (Equations 3.11 through 3.13) and Se-S species, as well as Se^0 nanostructures (Eswayah *et al.*, 2019):

$$GS{-}Se{-}SG + GSH \rightarrow GS{-}SeH + GSSG \tag{3.11}$$

$$GS{-}SeH + CH_3^{\ +} + H^+ \rightarrow CH_3{-}SeH + GSH \tag{3.12}$$

$$CH_3{-}SeH + CH_3^{\ +} \rightarrow (CH_3)_2{-}Se \tag{3.13}$$

3.3.2.2.2 Gram positive bacteria

A different mechanism of selenite reduction may occur in Gram-positive bacteria, which are unable to synthesize glutathione but can tolerate high levels of selenite. *Bacillus subtilis* was shown to reduce selenite to Se^0, which precipitated as Se granules between the cell wall and plasma membrane (Garbisu *et al.*, 1995). The authors confirmed that selenite reduction involves an inducible detoxification system rather than dissimilative electron transport. The reduction of selenite in

B. subtilis not only facilitates adaptation to high selenite concentrations but also induces a morphological change from rod-like to rounded cells. The mechanism of reduction appears to involve a dithiol system because thioredoxin and NADPH-thioredoxin reductase levels increased following exposure to selenite (Garbisu *et al.*, 1999).

The *Bacillus mycoides* strain SeITE01, isolated from the rhizosphere of *Astragalus bisulcatus* (Vallini *et al.*, 2005) completely reduces 2.0 mM selenite in the liquid culture after 24 h under aerobic conditions (Lampis *et al.*, 2014). Selenite reduction was associated with the formation of intracellular and extracellular Se^0 nanostructures, and because there was evidence of selenite reduction in the membrane protein fraction and spent medium, the authors proposed that the corresponding enzymes were not only found inside the cells but were also secreted to the extracellular space (Lampis *et al.*, 2014).

Selenite reduction may also have been facilitated by other thiol-rich molecules such as bacillithiol (BSH), a major low-molecular-weight thiol that plays a significant rôle in the cytosolic thiol-redox chemistry of Gram-positive bacteria together with thioredoxin (Gaballa *et al.*, 2010). BSH-synthesizing bacteria may contain enzymes analogous to those found in GSH-containing species, with bacilliredoxin (Brx) instead of glutaredoxin (Grx). In *B. mycoides* SeITE01 a Brx-like protein may play a similar role to Grx in an analogous pathway to the GSH system found in Gram-negative bacteria (Lampis *et al.*, 2014).

3.3.2.2.3 Reaction with sulfide

Selenite can also undergo abiotic reactions with biogenic sulfides to yield elemental Se and S (Equation 3.14). This has been observed in sulfate-reducing bacteria such as *Desulfomicrobium norvegicum*, where the reduction of sulfate is linked to the concomitant extracellular precipitation of elemental S and Se in cell cultures exposed to selenite (Hockin & Gadd, 2003).

$$SeO_3^{2-} + 2HS^- + 4H^+ \rightarrow Se^0 + 2S^0 + 3H_2O \qquad (3.14)$$

3.3.3 Siderophore driven detoxification

The reduction of selenite to Se^0 as a detoxification mechanism in *Pseudomonas stutzeri* KC is promoted by the siderophore PDTC, a broad-range metal chelator (Zawadzka *et al.*, 2006). Siderophores are iron-specific chelators excreted by microorganisms under iron-limiting conditions as a part of an iron acquisition system. The authors proposed that selenite could be reduced after binding to PDTC or its hydrolysis product dipicolinic acid-pyridine-2,6-bis(carboxylic acid) (DPA). The production and excretion of PDTC facilitates the extracellular reduction of metalloids and thus acts as an environmental detoxification mechanism that prevents the uptake of selenite into cells.

3.4 BIOTRANSFORMATION OF SELENIUM OXYANIONS BY *ARCHAEA*

A list of archaeal species associated with the aerobic or anaerobic reduction of Se oxyanions is provided in Table 3.2. One of the most interesting examples of selenate-respiring *Archaea* species is *Pyrobaculum arsenaticum* PZ6, a hyper-thermophilic, strictly anaerobic, facultative organotrophic strain isolated from a hot spring at Pisciarelli Solfatara (Naples, Italy). This strain can grow organotrophically in the presence of selenate, arsenate or elemental sulfur as inorganic electron acceptors (Huber *et al.*, 2000). Similarly, *Pyrobaculum ferrireducens* 1860(T) is a hyperthermophilic, anaerobic archaeon isolated from a terrestrial hot spring at Uzon Caldera, Kronotsky Nature Reserve (Kamchatka, Russia) (Slobodkina *et al.*, 2015). It can grow anaerobically using Fe(III), nitrate, arsenate, selenate and selenite as final electron acceptors. *Pyrobaculum aerophilum* is a hyperthermophilic strain belonging to the archaeon phylum *Crenarchaeota*. This strain is capable of chemolitho-autotrophic growth with selenate as the electron acceptor in the presence of H_2, and can also grow by the respiration of either selenate or selenite under organotrophic conditions in a basal salt medium supplied with yeast extract (Huber *et al.*, 2000; Völkl *et al.*, 1993). Interestingly, the reduction of selenite resulted in the formation of black colored Se^0 precipitates.

More recently, a microbial consortium capable of methane-dependent selenate reduction was enriched in a membrane biofilm batch reactor. Metagenomic analysis revealed that the consortium was mainly composed of methanotrophic archaeons belonging to the genus *Methanosarcina* and type II methanotrophic bacteria belonging to the genus *Methylocystis*, which carried out selenate reduction along with methane oxidation (Shi *et al.*, 2020). This agrees with recent insights into methane-metabolizing archaeons that either produce or consume methane (Evans *et al.*, 2019).

3.5 FUNGAL TRANSFORMATION OF SELENIUM OXYANIONS

3.5.1 Introduction

The aerobic reduction of selenite and selenate in the absence of respiration or assimilation reactions is used by many bacteria as a detoxification mechanism (Doran, 1982; Eswayah *et al.*, 2016; Gadd, 1993; Soudi *et al.*, 2009). This has considerable potential as a cost-effective approach for the bioremediation of seleniferous environments (Javed *et al.*, 2016; Maltman & Yurkov, 2018). Nevertheless, the ability to aerobically reduce selenite and/or selenate to elemental selenium (Se^0), or even to selenide (Se^{2-}) in the form of volatile methylated compounds such as dimethyl selenide (DMSe: CH_3SeCH_3), dimethyl

Table 3.2 *Archaea* capable of the selective reduction of selenium oxyanions.

Species	Taxonomic Classification	Reduction of Se Oxyanions	Reduction Conditions	Formation of Elemental Selenium Nano-structures	Reference
Halobacterium sp. SP1	*Euryarchaeota*, *Halobacteriales* (extreme halophiles)	10 mM selenite	AE	Yes – as amorphous red elemental Se precipitate	Naik *et al.*, 2017
Halococcus salifodinae BK18	*Euryarchaeota*, *Halobacteriales* (extreme halophiles)	2 mM selenite	AE	Yes – as amorphous red elemental Se nanoparticles	Srivastava *et al.*, 2014.
Halogeometricum borinquense E118	*Euryarchaeota*, *Halobacteriales* (extreme halophiles)	5 mM selenite	AE	Yes – intracellular amorphous red elemental Se nanoparticles	Abdollahnia *et al.*, 2020
Halorubrum xinjiangense	*Euryarchaeota*, *Halobacteriales* (extreme halophiles)	25 mM selenite	AE	Yes – as amorphous red elemental Se nanoparticles	Güven *et al.*, 2013
Methanosarcina sp.	*Euryarchaeota*, *Methanosarcinales* (methanogens)	~30 mg Se/L as selenate	ANA	Progressive disappearance of SeO_4^{2-} via transient SeO_3^{2-} accumulation due to the activity of a *Methanosarcina*/*Methylocystis* consortium in bioreactor; Formation of amorphous red Se^0 starting from selenite reduction	Shi *et al.*, 2020

(*Continued*)

Table 3.2 *Archaea* capable of the selective reduction of selenium oxyanions (*Continued*).

Species	Taxonomic Classification	Reduction of Se Oxyanions	Reduction Conditions	Formation of Elemental Selenium Nano-structures	Reference
Pyrobaculum aerophilum	*Crenarchaeota*, *Thermoproteales* (hyperthermophiles)	Organotrophic respiration of either selenate or selenite as final electron acceptors Chemolitho-autotrophic growth with selenate as electron acceptor in the presence of H_2	ANA	Yes – as hexagonal gray elemental Se, only during growth on selenite	Huber *et al.*, 2000
Pyrobaculum arsenaticum	*Crenarchaeota*, *Thermoproteales* (hyperthermophiles)	Anaerobic respiration of selenate	ANA	Yes – as a precipitate of hexagonal gray elemental Se	Huber *et al.*, 2000
Pyrobaculum ferrireducens 1860^T	*Crenarchaeota*, *Thermoproteales* (hyperthermophiles)	Anaerobic respiration of 10 mM selenite or 10 mM selenate	ANA	Yes – red precipitate with selenite, black precipitate with selenate	Slobodkina *et al.*, 2015

Reduction: AE (aerobic); ANA (anaerobic).

selenyl sulfide (DMSeS: CH_3SeSCH_3), and dimethyl diselenide (DMDSe: $CH_3SeSeCH_3$), is not unique to bacteria (Chasteen & Bentley, 2003; Herrero & Wellinger, 2015) (Figure 3.3). More recently, this property has been observed in a broad range of fungi (Morley *et al.*, 1996; Rosenfeld *et al.*, 2017; Sinharoy & Lens, 2020), including yeasts (Falcone & Nickerson, 1963; Kieliszek *et al.*, 2015; Soudi, 2003), filamentous fungi, and higher fungi (i.e., mushrooms) within the members of *Zygomycota* (Gharieb *et al.*, 1995), *Ascomycota* (Ramadan *et al.*, 1988; Sarkar *et al.*, 2011), and *Basidiomycota* (Espinosa-Ortiz *et al.*, 2015a; Vetchinkina *et al.*, 2016). A summary list of fungal species capable of selenite and/or selenate reduction is presented in Table 3.3.

Se tolerance and detoxification by fungi mainly involves the reduction of inorganic Se to less toxic and volatile derivatives such as DMSe (Gadd, 1993), or

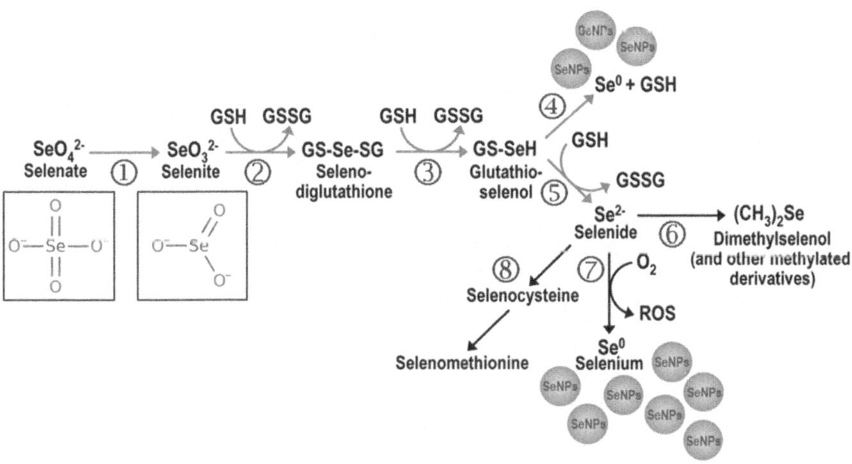

Figure 3.3 Pathways for the metabolic reduction of inorganic selenium (Se) and its conversion into organic species or the elemental form (Se^0) in *Saccharomyces cerevisiae* and other yeasts. Reductive reactions are shown as red arrows. Reaction 1 represents multiple reactions involving ATP sulfurylase and other enzymes from the initial steps of the sulfate assimilation pathway. Reactions 2–5 are non-enzymatic and result in the net conversion of reduced glutathione (GSH) to oxidized glutathione as glutathione disulfide (GSSG). Selenide can give rise to volatile methylated forms via reaction 6. Reaction 7 is also non-enzymatic, and results in the formation of diverse reactive oxygen species (ROS) along with Se^0 nanoparticles (which are also generated by reaction 4). In reaction 8, the Se-containing amino acids selenocysteine and selenomethionine are formed and incorporated into selenoproteins [adapted from Herrero & Wellinger, 2015].

Table 3.3 Fungal species capable of selenite and/or selenate reduction, associated with the accumulation of red amorphous Se^0 particulate matter.

Fungal Species	Selenite Reduction	Selenate Reduction	Reduction Products	Reference
Filamentous and Higher Fungi				
Acremonium strictum [A]	+	+	Se^0 nanoparticles; Se methylated volatile compounds	Rosenfeld *et al.*, 2017
Agaricus arvensis [B]	+		Spherical Se^0 nanoparticles	Vetchinkina *et al.*, 2019
Agaricus bisporus [B]	+		Spherical Se^0 nanoparticles	Vetchinkina *et al.*, 2019
Alternaria alternata [A]	+	+	Se^0 nanorods; Se^0 nanoparticles; Se methylated volatile compounds	Rosenfeld *et al.*, 2017; Sarkar *et al.*, 2012
Aspergillus clavatus [A]		+	Negligible zero-valent particles; Se methylated volatile compounds	Urík *et al.*, 2016
Aspergillus funiculosus [A]	+		Needle-like crystals of Se^0	Gharieb *et al.*, 1995
Aspergillus niger [A]	+		Amorphous red elemental Se^0	Gharieb *et al.*, 1995
Aspergillus oryzae [A]	+		Se^0 nanoparticles	Kimura *et al.*, 2014
Aspergillus parasiticus [A]	+	+	Se^0 nanoparticles	Moss *et al.*, 1987
Aspergillus sp. J2 [A]	+		Se^0 nanoparticles	Li *et al.*, 2018
Coriolus versicolor [B]	+		Amorphous red elemental Se^0	Gharieb *et al.*, 1995
Flammulina velutipes [B]	+		Se^0 nanoparticles	Wang *et al.*, 2016
Fusarium sp. [A]	+	+	Red elemental Se^0 particulate; Needle-like crystals of Se^0	Gharieb *et al.*, 1995; Ramadan *et al.*, 1988
Ganoderma lucidum [B]	+		Se^0 nanoparticles	Rosenfeld *et al.*, 2017
Grifila frondosa [B]	+		Se^0 nanoparticles	Rosenfeld *et al.*, 2017
Lentinula edodes [B]	+	+	Se^0 nanoparticles	Vetchinkina *et al.*, 2013

Mortierella humilis [M]	+		Se^0 nanoparticles	Liang *et al.*, 2019
Mortierella sp. [M]	+	+	Se^0 nanoparticles; Se methylated volatile compounds	Zieve *et al.*, 1985
Mucor hiemalis [Z]	+		Amorphous red elemental Se^0	Gharieb *et al.*, 1995
Mucor SK [Z]	+		Amorphous red elemental Se^0	Gharieb *et al.*, 1995
Paraconiothyrium sporulosum [A]	+		Stable Se^0 nanoparticles; Organo-selenium (Se^{-II}) compounds	Rosenfeld *et al.*, 2020
Penicillium chrysogenum [A]	+		Amorphous red elemental Se^0	Gharieb *et al.*, 1995
Penicillium funiculosum [A]	+		Amorphous red elemental Se^0	Gharieb *et al.*, 1995
Phanerochaete chrysosporium [B]	+	+	Se^0 nanoparticles	Espinosa-Ortiz *et al.*, 2015a, b
Phoma glomerata [A]	+		Se^0 nanoparticles	Liang *et al.*, 2019
Plectosphaerella cucumerina [A]	+	+	Se^0 nanoparticles; Se methylated volatile compounds	Rosenfeld *et al.*, 2017
Pleurotus ostreatus [B]	+		Se^0 nanoparticles	Vetchinkina *et al.*, 2016
Pyrenochaeta sp. [A]	+	+	Se^0 nanoparticles; Se methylated volatile compounds	Rosenfeld *et al.*, 2017
Rhizopus arrhizus [M]	+		Amorphous red elemental Se^0	Gharieb *et al.*, 1995
Stagonospora sp. [A]	+	+	Stable Se^0 nanoparticles; Organo-selenium (Se^{-II}) compounds	Rosenfeld *et al.*, 2020
Trichoderma harzianum [A]	+		Se^0 nanoparticles; Se oxide (SeO_2, downeyite) nanoparticles	Liang *et al.*, 2019
Trichoderma reesei [A]	+		Amorphous red elemental Se^0	Gharieb *et al.*, 1995

(Continued)

Table 3.3 Fungal species capable of selenite and/or selenate reduction, associated with the accumulation of red amorphous Se^0 particulate matter (*Continued*).

Fungal Species	Selenite Reduction	Selenate Reduction	Reduction Products	Reference
Yeast-like Fungi				
Aureobasidium pullulans [A]	+		Se^0 nanoparticles	Gharieb *et al.*, 1995; Liang *et al.*, 2019
Yeasts				
Candida albicans [A]	+		Elemental Se^0 precipitate	Falcone and Nickerson, 1963; Gharieb *et al.*, 1995
Candida glabrata [A]	+		Amorphous red elemental Se^0	Gharieb *et al.*, 1995
Candida humicola [A]	+	+	Se methylated volatile compounds; Elemental Se^0 precipitate	Cox and Alexander, 1974; Herrero and Wellinger, 2015
Candida lipolytica [A]	+		Amorphous red elemental Se^0	Gharieb *et al.*, 1995
Candida maltosa [A]		+	Elemental Se^0 precipitate	Golubev and Golubev, 2002
Cryptococcus sp. [B]		+	Elemental Se^0 precipitate	Golubev and Golubev, 2002
Hanseniaspora valbyensis [A]		+	Elemental Se^0 precipitate	Golubev and Golubev, 2002
Kluyveromyces marxianum [A]		+	Elemental Se^0 precipitate	Golubev and Golubev, 2002

Rhodotorula mucilaginosa [B]	+		Se^0 nanoparticles	Ruocco *et al.*, 2014
Rhodotorula rubra [B]	+		Amorphous red elemental Se^0	Gharieb *et al.*, 1995
Saccharomyces cerevisiae [A]	+		Amorphous red elemental Se^0	Gharieb *et al.*, 1995; Zhang *et al.*, 2012
Trichosporon sp. [B]		+	Elemental Se^0 precipitate	Golubev and Golubev, 2002
Yarrowia lipolytica [A]		+	Elemental Se^0 precipitate	Golubev and Golubev, 2002
Zygosaccharomyces rouxii [A]	+		Amorphous red elemental Se^0	Gharieb *et al.*, 1995

[A] = Ascomycota; [B] = Basidiomycota; [M] = Mucoromycota; [Z] = Zygomycota.

the reduction of Se oxyanions to Se^0 resulting in the formation of intracellular or extracellular orange-to-red deposits (Gharieb *et al.*, 1995; Konetzka, 1977) (Figure 3.4).

3.5.2 Yeasts

A broad review on the tolerance of ascomycetous and basidiomycetous yeasts to selenate was presented by Golubev and Golubev (2002). The ascomycetes *Candida maltosa*, *Hanseniaspora valbyensis*, *Kluyveromyces marxianus* and *Yarrowia lipolytica* show high tolerance to selenate, as do the basidiomycetes *Cryptococcus curvatus*, *Cr. humicola*, and *Trichosporon* spp. A few strains are able to grow at a selenate concentration of 0.1 M, although growth under these conditions is poor. Eventually, these yeasts produce pink-to-red colonies reflecting the reduction of selenate to Se^0, whereas the colonies remain colorless in the absence of selenate.

Yeasts were initially tested for their ability to reduce selenite by inoculating them onto MYGP (Malt extract, Yeast extract, Glucose and Peptone) agar supplemented with selenite at concentrations of 0.5–5 mM (Gharieb *et al.*, 1995). Several *Candida* spp. and *Saccharomyces* spp. were able to reduce selenite to Se^0 at all concentrations, resulting in the formation of pink colonies. In the baker's yeast *Saccharomyces cerevisiae*, Zhang *et al.* (2012) reported the biogenesis of red elemental selenium/protein nanoparticles following the reduction of selenate to Se^0 in a microaerophilic environment. Other strains, such as *Candida lypolitica* 37-1 and *Rhodotorula rubra* NCYC 797, produced bright red colonies due to their efficient reduction of selenite. Colonies of the polymorphic fungus *Aureobasidium pullulans* were – depending on the SeO_3^{2-} concentration – light pink in the presence of 1 mM selenite but red at a higher (5 mM) concentration (Gharieb *et al.*, 1995). The marine yeast strain *Rhodotorula mucilaginosa*-13B provides another clear example of the transformation of selenite into elemental Se in the form of intracellular or extracellular Se^0 nanoparticles (Ruocco *et al.*, 2014).

An illuminating contribution to the description of the interactions between yeast cells, mostly *S. cerevisiae* and *Candida* spp., and the different chemical species of Se, including Se oxyanions, is provided by Kieliszek *et al.* (2015). Inorganic Se forms such as selenite and selenate enter the yeast cells via promiscuous oxyanion transporters, and are reduced to avoid toxicity. They undergo a series of reduction reactions ultimately leading to the formation of selenide (H_2Se) using a pathway that, under physiological conditions, involves GSH. In many microbial species, selenide is a common intermediate used for the synthesis of selenoproteins or destined for conversion into methylated forms that are then eliminated through volatilization (Chasteen & Bentley, 2003). Actually, Cox and Alexander (1974) observed the formation of DMSe in cultures of *Candida humicola* supplemented with selenite and selenate. Another possibility is,

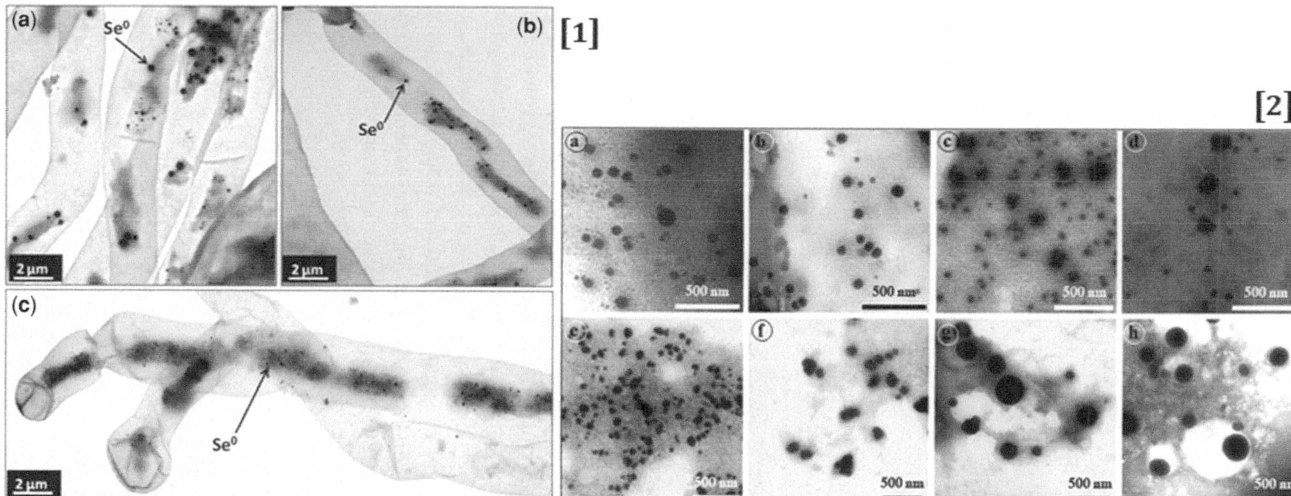

Figure 3.4 Intracellular and extracellular occurrence of SeNPs in fungi. [1] Transmission electron micrographs of elemental selenium nanoparticles produced in the biomass of the filamentous basidiomycete *P. chrysosporium*. A – distribution of Se^0 particles of different sizes within fungal biomass. B and C – localization of Se^0 particles within fungal biomass [from: Espinosa-Ortiz *et al.*, 2015a]. [2] TEM of Se (or even SeO_2) nanoparticles in a number of higher basidiomycetes. nanoparticles produced from Na_2SeO_3 with extracellular (a–d) and intracellular (e–h) extracts of *L. edodes* (a, e), *P. ostreatus* (b, f), *G. lucidum* (c, g), and *G. frondosa* (d, h). Bars mark 500 nm [from: Vetchinkina *et al.*, 2018].

however, the reverse oxidation of selenide to Se^0, which accumulates inside the cell (Equation 3.15)

$$H_2Se + 1/2O_2 \rightarrow Se^0 + H_2O \quad (3.15)$$

It is well known that SeO_3^{2-} can react spontaneously with GSH to initially produce selenodiglutathione (GS-Se-SG), (Equation 3.16). In the presence of excess GSH, GSSeSG is further reduced to glutathioselenol (GSSeH), (Equation 3.17). GSSeH either spontaneously dismutates into Se^0 and GSH, (Equation 3.18), or is further reduced by GSH to yield H_2Se (Equation 3.19). In oxic conditions, selenide is readily re-oxidized by O_2 into Se^0 (Equation 3.15) (Cupp-Sutton & Ashby, 2016).

$$SeO_3^{2-} + 4GSH + 2H^+ \rightarrow GSSeSG + GSSG + 3H_2O \quad (3.16)$$

$$GSSeSG + GSH \rightarrow GSSeH + GSSG \quad (3.17)$$

$$GSSeH \rightarrow Se^0 + GSH \quad (3.18)$$

$$GSSeH + GSH \rightarrow H_2Se + GSSG \quad (3.19)$$

3.5.3 Filamentous fungi

Filamentous fungi are also able to reduce selenium oxyanions, resulting in the formation of intracellular or extracellular Se^0 nanostructures. The pioneering study of Gharieb *et al.* (1995), discussed above for yeast, also revealed that a few species of filamentous fungi were able to carry out the reduction of selenite to Se^0. Increasing the selenite concentration from 1 to 10 mM generally led to more severe growth inhibition, although other factors affecting growth included the presence of sulfur compounds and/or ingredients that form selenite complexes in the medium. The species that successfully reduced selenite while growing on Czapek-Dox Agar (CDA) were *Fusarium* sp. and *Trichoderma reesei*, and – to a lesser extent – *Mucor hiemalis* and *Penicillium chrysogenum*, leading to the formation of red-to-orange colonies. Interestingly, electron microscopic images clearly showed that *Fusarium* sp. and *Aspergillus funiculosus* exposed to 50 mM selenite were able to form Se^0 crystals on the surface of the hyphae. Other species – namely *Aspergillus niger*, *Coriolus versicolor*, *Mucor* SK, and *Rhizopus arrhizus* – were instead able to reduce selenite only on Malt Extract Agar (MEA). These observations suggest that even different culture conditions such as nutrient supply, temperature, pH, and incubation time affect Se metabolism and the final characteristics of the Se nanoparticles. However, the effects of such parameters remain unclear, requiring further investigation for each fungal strain.

Importantly, not all the fungal species tested by Gharieb *et al.* (1995) were able to form Se^0. Some of these were able to reduce selenite via other mechanisms, such as methylation and/or the formation of Se^{2-} volatiles. Accordingly, the ability to grow on selenite does not necessarily reflect the ability to reduce selenite to Se^0. For example, Brady *et al.* (1996) showed that *Penicillium* sp. grown aerobically for 2 weeks in the presence of 1 mM selenite releases volatile organic derivatives (probably DMSe) during all four cultivation phases (lag, exponential growth, stationary and decline), whereas the reduction of selenite to amorphous Se^0 was only observed during the decline phase, as evidenced by the characteristic red color of both the fungal biomass and the culture substrate.

3.5.4 Higher fungi (mushrooms)

3.5.4.1 Ascomycetes

A deeper understanding of the role of fungi in the complex reactions characterizing the biogeochemical cycle of selenium has been powerfully reinforced by new insights into the ascomycetes *Paraconiothyrium sporulosum* and *Stagonospora* sp. (Rosenfeld *et al.*, 2020). Both species are capable of reducing selenate and selenite to Se^0 or even to Se^{2-} volatiles under aerobic conditions alongside the simultaneous oxidation of manganese (Mn^{II}). Rosenfeld and co-workers had previously compared *P. sporulosum* and *Stagonospora* sp. to four other ascomycetes (*Acremonium strictum*, *Alternaria alternata*, *Plectosphaerella cucumerina* and *Pyrenochaeta* sp.) isolated from sites contaminated with heavy metals (Rosenfeld *et al.*, 2017). All species showed a high tolerance of Se oxyanions and in most cases the ability to reduce selenite and selenate to Se^0, the exception being *P. sporulosum* which lacks the ability to reduce selenate. On the other hand, the ability of the yeast-like dematiaceous fungus *Aureobasidium pullulans* to reduce selenite has recently been confirmed, along with the zygomycete *Mortierella humilis* and the ascomycetes *Trichoderma harzianum* and *Phoma glomerata* (Liang *et al.*, 2019). Furthermore, *P. glomerata* was recently shown to precipitate intracellular and extracellular Se^0 nanoparticles when grown on medium supplemented with selenite (Liang *et al.*, 2020). *Aspergillus* strains are also known to reduce selenite (Kimura *et al.*, 2014) or selenate (Urík *et al.*, 2016) and produce Se^0 nanoparticles by the reduction of selenite, which accumulate mainly on the surface of the mycelial cell walls (Li *et al.*, 2018). Recently, pellets of the fungus *Aspergillus niger* KP were used in an airlift reactor to remove selenite from wastewater, demonstrating the potential of this species for environmental remediation (Negi *et al.*, 2020). Interestingly, field-emission transmission electron microscopy (FE-TEM) images revealed that Se^0 nanoparticles formed within the fungal cells, suggesting the intracellular conversion of selenite into these Se^0 nanostructures.

3.5.4.2 Basidiomycetes

Red coloring, indicating the accumulation of amorphous Se^0, has been reported in cultures of the basidiomycetes *Lentinula* (obsolete name, *Lentinus*) *edodes*, *Pleurotus ostreatus*, *Ganoderma lucidum* and *Grifila frondosa*, which form Se^0 nanoparticles when grown in media containing selenite (Vetchinkina *et al.*,

2016). In particular, *L. edodes* and *G. frondosa* accumulated Se^0 nanoparticles predominantly within the mycelia and – to a limited extent – on the hyphal surface. In contrast, *G. lucidum* and *P. ostreatus* formed Se^0 nanoparticles mostly in the growth medium. *L. edodes* reduced 0.3 mM selenite added to the growth medium, producing intracellular electron-dense spherical formations of Se^0, but cannot reduce 0.3 mM selenate under the same conditions (Vetchinkina *et al.*, 2013). Furthermore, the edible button mushroom *Agaricus bisporus* and the horse mushroom *Agaricus arvensis* reduced selenite and formed spherical Se^0 particles with diameters of 100–250 and 150–550 nm, respectively (Vetchinkina *et al.*, 2019). The widely cultivated basidiomycete mushroom *Flammulina velutipes* can reduce selenite to Se^0 when grown on a substrate containing optimal selenite concentrations of 0.03–0.1 mM to enrich the fungal biomass with dietary Se. In these conditions the strain is also able to form Se^0 nanoparticles, possibly as a detoxification mechanism (Wang *et al.*, 2016).

The ligninolytic basidiomycete *Phanerochaete chrysosporium* can also reduce selenite and selenate, but only forms Se^0 from selenite (Espinosa-Ortiz *et al.*, 2015a, b). Although the detailed mechanisms of selenate and selenite reduction are unknown, this species is likely to use a GSH-dependent mechanism for at least the reduction of selenite as proposed for bacteria (Debieux *et al.*, 2011; Xu *et al.*, 2014). Transmission electron microscopy (TEM) revealed that most Se nanoparticles were compartmentalized in the fungal cell but not equally distributed throughout the hyphae (Espinosa-Ortiz *et al.*, 2015a). *P. chrysosporium* has been tested as a means to bioreduce selenite from wastewater and recover non-toxic elemental Se in the form of Se^0 nanoparticles (Espinosa-Ortiz, 2016). There is evidence for synergy between *P. chrysosporium* and the bacterium *Delftia lacustris* in reducing selenite to Se^0 with contextual phenol degradation, offering another promising biotechnological approach for the bioremediation of polluted environmental matrices (Chakraborty *et al.*, 2019).

3.5.5 Selenium reduction by cell extracts

The synthesis of Se nanorods or nanoparticles was demonstrated in fungal extracts. In the plant pathogen *Alternaria alternata*, an ascomycete, Se^0 nanostructures synthesis was induced by adding 1 mM sodium selenate to the growth medium after removing the mycelium (Sarkar *et al.*, 2011, 2012). Extracellular and intracellular extracts, even from cultures of different basidiomycetes, were shown to generate Se^0 nanoparticles following sodium selenite addition (Vetchinkina *et al.*, 2018).

To conclude, focusing on the actual involvement of fungi in the biotransformation reactions of selenium compounds, with particular reference to the oxyanions selenite and selenate, assumes a prominent ecological and practical relevance where it is considered that fungal species isolated from the rhizosphere of selenium hyperaccumulator plants represent a significant fraction of the microbiome that interacts with the root systems of these botanical species (Wangeline *et al.*, 2011).

3.6 FUTURE PERSPECTIVES

We are now beginning to understand the mechanisms underlying the potential active transport of biogenic nanostructured Se^0 particles from the intracytoplasmic compartment to the outside environment, that not only provides new fundamental knowledge about microbial transport pathways and adaptations to stress, but could also be exploited for bioremediation of contaminated sites, mineral recovery in nanostructured forms and nutrient biofortification of food crops and feedstuff for livestock. The growing evidence for important interactions between oxidized species of selenium and diverse members of *Archaea* is of great impact in biotechnology. This topic is still in its infancy, but early results provide evidence that Se plays a key role in the biogeochemistry of thermo-acidophilic and halophilic extreme environments, and also suggest that methylotrophic methanogens could be used in anaerobic bioreactors intended for the treatment of different Se-laden waste streams to recover nanostructured elemental Se^0 following the transformation of Se oxyanions.

The burgeoning exploration of the relationship between diverse fungi – yeasts, molds or mushrooms – and Se oxyanions, not only represents an important theoretical advance but could also lead to the exploitation of fungi for the bioremediation of Se-contaminated environments or the biofortification of edible fungi with dietary Se. Microbes in the rhizosphere and inner tissues of plants can also facilitate the reduction of selenate or selenite in the soil, suggesting these microbes could be utilized for both phytoremediation (Lampis *et al.*, 2009; Staicu *et al.*, 2015) and the biofortification of crops (Acuña *et al.*, 2013). The endophytic bacterium *Pseudomonas moraviensis* ssp. *stanleyae*, isolated from the roots of the hyperaccumulator *Stanleya pinnata* growing on seleniferous soil in the Pine Ridge Natural Area (Colorado, USA), was shown to tolerate 150 mM selenate and 120 mM selenite, and to reduce selenite efficiently under aerobic and anaerobic conditions, producing extracellular Se^0 nanostructures (Staicu *et al.*, 2015). Moreover, *Bacillus mycoides* SeITE01 and *S. maltophilia* SeITE02 were shown to enhance the phytoextraction efficiency of *Brassica juncea* plants exposed to selenite or selenate in bio-augmented water-filtering artificial beds (Lampis *et al.*, 2009).

REFERENCES

Abdollahnia M., Makhdoumi A., Mashreghi M. and Eshghi H. (2020). Exploring the potentials of halophilic prokaryotes from a solar saltern for synthesizing nanoparticles: the case of silver and selenium. *PLoS ONE*, **15**(3), e0229886. doi:10.1371/journal.pone.0229886

Abin C. A. and Hollibaugh J. T. (2017). *Desulfuribacillus stibiiarsenatis* sp. nov., an obligately anaerobic, dissimilatory antimonate- and arsenate-reducing bacterium isolated from anoxic sediments, and emended description of the genus *Desulfuribacillus*. *International Journal of Systematic and Evolutionary Microbiology*, **67**(4), 1011–1017. doi:10.1099/ijsem.0.001732

Acuña J. J., Jorquera M. A., Barra P. J., Crowley D. E. and de la Luz Mora M. (2013). Selenobacteria selected from the rhizosphere as a potential tool for Se biofortification of wheat crops. *Biology and Fertility of Soil*, **49**(2), 175–185. doi:10.1007/s00374-012-0705-2

Antonioli P., Lampis S., Chesini I., Vallini G., Rinalducci S., Zolla L. and Righetti P. G. (2007). *Stenotrophomonas maltophilia* SeITE02, a new bacterial strain suitable for bioremediation of selenite-contaminated environmental matrices. *Applied and Environmental Microbiology*, **73**(21), 6854–6863. doi:10.1128/AEM.00957-07

Avazeri C., Turner R. J., Pommier J., Weiner J. H., Giordano G. and Vermeglio A. (1997). Tellurite and selenate reductase activity of nitrate reductases from *Escherichia coli*: correlation with tellurite resistance. *Microbiology*, **143**, 1181–1189. doi:10.1099/00221287-143-4-1181

Baesman S. M., Stolz J. F., Kulp T. R. and Oremland R. S. (2009). Enrichment and isolation of *Bacillus beveridgei* sp. nov., a facultative anaerobic haloalkaliphile from mono lake, California, that respires oxyanions of tellurium, selenium, and arsenic. *Extremophiles*, **13**(4), 695–705. doi:10.1007/s00792-009-0257-z

Basaglia M., Toffanin A., Baldan E., Bottegal M., Shapleigh J. P. and Casella S. (2007). Selenite-reducing capacity of the copper-containing nitrite reductase of *Rhizobium sullae*. *FEMS Microbiology Letters*, **269**(1), 124–130. https://doi.org/10.1111/j.1574-6968.2006.00617.x

Bébien M., Lagniel G., Garin J., Touati D., Verméglio A. and Labarre J. (2002). Involvement of superoxide dismutases in the response of *Escherichia coli* to selenium oxides. *Journal of Bacteriology*, **184**(6), 1556–1564. doi:10.1128/jb.184.6.1556-1564.2002

Bini E. (2010). Archaeal transformation of metals in the environment. *FEMS Microbiology Ecology*, **73**(1), 1–16. https://doi.org/10.1111/j.1574-6941.2010.00876.x

Björnstedt M., Kumar S. and Holmgren A. (1992). Selenodiglutathione is a highly efficient oxidant of reduced thioredoxin and a substrate for mammalian thioredoxin reductase. *Journal of Biological Chemistry*, **267**(12), 8030–8034. doi:10.1016/S0021-9258(18)42403-9

Blum J. S., Stolz J. F., Oren A. and Oremland R. S. (2001). *Selenihalanaerobacter shriftii* gen. nov., sp. nov., a halophilic anaerobe from dead sea sediments that respires selenate. *Archives of Microbiology*, **175**(3), 208–219. doi:10.1007/s002030100257

Böck A. (2001). Selenium metabolism in bacteria. In: Selenium, D. L. Hatfield (ed.) Springer, Boston, MA. https://doi.org/10.1007/978-1-4615-1609-5_2

Bock A., Forchhammer K., Heider J., Leinfelder W., Sawers G., Veprek B. and Zinoni F. (1991). Selenocysteine: the 21st amino acid. *Molecular Microbiology*, **5**, 515–520. https://doi.org/10.1111/j.1365-2958.1991.tb00722.x

Brady J. M., Tobin J. M. and Gadd G. M. (1996). Volatilization of selenite in aqueous medium by a *Penicillium* species. *Mycological Research*, **100**(8), 955–961. https://doi.org/10.1016/S0953-7562(96)80048-7

Butler C. S. (2012). Metals, non-metals and minerals: the complexity of bacterial selenate respiration. *Biochemist*, **34**(5), 23–27. doi: https://doi.org/10.1042/BIO03405023

Chakraborty S., Rene E. R. and Lens P. N. L. (2019). Reduction of selenite to elemental Se(0) with simultaneous degradation of phenol by co-cultures of *Phanerochaete chrysosporium* and *Delftia lacustris*. *Journal of Microbiology*, **57**, 738–747. doi:10.1007/s12275-019-9042-6

Chasteen T. G. and Bentley R. (2003). Biomethylation of selenium and tellurium: microorganisms and plants. *Chemical Reviews*, **103**(1), 1–25. doi:10.1021/cr010210+

Cox D. P. and Alexander M. (1974). Factors affecting trimethylarsine and dimethylselenide formation by *Candida humicola*. *Microbial Ecology*, **1**, 136–144. https://doi.org/10.1007/BF02512385

Cupp-Sutton K. A. and Ashby M. T. (2016). Biological chemistry of hydrogen selenide. *Antioxidants*, **5**, 42. https://doi.org/10.3390/antiox5040042

Debieux C. M., Dridge E. J., Mueller C. M., Splatt P., Paszkiewicz K., Knight I., Florance H., Love J., Titball R. W., Lewis R. J., Richardson D. J. and Butler C. S. (2011). A bacterial process for selenium nanosphere assembly. *Proceedings of the National Academy of Sciences*, **108**, 13480–13485. doi:10.1073/pnas.1105959108

DeMoll-Decker H. and Macy J. M. (1993). The periplasmic nitrite reductase of *Thauera selenatis* may catalyze the reduction of selenite to elemental selenium. *Archives of Microbiology*, **160**(3), 241–247. https://doi.org/10.1007/BF00249131

Di Gregorio S., Lampis S. and Vallini G. (2005). Selenite precipitation by a rhizospheric strain of *Stenotrophomonas* sp. Isolated from the root system of *Astragalus bisulcatus*: a biotechnological perspective. *Environment International*, **31**(2), 233–241. doi:10.1016/j.envint.2004.09.021

Doran J. W. (1982). Microorganisms and the biological cycling of selenium. In: Advances in Microbial Ecology, K. C. Marshall (ed.), Vol. **6**, Springer, Boston, MA, pp. 1–32. doi:10.1007/978-1-4615-8318-9

Dridge E. J., Watts C. A., Jepson B. J. N., Line K., Santini J. M., Richardson D. J. and Butler C. S. (2007). Investigation of the redox centres of periplasmic selenate reductase from *Thauera selenatis* by EPR spectroscopy. *Biochemical Journal*, **408**(1), 19–28. doi:10.1042/BJ20070669

Espinosa-Ortiz E. J. (2016). Bioreduction of selenite and tellurite by Phanerochaete chrysosporium. PhD Thesis, UNESCO-IHE Institute for Water Education, CRC Press/Balkema (Taylor & Francis Group), Leiden, The Netherlands.

Espinosa-Ortiz E. J., Gonzalez-Gil G., Saikaly P. E., van Hullebusch E. D. and Lens P. N. L. (2015a). Effects of selenium oxyanions on the white-rot fungus *Phanerochaete chrysosporium*. *Applied Microbiology and Biotechnology*, **99**, 2405–2418. https://doi.org/10.1007/s00253-014-6127-3

Espinosa-Ortiz E. J., Rene E. R., van Hullebusch E. D. and Lens P. N. L. (2015b). Removal of selenite from wastewater in a *Phanerochaete chrysosporium* pellet based fungal bioreactor. *International Biodeterioration Biodegradation*, **102**, 361–369. https://dx.doi.org/10.1016/j.ibiod.2015.04.014

Eswayah A. S., Smith T. J. and Gardiner P. H. E. (2016). Microbial transformations of Selenium Species of relevance to bioremediation. *Applied and Environmental Microbiology*, **82**(16), 4848–4859. https://doi.org/10.1128/AEM.00877-16

Eswayah A. S., Smith T. J., Scheinost A. C., Hondow N. and Gardiner P. H. E. (2017). Microbial transformations of selenite by methane-oxidizing bacteria. *Applied Microbiology and Biotechnology*, **101**, 6713–6724. https://doi.org/10.1007/s00253-017-8380-8

Eswayah A. S., Hondow N., Scheinost A. C., Merroun M., Romero-González M., Smith T. J. and Gardiner P. H. E. (2019). Methyl Selenol as a precursor in selenite reduction to Se/S species by methane-oxidizing bacteria. *Applied and Environmental Microbiology*, **85** (22), e01379–19. DOI:10.1128/AEM.01379-19

Evans P. N., Boyd J. A., Leu A. O., Woodcroft B. J., Parks D. H., Hugenholtz P. and Tyson G. W. (2019). An evolving view of methane metabolism in the Archaea. *Natural Reviews in Microbiology*, **17**, 219–232. https://doi.org/10.1038/s41579-018-0136-7

Falcone G. and Nickerson W. J. (1963). Reduction of selenite by intact yeast cells and cell-free preparations. *Journal of Bacteriology*, **85**, 754–762. https://doi.org/10.1128/jb.85.4.754-762

Flohe L., Günzler W. A. and Schock H. H. (1973). Glutathione peroxidase: a selenoenzyme. *FEBS Letters*, **32**(1), 132–134. doi:10.1016/0014-5793(73)80755-0

W. T. Frankenberger, Jr. and R. A. Engberg (eds.). (1998). Environmental chemistry of selenium. Marcel Dekker, Inc., New York, NY.

Fujita M., Ike M., Nishimoto S., Takahashi K and Kashiwa M. (1997). Isolation and characterization of a novel selenate-reducing bacterium, *Bacillus* sp SF-1. *Journal of Fermentation and Bioengineering*, **83**(6), 517–522. http://dx.doi.org/10.1016/S0922-338X(97)81130-0.

Fujita M., Ike M., Kashiwa M., Hashimoto R. and Soda S. (2002). Laboratory-scale continuous reactor for soluble selenium removal using selenate-reducing bacterium, *Bacillus* sp. SF-1. *Biotechnology and Bioengineering*, **80**, 755–761. doi:10.1002/bit.10425

Gaballa A., Newton G. L, Antelmann H., Parsonage D., Upton H., Rawat M., Claiborne A., Fahey R. C. and Helmann J. D. (2010). Biosynthesis and functions of bacillithiol, a major low-molecular-weight thiol in Bacilli. *Proceedings of the National Academy of Sciences USA*, **107**(14), 6482–6486. doi:10.1073/pnas.1000928107

Gadd G. M. (1993). Microbial formation and transformation of organometallic and organometalloid compounds. *FEMS Microbiology Reviews*, **11**(4), 297–316. https://www.sciencedirect.com/science/article/pii/016864459390003R

Ganther H. E. (1971). Reduction of the selenotrisulfide derivative of glutathione to a persulfide analog by gluthathione reductase. *Biochemistry*, **10**(22), 4089–4098. https://doi.org/10.1021/bi00798a013

Garbisu C., Gonzalez S., Yang W.-H., Yee B. C., Carlson D. E., Yee A., Smith N. R., Otero R., Buchanan B. B. and Leighton T. (1995). Physiological mechanisms regulating the conversion of selenite to elemental selenium by *Bacillus subtilis*. *BioFactors*, **5**, 29–37.

Garbisu C., Carlson D., Adamkiewicz M., Yee B. C., Wong J. H., Resto E., Leighton T. and Buchanan B. B. (1999). Morphological and biochemical responses of *Bacillus subtilis* to selenite stress. *Biofactors*, **10**(4), 311–319. doi:10.1002/biof.5520100401

Gharieb M. M., Wilkinson S. C. and Gadd G. M. (1995). Reduction of selenium oxyanions by unicellular, polymorphic and filamentous fungi: cellular location of reduced selenium and implications for tolerance. *Journal of Industrial Microbiology*, **14**, 300–311. https://doi.org/10.1007/BF01569943

Golubev V. I. and Golubev N. V. (2002). Selenium tolerance of yeasts. *Microbiology*, **71**(4), 386–390. https://doi.org/10.1023/A:1019829107239

Gromer S., Eubel J. K., Lee B. L. and Jacob J. (2005). Human selenoproteins at a glance. *Cellular and Molecular Life Sciences*, **62**, 2414–2437. https://doi.org/10.1007/s00018-005-5143-y

Güven K., Mutlu M. B., Çırpan C. and Kutlu H. M. (2013). Isolation and identification of selenite reducing archaea from Tuz (salt) lake In Turkey. *Journal of Basic Microbiology*, **53**(5), 397–401. doi:10.1002/jobm.201200008

Haroon M. F., Hu S. H., Shi Y., Imelfort M., Keller J., Hugenholtz P., Yuan Z. G. and Tyson G. W. (2013). Anaerobic oxidation of methane coupled to nitrate reduction in a novel archaeal lineage. *Nature*, **500**(7464), 567. doi:10.1038/nature12375

Herrero E. and Wellinger R. E. (2015). Yeast as a model system to study metabolic impact of selenium compounds. *Microbial Cell*, **2**(5):139–149.

Hockin S. L. and Gadd G. M. (2003). Linked redox precipitation of sulfur and selenium under anaerobic conditions by sulfate-reducing bacterial biofilms. *Applied and Environmental Microbiology*, **69**, 7063–7072. https://doi.org/10.1128/AEM.69.12.7063-7072.2003

Huber R., Sacher M., Vollmann A., Huber H. and Rose D. (2000). Respiration of arsenate and selenate by hyperthermophilic archaea. *Systematic and Applied Microbiology*, **23**, 305–314. http://dx.doi.org/10.1016/S0723-2020(00)80058-2.

Javed S., Sarwar A. and Tassawar M. (2016). Conversion of selenite to elemental selenium by indigenous bacteria isolated from polluted areas. *Chemical Speciation and Bioavailability*, **27**(4), 162–168. https://doi.org/10.1080/09542299.2015.1112751

Jones W. J., Leigh J. A., Mayer F., Woese C. F. and Wolfe R. S. (1983). *Methanococcus jannaschii* sp. Nov., an extremely thermophilic methanogen from a submarine hydrothermal vent. *Archives of Microbiology*, **136**, 254–261. https://doi.org/10.1007/BF00425213

Kashiwa M., Nishimoto S., Takahashi K., Ike M. and Fujita M. (2000). Factors affecting soluble selenium removal by a selenate reducing bacterium *Bacillus* sp. SF-1. *Journal of Biosciences and Bioengineering*, **89**, 528–533. https://doi.org/10.1016/S1389-1723(00)80051-1

Kessi J. and Hanselmann K. W. (2004). Similarities between the abiotic reduction of selenite with glutathione and the dissimilatory reaction mediated by *Rhodospirillum rubrum* and *Escherichia coli*. *Journal of Biological Chemistry*, **279**, 50662–50669. doi:10.1074/jbc.M405887200

Kessi J., Ramuz M., Wehrli E., Spycher M. and Bachofen R. (1999). Reduction of selenite and detoxification of elemental selenium by the phototrophic bacterium *Rhodospirillum rubrum*. *Applied and Environmental Microbiology*, **65**, 4734–4740. doi:10.1128/AEM.65.11.4734-4740

Kieliszek M., Błażejak S., Gientka I. and Bzducha-Wróbel A. (2015). Accumulation and metabolism of selenium by yeast cells. *Applied Microbiology and Biotechnology*, **99**, 5373–5382. doi:10.1007/s00253-015-6650-x

Kim H. Y. and Gladyshev V. N. (2007). Methionine sulfoxide reductases: selenoprotein forms and roles in antioxidant protein repair in mammals. *The Biochemical Journal*, **407**(3), 321–329. doi:10.1042/BJ20070929

Kimura H., Arima T.-H., Okamura Y., Oku T. and Sakaguchi T. (2014). Selenium recovery and conversion by a filamentous fungus, *Aspergillus oryzae* strain RIB40. *Asia Pacific Journal of Sustainable Agriculture Food and Energy (APJSAFE)*, **2**(2), 5–8.

Knight V. K., Nijenhuis I., Kerkhof L. J. and Mäggblom M. M. (2002). Degradation of aromatic compounds coupled to selenate reduction. *Geomicrobiology Journal*, **19**, 77–86. http://dx.doi.org/10.1080/01490450231724 6183

Konetzka WA.. (1977). Microbiology of metal transformations. In: "Microorganisms and Minerals" E. D. Weinberg (ed.), Marcel Dekker Inc., New York, pp. 317–342.

Kryukov G. V. and Gladyshev V. N. (2004). The prokaryotic selenoproteome. *EMBO Reports*, **5**(5), 538–543. doi:10.1038/sj.embor.7400126

Kumar S., Bjornstedt M. and Holmgren A. (1992). Selenite is a substrate for calf thymus thioredoxin reductase and thioredoxin elicits a large non-stoichiometric oxidation of NADPH in the presence of oxygen. *European Joirnal of Biochemistry*, **207**, 435–439. doi:10.1111/j.1432-1033.1992.tb17068.x

Kuroda M., Yamashita M., Miwa E., Imao K., Fujimoto N., Ono H., Nagano K., Sei K. and Ike M. (2011). Molecular cloning and characterization of the srdBCA operon, encoding the respiratory selenate reductase complex, from the selenate-reducing bacterium *Bacillus selenatarsenatis* SF-1. *Journal of Bacteriology*, **193**, 2141–2148. http://dx.doi.org/10.1128/JB.01197-10.

Lampis S., Ferrari A., Cunha-Queda A. C. F., Alvarenga P., Di Gregorio S. and Vallini G. (2009). Selenite resistant rhizobacteria stimulate seO_3^{2-} phytoextraction by *Brassica juncea* in bioaugmented water-filtering artificial beds. *Environmental Science and Pollution Research*, **16**(6), 663–670. doi:10.1007/s11356-008-0088-y

Lampis S., Zonaro E., Bertolini C., Bernardi P., Butler C. S. and Vallini G. (2014). Delayed formation of zero-valent selenium nanoparticles by *Bacillus mycoides* SeiTE01 as a consequence of selenite reduction under aerobic conditions. *Microbial Cell Factories*, **13**(1), 1–14. doi:10.1186/1475-2859-13-35

Lampis S., Zonaro E., Bertolini C., Cecconi F., Monti F., Micaroni M., Turner R. J., Butler C. S. and Vallini G. (2017). Selenite biotransformation and detoxification by *Stenotrophomonas maltophilia* SeITE02: novel clues on the route to bacterial biogenesis of selenium nanoparticles. *Journal of Hazardous Material*, **324**(Pt A), 3–14. doi:10.1016/j.jhazmat.2016.02.035

Lenz M. and Lens P. N. L. (2009). The essential toxin: the changing perception of selenium in environmental sciences, *Science of The Total Environment*, **407**(12), 3620–3633. doi:10.1016/j.scitotenv.2008.07.056 Abstract

Li D. B., Cheng Y. Y., Wu C., Li W. W., Li N., Yang Z. C., Tong Z. H. and Yu H. Q. (2014). Selenite reduction by *Shewanella oneidensis* MR-1 is mediated by fumarate reductase in periplasm. *Scientific Reports*, **4**, 3755. http://dx.doi.org/10.1038/srep03735.

Li Z., Li H. and Hu H. (2018). Selenite removal and reduction by growing *Aspergillus* sp. J2. *Biometals*, **31**, 45–50. doi:10.1007/s10534-017-0063-5

Liang X., Perez M. A. M-J., Nwoko K. C., Egbers P., Feldmann J., Csetenyi L. and Gadd G. M. (2019). Fungal formation of selenium and tellurium nanoparticles. *Applied Microbiology and Biotechnology*, **103**, 7241–7259. https://doi.org/10.1007/s00253-019-09995-6

Liang X., Perez M. A. M-J., Zhang S., Song W., Armstrong J. G., Bullock L. A., Feldmann J., Parnell J., Csetenyi L. and Gadd G. M. (2020). Fungal transformation of selenium and tellurium located in a volcanogenic sulfide deposit. *Environmental Microbiology*, **22**(6), 2346–2364. https://doi.org/10.1111/1462-2920.15012

Losi M. E. and Frankenberger W. T, Jr. (1997). Reduction of selenium oxyanions by *Enterobacter cloacae* SLD1a-1: isolation and growth of the bacterium and its expulsion of selenium particles. *Applied and Environmental Microbiology*, **63**(8), 3079–3084. doi:10.1128/aem.63.8.3079-3084

Lowe E. C., Bydder S., Hartshorne R. S., Tape H. L. U., Dridge E. J., Debieux C. M., Paszkiewicz K., Singleton I., Lewis R. J., Santini J. M., Richardson D. J. and Butler C. S. (2010). Quinol-cytochrome c oxidoreductase and cytochrome c4 mediate electron transfer during selenate respiration in *Thauera selenatis*. *Journal of Biological Chemistry*, **285**(24), 18433–18442. doi:10.1074/jbc.M110.115873

Macy J. M., Michel T. A. and Kirsch D. G. (1989). Selenate reduction by a *Pseudomonas* species: a new mode of anaerobic respiration. *FEMS Microbiology Letters*, **61**, 195–198. http://dx.doi.org/10.1111/j.1574-6968.1989.tb03577.x.

Macy J. M., Rech S., Auling G., Dorsch M., Stackebrandt E. and Sly L. I. (1993). *Thauera selenatis* gen. nov., sp. nov., a member of the beta subclass of Proteobacteria with a novel type of anaerobic respiration. *International Journal of Systematic Bacteriology*, **43**(1), 135–142. doi:10.1099/00207713-43-1-135

Maltman C. and Yurkov V. (2018). Bioremediation potential of bacteria able to reduce high levels of selenium and tellurium oxyanions. *Archives of Microbiology*, **200**, 1411–1417. doi:10.1007/s00203-018-1555-6

Morley G. F., Sayer J. A., Wilkinson S. C., Gharieb M. M. and Gadd G. M. (1996). Fungal sequestration, solubilisation and transformation of toxic metals. In: Fungi and environmental change, J. C. Frankland, N. Magan and G. M. Gadd (eds.), Cambridge University Press, Cambridge, pp. 235–256.

Moss M. O., Badii F. and Gibbs G. (1987). Reduction of biselenite to elemental selenium by *Aspergillus parasiticus*. *Transactions of the British Mycological Society*, **89**, 578–580.

Mustacich D. and Powis G. (2000). Thioredoxin reductase. *The Biochemical Journal*, **346**(Pt 1), 1–8. https://doi.org/10.1042/bj3460001

Myneni S. C. B., Tokunaga T. K. and Brown G. E., Jr. (1997). Abiotic selenium redox transformations in the presence of Fe(II, III) oxides. *Science*, **278**, 1106–1109. doi:10.1126/science.278.5340.1106

Naik M. M., Bagayat A. and Charya L. S. (2017). Extreme halophilic archaea *Halobacterium* sp. SP1 in bioremediation of hypersaline water polluted with metal, metalloid, organometal and aromatic compound. *Indian Journal of Applied Microbiology*, **20**(2), 42–54. doi:10.46798/ijam.2017.v20i02.003

Nakagawa T., Iino T., Suzuki K. I. and Harayama S. (2006). *Ferrimonas futtsuensis* sp. nov. And *Ferrimonas kyonanensis* sp. nov., selenate-reducing bacteria belonging to the Gammaproteobacteria isolated from Tokyo bay. *International Journal of Systematic and Evolutionary Microbiology*, **56**(Pt 11), 2639–2645. doi:10.1099/ijs.0.64399-0

Nancharaiah Y. V. and Lens P. N. L. (2015). Ecology and biotechnology of selenium-respiring Bacteria. *Microbiology and Molecular Biology Reviews*, **79**(1), 61–80. doi:10.1128/MMBR.00037-14

Narasingarao P. and Häggblom M. M. (2006). *Sedimenticola selenatireducens*, gen. nov., sp. nov., an anaerobic selenate-respiring bacterium isolated from estuarine sediment. *Systematic and Applied Microbiology*, **29**(5), 382–388. doi:10.1016/j.syapm.2005.12.011

Narasingarao P. and Häggblom M. M. (2007a). Identification of anaerobic selenate-respiring bacteria from aquatic sediments. *Applied and Environmental Microbiology*, **73**, 3519–3527. https://doi.org/10.1128/AEM.02737-06

Narasingarao P. and Häggblom M. M. (2007b). *Pelobacter seleniigenes* sp. nov., a selenate-respiring bacterium. *International Journal of Systematyc and Evolutionary Microbiology*, **57**(Pt 9), 1937–1942. doi:10.1099/ijs.0.64980-0

Negi B. B., Sinharoy A. and Pakshirajan K. (2020). Selenite removal from wastewater using fungal pelleted airlift bioreactor. *Environmental Science and Pollution Research*, **27**, 992–1003. doi:10.1007/s11356-019-06946-6

Newman D. K., Ahmann D. and Morel F. M. M. (1998). A brief review of microbial arsenate reduction. *Geomicrobiology Journal*, **15**, 255–268. http://dx.doi.org/10.1080/01490459809378082

Ojeda J. J., Merroun M. L., Tugarova A. V., Lampis S., Kamnev A. A. and Gardiner P. H. E. (2020). Developments in the study and applications of bacterial transformations of selenium species. *Critical Reviews in Biotechnology*. **40**(8), 1250–1264. doi:10.1080/07388551.2020.1811199

Oremland R. S. and Stolz J. F. (2000). Dissimilatory reduction of selenate and arsenate in nature. In Environmental microbe-metal interactions, Edited by D. R. Lovley (ed.), American Society for Microbiology, Washington, DC, pp. 199–224. https://doi.org/10.1128/9781555818098.ch9

Oremland R. S., Hollibaugh J. T., Maest A. S., Presser T. S., Miller L. G. and Culbertson CW.. (1989). Selenate reduction to elemental selenium by anaerobic bacteria in sediments and culture: biogeochemical significance of a novel sulfate-independent respiration. *Applied and Environmental Microbiology*, **55**, 2333–2343. https://doi.org/10.1128/aem.55.9.2333-2343.1989

Oremland R. S., Switzer Blum J., Culbertson C. W., Visscher P. T., Miller G., Dowdle P. and Strohmaier F. E. (1994). Isolation, growth, and metabolism of an obligately anaerobic, selenaterespiring bacterium, strain SES-3. *Applied and Environmental Microbiology*, **60**, 3011–3019. doi:10.1128/aem.60.8.3011-3019.1994

Orphan V. J., House C. H., Hinrichs K. U., McKeegan K. D. and DeLong E. F. (2001). Methane consuming archaea revealed by directly coupled isotopic and phylogenetic analysis. *Science*, **293**(5529), 484–487. doi:10.1126/science.1061338

Painter E. P. (1941). The chemistry and toxicity of selenium compounds, with special reference to the selenium problem. *Chemical Reviews*, **28**(2), 179–213. https://doi.org/10.1021/cr60090a001

Papp L. V., Lu J., Holmgren A. and Khanna K. K. (2007). From selenium to selenoproteins: synthesis, identity, and their role in human health. *Antioxidants and Redox Signaling*, **9** (7), 775–806. doi:10.1089/ars.2007.1528

Pearce C. I., Pattrick R. A., Law N., Charnock J. M., Coker V. S., Fellowes J. W., Oremland R. S. and Lloyd J. R. (2009). Investigating different mechanisms for biogenic selenite transformations: *Geobacter sulfurreducens*, *Shewanella oneidensis* and *Veillonella atypica*. *Environmental technology*, **30**(12), 1313–1326. https://doi.org/10.1080/09593330902984751

Piacenza E., Presentato A., Ambrosi E., Speghini A., Turner R. J., Vallini G. and Lampis S. (2018). Physical-chemical properties of biogenic selenium nanostructures produced by

Stenotrophomonas maltophilia SeITE02 and *Ochrobactrum* sp. MPV1. *Frontiers in Microbiology*, **9**, 3178. doi:10.3389/fmicb.2018.03178

Ramadan S. E., Razak A. A, Yousseff Y. A. and Sedky N. M. (1988). Selenium metabolism in a strain of *Fusarium*. *Biological Trace Element Research*, **18**, 161–170. https://doi.org/10.1007/BF02917500

Rauschenbach I., Yee N., Häggblom M. M. and Bini E. (2011). Energy metabolism and multiple respiratory pathways revealed by genome sequencing of *Desulfurispirillum indicum* strain S5. *Environmental Microbiology*, **13**, 1611–1621. doi:10.1111/j.1462-2920.2011.02473.x

Rayman M. P. (2012). Selenium and human health. *Lancet*, **379**, 1256–1268. doi:10.1016/S0140-6736(11)61452-9

Rech S. A. and Macy J. M. (1992). The terminal reductases for selenate and nitrate respiration in *Thauera selenatis* are two distinct enzymes. *Journal of Bacteriology*, **174**(22), 7316–7320. doi:10.1128/jb.174.22.7316-7320.1992

Ridley H., Watts C. A., Richardson D. J. and Butler C. S. (2006). Resolution of distinct membrane-bound enzymes from *Enterobacter cloacae* SLD1a-1 that are responsible for selective reduction of nitrate and selenate oxyanions. *Applied and Environmental Microbiology*; **72**(8), 5173–5180. doi:10.1128/AEM.00568-06

Rosenfeld C. E., Kenyon J. A., James B. R. and Santelli C. M. (2017). Selenium (IV,VI) reduction and tolerance by fungi in an oxic environment. *Geobiology*, **15**, 441–452. doi:10.1111/gbi.12224

Rosenfeld C. E., Sabuda M. C., Hinkle M. A. G., James B. R. and Santelli C. M. (2020). A fungal-mediated cryptic selenium cycle linked to manganese biogeochemistry. *Environmental Science and Technology*, **54**, 3570–3580. https://doi.org/10.1021/acs.est.9b06022

Ruocco M. H. W., Chan C. S., Hanson T. E. and Church T. M. (2014). Characterization and distribution of selenite reduction products in cultures of the marine yeast *rhodotorula mucilaginosa*-13B, *Geomicrobiology Journal*, **31**(9), 769–778. https://doi.org/10.1080/01490451.2014.888909

Sarkar J., Dey P., Saha S. and Acharya K. (2011). Mycosynthesis of selenium nanoparticles. *Micro and Nano Letters*, **6**(8), 599–602. doi:10.1049/mnl.2011.0227

Sarkar J., Saha S., Dey P. and Acharya K. (2012). Production of selenium nanorods by phytopathogen *Alternaria alternata*. *Advanced Science Letters*, **10**, 111–114. https://doi.org/10.1166/asl.2012.2137

Schrauzer G. N. (2000). Selenomethionine: a review of its nutritional significance, metabolism and toxicity. *The Journal of Nutrition*, **130**(7), 1653–1656. doi:10.1093/jn/130.7.1653

Schröder I., Rech S., Krafft T. and Macy J. M. (1997). Purification and characterization of the selenate reductase from *Thauera selenatis*. *Journal of Biological Chemistry*, **272**, 23765–23768. 10.1074/jbc.272.38.23765

Shi L. D., Lv P. L., Wang M., Lai C. Y. and Zhao H. P. (2020). A mixed consortium of methanotrophic archaea and bacteria boosts methane-dependent selenate reduction. *Science of the Total Environment*, **732**, 139310. doi:10.1016/j.scitotenv.2020.139310

Shisler J. L., Senkevich T. G., Berry M. J. and Moss B. (1998). Ultraviolet-induced cell death blocked by a selenoprotein from a human dermatotropic poxvirus. *Science*, **279**, 102–105. http://dx.doi.org/10.1126/science.279.5347.102.

Sinharoy A. and Lens P. N. L. (2020). Biological removal of selenate and selenite from wastewater: options for selenium recovery as nanoparticles. *Current Pollution Reports*, **6**, 230–249. https://doi.org/10.1007/s40726-020-00146-4

Slobodkina G. B., Lebedinsky A. V., Chernyh N. A., Bonch-Osmolovskaya E. A. and Slobodkin A. I. (2015). *Pyrobaculum ferrireducens* sp. nov., a hyperthermophilic Fe (III)-, selenate- and arsenate-reducing crenarchaeon isolated from a hot spring. *International Journal of Systematic and Evolutionary Microbiology*, **65**(Pt 3), 851–856. doi:10.1099/ijs.0.000027

Soudi M. R., Malekzadeh F. and Norouzi P. (2003). Screening yeast cells for absorption of selenium oxyanions. *World Journal of Microbiology and Biotechnology*, **19**, 181–184. https://doi.org/10.1023/A:1023290908937

Soudi M., Ghazvini P., Khajeh K. and Gharavi S. (2009). Bioprocessing of seleno-oxyanions and tellurite in a novel *Bacillus* sp. Strain STG-83: a solution to removal of toxic oxyanions in presence of nitrate. *Journal of Hazardous Materials*, **165**, 71–77. https://doi.org/10.1016/j.jhazmat.2008.09.065

Srivastava P., Braganca J. M. and Kowshik M. (2014). In vivo synthesis of selenium nanoparticles by *Halococcus salifodinae* BK18 and their anti-proliferative properties against HeLa cell line. *Biotechnology Progress*, **30**(6), 1480–1487. doi:10.1002/btpr.1992

Staicu L. C., Ackerson C. J., Cornelis P., Ye L., Berendsen R. L., Hunter W. J., Noblitt S. D., Henry C. S., Cappa J. J., Montenieri R. L., Wong A. O., Musilova L., Sura-de Jong M., van Hullebusch E. D., Lens P. N., Reynolds R. J. and Pilon-Smits E. A. (2015). *Pseudomonas moraviensis* subsp. *Stanleyae*, a bacterial endophyte of hyperaccumulator *Stanleya pinnata*, is capable of efficient selenite reduction to elemental selenium under aerobic conditions. *Journal of Applied Microbiology*, **119** (2), 400–410. doi: 10.1111/jam.12842.

Stolz J. F. and Oremland R. S. (1999). Bacterial respiration of arsenic and selenium, *FEMS Microbiology Reviews*, **23**(5), 615–627. https://doi.org/10.1111/j.1574-6976.1999.tb00416.x

Stolz J. F., Ellis D. J., Blum J. S., Ahmann D., Lovley D. R. and Oremland R. S. (1999). *Sulfurospirillum barnesii* sp. nov. And *Sulfurospirillum arsenophilum* sp. nov., new members of the *Sulfurospirillum* clade of the epsilon Proteobacteria. *International Journal of Systematic Bacteriology*, **49**(Pt 3), 1177–1180. doi:10.1099/00207713-49-3-1177

Stolz J. F., Basi P. and Oremland R. S. (2002). Microbial transformation of elements: the case of arsenic and selenium. *International Microbiology*, **5**, 201–207. http://dx.doi.org/10.1007/s10123-002-0091-y.

Switzer-Blum J., Burns Bindi A., Buzzelli J., Stolz J. F. and Oremland RS. . (1998). *Bacillus arsenicoselenatis*, sp. nov., and *Bacillus selenitireducens*, sp. nov.: two haloalkaliphiles from mono lake, California that respire oxyanions of selenium and arsenic. *Archives of Microbiology*, **171**, 19–30. https://doi.org/10.1007/s002030050673

Tan Y., Wang Y., Wang Y., Xu D., Huang Y., Wang D., Wang G., Rensing C. and Zheng S. (2018). Novel mechanisms of selenate and selenite reduction in the obligate aerobic bacterium *Comamonas testosteroni* s44. *Journal of Hazardous Materials*, **359**, 129–138, https://doi.org/10.1016/j.jhazmat.2018.07.014.

Tastet L., Schaumlöffel D., Bouyssiere B. and Lobinski R. (2008). Identification of selenium-containing proteins in selenium-rich yeast aqueous extract by 2D gel electrophoresis, nanoHPLC–ICP MS and nanoHPLC–ESI MS/MS. *Talanta*, **75**(4), 1140–1145. https://doi.org/10.1016/j.talanta.2008.01.003

Tejo Prakash N., Sharma N., Prakash R., Raina K. K., Fellowes J., Pearce C. I., Lloyd J. R. and Pattrick R. A. (2009). Aerobic microbial manufacturing of nanoscale selenium: exploiting nature's bio-nanomineralization potential. *Biotechnology Letters*, **31**, 1857–1862. http://dx.doi.org/10.1007/s10529-009-0096-0.

Theisen J. and Yee N. (2014). The molecular basis for selenate reduction in *Citrobacter freundii*. *Geomicrobiology Journal*, **31**, 875–883. doi:10.1080/01490451.2014.907377.

Tomei F. A., Barton L. L., Lemanski C. L. and Zocco T. G. (1992). Reduction of selenate and selenite to elemental selenium by *Wolinella succinogenes*. *Canadian Journal of Microbiology*, **38**, 1328–1333. https://doi.org/10.1139/m92-219

Tomei F. A., Barton L. L., Lemanski C. L., Zocco T. G., Fink N. H. and Sillerud L. O. (1995). Transformation of selenate and selenite to elemental selenium by *Desulfovibrio desulfuricans*. *Journal of Industrial Microbiology*, **14**, 329–336. https://doi.org/10.1007/BF01569947

Urík M., Boriová K., Bujdoš M. and Matúš P. (2016). Fungal selenium (VI) accumulation and biotransformation—filamentous fungi in selenate contaminated aqueous media remediation. *CLEAN – Soil Air Water*, **44**(6), 610–614. https://doi.org/10.1002/clen.201500100

Vallini G., Di Gregorio S. and Lampis S. (2005). Rhizosphere-induced selenium precipitation for possible applications in phytoremediation of Se polluted effluents. *Zeitschrift fur Naturforschung C Journal of Bioscience*, **60**(3–4), 349–356. https://doi.org/10.1515/znc-2005-3-419

Vetchinkina E., Loshchinina E., Kursky V. and Nikitina V. (2013). Reduction of organic and inorganic selenium compounds by the edible medicinal basidiomycete *lentinula edodes* and the accumulation of elemental selenium nanoparticles in its mycelium. *Journal of Microbiology*, **51**(6), 829–835. doi:10.1007/s12275-013-2689-5

Vetchinkina E., Loshchinina E., Kurskyi V. and Nikitina V. (2016). Biological synthesis of selenium and germanium nanoparticles by xylotrophic basidiomycetes. *Applied Biochemistry and Microbiology*, **52**(1), 87–97. https://doi.org/10.1134/S0003683816010130

Vetchinkina E., Loshchinina E., Kupryashina M., Burov A., Pylaev T. and Nikitina V. (2018). Green synthesis of nanoparticles with extracellular and intracellular extracts of basidiomycetes. *Peer Journal - Life and Environment*, **6**, e5237. https://doi.org/10.7717/peerj.5237

Vetchinkina E., Loshchinina E., Kupryashina M., Burov A. and Nikitina V. (2019). Shape and size diversity of gold, silver, selenium, and silica nanoparticles prepared by green synthesis using fungi and Bacteria. *Industrial and Engineering Chemistry Research*, **58**, 17207–17218. https://doi.org/10.1021/acs.iecr.9b03345

Völkl P., Huber R., Drobner E., Rachel R., Burggraf S., Trincone A. and Stetter O. K. (1993). *Pyrobaculum aerophilum* sp. nov., a novel nitrate-reducing hyperthermophilic archaeum. *Applied and Environmental Microbiology*, **59**(9), 2918–2926. doi:10.1128/aem.59.9.2918-2926.1993

Vorholt J. A., Vaupel M. and Thauer R. K. (1997). A selenium-dependent and a selenium-independent formylmethanofuran dehydrogenase and their transcriptional regulation in the hyperthermophilic *Methanopyrus kandleri*. *Molecular Microbiology*, **23**, 1033–1042. doi:10.1046/j.1365-2958.1997.2931653.x

Wang J., Wang B., Zhang D. and Wu Y. (2016). Selenium uptake, tolerance and reduction in *Flammulina velutipes* supplied with selenite. *Peer Journal - Life and Environment*, **4**: e1993. doi:10.7717/peerj.1993

Wangeline A. L., Valdez J. R., Lindblom S. D., Bowling K. L., Reeves F. B. and Pilon-Smits E. A. H. (2011). Characterization of rhizosphere fungi from selenium hyperaccumulator and nonhyperaccumulator plants along the eastern rocky mountain front range. *American Journal of Botany*, **98**(7), 1139–1147. doi:10.3732/ajb.1000369

Watts C. A., Ridley H., Condie K. L., Leaver J. T., Richardson D. J. and Butler C. S. (2003). Selenate reduction by *Enterobacter cloacae* SLD1a-1 is catalyzed by a molybdenum-dependent membrane-bound enzyme that is distinct from the membrane-bound nitrate reductase. *FEMS Microbiology Letters*, **228**(2), 273–279. doi:10.1016/S0378-1097(03)00782-1

Wells M., McGarry J., Gaye M. M., Basu P., Oremland R. S. and Stolz J. F. (2019). Respiratory selenite reductase from *Bacillus selenitireducens* strain MLS10. *Journal of Bacteriology*, **201**(7), e00614–18. Published 2019 Mar 13. doi:10.1128/JB.00614-18

Winkel L. H., Vriens B., Jones G. D., Schneider L. S., Pilon-Smits E. and Bañuelos G. S. (2015). Selenium cycling across soil-plant-atmosphere interfaces: a critical review. *Nutrients*, **7**(6), 4199–4239. doi:10.3390/nu7064199

Whitham G. H. (1995). Organosulfur chemistry. Oxford University Press, Oxford.

Xia X., Wu S., Li N., Wang D., Zheng S. and Wang G. (2018). Novel bacterial selenite reductase Csrf responsible for Se(IV) and Cr(VI) reduction that produces nanoparticles in *Alishewanella* sp. WH16–1. *Journal of Hazardous Materials*, **342**, 499–509. https://doi.org/10.1016/j.jhazmat.2017.08.051.

Xu P., Liu L., Zeng G., Huang D., Lai C., Zhao M., Huang C., Li N., Wei Z., Wu H., Zhang C., Lai M. and He Y. (2014). Heavy metal-induced glutathione accumulation and its role in heavy metal detoxification in *Phanerochaete chrysosporium*. *Applied Microbiology and Biotechnology*, **98**, 6409–6418. doi:10.1007/s00253-014-5667-x

Yamamura S., Yamashita M., Fujimoto N., Kuroda M., Kashiwa M., Sei K., Fujita M. and Ike M. (2007). *Bacillus selenatarsenatis* sp. nov., a selenate- and arsenate-reducing bacterium isolated from the effluent drain of a glass-manufacturing plant. *International Journal of Systematic and Evolutionary Microbiology*, **57**, 1060–1064. DOI 10.1099/ijs.0.64667-0

Yee N., Ma J., Dalia A., Boonfueng T. and Kobayashi D. Y. (2007). Se(VI) reduction and the precipitation of Se(0) by the facultative bacterium *Enterobacter cloacae* SLD1a-1 are regulated by FNR. *Applied and Environmental Microbiology*, **73**, 1914–1920. doi:10.1128/AEM.02542-06.

Zannoni D., Borsetti F., Harrison J. J. and Turner R. J. (2008). The bacterial response to the chalcogen metalloids Se and Te. *Advances in Microbial Physiology*, **53**, 1–72. doi:10.1016/S0065-2911(07)53001-8

Zawadzka A. M., Crawford R. L. and Paszczynski A. J. (2006) Pyridine-2,6-bis (thiocarboxylic acid) produced by *Pseudomonas stutzeri* KC reduces and precipitates selenium and tellurium oxyanions. *Applied and Environmental Microbiology*, **72**, 3119–3129. doi:10.1128/AEM.72.5.3119-3129.2006

Zhang Y. Q., Siddique T., Wang J. and Frankenberger W. T. (2004a). Selenate reduction in river water by *Citerobacter freundii* isolated from a selenium-contaminated sediment. *Journal of Agricultural and Food Chemistry*, **52**, 1594–1600. https://doi.org/10.1021/jf0306789

Zhang Y., Zahir Z. A. and Frankenberger W. T., Jr. (2004b). Fate of colloidal particulate elemental selenium in aquatic systems. *Journal of Environmental Quality*, **33**, 559–564. https://doi.org/10.2134/jeq2004.5590

Zhang L., Li D. and Gao P. (2012). Expulsion of selenium/protein nanoparticles through vesicle-like structures by *Saccharomyces cerevisiae* under microaerophilic environment. *World Journal of Microbiology and. Biotechnology*, **28**(12), 3381–3386. doi:10.1007/s11274-012-1150-y

Zheng S., Su J., Wang L., Yao R., Wang D., Deng Y., Wang R., Wang G. and Rensing C. (2014). Selenite reduction by the obligate aerobic bacterium *Comamonas testosteroni* s44 isolated from a metal-contaminated soil. *BMC Microbiology*, **14,** 204 https://doi.org/10.1186/s12866-014-0204-8

Zieve R., Ansell P. J., Young T. W. K. and Peterson P. J. (1985). Selenium volatilization by *Mortierella* species. *Transactions of the British Mycological Society*, **84**, 177–179.

Zoidis E., Seremelis I., Kontopoulos N. and Danezis G. P. (2018). Selenium-dependent antioxidant enzymes: actions and properties of selenoproteins. *Antioxidants*, **7**(5), 66. doi:10.3390/antiox7050066

Zonaro E., Piacenza E., Presentato A., Monti F., Dell'Anna R., Lampis S. and Vallini G. (2017). *Ochrobactrum* sp. MPV1 from a dump of roasted pyrites can be exploited as bacterial catalyst for the biogenesis of selenium and tellurium nanoparticles. *Microbial Cell Factories*, **16**(1), 215. doi:10.1186/s12934–017-0826-2

Chapter 4

Microbial ecology of selenium-respiring bacteria

Joshua P. Boltz and Bruce E. Rittmann

4.1 SELENIUM, SULFUR, AND NITROGEN IN A COMMON AQUATIC ENVIRONMENT

Irrigated agriculture, steam-power generation, mining, and other human activities result in water that is co-contaminated by selenium (Se), sulfur (S), and nitrogen (N) that typically exist as selenate (SeO_4^{2-}) and/or selenite ($HSeO_3^{-}$), sulfate (SO_4^{2-}), and nitrate (NO_3^{-}), respectively. Usually, their concentrations are very different, whether in irrigated agriculture run-off or in wastewater. The S-to-Se mass ratio (S:Se) is typically in the order of 1000:1 (g S:g Se) and the N-to-Se mass ratio (N:Se) is typically in the order of 50:1 (g N:g Se). The target contaminant concentrations in treated effluent also show great disparity. For example, the United States Environmental Protection Agency (EPA, 2020) requires existing steam-power-generation facilities to discharge water having less than 3 g N/m^3 and 0.029 g Se/m^3 (average daily concentrations over a consecutive 30-day period); this is a N:Se ratio ~100:1 (g N:g Se).

Selenium is among the first micro-pollutants that, according to regulation (EPA, 2020), require biological wastewater treatment and have regulated surface-water discharge standards. When the contaminated water has a pH of 6 to 8 and a temperature of 15 to 30°C, bacteria can anaerobically reduce these oxyanions at a rate that makes bioreactors an economically viable treatment alternative (Boltz &

doi: 10.2166/9781789061055_0145

Rittmann, 2019). While much is known about denitrifying heterotrophic bacteria (denoted X_H) and sulfate-reducing bacteria (denoted X_{SRB}) (see Grady *et al.*, 2011; Rittmann & McCarty, 2020), the metabolic pathways of selenium-respiring bacteria (SeRB) are not as well understood.

Selenium-respiring bacteria are phylogenetically and physiologically diverse, and they include the selenate-reducing bacterium *Thauera selenatis* (Macy *et al.*, 1993) and selenite-reducing bacterium *Bacillus selenitireducens* (Switzer Blum *et al.*, 1998). Much like denitrifying heterotrophic and sulfate-reducing bacteria, SeRB carry out respiration, synthesis, endogenous, and detoxification processes. These bacteria respire selenium, synthesize essential macro- (e.g., C and N) and micro-nutrients (e.g., Se and S), release and/or utilize their internal components through endogenous processes, and emit organic selenium compounds (e.g., selenomethionine, or SeMet) to prevent accumulating an internal selenide mass that is toxic to the bacteria. Selenate-reducing bacteria (denoted X_{SeO_4}) biologically transform selenate into selenite during respiration. Similarly, selenite-reducing bacteria (denoted X_{SeO_3}) biologically transform selenite into elemental selenium (Se^0) during respiration. Elemental selenium is a stable product that predominantly accumulates in the cells' extracellular polymeric substances (EPS). Intra-cellular Se generally exists as selenide (Se(-II)). Selenium-respiring bacteria can reduce selenium oxyanions directly to intra-cellular selenide (Se(-II)) during synthesis. Figure 4.1 illustrates the modeled processes of respiration, synthesis, and endogenous decay for selenite-reducing bacteria. The reduction of selenium oxyanions to selenide and detoxification are not modeled.

Energetically, SeRB preferentially utilize selenide as their source of Se for biomass synthesis (Eswayah *et al.*, 2016). However, Se in oxyanions also can be used for this purpose, but with an associated energy cost in the form of electron donor demand. Endogenous decay processes re-introduce selenide to the bulk of the liquid, where it can be transformed into metal selenides or re-used for biomass synthesis. In addition, selenium resulting from endogenous processes and detoxification may volatilize (see Chapter 3). Factors influencing bacterial selenium volatilization include macro-nutrient availability (particularly C and N), exposure to aerobic conditions, temperature (>30 to 35°C), pH (>8), selenium speciation, and microbial community composition (De Souza *et al.*, 2001).

This chapter begins by describing interactions among denitrifying heterotrophic bacteria, SeRB, and sulfate-reducing bacteria. Figure 4.2 illustrates that all these bacteria carry out respiration, synthesis, and endogenous decay while competing for a common electron donor and nutrient pool. The chapter then provides an energetic analysis and mathematical model of these processes, and the chapter compares model results with published data. Used together, they identify the factors controlling how SeRB occur along with denitrifying heterotrophic and sulfate-reducing bacteria in bioreactors treating Se-laden waters.

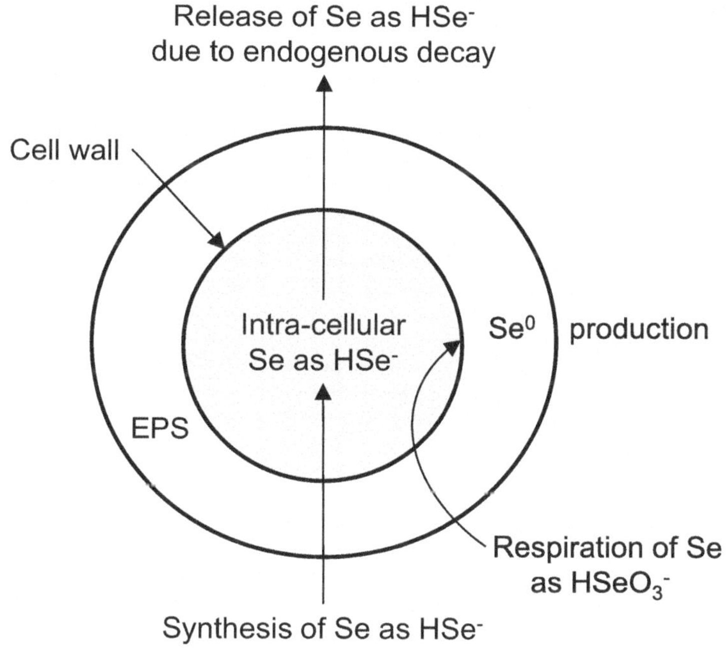

Figure 4.1 Simulated metabolic processes for selenite-reducing bacteria (X_{SeO_3}). Selenite-reducing bacteria respire selenite ($HSeO_3^-$) to produce elemental selenium (Se^0), which may accumulate inside and outside the cell wall. These bacteria synthesize selenium as selenide (HSe^-). An intra-cellular pool of nutrients, including HSe^-, exists inside a cell wall. Extracellular polymeric substances (EPS) exist outside the cell wall.

4.2 SUBSTRATE PARTITIONING, ENERGETICS, AND BIOMASS YIELD

Microorganisms carry out oxidation-reduction reactions to obtain energy for synthesis and cell maintenance (Rittmann & McCarty, 2020). The amount of energy released per electron equivalent depends on the reaction. Consequently, the amount of synthesis resulting from an equivalent of electron donor oxidized also depends on the reaction. When microorganisms utilize an electron donor, they transfer a portion of their electrons to the electron acceptor (fraction f_e^0) to supply the energy required for converting the remaining electrons (fraction f_s^0) into microbial cells; the fractions add to one, or $f_e^0 + f_s^0 = 1$. We apply a thermodynamic electron-equivalent model (McCarty, 2007; Rittmann & McCarty, 2020) to relate synthesis and reaction energetics. We begin with the four electron acceptors used in respiration, consider a common electron donor, develop the energy reactions, and then factor in biomass synthesis.

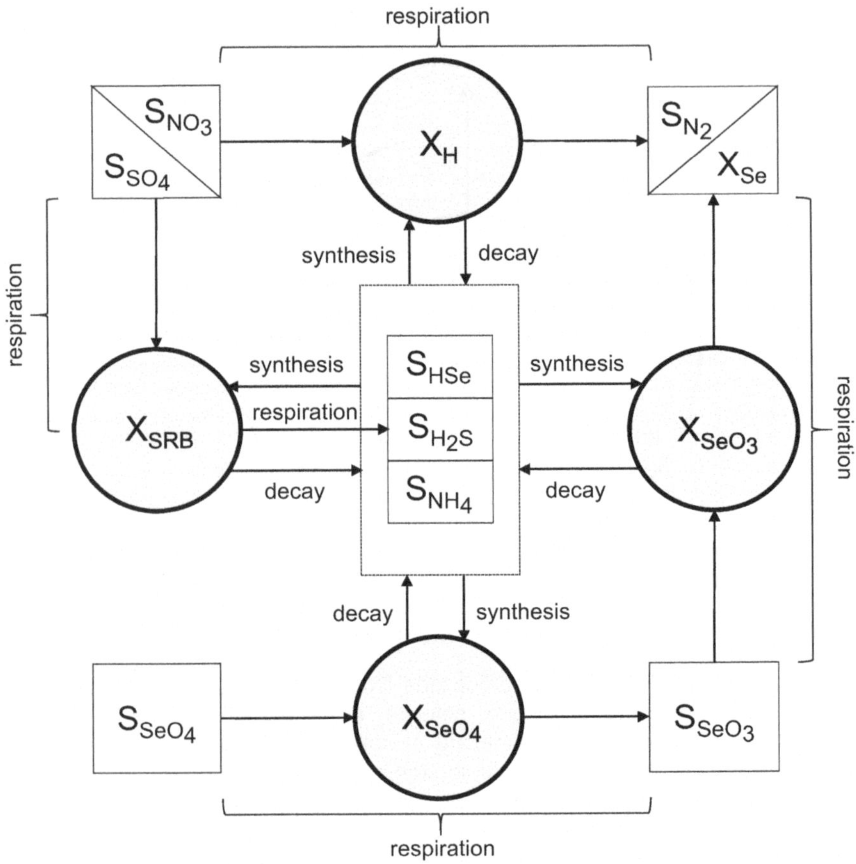

Figure 4.2 Interactions among denitrifying heterotrophic bacteria (X_H), selenate-reducing bacteria (X_{SeO_4}), selenite-reducing bacteria (X_{SeO_3}), and sulfate-reducing bacteria (X_{SRB}) undergoing respiration, synthesis, and endogenous decay. Symbols: nitrate (S_{NO_3}), di-nitrogen (S_{N_2}), ammonium (S_{NH_4}), selenate (S_{SeO_4}), selenite (S_{SeO_3}), elemental selenium (X_{Se}), selenide (S_{HSe}), sulfate (S_{SO_4}), and hydrogen sulfide (S_{H_2S}).

4.2.1 Electron-acceptor reductions

Equation (4.1) is the half reaction for nitrate reduction to di-nitrogen (N_2) and its free energy (Grady *et al.*, 2011; Rittmann & McCarty, 2020):

$$\frac{1}{5}NO_3^- + \frac{6}{5}H^+ + e^- \rightarrow \frac{1}{10}N_2 + \frac{3}{5}H_2O \quad \Delta G^{0'}_{EA} = -72.2\frac{kJ}{e^-eq_{EA}} \tag{4.1}$$

Equation (4.2) is the half reaction for selenate reduction to selenite and its free energy (Olin *et al.*, 2020; PSI, 2007):

$$\frac{1}{2}SeO_4{}^{2-} + \frac{3}{2}H^+ + e^- \rightarrow \frac{1}{2}HSeO_3{}^- + \frac{1}{2}H_2O \qquad \Delta G^{0'}_{EA} = -43.9\frac{kJ}{e^-eq_{EA}} \tag{4.2}$$

Equation (4.3) is the half reaction for selenite reduction to elemental selenium and its free energy (Olin *et al.*, 2020; PSI, 2007):

$$\frac{1}{4}HSeO_3{}^- + \frac{5}{4}H^+ + e^- \rightarrow \frac{1}{4}Se^0 + \frac{3}{4}H_2O \qquad \Delta G^{0'}_{EA} = -25.4\frac{kJ}{e^-eq_{EA}} \tag{4.3}$$

Finally, Equation (4.4) is the half reaction for sulfate reduction to hydrogen sulfide (H_2S) and its free energy (Grady *et al.*, 2011; Rittmann & McCarty, 2020):

$$\frac{1}{8}SO_4{}^{2-} + \frac{5}{4}H^+ + e^- \rightarrow \frac{1}{8}H_2S + \frac{1}{2}H_2O \qquad \Delta G^{0'}_{EA} = +20.1\frac{kJ}{e^-eq_{EA}} \tag{4.4}$$

Standard free energies have been adjusted to pH = 7 (i.e., $\Delta G^{0'}$) for all reactions, following Rittmann and McCarty (2020).

4.2.2 Oxidation of a common electron donor

Acetate (CH_3COO^-) is the common electron donor for all electron acceptors. Equation (4.5) is the half reaction for acetate reduction and its free energy (Rittmann & McCarty, 2020):

$$\frac{1}{8}CH_3COO^- + \frac{1}{8}H_2O \rightarrow \frac{1}{8}CO_2 + \frac{1}{8}HCO_3{}^- + H^+ + e^-$$
$$\Delta G^{0'}_{ED} = +27.4\frac{kJ}{e^-eq_{ED}} \tag{4.5}$$

4.2.3 Energy reactions

Equations (4.6) through (4.9) are the energy reactions for each of the four electron acceptors when acetate is the common electron donor. Also shown are the free energy values for each energy reaction ($\Delta G^{0'}$) kJ/e$^-$eq.

$$\frac{1}{8}CH_3COO^- + \frac{1}{5}NO_3{}^- + \frac{1}{5}H^+ \rightarrow \frac{1}{8}CO_2 + \frac{1}{10}N_2 + \frac{1}{8}HCO_3{}^- + \frac{19}{40}H_2O$$
$$\Delta G^{0'}_{ED} = -99.6\frac{kJ}{e^-eq_{ED}} \tag{4.6}$$

$$\frac{1}{8}CH_3COO^- + \frac{1}{2}SeO_4^{2-} + \frac{1}{2}H^+ \rightarrow \frac{1}{8}CO_2 + \frac{1}{2}HSeO_3^- + \frac{1}{8}HCO_3^- + \frac{3}{8}H_2O$$
$$\Delta G^{0'}_{ED} = -71.3\frac{\text{kJ}}{\text{e}^-\text{eq}_{\text{ED}}} \quad (4.7)$$

$$\frac{1}{8}CH_3COO^- + \frac{1}{4}HSeO_3^- + \frac{1}{4}H^+ \rightarrow \frac{1}{8}CO_2 + \frac{1}{4}Se^0 + \frac{1}{8}HCO_3^- + \frac{5}{8}H_2O$$
$$\Delta G^{0'}_{ED} = -49.8\frac{\text{kJ}}{\text{e}^-\text{eq}_{\text{ED}}} \quad (4.8)$$

$$\frac{1}{8}CH_3COO^- + \frac{1}{8}SO_4^{2-} + \frac{1}{4}H^+ \rightarrow \frac{1}{8}CO_2 + \frac{1}{8}H_2S + \frac{1}{8}HCO_3^- + \frac{3}{8}H_2O$$
$$\Delta G^{0'}_{ED} = -7.3\frac{\text{kJ}}{\text{e}^-\text{eq}_{\text{ED}}} \quad (4.9)$$

Equations (4.6) through (4.9) show significant differences in $\Delta G^{0'}$ values, which vary from −99.6 to −7.3 kJ/e$^-$eq$_{ED}$. These differences stem from the free energies for the electron acceptors (Equations (4.1) through (4.4)), which range from −72.2 to +20.1 kJ/e$^-$eq$_{EA}$. These free-energy differences have strong implications on the competitiveness of bacteria carrying out different respirations. Denitrifying heterotrophic bacteria have a significant energy advantage over the other bacteria considered here, while sulfate-reducing bacteria have a significant energy disadvantage. Noteworthy is that the conversion of selenate into selenite (Equation (4.2)) has a $\Delta G^{0'}$ value that is more negative by 18.5 kJ/e$^-$$_{ED}$ when compared with the conversion of selenite into elemental selenium (Equation (4.3)). The more negative free energy indicates a competitive advantage for selenate-reducing bacteria when they are competing with selenite-reducing bacteria for a common electron donor.

4.2.4 Considering biomass synthesis

Rittmann and McCarty (2020) presented a method for quantifying biomass synthesis based on balancing energy gained in the energy reactions (i.e., Equations (4.6) through (4.9)) with the demand that is associated with synthesizing biomass from available forms of the essential macro-nutrients, of which C and N are dominating. Because all of the bacteria in our analysis have the same sources of C (acetate) and N (ammonium), the energy associated with synthesis is the same in all cases: 37.2 kJ/e$^-$ eq$_X$. This value is combined with ΔG_r to compute A, which is the ratio of electron equivalents used for the energy reaction to the electron equivalents used for biomass synthesis (from the energy balance):

$$A = \frac{-37.2}{\varepsilon \cdot \Delta G_r} \tag{4.10}$$

Here, ε is the energy-transfer efficiency, which we take as 0.68, a value within the typical range of 0.55 to 0.70 (Rittmann & McCarty, 2020). The ΔG_r values are taken from Equations (4.6) through (4.9).

Then, f_s^0 can be calculated by:

$$f_s^0 = \frac{1}{1 + A} \tag{4.11}$$

Since one electron equivalent equals 8 g O_2, the true yield for any of the bacteria (Y_k) can be calculated by:

$$Y_k = \frac{f_s^0 \cdot M_c}{8 \cdot n_e} \cdot 1.42 \frac{\text{g COD}_\text{X}}{\text{g VSS}} \tag{4.12}$$

Here, M_c is the molecular weight of bacterial cells (113 g volatile suspended solids (VSS)/mol VSS), n_e is the number of electron equivalents in an empirical mole of bacterial cells, i.e., 20 e^-eq_{VSS}/mol VSS when ammonium is the N source, and 1.42 is a conversion factor for biomass chemical oxygen demand (COD) and VSS (Rittmann & McCarty, 2020).

The maximum specific rate of electron-donor utilization, or q_k, can be calculated by:

$$q_k = \frac{8}{f_e^0} \cdot \frac{1}{1.42} \frac{\text{g VSS}}{\text{g COD}_\text{X}} \tag{4.13}$$

Here, q_k has units $\left(\frac{\text{g COD}_\text{ED}}{\text{g COD}_\text{X} \cdot \text{d}}\right)$. The maximum specific growth rate of any biomass k ($\mu_{\max,k}$) can be calculated by:

$$\mu_{\max,k} = Y_k \cdot q_k \tag{4.14}$$

Table 4.1 summarizes the results of this analysis for the respirations described by Equations (4.6) through (4.9). Striking about Table 4.1 are the significant differences in Y_k and $\mu_{\max,k}$ values. Reflecting the trends in the $\Delta G^{0'}$ values associated with Equations (4.6) through (4.9), the Y_k and $\mu_{\max,k}$ values indicate a clear advantage for denitrifying heterotrophic bacteria when compared with the other bacteria considered. The advantages indicated by both of these parameters suggest a kinetic hierarchy for the specific growth rate: denitrifying heterotrophic bacteria > selenate-reducing bacteria > selenite-reducing bacteria > sulfate-reducing bacteria. The differences among Y_k and $\mu_{\max,k}$ are substantial enough to suggest that it is possible to favor or suppress the accumulation of different bacterial types in a well-designed and operated bioreactor.

Table 4.1 Stoichiometry and energetics for the four respiration reactions.

Electron Donor (ED)	Electron Acceptor (EA)	Reduced Product	Energy Equivalents Required for Cell Production, *A*	Fraction of Electrons Required for Cell Synthesis, f_s^0	Fraction of Electrons Required for Energy Production, f_e^0	Biomass Yield, *Y*	Maximum Specific Rate of Substrate Utilization, q_{max}	Maximum Specific Growth Rate of Biomass, μ_{max}
			$\left(\frac{e^- eq_s}{e^- eq_{ED}}\right)$	$\left(\frac{e^- eq_X}{e^- eq_{ED}}\right)$	$\left(\frac{e^- eq_{EA}}{e^- eq_{ED}}\right)$	$\left(\frac{g_{COD_X}}{g_{COD_{ED}}}\right)$	$\left(\frac{g_{COD_{ED}}}{g_{COD_X} \cdot d}\right)$	$\left(\frac{1}{d}\right)$
Acetate	NO_3^-	N_2	0.55	0.65	0.35	0.65	15.9	10.3
Acetate	SeO_4^{2-}	$HSeO_3^-$	0.77	0.57	0.43	0.57	13.0	7.4
Acetate	$HSeO_3^-$	Se^0	1.04	0.49	0.51	0.49	11.1	5.5
Acetate	SO_4^{2-}	H_2S	7.49	0.12	0.88	0.12	6.4	0.8

4.3 MATHEMATICAL MODEL OF DENITRIFYING HETEROTROPHIC BACTERIA, SELENIUM-RESPIRING BACTERIA, AND SULFATE-REDUCING BACTERIA

The mathematical model presented in this chapter explicitly simulates electron-donor oxidation, electron-acceptor respiration, biomass synthesis, macro- and micro-nutrient assimilation for synthesis, and endogenous decay for denitrifying heterotrophic bacteria, selenate-reducing bacteria, selenite-reducing bacteria, and sulfate-reducing bacteria. These bacteria compete for essential nutrients that include the electron donor acetate (denoted S_{VFA}), macro-nutrient ammonium (denoted S_{NHx}), and the micro-nutrients hydrogen sulfide (denoted S_{H_2S}) and selenide (denoted S_{HSe}). Grady *et al.* (2011) and Rittmann and McCarty (2020) present comprehensive discussions about biological process models and their applications.

Table 4.2 presents the process, kinetic, and stoichiometric matrix for denitrifying heterotrophic bacteria. Simulated processes include the anaerobic biological transformation of nitrate (denoted S_{NO_3}) to di-nitrogen (denoted S_{N_2}) and anaerobic endogenous decay. Table 4.3 presents the process, kinetic, and stoichiometric matrix for selenate-reducing bacteria. Simulated processes include the anaerobic biological transformation of selenate (denoted S_{SeO_4}) to selenite (denoted S_{SeO_3}) and anaerobic endogenous decay. Table 4.4 presents the process, kinetic, and stoichiometric matrix for selenite-reducing bacteria. Simulated processes include the anaerobic biological transformation of selenite to elemental selenium (denoted X_{Se}) and anaerobic endogenous decay. Finally, Table 4.5 presents the process, kinetic, and stoichiometric matrix for sulfate-reducing bacteria. Simulated processes include the anaerobic biological transformation of sulfate (denoted S_{SO_4}) to hydrogen sulfide and anaerobic endogenous decay.

Kinetic expressions for each of the simulated processes are listed in Table 4.6. The rates of electron-donor oxidation, electron-acceptor reduction, biomass synthesis, and macro- and micro-nutrient assimilation are represented as the product of a maximum specific growth rate of biomass ($\mu_{\max,k}$), Monod-type hyperbolic functions for each potentially rate-limiting substrate i $\left(\frac{S_i}{S_i + K_i}\right)$, and the concentration of biomass k (X_k). The rate expressions may include one or more functions that inhibit bacterial growth when other electron acceptors are present in sufficient quantity $\left(\frac{K_i}{S_i + K_i}\right)$. For example, a sufficient nitrate concentration will inhibit selenate, selenite, and sulfate reductions. Endogenous decay kinetics are simulated as first-order expressions that multiply a biomass decay rate (b_k) and the concentration of biomass k (X_k).

Table 4.7 defines kinetic parameters, and Table 4.8 defines conversion factors and stoichiometric parameters for each of the modeled bacteria. A parameter of

Table 4.2 Process, kinetic, and stoichiometric matrix for denitrifying heterotrophic bacteria (X_H).

j	Name	S_{VFA}	S_{NHx}	S_{N_2}	S_{NO_3}	S_{H_2S}	S_{HSe}	X_H	X_B	X_I	Rate
1	Anaerobic growth of X_H on S_{VFA} ($S_{NO_3} \rightarrow S_{N_2}$)	$-\frac{1}{Y_H}$	$-i_{N,BM}$	$\frac{1-Y_H}{Y_H \cdot \frac{40}{14}}$	$-\frac{1-Y_H}{Y_H \cdot \frac{40}{14}}$	$-i_{S,BM}$	$-i_{Se,BM}$	1			R1
2	Anaerobic endogenous decay of X_H		$i_{N,2}$			$i_{S,2}$	$i_{Se,2}$	-1	$1-f_{I,H}$	$f_{I,H}$	R2
	***Composition*:**										
	COD	1	0	$-\frac{24}{14}$	$-\frac{64}{14}$	0	0	1	1	1	
	N	0	1	1	1	0	0	$i_{N,BM}$	$i_{N,XB}$	$i_{N,I}$	
	S	0	0	0	0	1	0	$i_{S,BM}$	$i_{S,XB}$	$i_{S,I}$	
	Se	0	0	0	0	0	1	$i_{Se,BM}$	$i_{Se,XB}$	$i_{Se,I}$	

$i_{N,2} = i_{N,BM} - i_{N,I} \cdot f_{I,H} - i_{N,XB} \cdot (1 - f_{I,H})$
$i_{S,2} = i_{S,BM} - i_{S,I} \cdot f_{I,H} - i_{S,XB} \cdot (1 - f_{I,H})$
$i_{Se,2} = i_{Se,BM} - i_{Se,I} \cdot f_{I,H} - i_{Se,XB} \cdot (1 - f_{I,H})$

Table 4.3 Process, kinetic, and stoichiometric matrix for selenate-reducing bacteria (X_{SeO_4}).

j	Name	S_{VFA}	S_{NHx}	S_{H_2S}	S_{HSe}	S_{SeO_3}	S_{SeO_4}	X_{SeO_4}	X_B	X_I	Rate
3	Anaerobic growth of X_{SeO_4} on S_{VFA} ($S_{SeO_4} \rightarrow S_{SeO_3}$)	$-\frac{1}{Y_{SeO_4}}$	$-i_{N,BM}$	$-i_{S,BM}$	$-i_{Se,BM}$	$\frac{1-Y_{SeO_4}}{Y_{SeO_4}\cdot\frac{16}{79}}$	$-\frac{1-Y_{SeO_4}}{Y_{SeO_4}\cdot\frac{16}{79}}$	1			R3
4	Anaerobic endogenous decay of X_{SeO_4}		$i_{N,4}$	$i_{S,4}$	$i_{Se,4}$			-1	$1-f_{I,SeO_4}$	f_{I,SeO_4}	R4
	Composition:										
	COD	1	0	0	0	$-\frac{48}{79}$	$-\frac{64}{79}$	1	1	1	
	N	0	1	0	0	0	0	$i_{N,BM}$	$i_{N,XB}$	$i_{N,I}$	
	S	0	0	1	0	0	0	$i_{S,BM}$	$i_{S,XB}$	$i_{S,I}$	
	Se	0	0	0	1	1	1	$i_{Se,BM}$	$i_{Se,XB}$	$i_{Se,I}$	

$i_{N,4} = i_{N,BM} - i_{N,I}\cdot f_{I,SeO_4} - i_{N,XB}\cdot(1 - f_{I,SeO_4})$

$i_{S,4} = i_{S,BM} - i_{S,I}\cdot f_{I,SeO_4} - i_{S,XB}\cdot(1 - f_{I,SeO_4})$

$i_{Se,4} = i_{Se,BM} - i_{HSe,I}\cdot f_{I,SeO_4} - i_{Se,XB}\cdot(1 - f_{I,SeO_4})$

Table 4.4 Process, kinetic, and stoichiometric matrix for selenite-reducing bacteria (X_{SeO_3}).

j	Name	S_{VFA}	S_{NHx}	S_{H_2S}	S_{HSe}	X_{Se}	S_{SeO_3}	X_{SeO_3}	X_B	X_I	Rate
5	Anaerobic growth of X_{SeO_3} on S_{VFA} ($S_{SeO_3} \rightarrow X_{Se}$)	$-\frac{1}{Y_{SeO_3}}$	$-i_{N,BM}$	$-i_{S,BM}$	$-i_{Se,BM}$	$\frac{1-Y_{SeO_3}}{Y_{SeO_3} \cdot \frac{32}{79}}$	$-\frac{1-Y_{SeO_3}}{Y_{SeO_4} \cdot \frac{32}{79}}$	1			R5
6	Anaerobic endogenous decay of X_{SeO_3}		$i_{N,6}$	$i_{S,6}$	$i_{Se,6}$			−1	$1 - f_{I,SeO_3}$	f_{I,SeO_3}	R6
	Composition:										
	COD	1	0	0	0	$-\frac{16}{79}$	$-\frac{48}{79}$	1	1	1	
	N	0	1	0	0	0	0	$i_{N,BM}$	$i_{N,XB}$	$i_{N,I}$	
	S	0	0	1	0	0	0	$i_{S,BM}$	$i_{S,XB}$	$i_{S,I}$	
	Se	0	0	0	1	1	1	$i_{Se,BM}$	$i_{Se,XB}$	$i_{Se,I}$	

$i_{N,6} = i_{N,BM} - i_{N,I} \cdot f_{I,SeO_3} - i_{N,XB} \cdot (1 - f_{I,SeO_3})$
$i_{S,6} = i_{S,BM} - i_{S,I} \cdot f_{I,SeO_3} - i_{S,XB} \cdot (1 - f_{I,SeO_3})$
$i_{Se,6} = i_{Se,BM} - i_{Se,I} \cdot f_{I,SeO_3} - i_{Se,XB} \cdot (1 - f_{I,SeO_3})$

Table 4.5 Process, kinetic, and stoichiometric matrix for sulfate-reducing bacteria (X_{SRB}).

j	Name	S_{VFA}	S_{NHx}	S_{H_2S}	S_{SO_4}	S_{HSe}	X_{SRB}	X_B	X_I	Rate
7	Anaerobic growth of X_{SRB} on S_{VFA} ($S_{SO_4} \rightarrow S_{H_2S}$)	$-\frac{1}{Y_{SRB}}$	$-i_{N,BM}$	$\frac{1-Y_{SRB}}{Y_{SRB}\cdot\frac{64}{32}} - i_{S,BM}$	$-\frac{1-Y_{SRB}}{Y_{SRB}\cdot\frac{64}{32}}$	$-i_{Se,BM}$	1			R7
8	Anaerobic endogenous decay of X_{SRB}		$i_{N,8}$	$i_{S,8}$		$i_{Se,8}$	−1	$1-f_{I,SRB}$	$f_{I,SRB}$	R8
	Composition:									
	COD	1	0	0	$-\frac{64}{32}$	0	1	1	1	
	N	0	1	0	1	0	$i_{N,BM}$	$i_{N,XB}$	$i_{N,I}$	
	S	0	0	1	0	0	$i_{S,BM}$	$i_{S,XB}$	$i_{S,I}$	
	Se	0	0	0	0	1	$i_{Se,BM}$	$i_{Se,XB}$	$i_{Se,I}$	

$i_{N,8} = i_{N,BM} - i_{N,I}\cdot f_{I,SRB} - i_{N,XB}\cdot(1-f_{I,SRB})$
$i_{S,8} = i_{S,BM} - i_{S,I}\cdot f_{I,SRB} - i_{S,XB}\cdot(1-f_{I,SRB})$
$i_{Se,8} = i_{Se,BM} - i_{Se,I}\cdot f_{I,SRB} - i_{Se,XB}\cdot(1-f_{I,SRB})$

Table 4.6 Process rate equations for denitrifying heterotrophic bacteria (X_H), selenate-reducing bacteria (X_{SeO_4}), selenite-reducing bacteria (X_{SeO_3}), and sulfate-reducing bacteria (X_{SRB}) growth and endogenous decay.

	Name	Process Rate Equation (g/m³ · d)
R1	Anaerobic growth of X_H on S_{VFA} ($S_{NO_3} \rightarrow S_{N_2}$)	$\mu_{max,H} \cdot \frac{S_{VFA}}{K_{VFA,H} + S_{VFA}} \cdot \frac{S_{NO_3}}{K_{NO_3,H} + S_{NO_3}} \cdot \frac{S_{NHx}}{K_{NHx,H} + S_{NHx}} \cdot \frac{S_{H_2S}}{K_{H_2S,H} + S_{H_2S}} \cdot \frac{S_{HSe}}{K_{HSe,H} + S_{Hse}} \cdot X_H$
R2	Anaerobic endogenous decay of X_H	$b_H \cdot X_H$
R3	Anaerobic growth of X_{SeO_4} on S_{VFA} ($S_{SeO_4} \rightarrow S_{SeO_3}$)	$\mu_{max,SeO_4} \cdot \frac{K_{NO_3,SeO_4}}{K_{NO_3,SeO_4} + S_{NO_3}} \cdot \frac{S_{VFA}}{K_{VFA,SeO_4} + S_{VFA}} \cdot \frac{S_{SeO_4}}{K_{SeO_4,SeO_4} + S_{SeO_4}} \cdot \frac{S_{NHx}}{K_{NHx,SeO_4} + S_{NHx}} \cdot \frac{S_{H_2S}}{K_{H_2S,SeO_4} + S_{H_2S}} \cdot \frac{S_{HSe}}{K_{HSe,SeO_4} + S_{Hse}} \cdot X_{SeO_4}$
R4	Anaerobic endogenous decay of X_{SeO_4}	$b_{SeO_4} \cdot X_{SeO_4}$
R5	Anaerobic growth of X_{SeO_3} on S_{VFA} ($S_{SeO_3} \rightarrow X_{Se}$)	$\mu_{max,SeO_3} \cdot \frac{K_{NO_3,SeO_3}}{K_{NO_3,SeO_3} + S_{NO_3}} \cdot \frac{K_{SeO_4,SeO_3}}{K_{SeO_4,SeO_3} + S_{SeO_4}} \cdot \frac{S_{VFA}}{K_{VFA,SeO_3} + S_{VFA}} \cdot \frac{S_{SeO_3}}{K_{SeO_3,SeO_3} + S_{SeO_3}} \cdot \frac{S_{NHx}}{K_{NHx,SeO_3} + S_{NHx}} \cdot \frac{S_{H_2S}}{K_{H_2S,SeO_3} + S_{H_2S}} \cdot \frac{S_{HSe}}{K_{HSe,SeO_3} + S_{Hse}} X_{SeO_3}$
R6	Anaerobic endogenous decay of X_{SeO_3}	$b_{SeO_3} \cdot X_{SeO_3}$
R7	Anaerobic growth of X_{SRB} on S_{VFA} ($S_{SO_4} \rightarrow S_{H_2S}$)	$\mu_{max,SRB} \cdot \frac{K_{NO_3,SRB}}{K_{NO_3,SRB} + S_{NO_3}} \cdot \frac{K_{SeO_4,SRB}}{K_{SeO_4,SRB} + S_{SeO_4}} \cdot \frac{K_{SeO_3,SRB}}{K_{SeO_3,SRB} + S_{SeO_3}} \cdot \frac{S_{VFA}}{K_{VFA,SRB} + S_{VFA}} \cdot \frac{S_{SO_4}}{K_{SO_4,SRB} + S_{SO_4}} \cdot \frac{S_{NHx}}{K_{NHx,SRB} + S_{NHx}} \cdot \frac{S_{H_2S}}{K_{H_2S,SRB} + S_{H_2S}} \cdot \frac{S_{HSe}}{K_{HSe,SRB} + S_{Hse}} \cdot X_{SRB}$
R8	Anaerobic endogenous decay of X_{SRB}	$b_{SRB} \cdot X_{SRB}$

Table 4.7 Kinetic parameters for denitrifying heterotrophic bacteria (X_H), selenate-reducing bacteria (X_{SeO_4}), selenite-reducing bacteria (X_{SeO_3}), and sulfate-reducing bacteria (X_{SRB}) growth and endogenous decay.

Symbol	Description	Value	Unit
Denitrifying heterotrophic bacteria (X_H):			
$\mu_{max,H}$	Maximum growth rate of X_H on S_{VFA}	10.30	1/d
b_H	Anaerobic endogenous decay rate, X_H	0.40	1/d
$K_{VFA,H}$	Half-saturation concentration, S_{VFA}	4.00	g COD_S/m^3
$K_{NO_3,H}$	Half-saturation concentration, S_{NO_3}	0.50	g N/m^3
$K_{NHx,H}$	Half-saturation concentration, S_{NHx}	0.05	g N/m^3
$K_{H_2S,H}$	Half-saturation concentration, S_{H_2S}	0.01	g S/m^3
$K_{HSe,H}$	Half-saturation concentration, S_{HSe}	0.001	g Se/m^3
Selenate-reducing bacteria (X_{SeO_4}):			
μ_{max,SeO_4}	Maximum growth rate of X_{SeO_4} on S_{VFA}	7.40	1/d
b_{SeO_4}	Anaerobic endogenous decay rate, X_{SeO_4}	0.29	1/d
K_{VFA,SeO_4}	Half-saturation concentration, S_{VFA}	4.00	g COD_S/m^3
K_{NO_3,SeO_4}	Inhibition concentration, S_{NO_3}	0.20	g N/m^3
K_{NHx,SeO_4}	Half-saturation concentration, S_{NHx}	0.05	g N/m^3
K_{H_2S,SeO_4}	Half-saturation concentration, S_{H_2S}	0.01	g S/m^3
K_{SeO_4,SeO_4}	Half-saturation concentration, S_{SeO_4}	0.20 *	g Se/m^3
K_{HSe,SeO_4}	Half-saturation concentration, S_{HSe}	0.001	g Se/m^3
Selenite-reducing bacteria (X_{SeO_3}):			
μ_{max,SeO_3}	Maximum growth rate of X_{SeO_3} on S_{VFA}	5.50	1/d
b_{SeO_3}	Anaerobic endogenous decay rate, X_{SeO_3}	0.21	1/d
K_{VFA,SeO_3}	Half-saturation concentration, S_{VFA}	4.00	g COD_S/m^3
K_{NO3,SeO_3}	Inhibition concentration, S_{NO_3}	0.20	g N/m^3
K_{NHx,SeO_3}	Half-saturation concentration, S_{NHx}	0.05	g N/m^3
K_{H_2S,SeO_3}	Half-saturation concentration, S_{H_2S}	0.01	g S/m^3
K_{SeO_3,SeO_3}	Half-saturation concentration, S_{SeO_3}	0.20 *	g Se/m^3
K_{HSe,SeO_3}	Half-saturation concentration, S_{HSe}	0.001	g Se/m^3
K_{SeO_4,SeO_3}	Inhibition concentration, S_{SeO_4}	0.02	g Se/m^3
Sulfate-reducing bacteria (X_{SRB}):			
$\mu_{max,SRB}$	Maximum growth rate of X_{SRB} on S_{VFA}	0.80	1/d
b_{SRB}	Anaerobic endogenous decay rate, X_{SRB}	0.03	1/d
$K_{VFA,SRB}$	Half-saturation concentration, S_{VFA}	4.00	g COD_S/m^3
$K_{NO_3,SRB}$	Inhibition concentration, S_{NO_3}	0.20	g N/m^3

(*Continued*)

Table 4.7 Kinetic parameters for denitrifying heterotrophic bacteria (X_H), selenate-reducing bacteria (X_{SeO_4}), selenite-reducing bacteria (X_{SeO_3}), and sulfate-reducing bacteria (X_{SRB}) growth and endogenous decay (*Continued*).

Symbol	Description	Value	Unit
$K_{NHx,SRB}$	Half-saturation concentration, S_{NHx}	0.05	g N/m^3
$K_{SO_4,SRB}$	Half-saturation concentration, S_{SO_4}	0.20	g S/m^3
$K_{H_2S,SRB}$	Half-saturation concentration, S_{H_2S}	0.01	g S/m^3
$K_{HSe,SRB}$	Half-saturation concentration, S_{HSe}	0.001	g Se/m^3
$K_{SeO_4,SRB}$	Inhibition concentration, S_{SeO_4}	0.02	g Se/m^3
$K_{SeO_3,SRB}$	Inhibition concentration, S_{SeO_3}	0.02	g Se/m^3

* If $S_{SeOx} > 1$, then $K_{SeOx,SeOx} = 0.04$ ($S_{SeOx,INF}$); $S_{SeOx} = S_{SeO_4} + S_{SeO_3}$

particular interest is the mass of selenium consumed for biomass synthesis ($i_{Se,BM}$ = 0.0002 g Se/g COD$_X$). Selenite-reducing bacteria have an intra-cellular selenium content that depends on their activity: it approaches 0.04 g Se/g COD$_X$ when they are respiring selenium (Tan, 2018). Mörschbächer *et al.* (2018) reported that the respiration-associated content of selenium in selenite-reducing bacteria is 0.01 to 0.02 g Se/g COD$_X$, but also demonstrated that selenite-reducing bacteria have an intra-cellular selenium content of 0.0001 to 0.0002 g Se/g COD$_X$ prior to respiring selenite and accumulating elemental selenium. De Souza *et al.* (2001) claimed 6% Se for selenate reduction by a bacterial consortium in an aerobic, hypersaline pond. Selenium-oxyanion reduction predominantly takes place under anaerobic conditions; therefore, non-selenium respiring bacteria reduced selenate to acquire selenium for cell synthesis. The researchers also reported that the cellular selenium content resulting from assimilation was 0.0001 to 0.0003 g Se/g COD$_X$. The amount of selenium is less than the amount of nitrogen required for assimilation ($i_{N,BM}$ = 0.07 g N/g COD$_X$). Comparing the mass of nitrogen and selenium consumed for biomass synthesis, bacteria require approximately 350 times more nitrogen than selenium.

Because the mass of selenium that needs to be removed from wastewater is typically significantly less than the mass of nitrogen (recall the first paragraph), it is reasonable to consider whether or not significant selenium removal can be achieved by selenium oxyanion reduction for selenium assimilation into denitrifying heterotrophic bacteria. We consider denitrifying heterotrophic bacteria (Y_H = 0.65 g COD$_x$/g COD$_{ED}$) in a bioreactor treating wastewater containing 50 g N/m^3 as nitrate. Since denitrification requires 2.86 g COD$_{ED}$/g N, [50 g N/m^3 · 2.86 g COD$_{ED}$/g N · 0.65 g COD$_X$/g COD$_{ED}$ · 0.0002 g Se/g COD$_X$] = 0.019 g Se/m^3 is assimilated by synthesis of denitrifying heterotrophic bacteria when all of the influent nitrate is respired. If the influent selenium

Table 4.8 Conversion factors and stoichiometric parameters.

Symbol	Description	Value	Unit
Nitrogen:			
$i_{N,BM}$	Nitrogen content of biomass: X_H, X_{SeO_4}, X_{SeO_3}, X_{SRB}	0.07	g N/g COD_X
$i_{N,XB}$	Nitrogen content of slowly biodegradable COD, X_B	0.04	g N/g COD_X
$i_{N,I}$	Nitrogen content of particulate inert COD, X_I	0.03	g N/g COD_X
Selenium:			
$i_{Se,BM}$	Selenium content of biomass: X_H, X_{SeO_4}, X_{SeO_3}, X_{SRB}	0.0002	g Se/g COD_X
$i_{Se,XB}$	Selenium content of particulate COD, X_B	0.0000	g Se/g COD_X
$i_{Se,I}$	Selenium content of particulate inert COD, X_I	0.0000	g Se/g COD_X
Sulfur:			
$i_{S,BM}$	Sulfur content of biomass: X_H, X_{SeO_4}, X_{SeO_3}, X_{SRB}	0.006	g S/g COD_X
$i_{S,XB}$	Sulfur content of slowly biodegradable COD, X_B	0.000	g S/g COD_X
$i_{S,I}$	Sulfur content of particulate inert COD, X_I	0.000	g S/g COD_X
Stoichiometric:			
$f_{I,H}$	Fraction of particulate inert COD by X_H decay	0.10	g COD_X /g COD_X
f_{I,SeO_4}	Fraction of particulate inert COD by X_{SeO_4} decay	0.10	g COD_X /g COD_X
f_{I,SeO_3}	Fraction of particulate inert COD by X_{SeO_3} decay	0.10	g COD_X /g COD_X
$f_{I,SRB}$	Fraction of particulate inert COD by X_{SRB} decay	0.10	g COD_X /g COD_X

concentration is 1 g Se/m^3, then a maximum 1.9% of the selenium removal can be achieved by assimilation into denitrifying heterotrophic bacteria. Hence, sufficient removal of selenium oxyanions by assimilation for synthesis of denitrifying heterotrophs can be feasible only when the N:Se ratio is very much greater than 50 g N:1 g Se. Thus, we conclude that a substantial removal of selenium oxyanions can occur only through their respiration.

4.4 MINIMUM SRT AND DONOR-SUBSTRATE CONCENTRATION

A reliable diagnostic parameter is the minimum solids residence time ($SRT_{\text{lim}}^{\text{min}}$) required to accumulate steady-state biomass. It can be calculated by:

$$SRT_{\text{lim}}^{\text{min},k} = \frac{1}{\mu_{\text{max},k} - b_k} \tag{4.15}$$

The minimum solids residence time (*SRT*) tells us how long the actual *SRT* must be to avoid the washout of a biomass *k*. Reliable performance requires an actual *SRT* that is greater than the minimum value calculated using Equation (4.15). A comparison of the actual *SRT* to the minimum *SRT* applies equally to suspended-growth and biofilm processes, although the method to compute the actual *SRT* differs (Rittmann & McCarty, 2020). As *SRT* increases well above the minimum *SRT*, the concentration of S_i approaches a limiting value that represents the minimum concentration of substrate *i* that is capable of supporting steady-state biomass ($S_{\text{min},i}$). $S_{\text{min},i}$ can be calculated by:

$$S_{\text{min},i} = K_i \frac{b_k}{\mu_{\text{max},k} - b_k} \tag{4.16}$$

Here, K_i is the half-saturation concentration of substrate *i* (g/m^3), which can be the electron donor or acceptor, depending on which one is rate-limiting.

Applying Equations (4.15) and (4.16), along with the appropriate $\mu_{\text{max},k}$, b_k, and K_i parameter values listed in Table 4.7, makes it possible to compute $SRT_{\text{lim}}^{\text{min},k}$ and $S_{\text{min},i}$ for the electron acceptor and donor used by each bacterial group considered. These values are listed in Table 4.9 and provide two very important insights into anaerobic bioreactors exposed to wastewater containing nitrate, selenate/selenite, and sulfate. First, all of the $S_{\text{min},i}$ values are much less than 1 g/m^3, and the values for selenate and selenite are less than 0.0001 g/m^3. This means that a biological process with a long-enough SRT can drive the concentrations of all oxyanions, as well as acetate, to very low concentrations. Second, the $SRT_{\text{lim}}^{\text{min}}$ values span a large range. For example, the $SRT_{\text{lim}}^{\text{min}}$ for sulfate-reducing bacteria is nearly 14-fold greater than that for denitrifying heterotrophic bacteria. Likewise, the bacteria respiring selenium oxyanions have $SRT_{\text{lim}}^{\text{min}}$ values that are

Table 4.9 Comparison of estimated $SRT_{\lim,i}^{\min,k}$ and $S_{\min,i}$ values for the four types of respiration.

Variable	Symbol	$SRT_{\lim,i}^{\min,k}$ (days)	$S_{\min,i}$ (g/m^3)
1. Minimum *SRT*, Denitrifying heterotrophic bacteria	X_H	0.10	
2. Concentration at a long *SRT*			
a. Nitrate	S_{NO_3}		0.02
b. Acetate	S_{VFA}		0.16
3. Minimum *SRT*, Selenate-reducing bacteria	X_{SeO_4}	0.14	
4. Concentration at a long *SRT*			
a. Selenate	S_{SeO_4}		0.00004
b. Acetate	S_{VFA}		0.16
5. Minimum *SRT*, Selenite-reducing bacteria	X_{SeO_4}	0.19	
6. Concentration at a long *SRT*			
a. Selenite	S_{SeO_3}		0.00004
b. Acetate	S_{VFA}		0.16
7. Minimum *SRT*, Sulfate-reducing bacteria	X_{SRB}	1.38	
8. Concentration at a long *SRT*			
a. Sulfate	S_{SO_4}		0.01
b. Acetate	S_{VFA}		0.16

7 to 10-fold less than that for sulfate-reducing bacteria. These differences mean that – at least in principle – sulfate-reduction can be suppressed or minimized by selecting an operating *SRT* that supports the accumulation of denitrifying heterotrophic bacteria and SeRB, but prevents the accumulation of sulfate-reducing bacteria.

4.5 SIMULATION OF SeRB POPULATION DYNAMICS

We carried out simulations using a commercially available wastewater treatment plant simulator (SUMO®, Dynamita, France). In both cases, a continuous flow stirred tank reactor (CFSTR) was modeled which has a hydraulic retention time (*HRT*) equal to the *SRT*.

4.5.1 Model comparison with observed selenium oxyanion reduction

A focused evaluation of our SeRB model was carried out by simulating the experimental system reported by Fujita *et al.* (2002) and then comparing the model output with data reported by the researchers. They operated a laboratory-scale CFSTR ($V_R = 0.0005\ m^3$) inoculated with SeRB. The CFSTR was fed synthetic wastewater having concentrations of 42 g Se/m^3 as selenate and 2200 g COD/m^3 as lactate. The influent flow rate was adjusted to achieve SRT values of 2.9, 6.0, 8.9, 12.0, 23.8, 48.0, and 95.2 hours. A biomass concentration of 28.4 g COD_X/m^3 in the CFSTR was reported by Fujita *et al.* (2002).

Since lactate was the electron donor instead of acetate, the energy equation should consider lactate ($CH_3CHOHCOO^-$) as the common electron donor for the electron acceptors selenate and selenite. Equation (4.17) is the half reaction for lactate reduction and its free energy (Rittmann & McCarty, 2020).

$$\frac{1}{12}CH_3CHOHCOO^- + \frac{1}{3}H_2O \rightarrow \frac{1}{6}CO_2 + \frac{1}{12}HCO_3^- + H^+ + e^-$$

$$\Delta G_{ED}^{0'} = +32.3\frac{kJ}{e^- eq_{ED}} \tag{4.17}$$

Combining Equations (4.2) and (4.17) and combining Equations (4.3) and (4.17) result in the energy reactions for selenate and selenite reduction, respectively, when lactate is the common electron donor (Equations (4.18) and (4.19)). Also shown are the free energy values for each energy reaction ($\Delta G^{0'}$) in kJ/e$^-$eq.

$$\frac{1}{12}CH_3CHOHCOO^- + \frac{1}{2}SeO_4^{2-} + \frac{1}{2}H^+ \rightarrow \frac{1}{12}CO_2 + \frac{1}{2}HSeO_3^- + \frac{1}{12}HCO_3^- + \frac{5}{24}H_2O \quad \Delta G_{ED}^{0'} = -76.2\frac{kJ}{e^- eq_{ED}} \tag{4.18}$$

$$\frac{1}{12}CH_3CHOHCOO^- + \frac{1}{4}HSeO_3^- + \frac{1}{4}H^+ \rightarrow \frac{1}{12}CO_2 + \frac{1}{4}Se^0 + \frac{1}{12}HCO_3^- + \frac{5}{25}H_2O \quad \Delta G_{ED}^{0'} = -57.7\frac{kJ}{e^- eq_{ED}} \tag{4.19}$$

Applying Equations (4.10) through (4.14), the following parameter values were calculated for:

(1) Selenate-reducing bacteria:
   - $Y_{SeO_4,\text{lactate}} = 0.58$ g COD_X/g COD_{ED}
   - $\hat{q}_{SeO_4,\text{lactate}} = 13.5$ g COD_{ED}/(g $COD_X \cdot$ d)
   - $\mu_{SeO_4,\text{lactate}} = 7.9$ 1/d

(2) Selenite-reducing bacteria:
   - $Y_{SeO_3,\text{lactate}} = 0.51$ g COD_X/g COD_{ED}
   - $\hat{q}_{SeO_3,\text{lactate}} = 11.6$ g COD_{ED}/(g $COD_X \cdot$ d)
   - $\mu_{SeO_3,\text{lactate}} = 6.0$ 1/d

Based on these parameter values, Figure 4.3 presents the selenate, selenite, and elemental selenium concentrations observed by Fujita *et al.* (2002) and those

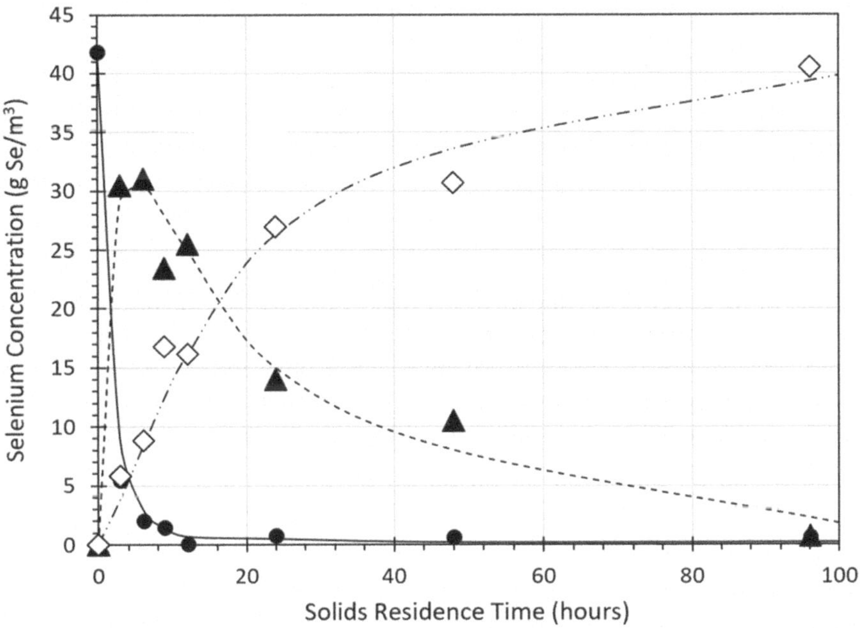

Figure 4.3 Observed selenate (●), selenite (▲), and elemental selenium (◇) concentrations in the effluent of a bench-scale continuous-flow stirred-tank reactor (CFSTR) versus solids residence time. Data reported by Fujita *et al.* (2002). Observations compared to simulated selenate (—), selenite (----), and elemental selenium (–··–) concentrations.

resulting from the simulation of their experimental system using the SeRB model described in this chapter (i.e., neglecting denitrifying heterotrophic and sulfate-reducing bacteria). The experimental data and model results agree well in terms of all trends and absolute concentrations, and the coefficients of determination (R^2) for selenate, selenite, and elemental selenium are 0.99, 0.97, and 0.97, respectively. The observed rate of selenium-oxyanion ($S_{SeO_4} + S_{SeO_3}$) reduction was 0.37 g Se/g $COD_X \cdot d$, which is similar to the simulated selenium-oxyanion reduction rate of 0.35 g Se/g $COD_X \cdot d$.

4.5.2 Ecology of denitrifying heterotrophic bacteria, selenium-respiring bacteria, and sulfate-reducing bacteria

To evaluate the microbial ecology of a reactor treating wastewater containing typical concentrations of nitrate, selenate, and sulfate, we simulated a 1 m^3 CFSTR. The simulated influent concentration of nitrate was 50 g N/m^3, selenate was 1 g Se/m^3, and sulfate was 1000 g S/m^3. The simulated wastewater flow rate was adjusted to give SRTs that were incrementally increased from 0 to 8.5 days. Each simulated wastewater flow rate equals the CFSTR volume (1 m^3) multiplied by the *SRT*. Initially, we used influent electron donor, macro- and micro-nutrient concentrations that were non-rate limiting: acetate was 3000 g COD/m^3, ammonium was 100 g N/m^3, hydrogen sulfide was 1 g S/m^3, and hydrogen selenide was 0.1 g Se/m^3. Subsequently we lowered the influent electron donor concentration to 500 g COD/m^3, while the other essential nutrient concentrations remained the same.

Figure 4.4 illustrates acetate (as COD), nitrate (as N), selenate (as Se), selenite (as Se), elemental selenium (as Se), sulfate (as S), and hydrogen sulfide (as S) as a function of SRT. The first thing to notice is that the effluent contains a significant concentration of acetate (at least 500 g COD/m^3) because the simulations were set up with an acetate-input concentration greater than what is needed for the stoichiometric reduction of all oxyanions. This approach emphasizes the relative specific growth rates impacts, but not competition for the electron donor (i.e., acetate).

Figure 4.4 shows significant nitrate reduction to 0.20 g N/m^3, after a 0.4-day *SRT*, which is greater than its $SRT_{\text{lim}}^{\text{min},H}$ of 0.10 days. The simulated nitrate concentration for *SRTs* greater than 0.4 days approaches the S_{min,NO_3} value 0.02 g N/m^3. Similarly, Figure 4.4 illustrates significant selenate reduction to 0.0075 g Se/m^3 after a 1.5-day *SRT*, which is greater than its $SRT_{\text{lim}}^{\text{min},SeO_4}$ of 0.14 days. When the *SRT* is greater than 1.5 days, the effluent selenate concentration approaches its S_{min,SeO_4} value of 0.00004 g Se/m^3. For selenite, Figure 4.4 illustrates its significant reduction to 0.0093 g Se/m^3 after a 4.0-day *SRT*, which is greater than its $SRT_{\text{lim}}^{\text{min},SeO_3}$ of 0.19 days. For *SRTs* greater than 4.0 days, the selenite concentration approaches its S_{min,SeO_3} value of 0.00004 g Se/m^3. Finally, Figure 4.4 illustrates that sulfate reduction begins when the *SRT* is around 2.0 days, which is greater than its $SRT_{\text{lim}}^{\text{min},SRB}$ of 1.38 days. Significant sulfate reduction, to 0.28 g S/m^3, occurs at an 8.5-day *SRT*. These results illustrate the generalizable trend that the actual *SRT* must be roughly three-fold greater than $SRT_{\text{lim}}^{\text{min},k}$ to drive the concentration of substrate *i* to low levels.

Reducing the influent acetate concentration to a value that is substantially less than what is required to stoichiometrically convert all of the oxyanions to their most reduced state has a profound impact on simulation results. Figure 4.5 shows the effects of lessening the influent acetate concentration to 500 g COD/m^3, which provides only 17% of the stoichiometric demand to reduce all of the oxyanions. While denitrification, selenate reduction, and selenite reduction are hardly affected, sulfate reduction is almost entirely suppressed. Because sulfate-reducing bacteria are the slowest growing bacteria, they can be out-competed for acetate if its supply cannot meet the demand of all oxyanions. In this example, sulfate has the greatest electron-donor demand, which means that only 17% of the electron donor needed for all oxyanions was sufficient for the reduction of nitrate, selenate, and selenite.

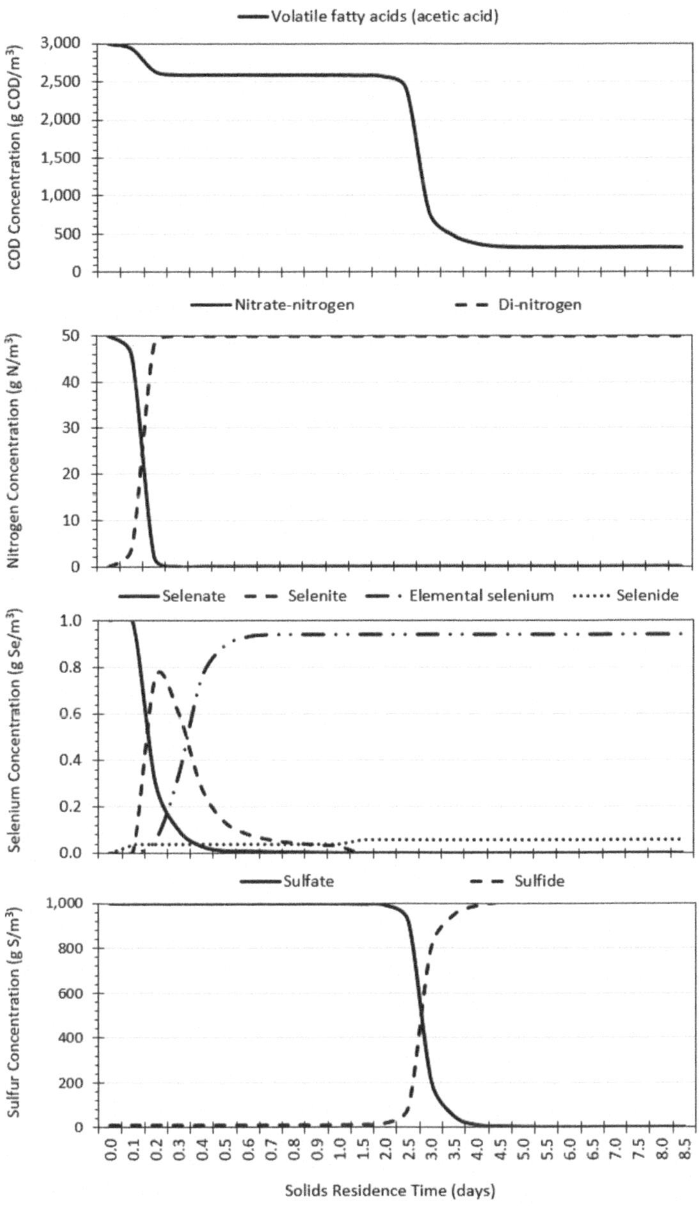

Figure 4.4 Impact of *SRT* on acetate, nitrate, selenate, selenite, elemental selenium, hydrogen selenide, sulfate, and hydrogen sulfide concentrations in a continuous-flow stirred-tank reactor (CFSTR) treating wastewater containing concentrations of acetate = 3000 g COD/m^3, nitrate = 50 g N/m^3, selenate = 1 g Se/m^3, and sulfate = 1000 g S/m^3.

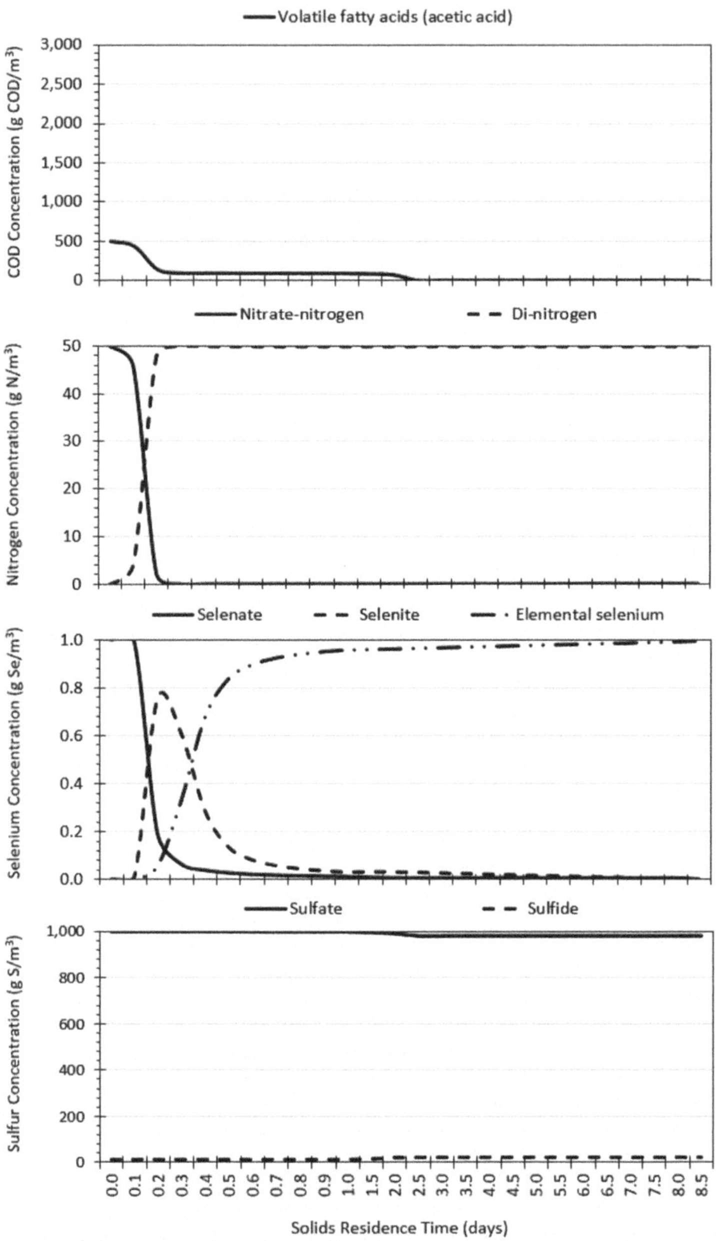

Figure 4.5 Impact of *SRT* on acetate, nitrate, selenate, selenite, elemental selenium, sulfate, and hydrogen sulfide concentrations in a continuous-flow stirred-tank reactor (CFSTR) treating wastewater containing concentrations of acetate = 500 g COD/m^3, nitrate = 50 g N/m^3, selenate = 1 g Se/m^3, and sulfate = 1000 g S/m^3.

4.6 KEY POINTS

- The thermodynamics of the respiration reactions, manifested with values of Y_k and $\mu_{\mathrm{max},k}$ give major growth advantages in the order: denitrification > selenate reduction > selenite reduction > sulfate reduction.
- As sulfate reduction normally is undesired, it can be suppressed by maintaining a sufficiently short *SRT*, controlling the electron-donor supply, or a combination thereof.
- In contrast, using a too long *SRT* with a too great input of electron donor will allow sulfate reduction.
- Using a too low *SRT* also can suppress selenite reduction, even if selenate reduction occurs.

REFERENCES

Boltz J. P. and Rittmann B. E. (2019). A model of Selenium, Sulfur, and Nitrogen Species (SeSANS) conversion in a biofilm reactor. EPRI, Palo Alto, CA. 3002017266.

De Souza M. P., Amini A., Dojka M. A., Pickering I. J., Dawson S. C., Pace N. R. and Terry N. (2001). Identification and characterization of bacteria in a selenium-contaminated hypersaline evaporation pond. *Applied and Environmental Microbiology*, **67**(9), 3785–3794.

EPA (2020). Steam electric reconsideration rule. 40 CFR Part 23; 6560-50-P. RIN 2040-AF77.

Eswayah A. S., Smith T. J. and Gardiner P. H. E. (2016). Microbial transformations of selenium species of relevance to bioremediation. *Applied and Environmental Microbiology*, **82**(16), 4848–4859.

Fujita M., Ike M., Kashiwa M., Hashimoto R. and Soda S. (2002). Laboratory-scale continuous reactor for soluble selenium removal using selenate-reducing bacterium, *Bacillus* sp. SF-1. *Biotechnology and Bioengineering*, **80**(7), 755–761.

Grady C. P., Daigger G. T., Love N. G. and Filipe C. D. M. (2011). Biological wastewater treatment, 3rd edn, IWA Publishing, CRC Press, Boca Raton, FL.

Macy J. M., Lawson S. and Decker H. D. (1993). Bioremediation of selenium oxyanions in San Joaquin drainage water using *Thauera selenatis* in a biological reactor system. *Applied Microbiology and Biotechnology*, **40**, 588–594.

McCarty P. L. (2007). Thermodynamic electron equivalents model for bacterial yield prediction: modifications and comparative evaluations. *Biotechnology and Bioengineering*, **97**(2), 377–388.

Mörschbächer A. P., Dullius A., Dullius C. H., Bandt C. R., Kuhn D., Brietzke D. T., Kuffel F. J. M., Etgeton H. P., Altmayer T., Gonçalves T. E., Oreste E. Q., Ribeiro A. S., de Souza C. F. V. and Hoehne L. (2018). Validation of an analytical method for the quantitative determination of selenium in bacterial biomass by ultraviolet-visible spectrophotometry. *Food Chemistry*, **255**(30), 1820186.

Olin Å., Noläng B., Osadchii E. G., Öhman L.-O. and Rosén E. (2020). Chemical thermodynamics of selenium. OECD Nuclear Energy Agency, Data Bank.

PSI (Paul Scherrer Institute) (2007). The PSI/Nagra chemical thermodynamic database: data selection of selenium. TM-44-14-07.

Rittmann B. E. and McCarty P. L. (2020). *Environmental Biotechnology: Principles and Applications*, 2nd edn, McGraw-Hill, U.S.A.

Switzer Blum J., Burns Bindi A., Buzzelli J., Stolz J. F. and Oremland R. S. (1998). *Bacillus arsenicoselenatis*, sp. nov., and *Bacillus selenitireducens*, sp. nov.: two haloalkaliphiles from Mono Lake, California that respire oxyanions of selenium and arsenic. *Archives of Microbiology*, **171**, 19–30.

Tan L. C. (2018). Anaerobic treatment of mine wastewater for the removal of selenate and its co-contaminants. Doctoral dissertation. Université Paris-est and UNESCO-IHE.

Part II

Remediation of Selenium Contamination

Chapter 5

Reactivity and selectivity of zerovalent iron toward selenium oxyanions under aerobic conditions

Jinxiang Li, Yuankui Sun and Xiaohong Guan

5.1 AQUEOUS CHEMISTRY OF ZVI WITH SELENIUM

Zerovalent iron (ZVI, Fe^0) is readily available, inexpensive, reactive and environmentally friendly and thus has been widely used in sequestering various contaminants (Guan *et al.*, 2015; Johnson *et al.*, 1996; Li *et al.*, 2015a; Liang *et al.*, 2013, 2014a; Scherer *et al.*, 1998). It can undergo a series of corrosion reactions in the H_2O/O_2 system, including transformation of the released Fe^{2+}/Fe^{3+} (Figure 5.1), which are accompanied by the liberation of hydrogen and/or the reduction of oxygen (Guan *et al.*, 2015; Noubactep, 2008). Therefore, ZVI can act as not only a generator of adsorbing agents (i.e., iron (hydr)oxides) but also as an electron donor for the reduction of the selenium oxyanions, i.e. selenite (Se(IV)) and selenate (Se(VI)) (Fan *et al.*, 2019b; Noubactep, 2015).

Considering the redox-dependent solubility of selenium, as shown in Figure 5.1, the reduction of the soluble Se(IV) and Se(VI) to the insoluble Se(-II)/Se(0) by ZVI is a wise choice and of high environmental significance (Liang *et al.*, 2013; Shao *et al.*, 2018). Nevertheless, granular ZVI particles generally have low reactivity and utilization ratio toward Se(IV) and Se(VI) due to the resistance of the inherent or newly formed iron (hydr)oxides (Fan *et al.*, 2019a, b; Guo *et al.*, 2016; Li *et al.*, 2018, 2019, 2020a; Liang *et al.*, 2013, 2014a, b, 2015a, b; Qiao *et al.*, 2018; Tang *et al.*, 2014, 2016; Xu *et al.*, 2016; Zhang *et al.*, 2018, 2020),

doi: 10.2166/9781789061055_0173

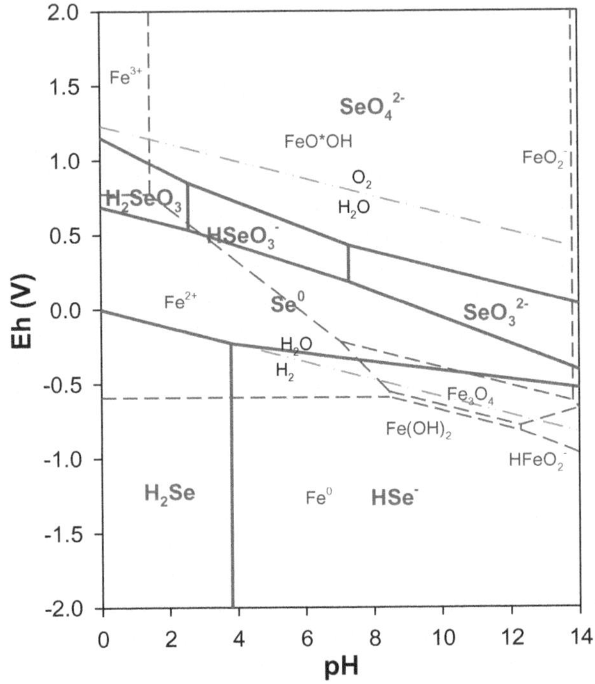

Figure 5.1 Eh-pH diagram for selenium in Fe^0/H_2O system at 298.15 K and 101.3 kPa (Qin *et al.*, 2016).

which can passivate ZVI and concurrently make the iron core (i.e., Fe^0) inaccessible. When applying ZVI to reduce Se(IV)/Se(VI) in the H_2O/O_2 system, the coexisting oxidizing compounds including H_2O, H^+, O_2, as well as other reducible species can compete with Se(IV)/Se(VI) for the electrons of ZVI, resulting in a low electron efficiency (EE) of ZVI. The parameter EE is defined as the proportion of electrons utilized by the reduction of the target contaminant to all available electrons provided by ZVI (Fan *et al.*, 2019a, b; Gu *et al.*, 2017; He *et al.*, 2020; Li *et al.*, 2017b, 2018, 2020a, b; Qiao *et al.*, 2018; Qin *et al.*, 2017).

In view of the limitations of ZVI-based technology, this chapter will systematically overview the performance of employing some promising strategies, such as imposing a weak magnetic field (WMF, see Section 5.2), dosing ferrous iron (see Section 5.3), and pre-treatment with sulfidation (see Section 5.4), to improve the reactivity and selectivity of ZVI toward Se(IV)/Se(VI). All the above-mentioned methods have gained great interest in recent years, while there are still some limitations in real practice. Thus, the corresponding suggestions for these three enhanced-ZVI technologies are further provided in Section 5.5.

5.2 WMF ENHANCES THE REACTIVITY AND SELECTIVITY OF ZVI TOWARD Se(IV) AND Se(VI)

5.2.1 Effect of WMF on the reactivity of ZVI toward Se(IV)/Se(VI)

The improvement of ZVI reactivity induced by a weak magnetic field (WMF) was first reported by Liang *et al.* (2014a). A WMF was found to significantly accelerate Se(IV) removal and extend the working pH range of ZVI from 4.0–6.0 (without WMF) to 4.0–7.2 (with WMF). Since the WMF can be applied by two pieces of permanent magnet, the WMF-assisted ZVI technology is environmentally friendly and thus promising for real applications such as groundwater/wastewater remediation/treatment (Guan *et al.*, 2015; Liang *et al.*, 2014b; Reinsch *et al.*, 2010; Sarathy *et al.*, 2008; Wang *et al.*, 2010).

The positive effect of a WMF on Se(IV)/Se(VI) removal by ZVI under various reaction conditions has been extensively reported (Li *et al.*, 2020a; Liang *et al.*, 2014b; Xu *et al.*, 2016; Zhang *et al.*, 2018), as shown in Figure 5.2. Besides pristine ZVI (Pri-ZVI), the performance of aged ZVI (AZVI) with WMF has also been explored (Guan *et al.*, 2015; Liang *et al.*, 2014b; Reinsch *et al.*, 2010; Sarathy *et al.*, 2008; Wang *et al.*, 2010). A WMF could significantly enhance the reactivity of all AZVI samples prepared under different background matrices and aging durations. Moreover, the WMF enhancing effect could enable a high reactivity of ZVI collected from different origins toward both Se(IV) and Se(VI) removal to various extents.

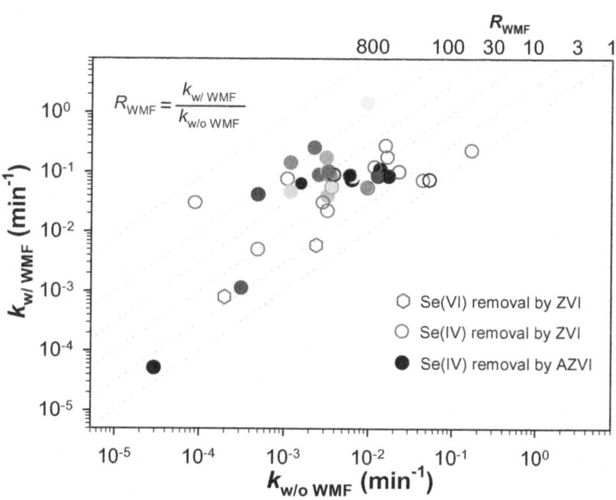

Figure 5.2 Rate constants for sequestration of Se(IV) and Se(VI) with and without WMF by ZVI and AZVI (Zhang *et al.*, 2018).

In order to better illustrate the enhancing effect of WMF on the reactivity of the ZVI/AZVI, the wide spectrum of rate constants (min^{-1}) of the pseudo-first-order kinetics for sequestration of Se(IV) and Se(VI) were compared by the ratio R_{WMF} (formulated in Figure 5.2), where $k_{w/\ WMF}$ and $k_{w/o\ WMF}$ are the corresponding observed rate constants with and without WMF, respectively. Figure 5.2 shows that the R_{WMF} mainly falls in the range of 3.0–100.0, with only a few cases outside of this range. As compared to the R_{WMF} values (i.e., 2.5–4.0) for Se(VI) removal, a WMF induced greater enhancement in Se(IV) removal by ZVI with R_{WMF} values ranging from 1.3 to 330.4. Furthermore, it should be noted that the enhancing effects of a WMF on AZVI samples were more outstanding than those of Pri-ZVI based on a wide-spectrum analysis (Li *et al.*, 2015a, b, 2016, 2017a; Sun *et al.*, 2014, 2017; Zhang *et al.*, 2018), which may be associated with the promotion of mass transfer and the enrichment of reactants by the synergistic effects of a WMF and iron (hydr)oxides.

5.2.2 Effect of WMF on the selectivity of ZVI toward Se(IV)/Se(VI)

After elucidating the improved reactivity of ZVI by the WMF, the undesired reactions of ZVI with the non-target compounds (e.g., O_2 and H_2O/H^+) may be concurrently inhibited (Gu *et al.*, 2017; He *et al.*, 2020; Li *et al.*, 2017b, 2018, 2020b; Qiao *et al.*, 2018; Zhang *et al.*, 2020). There may be a trade-off between the reactivity and electron selectivity of ZVI in the presence of WMF (Qiao *et al.*, 2018). As such, this section mainly focused on the effect of a WMF on the electron utilization (EU) and electrical efficiency (EE) of ZVI toward Se(IV) and Se(VI). Based on the detailed studies of selenium removal by ZVI/WMF via diverse spectral protocols, it was revealed that WMF could not only accelerate the corrosion of ZVI during Se decontamination, but also facilitate the transformation of Se(IV)/Se(VI) to Se(0) (Li *et al.*, 2020a; Liang *et al.*, 2014a, b; Xu *et al.*, 2016). However, it was found that the corresponding EU values were almost identical when the reactions between Se(IV)/Se(VI) and ZVI reach equilibrium, as shown in Figure 5.3a.

Since applying a WMF can tune the mechanism of Se(IV) removal by ZVI from adsorption followed by reduction to direct reduction (Liang *et al.*, 2014a, b), it suggests that a WMF can improve the electron transport in the ZVI, thus enhancing the EE of ZVI for selenium reduction. As plotted in Figure 5.3b, the EE of ZVI with WMF for reduction of Se(IV)/Se(VI) was 3.7- to 14.1-fold greater than that without WMF, regardless of the 'single' or 'multiple' dosing mode (Li *et al.*, 2020a). In addition, the EE of ZVI/WMF toward Se(IV) was higher than that toward Se(VI), which was likely due to the different involvement of the redox ability and affinity between Se(IV) and Se(VI) (Li *et al.*, 2020a).

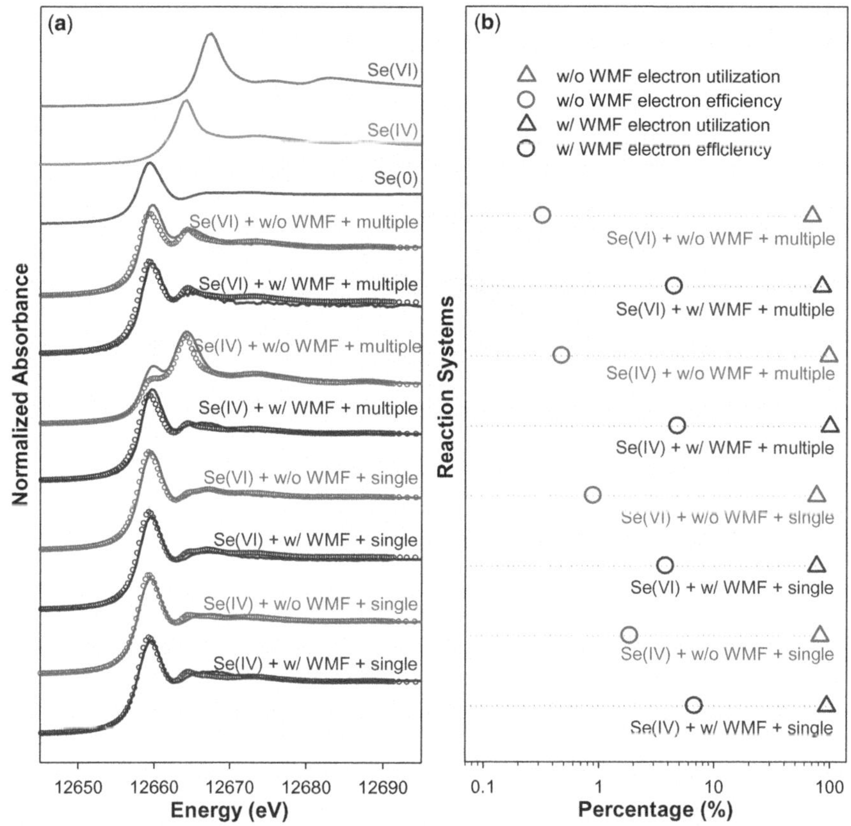

Figure 5.3 Se *K*-edge XANES spectra (a) of the Se-treated ZVI samples at sequestration equilibrium, along with the corresponding active parameters (b) of ZVI. The circles and the thick lines represent the linear combination fits and the experimental data, respectively (Li *et al.*, 2020a).

5.2.3 Contributions of WMF to the improved reactivity and selectivity of ZVI toward Se(IV)/Se(VI)

As for the genesis of enhanced reactivity and selectivity, Liang *et al.* (2014a) and Sun *et al.* (2014) showed that a WMF could accelerate the corrosion of ZVI and the transformation of amorphous iron (hydr)oxides to lepidocrocite, which favored the contaminant sequestration. Besides, under aerobic conditions, Li *et al.* (2016) further confirmed that the enhancing effect of WMF was mainly ascribed to the magnetic field gradient force ($F_{\Delta B}$), which can drive the movement of paramagnetic Fe^{2+} toward the site with higher induced magnetic field. Following this study, Sun *et al.* (2017) further demonstrated the coupled effect of coexisting anions and WMF on the enhanced reactivity of ZVI, which

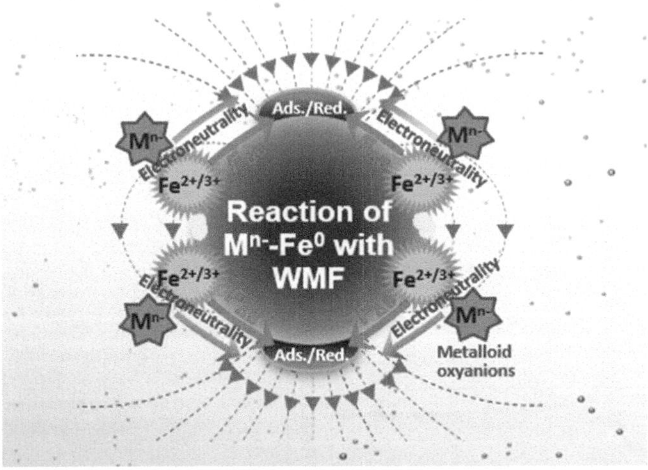

Figure 5.4 Proposed schematic illustration of the reactions of ZVI with Se(IV)/Se(VI) in the presence of a WMF (Li *et al.*, 2020a).

was associated with the simultaneous movement of anions and paramagnetic Fe^{2+} to keep local electroneutrality in the solution. Likewise, the movement of an anionic target should be also reflected by the transportation of paramagnetic Fe^{2+} in the ZVI/WMF system. As such, the $F_{\Delta B}$ induced by a WMF can selectively drive the metalloid oxyanions, including Se(IV) and Se(VI), toward the *in situ* formed iron (hydr)oxides, thereby leading to an enhanced reactivity and specific removal capacity (SRC) of ZVI toward the metalloid oxyanions (Li *et al.*, 2020a).

Considering that enrichment is the first step of electron transfer from the Fe^0 core to the target, the efficient incorporation of a contaminant into the iron corrosion products in the presence of a WMF is beneficial to the EE of ZVI for the reduction of Se(IV) and Se(VI). In general, it could be reasonably inferred that the improved reactivity and selectivity of ZVI toward Se(IV) and Se(VI) induced by a WMF is mainly ascribed to the $F_{\Delta B}$-derived movement of paramagnetic Fe^{2+} with Se(IV)/Se(VI) for the local electroneutrality at the ZVI reaction interface, as depicted in Figure 5.4. In summary, WMF is a promising and environmentally friendly method to enhance the reactivity and selectivity of ZVI toward the Se(IV) and Se(VI) under aerobic conditions.

5.3 FERROUS ION ENHANCES THE REACTIVITY AND SELECTIVITY OF ZVI TOWARD Se(VI)

5.3.1 Influence of Fe(II) on the reactivity of ZVI towards Se(VI)

Recently, the research on contaminant removal by ZVI has greatly increased the dimensions of complexity in this area (Shao *et al.*, 2018). Since Fe(II) and aerobic conditions are two of the most broadly relevant parameters, this section introduces the dynamic interactions between Fe(II) and O_2 on the reactivity and selectivity of ZVI towards Se(IV). As shown in Figure 5.5, the removal capacity and rate of Se(VI) by ZVI increased with an increase in the Fe(II) concentration from 0 to 1.0 mM and the enhancement was greater at higher oxygen concentration (He *et al.*, 2020; Li *et al.*, 2020a; Qiao *et al.*, 2018; Qin *et al.*, 2017). It was found that oxygen is essential for the sequestration of Se(VI) by ZVI. On the one hand, specifically, the SRC of ZVI towards Se(VI) increased progressively from 17.7–24.7 to 49.2–49.9 mg/g with an increase in the initial concentration of Fe(II) from 0 to 1.0 mM under aerobic conditions (Figure 5.5a). On the other hand, oxygen and Fe(II) showed a synergistic effect on the removal rate of Se(VI) by ZVI, which implies that the presence of Fe(II) had a greater accelerating effect on Se(VI) removal by ZVI at high oxygen concentrations (Qin *et al.*, 2017), and vice versa.

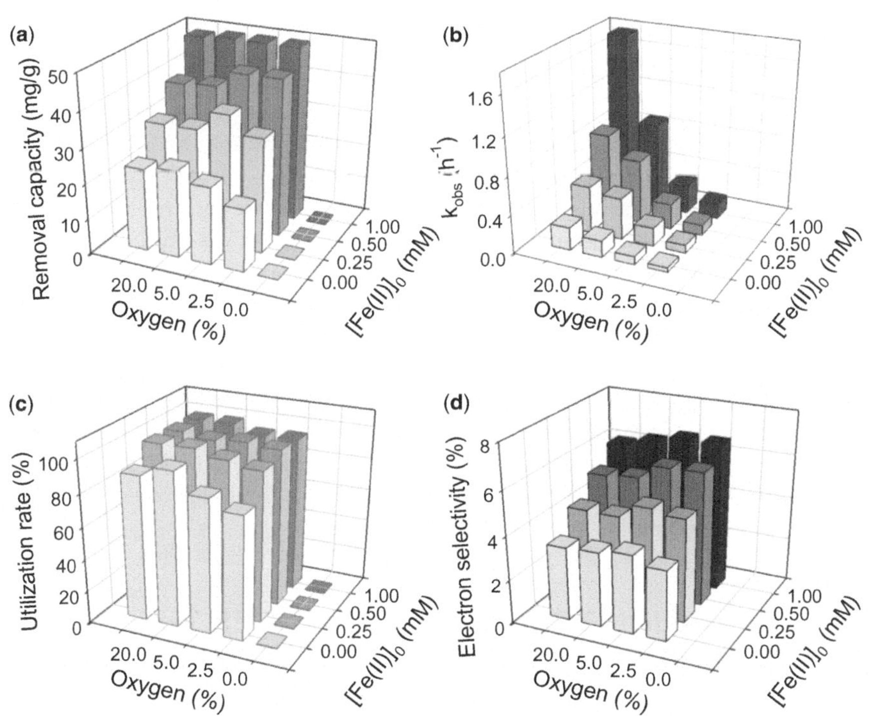

Figure 5.5 Influence of oxygen volume fraction in the purging gas and Fe(II) concentration on (a) capacity and (b) pseudo-first order rate constants (k_{obs}) of Se (VI) removal by ZVI as well as on utilization rate (c), and electron selectivity (d) of ZVI (Qin *et al.*, 2017).

5.3.2 Influence of Fe(II) on the selectivity of ZVI towards Se(VI)

Figure 5.5c shows Fe(II) dosing enhances the utilization ratio (UR) of ZVI to a greater extent at a lower oxygen fraction in the sparging gas (i.e., 2.5–5.0%) than at a higher oxygen fraction (i.e., 20.0–50.0%). Correspondingly, the UR values of ZVI for Se(VI) sequestration at oxygen fractions of 2.5, 5.0, 20, and 50% were increased by 19.4, 12.7, 5.0, and 11.0%, respectively, when the initial Fe(II) concentration increased from 0 to 1.0 mM. Furthermore, there are two obvious trends in the electron selectivity or EE of ZVI toward Se(VI) under different conditions. First, the EE of ZVI for Se(VI) reduction under aerobic conditions increased progressively from 3.2–3.6% to 6.2–6.8% as the initial Fe(II) concentration increased from 0 to 1.0 mM. Second, regardless of the initial Fe(II) concentration, the EE of ZVI improved slightly by increasing the oxygen fraction in the sparging gas from 2.5% to 5.0%. However, a further increase in the oxygen fraction resulted in only a slight drop in the EE of ZVI.

5.3.3 Role of Fe(II) in improving the reactivity and selectivity of ZVI for Se(VI) reduction

The sequestration of Se(VI) by ZVI is a heterogeneous redox reaction at the ZVI-H_2O interface, thus the mass transfer of Se(VI) toward the ZVI surface is a prerequisite of Se(VI) reduction by the electrons from the Fe^0 core. Characterization results suggested that with the increase of the initial Fe(II) concentration, the fractions of Fe^0 and γ-Fe_2O_3 in the Se(VI)-treated ZVI under aerobic conditions dropped, while that of lepidocrocite (γ-FeOOH) and magnetite (Fe_3O_4) increased (Qin *et al.*, 2017), which enhanced Se(VI) adsorption and the subsequent electron transfer between the underlying Fe^0 and the surface-decorated Se(VI). As shown in Figure 5.6, the Fe(II)-induced improvement in the rate constants of Se(VI) sequestration by Fe^0 and the electron selectivity of Fe^0 towards Se(VI) under aerobic conditions were attributed to the weak acidity arising from the Fe(II) addition (Huang and Zhang, 2005; Liu *et al.*, 2013) and the facilitated Se VI) enrichment by the γ-FeOOH, along with the favored electron conductivity by Fe_3O_4 generated on the surface of the ZVI particle.

5.4 SULFIDATION TREATMENT ENHANCES THE REACTIVITY AND SELECTIVITY OF ZVI TOWARD Se(VI)

Sulfidation of ZVI has gained increased attention due to its positive role in enhancing both the reactivity and selectivity of ZVI under anaerobic conditions (Fan *et al.*, 2016, 2017; He *et al.*, 2018; Li *et al.*, 2017b; Qin *et al.*, 2019; Rajajayavel and Ghoshal, 2015). Hydrogen production during reaction of ZVI

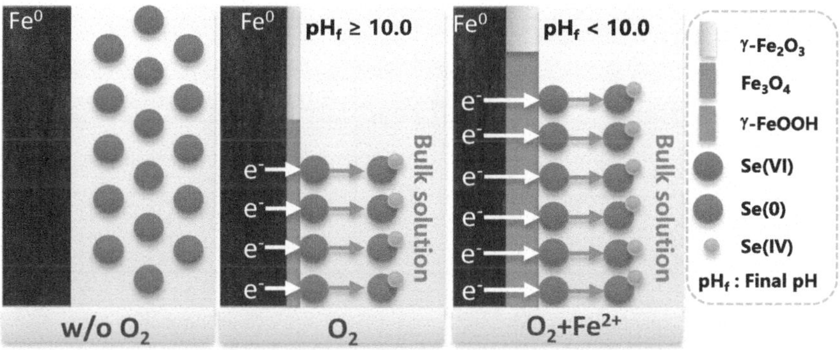

Figure 5.6 Proposed mechanistic model for the influence of oxygen and Fe(II) on Se(VI) removal by ZVI (w/o O_2: without O_2) (Qin *et al.*, 2017).

with water is suppressed by sulfidation under anaerobic conditions, thus leading to an increase in the selectivity of ZVI toward contaminant reduction. Considering the potential application of sulfidated ZVI (S-ZVI) in wastewater treatment, where dissolved oxygen is commonly present and is a much stronger electron acceptor than water, the influence of sulfidation on the reactivity and selectivity of ZVI under aerobic conditions needs to be understood. Hence, this section discusses the impacts of sulfidation on the reactivity and selectivity of ZVI toward Se (VI) reduction.

5.4.1 Influence of sulfidation on the reactivity of ZVI toward Se(VI)

With regard to the reactivity of S-ZVI, Qiao *et al.* (2018) systematically compared the Se(VI) reduction performance of unamended ZVI and sulfidated ZVI in the presence of different coexisting ions including Cl^-, SO_4^{2-}, PO_4^{3-}, Ca^{2+}, and Mg^{2+} under aerobic conditions. Compared with the ball-milled ZVI without elemental S (ZVI^{bm}), S-ZVI synthesized by ball-milling with elemental sulfur (S-ZVI^{bm}) enhanced the reactivity of ZVI toward Se(VI), whereas the positive effect was not very pronounced in the presence of some specific background electrolytes. For example, at reaction equilibrium, sulfidation increased the Se (VI) removal capacity of ZVI from ~15.6 to 30.4 mg/g in the presence of 1 mM Cl^-, whereas there was no noticeable influence when further increasing the Cl^- concentration to 10 and 20 mM (Qiao *et al.*, 2018). Taking the 10 mM NaCl case as a benchmark, it was further found that the presence of SO_4^{2-} or PO_4^{3-} inhibits the Se(VI) removal capacity of S-ZVI^{bm} (still higher than that of ZVI^{bm}), which became more remarkable with further increasing of the anion concentration (Qiao *et al.*, 2018). On the contrary, the hardness ions could increase the removal

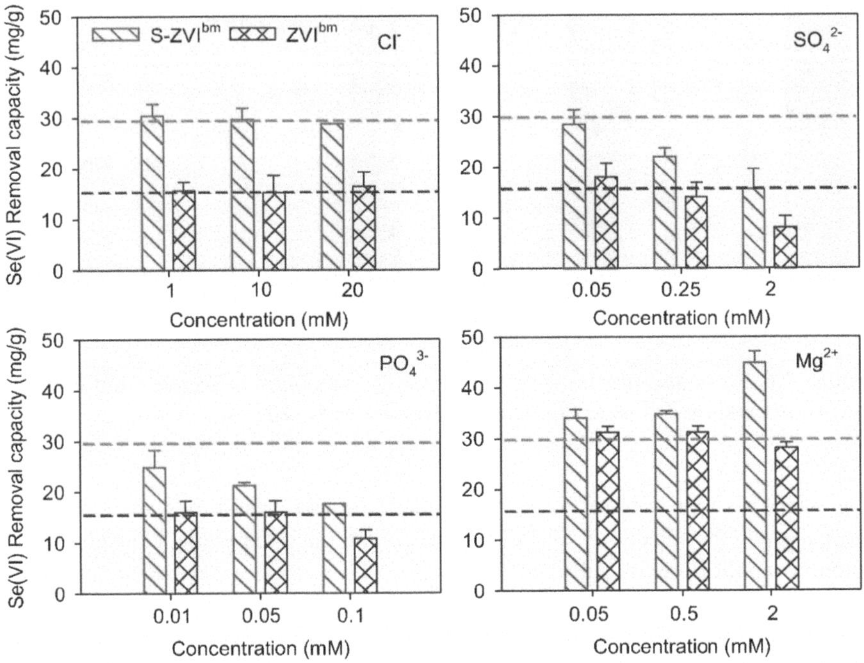

Figure 5.7 Comparison of the removal capacity of ZVI^{bm} and $S\text{-}ZVI^{bm}$ toward Se(VI) in the presence of different coexisting ions. Reaction conditions: $[ZVI]_0 = 0.5$ g/L, $[Se(VI)]_0 = 40$ mg/L, $pH_{ini} = 5.0$, T = 25 °C, reaction time = 24 h. The horizontal lines reflect the Se(VI) removal capacity of ZVI^{bm} (blue) and $S\text{-}ZVI^{bm}$ (red) obtained in 10 mM NaCl benchmark solutions (Qiao *et al.*, 2018).

capacity from 29.6 mg/g to as high as 44.8 mg/g with 2.0 mM Mg^{2+}, as depicted in Figure 5.7.

5.4.2 Influence of sulfidation on the selectivity of ZVI toward Se(VI)

With respect to the selectivity of S-ZVI toward Se(VI), the different coexisting ions could either negligibly, positively or negatively impact the EE of ZVI toward Se (VI), depending on their types and concentrations (Qiao *et al.*, 2018). On the one hand, NaCl did not greatly affect the EE of $S\text{-}ZVI^{bm}$, whereas the presence of SO_4^{2-} and PO_4^{3-} suppressed it, especially at relatively high concentrations (Figure 5.8). On the other hand, Mg^{2+} improved the selectivity of ZVI and a maximum EE value of 8.8% was obtained with the introduction of 2 mM Mg^{2+}. In general, this study sheds some light on evaluating the broader applicability of sulfidation and designing optimal operating conditions for the desirable ZVI performance (Qiao *et al.*, 2018).

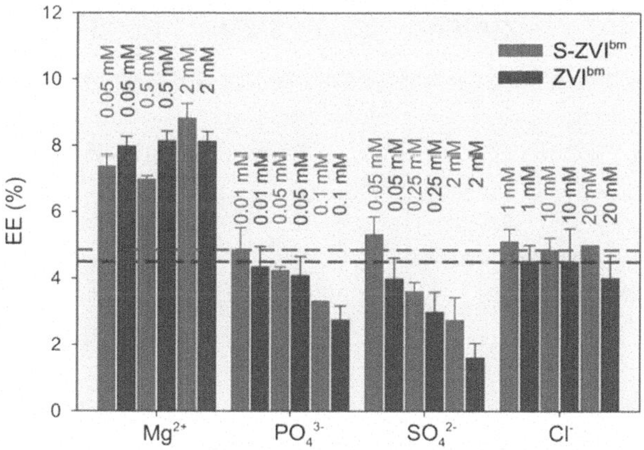

Figure 5.8 Effect of coexisting ions on the EE of ZVI^{bm} and S-ZVI^{bm} toward Se(VI) (Qiao *et al.*, 2018).

5.4.3 Coupled effects of sulfidation and ferrous dosing on Se(VI) removal by ZVI

In light of the improvements of ZVI selectivity toward Se(VI) (Li *et al.*, 2020a; Qin *et al.*, 2017), it is likely that the EE of the aerobic ZVI system can be enhanced via regulating its corrosion behavior. Sulfidation may impact the selectivity of ZVI since it can influence the Fe^0 corrosion behavior. All of Fe^0 in S-ZVI^{bm} was exhausted within 24 h with lepidocrocite and magnetite being the primary corrosion products (Qiao *et al.*, 2018). This suggested that the coexisting ions may influence the EE by affecting Se(VI) adsorption and the subsequent electron transfer from the Fe^0 core to Se(VI) at the water-particle interface. Given these facts, Fan *et al.* (2019b) further confirmed that sulfidation coupled with Fe (II) dosing could synergistically improve the reactivity and selectivity of ZVI toward Se(VI) under aerobic conditions (Figure 5.9). The promoting effect on Se (VI) sequestration by S-ZVI/Fe^{2+} is mainly associated with the following aspects: (1) Fe^{2+} could maintain a relatively low pH level during Se(VI) removal by S-ZVI; (2) S-ZVI/Fe^{2+} could retard the consumption of Fe^0 by the non-target O_2/H^+, thus facilitating the EE of S-ZVI; and (3) Fe(II) dosing could enable the electron transfer by forming semiconductive Fe_3O_4.

5.5 OUTLOOK

The superposition of WMF, Fe(II) dosing, and sulfidation pre-treatment can greatly improve the reactivity and selectivity of ZVI toward Se(IV)/Se(VI) under aerobic conditions, but to diverse extents. Among these three enhanced-ZVI technologies, ZVI/Fe^{2+} attained the largest amount of Se(VI) sequestration, while S-ZVI had the highest k_{obs} for removing Se(VI) (Figure 5.10). As for the removal

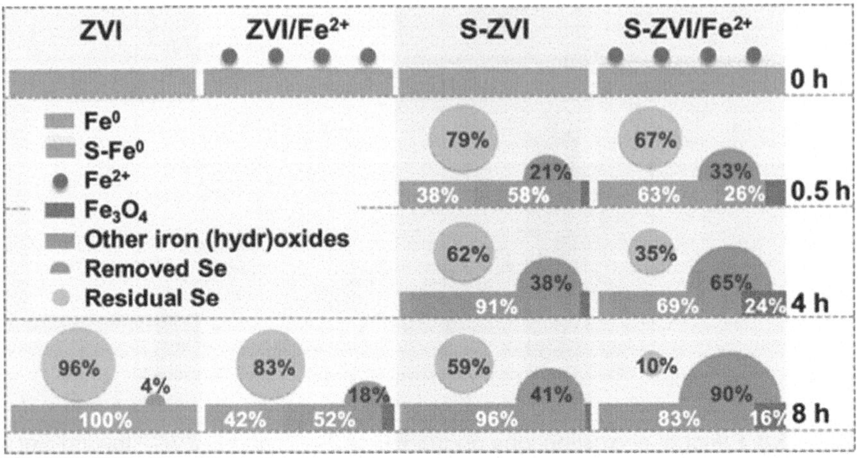

Figure 5.9 Coupled effects of sulfidation and ferrous dosing on Se(VI) removal by ZVI (Fan *et al.*, 2019b).

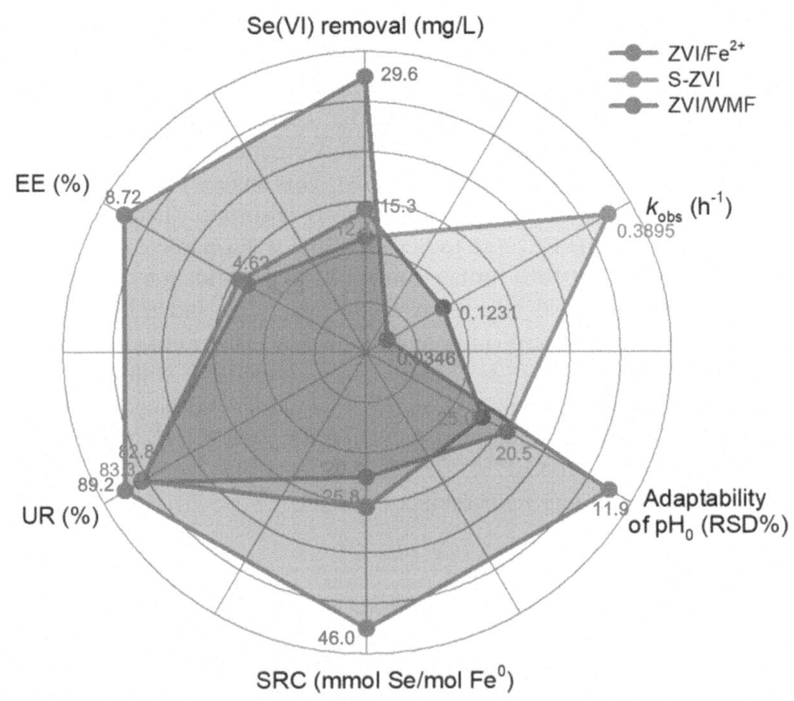

Figure 5.10 Comprehensive comparison of the performance of Se(VI) sequestration by ZVI/WMF, ZVI/Fe^{2+} and S-ZVI systems (Fan *et al.*, 2019a).

mechanism of Se(VI), these three enhancing techniques did not change the reaction mechanism of Se(VI) with ZVI, but only affected the rates and amounts of Se(VI) removal by ZVI. In theory, these enhancing techniques accelerate the corrosion of ZVI and tune the corresponding products transformation for better mass transfer of Se(IV)/Se(VI) (Fan *et al.*, 2019a, b; Li *et al.*, 2020a; Qiao *et al.*, 2018; Qin *et al.*, 2017), confirming that the corrosion of ZVI is a necessity and induces the removal of contaminants by ZVI. Considering the amount of removed Se(VI), the ZVI/Fe^{2+} system seems to be superior to the two other enhanced-ZVI technologies in terms of the utilization efficiency of ZVI at the end of reaction, and the adaptability over a wide pH range, as well as the chemical cost (Fan *et al.*, 2019a). Therefore, future research should focus on illustrating the performance of the ZVI/Fe^{2+} system toward Se(VI) on a pilot-scale (Fan *et al.*, 2019a). Moreover, the feasibility of employing Fe^{2+} to enhance the performance of ZVI should be examined with ZVI of different origins, different target contaminants, and in the presence of different background matrices (Fan *et al.*, 2019a), as well as coupled with WMF or sulfidation (Fan *et al.*, 2019b).

REFERENCES

Fan D., O'Brien Johnson G. S., Tratnyek P. G. and Johnson R. L. (2016). Sulfidation of nano zerovalent iron (nZVI) for improved selectivity during in-situ chemical reduction (ISCR). *Environmental Science Technology*, **50**(17), 9558–9565.

Fan D., Lan Y., Tratnyek P. G., Johnson R. L., Filip J., O'Carroll D. M., Nunez Garcia A. and Agrawal A. (2017). Sulfidation of iron-based materials: A review of processes and implications for water treatment and remediation. *Environmental Science Technology*, **51**, 13070–13085.

Fan P., Li L., Sun Y., Qiao J., Xu C. and Guan X. (2019a). Selenate removal by Fe^0 coupled with ferrous iron, hydrogen peroxide, sulfidation, and weak magnetic field: a comparative study. *Water Research*, **159**, 375–384.

Fan P., Sun Y., Zhou B. and Guan X. (2019b). Coupled effect of sulfidation and ferrous dosing on selenate removal by zerovalent iron under aerobic conditions. *Environmental Science Technology*, **53**(24), 14577–14585.

Gu Y., Wang B., He F., Bradley M. J. and Tratnyek P. G. (2017). Mechanochemically sulfidated microscale zero valent iron: pathways, kinetics, mechanism, and efficiency of trichloroethylene dechlorination. *Environmental Science Technology*, **51**(21), 12653–12662.

Guan X. H., Sun Y. K., Qin H. J., Li J. X., Lo I. M., He D. and Dong H. R. (2015). The limitations of applying zero-valent iron technology in contaminants sequestration and the corresponding countermeasures: the development in zero-valent iron technology in the last two decades (1994–2014). *Water Research*, **75**, 224–248.

Guo X. J., Yang Z., Dong H. Y., Guan X. H., Ren Q. D., Lv X. F. and Jin X. (2016). Simple combination of oxidants with zero-valent-iron (ZVI) achieved very rapid and highly efficient removal of heavy metals from water. *Water Research*, **88**, 671–680.

He F., Li Z., Shi S., Xu W., Sheng H., Gu Y., Jiang Y. and Xi B. (2018). Dechlorination of excess trichloroethene by bimetallic and sulfidated nanoscale zero-valent iron. *Environmental Science Technology*, **52**(15), 8627–8637.

He F., Gong L., Fan D., Tratnyek P. G. and Lowry G. V. (2020). Quantifying the efficiency and selectivity of organohalide dechlorination by zerovalent iron. *Environmental Science Processes & Impacts*, **22**, 528–542.

Huang Y. H. and Zhang T. C. (2005). Effects of dissolved oxygen on formation of corrosion products and concomitant oxygen and nitrate reduction in zero-valent iron systems with or without aqueous Fe^{2+}. *Water Research*, **39**(9), 1751–1760.

Johnson T. L., Scherer M. M. and Tratnyek P. G. (1996). Kinetics of halogenated organic compound degradation by iron metal. *Environmental Science Technology*, **30**(8), 2634–2640.

Li J., Qin H. and Guan X. (2015a). Premagnetization for enhancing the reactivity of multiple zerovalent iron samples toward various contaminants. *Environmental Science Technology*, **49**(24), 14401–14408.

Li J., Shi Z., Ma B., Zhang P., Jiang X., Xiao Z. and Guan X. (2015b). Improving the reactivity of zerovalent iron by taking advantage of its magnetic memory: implications for arsenite removal. *Environmental Science Technology*, **49**(17), 10581–10588.

Li J., Qin H., Zhang W. X., Shi Z., Zhao D. and Guan X. (2016). Enhanced Cr(VI) removal by zero-valent iron coupled with weak magnetic field: role of magnetic gradient force. *Separation and Purification Technology*, **176**, 40–47.

Li J., Qin H., Zhang X. and Guan X. (2017a). Improving the reactivity of zerovalent iron toward various contaminants by weak magnetic field: Performances and Mechanisms. *Acta Chimica Sinica*, **75**(6), 544.

Li J., Zhang X., Sun Y., Liang L., Pan B. C., Zhang W. and Guan X. (2017b). Advances in sulfidation of zerovalent iron for water decontamination. *Environmental Science Technology*, **248**(23), 173–182.

Li J., Zhang X., Liu M., Pan B. C., Zhang W., Shi Z. and Guan X. (2018). Enhanced reactivity and electron selectivity of sulfidated zerovalent iron toward chromate under aerobic conditions. *Environmental Science Technology*, **52**(5), 2988–2997.

Li J., Dou X. M., Qin H., Sun Y. K., Yin D. Q. and Guan X. (2019). Characterization methods of zerovalent iron for water treatment and remediation. *Water Research*, **148**, 70–85.

Li J., Sun Y., Zhang X. and Guan X. (2020a). Weak magnetic field enables high selectivity of zerovalent iron toward metalloid oxyanions under aerobic conditions. *Journal of Hazardous Materials*, **400**, 123330.

Li M., Mu Y., Shang H., Mao C., Cao S., Ai Z. and Zhang L. (2020b). Phosphate modification enables high efficiency and electron selectivity of nZVI toward Cr(VI) removal. *Applied Catalysis B-Environmental*, **263**, 118364.

Liang L., Yang W., Guan X., Li J., Xu Z., Wu J., Huang Y. and Zhang X. (2013). Kinetics and mechanisms of pH-dependent selenite removal by zero valent iron. *Water Research*, **47**(15), 5846–5855.

Liang L., Sun W., Guan X., Huang Y., Choi W., Bao H., Li L. and Jiang Z. (2014a). Weak magnetic field significantly enhances selenite removal kinetics by zero valent iron. *Water Research*, **49**, 371–380.

Liang L. P., Guan X. H., Shi Z., Li J. L., Wu Y. N. and Tratnyek P. G. (2014b). Coupled effects of aging and weak magnetic fields on sequestration of selenite by zero-valent iron. *Environmental Science Technology*, **48**(11), 6326–6334.

Liang L., Guan X., Huang Y., Ma J., Sun X., Qiao J. and Zhou G. (2015a). Efficient selenate removal by zero-valent iron in the presence of weak magnetic field. *Separation and Purification Technology*, **156**, 1064–1072.

Liang L., Jiang X., Yang W., Huang Y., Guan X. and Li L. (2015b). Kinetics of selenite reduction by zero-valent iron. *Desalination and Water Treatment*, **53**(9), 2540–2548.

Liu T., Li X. and Waite T. D. (2013). Depassivation of aged Fe^0 by ferrous ions: implications to contaminant degradation. *Environmental Science Technology*, **47**(23), 13712–13720.

Noubactep C. (2008). A critical review on the process of contaminant removal in Fe^0-H_2O systems. *Environmental Technology*, **29**(8), 909–920.

Noubactep C. (2015). Metallic iron for environmental remediation: a review of reviews. *Water Research*, **85**, 114–123.

Qiao J., Song Y., Sun Y. and Guan X. (2018). Effect of solution chemistry on the reactivity and electron selectivity of zerovalent iron toward Se(VI) removal. *Chemical Engineering Journal*, **353**, 246–253.

Qin H. J., Li J. X., Bao Q. Q., Li L. N. and Guan X. H. (2016). Role of dissolved oxygen in metal(loid) removal by zerovalent iron at different pH: its dependence on the removal mechanisms. *RSC Advances*, **6**, 50144–50152.

Qin H. J., Li J. X., Yang H. Y., Pan B. C., Zhang W. M. and Guan X. H. (2017). Coupled effect of ferrous ion and oxygen on the electron selectivity of zerovalent iron for selenate sequestration. *Environmental Science Technology*, **51**(9), 5090–5097.

Qin H., Guan X. and Tratnyek P. G. (2019). Effects of sulfidation and nitrate on the reduction of N-Nitrosodimethylamine by zerovalent iron. *Environmental Science Technology*, **53**(16), 9744–9754.

Rajajayavel S. R. and Ghoshal S. (2015). Enhanced reductive dechlorination of trichloroethylene by sulfidated nanoscale zerovalent iron. *Water Research*, **78**, 144–153.

Reinsch B. C., Forsberg B., Penn R. L., Kim C. S. and Lowry G. V. (2010). Chemical transformations during aging of zerovalent iron nanoparticles in the presence of common groundwater dissolved constituents. *Environmental Science Technology*, **44**(9), 3455–3461.

Sarathy V., Tratnyek P. G., Nurmi J. T., Baer D. R., Amonette J. E., Chun C. L., Penn R. L. and Reardon E. J. (2008). Aging of iron nanoparticles in aqueous solution: effects on structure and reactivity. *Journal of Physical Chemistry C*, **112**(7), 2286–2293.

Scherer M. M., Balko B. A., Gallagher D. A. and Tratnyek P. G. (1998). Correlation analysis of rate constants for dechlorination by zero-valent iron. *Environmental Science Technology*, **32**(19), 3026–3033.

Shao Q., Xu C., Wang Y., Huang S., Zhang B., Huang L., Fan D. and Tratnyek P. G. (2018). Dynamic interactions between sulfidated zerovalent iron and dissolved oxygen: mechanistic insights for enhanced chromate removal. *Water Research*, **135**, 322–330.

Sun Y. K., Guan X. H., Wang J. M., Meng X. G., Xu C. H. and Zhou G. M. (2014). Effect of weak magnetic field on arsenate and arsenite removal from water by zerovalent iron: an XAFS investigation. *Environmental Science Technology*, **48**(12), 6850–6858.

Sun Y., Hu Y., Huang T., Li J., Qin H. and Guan X. (2017). Combined effect of weak magnetic field and anions on arsenite sequestration by zero-valent iron: kinetics and mechanisms. *Environmental Science Technology*, **51**(7), 3742–3750. doi:10.1021/acs.est.6b06117

Tang C. L., Huang Y. H., Zeng H. and Zhang Z. Q. (2014). Reductive removal of selenate by zero-valent iron: the roles of aqueous Fe^{2+} and corrosion products, and selenate removal mechanisms. *Water Research*, **67**, 166–174.

Tang C. L., Huang Y. P., Zhang Z. Q., Chen J. J., Zeng H. and Huang Y. H. (2016). Rapid removal of selenate in a zero-valent iron/Fe_3O_4/Fe^{2+} synergetic system. *Applied Catalysis B-Environmental*, **184**, 320–327.

Wang Q., Lee S. and Choi H. (2010). Aging study on the structure of Fe 0 nanoparticles: stabilization, characterization, and reactivity. *Journal of Physical Chemistry, C* **114**, 2027–2033.

Xu H., Sun Y., Li J., Li F. and Guan X. (2016). Aging of zerovalent iron in synthetic groundwater: X-ray photoelectron spectroscopy depth profiling characterization and depassivation with uniform magnetic field. *Environmental Science Technology*, **50**(15), 8214–8222.

Zhang X., Li J., Sun Y., Li L., Pan B., Zhang W. and Guan X. (2018). Aging of zerovalent iron in various coexisting solutes: characteristics, reactivity toward selenite and rejuvenation by weak magnetic field. *Separation and Purification Technology*, **191**, 94–100.

Zhang J., Cheng Z., Yang X., Luo J., Li H., Chen H., Zhang Q. and Li J. (2020). Mediating the reactivity and selectivity of nanoscale zerovalent iron toward nitrobenzene under porous carbon confinement. *Chemical Engineering Journal*, **393**, 124779.

Chapter 6

Biological treatment technologies

Robert Simm

6.1 INTRODUCTION

Selenium contaminated waters are produced by various industrial activities such as mining (coal, hard rock, uranium and phosphate), refineries (metal and oil), power generation (coalfired power plants), and agriculture (irrigation waters and selenium fortification) (Chapman *et al.*, 2010). Selenium can also be present in municipal wastewater, due to inflow and infiltration of groundwater into sewers emanating from alluvium or sedimentary deposits that are high in selenium (Pontarolo *et al.*, 2017).

Background concentrations of Se in uncontaminated surface waters are generally below 1 μg/L (Seiler *et al.*, 1965), but in areas where the weathering and erosion of seleniferous soils is augmented by anthropogenic activities, environmental Se concentrations have been shown to be significantly elevated. Elevated intake of selenium, particularly in oviparous animals such as predatory fish and waterfowl, can result in teratogenic effects on entire populations in contaminated ecosystems (Janz *et al.*, 2010). The maximum permissible concentration of Se in the aquatic environment is, on average, 10 times lower than that in drinking water, at values of between 1 to 5 μg/L and 10 to 50 μg/L, respectively (Lemly, 2007). While overall, guidelines for Se in freshwater for aquatic life are very low level, they do vary by country.

An excellent review of global selenium regulations by Kumkrong *et al.* (2018) suggests Se is only regulated under Directive 98/83/EEC on the quality of water

doi: 10.2166/9781789061055_0189

intended for human consumption in the EU (European Union) which controls the maximum level at 10 µg/L. This same reference suggests there is little English information available relating to standards for Se in waters in Asia, though the drinking water standard in many Asian countries is set at 10 µg/L. This is despite documented negative impacts on fish and aquatic birds in locations around the world (Lemly, 2004).

The US and Canada have some of the world's most stringent aquatic water quality guidelines limiting the concentration of Se in fresh water to protect aquatic life at low levels of between 1 and 5 µg/L. Increasing public awareness in specific geographies in North America and progressively more stringent guidelines have increased demand for selenium management and treatment. Coincidentally, most full-scale treatment facilities are located in North America.

Biological treatment has been identified by the United States Environmental Protection Agency (US EPA) and others (CH2M Hill, 2010) as the best available technology (BAT) for selenium reduction to achieve ultra-low effluent selenium concentrations of less than 10 µg/L. A recent review by Golder (2020) for the North American Metals Council – Selenium Working Group (NAMC-SWG) suggests that approximately 30 full-scale selenium treatment systems with capacities between 410 m^3/d and 15,260 m^3/d were constructed in North America between 2007 and 2018, 70% of which rely on biological treatment. A recent review by Simm (2018) lists the limited number of vendors with full-scale experience, the high capital and operating and maintenance costs of installing these systems, and required system optimization as key issues to be addressed with biological treatment technologies.

This chapter provides an overview of currently available selenium bioremediation technologies including their advantages and disadvantages and summarizes current challenges with the technology based upon recently implemented projects. The primary focus of this review will be on technologies currently applied at full scale. Finally, a summary of fundamental and applied research needs will be identified to improve system performance and reliability.

6.2 PRINCIPLES OF SELENIUM BIOREMEDIATION IN BIOREACTOR SYSTEMS

A general overview of the principles of bioremediation in bioreactor systems is provided here for completeness. The reader is referred to Chapter 3 and the excellent review by Nancharaiah and Lens (2015a, 2015b) for a more detailed treatise.

In industrial and municipal wastewaters, selenium is typically available in the form of the selenium oxyanions selenate (SeO_4^{2-}) and selenite (SeO_3^{2-}). Both are toxic to living systems, although SeO_3^{2-} is potentially more toxic than SeO_4^{2-} (Nancharaiah & Lens, 2015a, 2015b).

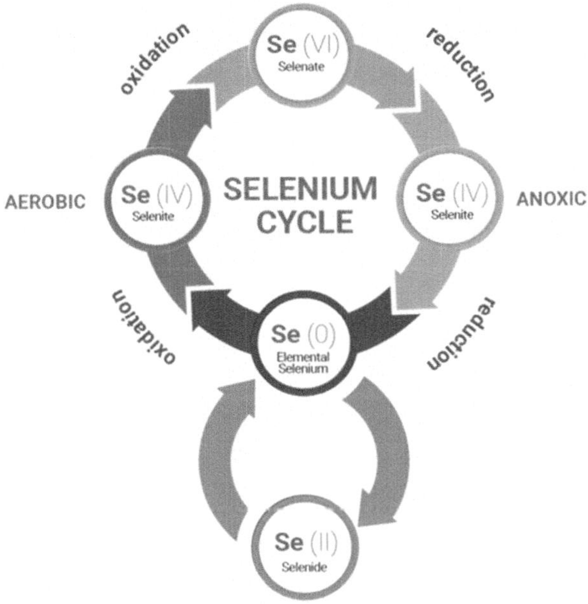

Figure 6.1 Bioconversions of Se compounds in the natural selenium cycle (*Source*: adapted from Simm, 2018).

Selenium bioremediation is based upon reactions carried out in nature as part of the global selenium cycle first identified by Shrift (1964). Natural bacteria and archaea readily metabolize selenium and are involved in a range of metabolic functions including assimilation, methylation, detoxification and anaerobic respiration. The reactions of primary importance in biological selenium removal systems are presented in Figure 6.1. The figure does not include all known reactions such as the aerobic conversion of Se(VI) to Se(IV) and to Se(0).

Bioremediation of selenate can occur by two processes: assimilatory and dissimilatory reduction, the latter being more common in biological treatment. In assimilatory reduction, SeO_4^{2-} is taken up and chemically reduced by bacteria and subsequently used in the synthesis of selenium-containing amino acids, such as selenomethionine and selenocysteine, both of which are classified as organo-selenium compounds (Nancharaiah & Lens, 2015a, 2015b).

Because of the small amounts of SeO_4^{2-} required in the formation of selenomethionine and selenocysteine, assimilatory reduction is not a major process in wastewater selenium treatment. The basis of virtually all biological Se removal techniques is the dissimilatory reduction pathway in which inorganic soluble selenium oxyanions (SeO_4^{2-} and SeO_3^{2-}) are reduced to inorganic, insoluble and less toxic biogenic elemental selenium – Se^0 – with temperature,

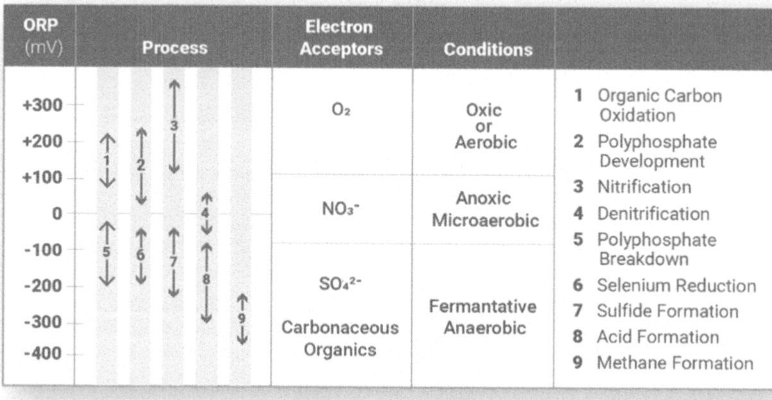

Figure 6.2 Typical ORP ranges for specific biological treatment objectives (*Source*: Electrical Power Research Institute, EPRI, 2009).

pH, and electron donor concentrations being among the controlling factors (Nancharaiah & Lens, 2015a, 2015b). Elemental selenium is removed from a biological treatment system with the treatment process residuals in an active treatment process.

Selenate can be reduced to selenite and both selenate and selenite can be reduced to elemental selenium or alkyl selenides. Selenate reduction is associated with energy production in bacteria whereas selenite reduction can be associated with either energy production or selenite detoxification. Although reduction of SeO_4^{2-} to elemental selenium was shown to be an environmentally significant process, only a few SeO_3^{2-} respiring bacteria have been isolated (Nancharaiah & Lens, 2015a, 2015b). Certain SeO_4^{2-} reducing bacteria have been shown to perform dissimilatory SeO_3^{2-} reduction as well.

Elemental selenium can be reduced further microbiologically to soluble selenide, which in combination with metal ions forms insoluble metal selenides. Selenide can also be emitted as the volatile and highly reactive H_2Se gas, but this is spontaneously and rapidly oxidized to elemental selenium in the presence of oxygen. Organo-selenium and methylated selenium species also contain selenide.

Oxidation of elemental selenium and selenide back to selenite or selenate by selenium oxidizing bacteria completes the selenium cycle. In general, the oxidation rates are 3 to 4 orders of magnitude lower than those in the reductive part of the selenium cycle.

Numerous microorganisms capable of reducing selenium oxyanions to Se^0 have been found in biological wastewater treatment processes such as activated sludge, denitrifying sludge, as well as sulfate-reducing and methanogenic sludge (Soda *et al.*, 2011). Nancharaiah and Lens (2015a, 2015b) have indicated that not only are selenate reducing bacteria phylogenetically diverse, but they are capable of coupling growth to a wide range of electron acceptors.

Nitrates compete as an electron acceptor with selenate and selenite reduction. Therefore, nitrates need to be essentially removed for complete reduction of selenium. The presence and/or absence of specific electron acceptors is typically reflected in the bulk solution oxidation-reduction potential (ORP). Figure 6.2 suggests selenate and selenite reduction take place at ORP values of between −50 mV and −200 mV. Sulfate can also begin to compete with the selenium oxyanions as an electron donor, but at the lower end of the optimal ORP range for selenium reduction.

6.3 HISTORY AND CURRENT PRACTICE OF SELENIUM BIOREMEDIATION

Selenium bioremediation technologies include active, passive, and *in situ* technologies with most biological treatment processes relying on the anaerobic-anoxic reduction processes described in Section 6.2.

The North American Metals Council – Selenium Working Group (NAMC-SWG) has commissioned several state-of-the-art reviews of selenium removal technology over the last 10 years (CH2M Hill, 2010, 2013; Golder, 2020). The NAMC-SWG comprises professionals from industry and consulting engaged in sharing and commissioning technical research on issues pertaining to ecological and human health effects, regulation, and water treatment of selenium in the context of industrial discharges. The Golder (2020) study indicated that 30 full-scale selenium treatment systems were installed predominantly in North America between 2007 and 2018 with design flows between 410 m^3/d and 15,260 m^3/d with biological systems accounting for 70% of the systems surveyed.

Several promising selenium treatment technologies have been studied but have not been implemented extensively at full scale for selenium removal. These include upflow anaerobic sludge blanket (UASB), the hydrogen-based hollow fibre biofilm reactor (MBfR), anaerobic membrane bioreactor (MBR), and the hybrid electro-biochemical reactor. These process concepts will be presented here but will not be discussed in detail.

Implementation of suspended growth systems for selenium bioremediation at full-scale are not widespread. The iBIO® process developed by Degremont is a suspended growth system that has been applied on wet flue gas desulfurization (WFGD) wastewaters and is discussed in Section 6.5. There are very few studies on the use of activated sludge for treatment of selenium wastewaters (Mal *et al.*, 2017). These authors have also suggested the effect of alternating aerobic – anoxic or anaerobic conditions on selenite bioreduction, whereas the fate of biogenic selenium nanoparticles in the activated sludge wastewater treatment system are unknown.

Passive treatment options which include wetlands, biochemical reactors (BCRs), gravel bed reactors, and submerged rock fills (SRFs) are receiving increased attention particularly by the North American mining industry. This is primarily a result of the high capital and operation and maintenance (O&M) costs associated with active treatment.

Full-scale biological selenium treatment practice is currently dominated by attached growth biofilm reactors, particularly packed beds, fluidized bed reactors (FBRs), and a combination of expanded bed biofilm reactors (EBBRs)-packed bed reactors. These reactor types dominate in the power and mining industries. The reason for this is partly historical as indicated below.

There are four primary vendors in the active treatment market in North America: Suez, Frontier, Envirogen, and Veolia (Simm, 2018). These four vendors supply the packed bed, expanded bed-packed bed, fluidized bed, and moving bed biofilm reactor (MBBR) processes for biological selenium removal, respectively. The key project milestones in the development of biological selenium removal at full scale in North America between 1999 and 2018 are presented in Figure 6.3.

The ABMet® process was developed in the late 1990s by Applied Biosciences out of Utah (USA). ABMet® is a packed bed fixed film process for selenium reduction. The process was employed at Kennecott Copper in 2000 and then at the Landusky Mine in Montana in 2002. In 2006, Applied Biosciences was acquired by Zenon (which eventually was acquired by GE). Zenon applied for a patent for an apparatus and method of treating flue gas desulfurization (FGD) blowdown or similar liquids in July 2006 (US Patent No. US 7,790,034 B2). GE Water & Process Technologies, a unit of General Electric Company, completed its acquisition of Zenon that same year. The Zenon patent included the ABMet® process and indicated it could be operated in either an upflow or downflow configuration. GE Water & Process Technologies was subsequently purchased by Suez in 2017.

The first ABMet® projects for FGD blowdown commenced at Duke Energy's Roxboro and Mayo Power Generating Stations in 2006 and 2007, respectively. Interestingly, the original design for the Roxboro system was based upon an upflow configuration which was changed to a downflow configuration during the planning process (Kennedy, personal communications). The ABMet® installations at Roxboro and Mayo were both reinforced concrete constructions with equal volume for the first and second stage reactors and using first stage effluent for system backwash. These were followed by ABMet® facilities at Bellows Creek and Allen stations (planned with downflow filters and similar configuration to Roxboro and Mayo), American Electrical Power's Mountaineer Station (planned with downflow filters and similar configuration to the previous reactors but adding ORP and pH monitoring and using second stage effluent for backwash), and finally the most recent upgrade at Roxboro station which was similar to Mountaineer in many respects, but used a fibreglass tank construction (Kennedy, personal communications).

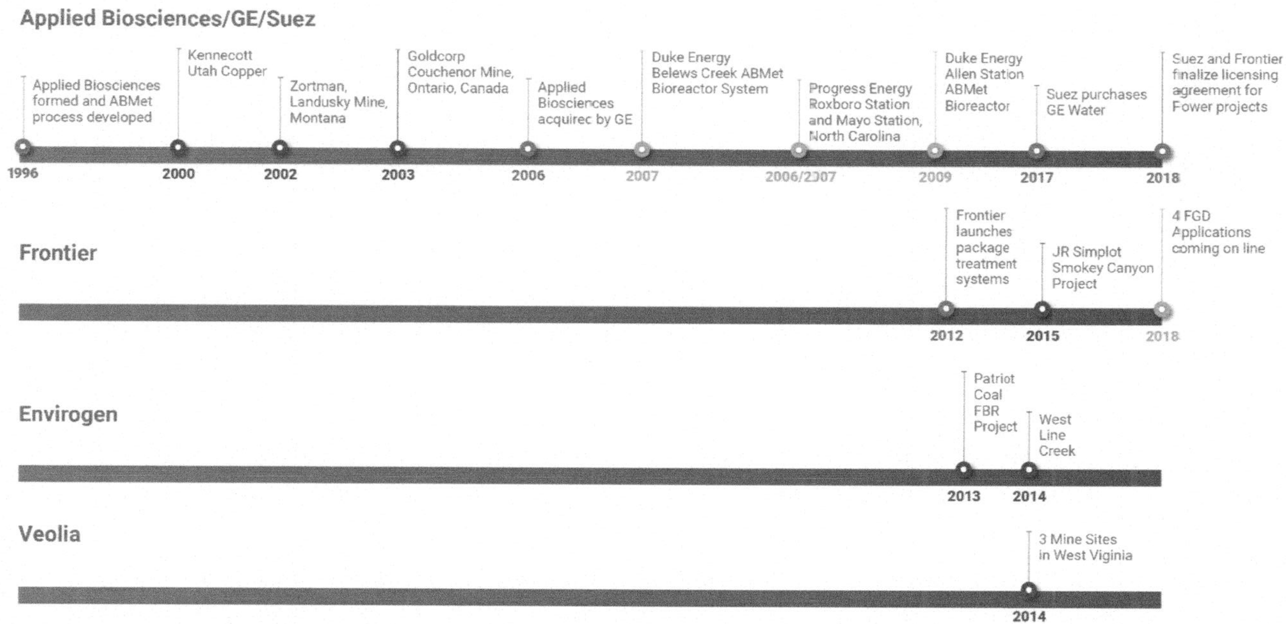

Figure 6.3 Key project milestones in the development of biological selenium treatment at full scale in North America (Simm, 2018).

In 2012, one of the original founders of Applied Biosciences left GE Water & Process and started Frontier Water Systems which specializes in modular biological selenium treatment systems. The Frontier system is based upon a two-stage reactor process with the first stage operated in an upflow mode and the second stage operated as a downflow packed bed. Evoqua acquired a majority stake in Frontier Water Systems in 2019.

The key features of each of the major fixed film processes used for selenium bioremediation are discussed in the following section.

6.4 ATTACHED BIOFILM REACTORS

Attached growth reactors rely on biochemical transformations performed by a biofilm on a surface. Fundamentally, attached growth systems are applied to wastewater where the parameters of primary interest are relatively dilute and variable in concentration. The most widely used processes are packed bed, combined expanded bed-packed bed reactors, and fluidized bed bioreactors. Moving bed biofilm reactors (MBBRs) have been used for selenium bioremediation at full scale, but to a lesser degree to date.

6.4.1 Packed bed reactor

Interest in packed bed reactors stems from the high biomass concentrations that can be achieved relative to suspended growth processes for the same solids retention time (SRT). This results in shorter hydraulic retention time (HRT) and more compact system designs. The primary components of a packed bed reactor include the reactor vessel, support media for biofilm growth, influent distribution system and effluent withdrawal system. Some of the advantages and disadvantages of the packed bed reactor system are summarized in Table 6.2.

The ABMet® packed bed reactor system provided by Suez has been used for simultaneous nitrate and selenium removal since 2000. ABMet® stands for advanced biological metals removal process. A simplified schematic of an ABMet® process flowsheet is presented in Figure 6.4.

There were 12 ABMet® plants installed around the world as of September 2018. A summary of full-scale ABMet® plants is presented in Table 6.1. Most ABMet® applications are in the power industry treating blowdown from flue gas desulfurization (FGD) systems.

Suez has completed more than 20 pilots of the ABMet® on FGD blowdown in addition to the full-scale facilities presented in Table 6.1 (this is as of September 2018).

A typical ABMet® process flowsheet includes pre-treatment, nutrient addition, two-stage ABMet® tanks with intermediate break tank between stages 1 and 2, backwash tank, backwash pumps and effluent storage tank. The process flowsheet can include post-treatment operations depending upon the application and effluent requirements. The need for pre-treatment will depend upon influent

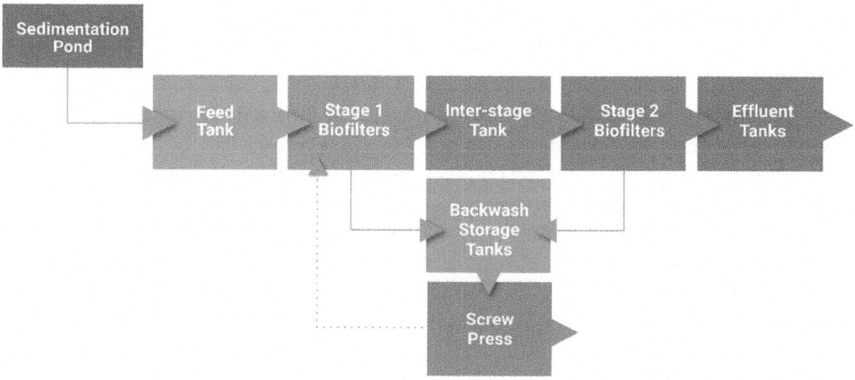

Figure 6.4 Simplified process flow diagram for an ABMet® process (*Source*: adapted from Patterson, 2017).

total suspended solids (TSS) loadings. Pre-treatment, such as filtration, should be considered for influent TSS concentrations exceeding 50 mg/L.

The ABMet® process uses granular activated carbon (GAC) media with a typical media effective size of 0.8 to 1.0 mm. Media are designed to provide a high surface area for biological growth and attachment and have uniformity in size, while allowing water, suspended solids, and sloughing biological growth to pass through the media without plugging during backwashes.

Influent wastewater is distributed across the top of the system and flows downward through the interstitial space between media and over the biomass. Periodically, the flow is reversed to partially remove suspended solids that accumulate at the biofilm surface, promote washing the dead biomass away from

Table 6.1 Full-scale ABMet® plants as of September 2018 (Simm, 2018 – produced with permission of Suez).

Plant	Region	Industry	Start Up Year	Design Flow (m^3/h)
Site 1	United States	Power (FGD)	2008	318
Site 2	United States	Power (FGD)	2008	146
Site 3	United States	Power (FGD)	2009	59
Site 4	United States	Power (FGD)	2009	100
Site 5	United States	Power (FGD)	2011	136
Site 6	Europe	Metal Refining	2015	230
Site 7	Canada	Coal Mining	2015	86
Site 8	United States	Agricultural Runoff	2015	23
Site 9	United States	Power (Ash Landfill Leachate)	2016	6

Table 6.2 Advantages and disadvantages of ABMet® process.

Advantages	Potential Disadvantages
• Longest track record and most full-scale selenium treatment applications (as of September 2018). • May not need downstream solids removal depending on effluent Se requirements. • Longer HRT and lower ORP conditions (less than −350 mV) could potentially provide more complete Se reduction. • Likely most cost-effective approach to large custom-built systems.	• Low tolerance for influent TSS levels >50 mg/L. • Long HRT and larger bioreactors than other technologies. • Potential short-circuiting from solids channelling and gas accumulation. • Lower tolerance for higher influent nitrate loadings than other technologies (example FBR).

the biofilm, and/or remove gases entrained in the media. Backwashing is used to remove accumulated solids and slough biomass from the system in a controlled manner on a periodic basis. The backwash flow rates are typically 4 to 5 times the design forward flow rate for 10- to 20-minute durations to remove TSS and biomass and lift and separate the media. Backwash solids are directed to the backwash tank for settling. Settled solids are directed to further thickening and ultimate dewatering. The backwash supernatant is directed back to the head end of the process. Backwashing is application dependent but is typically done every 2 to 6 weeks.

Degassing operations are typically done once per day to once per week. Degassing requires flow to be stopped to the treatment unit, followed by a back pulse of water supplied by the backwash system at a flow rate in the order of 163–570 L/m^2/min and a duration of approximately 1 to 10 minutes (EPRI, 2019). Degassing typically is required when there is reduced flow through the bed, a higher head loss, or by a routine schedule established by operational experience.

The ABMet® system includes a provision for nutrient addition upstream of both the first and second stage reactors. The original systems used molasses as a carbon source. However, later systems (Site 8 in Table 6.1) use alternative carbon sources such as acetate.

The key design criteria for packed bed reactors in addition to influent TSS are temperature, influent nitrate-nitrogen and empty bed contact time (EBCT). Influent temperature is a key design consideration particularly for industrial applications (potentially too hot for power or too cold for mining applications). Empty bed contact time (EBCT) is defined as the time required for the influent flow to displace the equivalent volume of media in the reactor. The design EBCT for each packed bed reactor currently deployed for FGD wastewater is between 4 and 8 hours or 8 to 16 hours total, respectively, as two reactors in series are used (EPRI, 2019).

The majority of wastewaters treated via packed bed reactors contain a mixture of nitrates, sulfates, and selenium oxyanions. Nitrate is typically present in the parts per million (ppm) range whereas selenium is present in the parts per billion (ppb) range. The majority of the influent nitrate-nitrogen must be reduced in order to get good selenium reduction. Nitrate reduction is typically responsible for the majority of the biomass production in the system with higher influent nitrate concentrations resulting in more biomass and more gaseous nitrogen production with the associated increase in backwash and degassing frequencies.

Packed bed reactors do not deal with high influent nitrate concentrations as well as other reactor configurations. EPRI (2019) has suggested influent nitrate concentrations in the feed water are ideally less than 25 mg/L for a packed bed. Higher influent concentrations may require pre-denitrification or an alternative process selection. Typical applied mass loading of nitrate-nitrogen is in the order of 0.32 to 3.2 $kg/d/m^3$ of media, but additional volume may be required for reaction zones for dissolved oxygen utilization and selenium reduction (EPRI, 2019).

6.4.2 Fluidized bed reactor

A fluidized bed reactor (FBR) is one in which the biofilm grows attached to small carrier particles or media that remain suspended in the fluid. The FBRs are configured so that water flows upward through a vessel containing the media (typically sand or GAC) to promote media fluidization. The velocity of the water flowing upward is typically fixed such that the media bed is expanded by 50% to 70% of the resting volume. Most FBRs are two-phase systems containing only water and bioparticles. Three-phase systems allow for the incorporation of a gas phase. Denitrifying systems can actually be considered as two-phase systems provided the gas flow rate is relatively small compared to the liquid flow (Grady *et al.*, 1999).

The advantages and disadvantages of the FBR system for selenium removal are summarized in Table 6.3.

Table 6.3 Advantages and disadvantages of FBR process.

Advantages	Potential Disadvantages
• Ability to deal with higher influent TSS than packed bed. • Ability to deal with higher influent nitrate loadings than packed bed. • Shorter retention time than packed bed and smaller reactor sizes. • Cost effective for custom built applications.	• Downstream solids removal required. • No full-scale track record on power applications like FGD. • Fewer full-scale operations on selenium removal than either packed bed or combined expanded bed and packed bed configuration.

The main advantage of FBRs relative to other attached growth bioreactors is the small size of the carrier particles which provide a very large surface area for biomass growth. This high specific surface area allows for the maintenance of very high biomass concentrations resulting in shorter hydraulic retention times (HRTs), of minutes, relative to other treatment processes. Periodic deep bed cleaning of the media ensures uniform biofilm and elimination of excess biomass growth, which may cause mounding or poor fluidization of media within the system leading to poor mass transfer. This is typically accomplished via daily automated air lift sparging and periodically with a deep bed hydraulic eductor-based solids separation system or air sparging apparatus.

FBR staging should be considered if the parameters degraded at higher ORP conditions (e.g., oxygen, nitrate) are 15 to 20 times higher in concentration than the other parameters (such as selenium). Staging, in this case, would provide removal of oxygen and nitrate in the first stage, followed by treatment of selenium in the second stage. If a large reaction zone is required, sequential systems may offer a more energy-efficient alternative to a taller or wider system (EPRI, 2019).

The primary design parameters for the FBR include minimum recirculation rate, influent temperature, influent HRT, and TSS. The typical HRT in an Envirogen FBR reactor is 30 to 60 minutes for example compared to the 4 to 8 hours required for a packed bed reactor. Although it is less sensitive to influent TSS than a packed bed the FBR generally cannot tolerate TSS levels greater than 100 mg/L on average. An FBR will typically require a post-treatment solids removal process to meet a low effluent selenium concentration.

The FBR system provided by Envirogen Technologies Inc. uses fine sand and/or activated carbon as the carrier particles. Selenium-containing wastewater is pumped into the FBR in an upflow direction. A recirculation line returns flow to the suction of the influent pump. The combined influent plus recirculation flow is controlled to maintain carrier particle fluidization. The Envirogen FBR system is typically based upon a dual stage design using sand as the carrier particle in the first stage and a GAC carrier in the second stage. The bulk of the denitrification and some of the conversion of selenate to selenite occurs in the Stage 1 reactor. The remaining selenium conversion to elemental selenium occurs in Stage 2. A simple schematic of the Envirogen system is presented as Figure 6.5.

There were two full-scale Envirogen FBR systems installed for biological selenium removal as of September 2018. Both systems were installed to treat mine impacted water. The first system was installed at Patriot Coal in the United States in 2013. The second system was installed at Teck Coal's West Line Creek (WLC) metallurgical coal operations in the Elk Valley of British Columbia in 2014.

McKevitt (2019) presented information on the West Line Creek (WLC) FBR facility at the Metal Leaching/Acid Rock Drainage Workshop in Vancouver in December 2019. The two-stage FBR system includes a ballasted sand clarifier (BSC), moving bed biofilm reactor (MBBR), and sand filtration for post

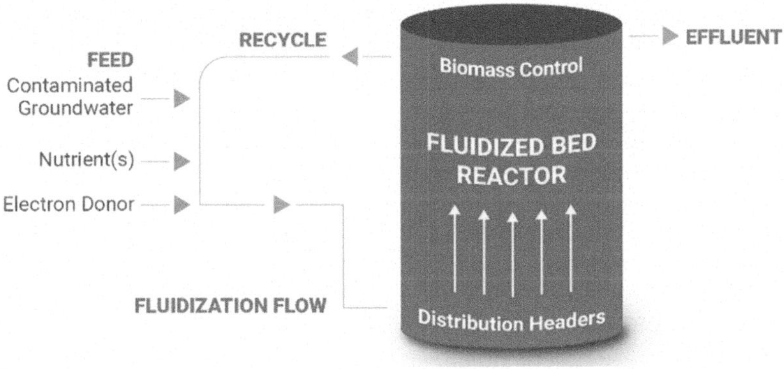

Figure 6.5 Simplified process schematic for fluidized bed reactor (*Source*: adapted from Simm, 2018).

treatment. The BSC and sand filtration units are required primarily for effluent solids and selenium control. This contrasts with a typical packed bed reactor system that can be operated with minimal, if any, post treatment processing depending on the effluent limits, given the packed bed acts as a filter.

Patterson (2017) presented the details of a two-stage ABMet® system treating mine impacted water in North Eastern British Columbia and with similar effluent TSS and selenium limits to WLC. The ABMet® system is able to meet these limits without post treatment downstream of the second stage packed bed.

6.4.3 Combination of expanded bed and packed bed reactor configuration

A hybrid system, using an expanded bed biofilm reactor (EBBR) and packed bed can potentially offer advantages over either of the packed bed or FBR. This system takes advantage of the high mass transfer capability of the EBBR and solids capturing and biodegradation capacity of the packed bed. An EBBR is fed from the bottom of the reactor similar to an FBR but without the recycle. The velocity of the water flowing upward is typically fixed such that the media bed is expanded by 25% to 40%. The advantages and disadvantages of the combined EBBC-packed bed system for selenium removal are summarized in Table 6.4.

Supplemental carbon is typically added both before and after the EBBR system to achieve reduction of nitrate and selenium. The system theory is based on having most of the denitrification and therefore gas formation occurring in the EBBR stage of the process, where it will not create short-circuiting through the reactor or require degassing. The packed bed stage of the system then does not need to be increased in size to allow for retention of this gas volume and the downtime associated with expelling this gas. Effluent solids from the EBBR are collected at the top of the packed bed. Cleaning of the EBBR may involve the use of periodic

Table 6.4 Advantages and disadvantages of combined EBBC-packed bed process.

Advantages	Potential Disadvantages
• Ability to deal with higher influent TSS than packed bed. • Ability to deal with higher influent nitrate loadings than packed bed. • Shorter retention time than packed bed and smaller reactor sizes. • Cost effective for smaller capacities where modular reactor configuration can be used. • Large number of installations relative to other technologies.	• Cost effective capacity potentially constrained by size of largest modular unit.

air scouring followed by a forward flush with the waste transferred along with the packed bed backwash water to solids handling.

The SeHAWK® system supplied by Frontier Water Systems supplies a combination of an EBBR and a packed bed system. The SeHAWK® system is based on a modular bioreactor design to reduce the overall capital cost. A simplified schematic of the SeHAWK® system is presented in Figure 6.6. Frontier had nine full-scale systems installed in mining applications and four full-scale FGD systems under construction as of September 2018 (Simm, 2018). The SeHAWK® systems at Duke Energy's Marshall, Crystal River, and Miller Generating Stations were commissioned in 2019 (Kennedy & Henderson, 2020). All of the SeHAWK® installations as of September 2018 are located in North America.

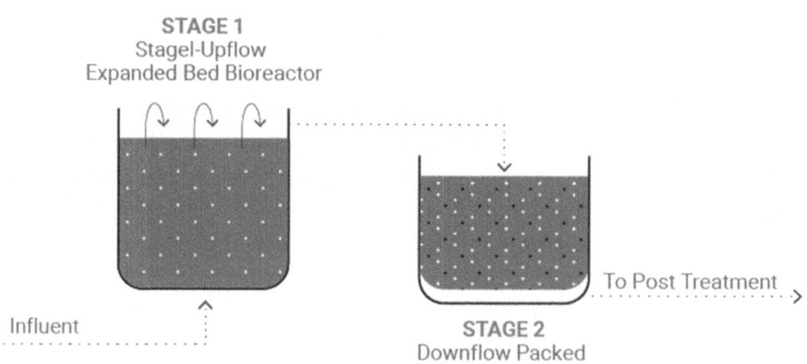

Figure 6.6 Simplified schematic of Frontier SeHawk® system (adapted from Pickett, 2020).

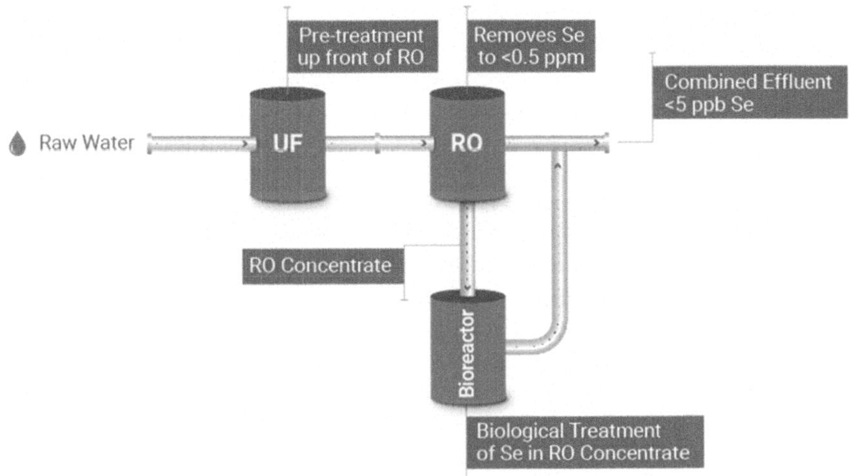

Figure 6.7 Se bioremediation with UF-RO concentration step (from Simm, 2018).

The cost effectiveness of the SeHAWK® system will likely be limited by available module size although Frontier continues to develop larger capacity modular units. In addition, they have employed ultrafiltration (UF) – reverse osmosis (RO) on some projects as a concentration step to reduce the size of the biological treatment step. A simplified process schematic of the UF-RO-Bioremediation configuration is presented in Figure 6.7. This approach is limited by the total dissolved solids (TDS) concentration in the RO concentrate. High TDS concentrates could potentially precipitate on the bioreactor media, thus negatively impacting treatment efficacy. The approach illustrated in Figure 6.7 was used for a selenium treatment system at a phosphate mine in Idaho and is described by Witt *et al.* (2019).

6.4.4 Moving bed biofilm reactor (MBBR)

An MBBR uses specifically designed plastic carrier elements for biofilm attachment that are neutrally buoyant (specific gravity less than/or equal to water). The carrier elements, or media, are held in suspension throughout the reactor by turbulent energy created by mechanical mixing or liquid recirculation, thereby resulting in improved mass transfer through a completely mixed regime. The medium takes up between one-third and two-thirds of the reactor volume. To allow treated water to pass through to the next treatment step while retaining the media in the reactor, a perforated plate or screen is installed on the effluent end of the reactor.

MBBRs combine the advantages of attached growth with many of the advantages of suspended growth systems. One advantage of the MBBR in selenium treatment applications is the ability to handle high influent solids levels. The completely mixed media and continuous flow through process of an MBBR eliminates the need for backwashing to maintain throughput and performance because there is no potential for clogging or plugging. Solids separation will, however, be required following the MBBR to remove solids like reduced selenium particles. MBBR systems can suffer performance problems if precipitates form on the media hampering media neutral buoyancy. This is a particular concern with industrial applications with a high influent TDS.

Veolia is one of several suppliers of MBBR systems. Veolia had provided three MBBR systems for selenium removal in mine water treatment applications with capacities of between 757 litres/min and 3,028 litres/min, as of September 2018 (Simm, 2018). These systems have since been shut down and replaced with passive treatment systems having a lower operation and maintenance cost.

6.5 SUSPENDED GROWTH SYSTEMS

6.5.1 Biofloc systems

Suspended growth systems are distinctly different from attached growth systems in that biological growth is not supported by a substrate, but rather aggregated microorganisms are suspended freely within a water matrix, where they both naturally coagulate and flocculate into bioflocs. A method of solid-liquid separation, such as sedimentation or membrane filtration, is required to concentrate and remove the bioflocs from the treated effluent. Larger tanks are typically required for suspended growth systems relative to fixed film systems because of the higher biomass inventory carried by fixed film processes.

6.5.1.1 Continuous stirred tank system

Lau *et al.* (2012) reported on the design, start-up and commissioning of a full-scale biological treatment system that was installed at a coal-fired power generating station to remove selenium and nitrates from a flue gas desulfurization (FGD) blowdown stream. The patented iBIO® wastewater treatment (WWT) system, which was first pilot-tested at power generating stations, is based on a suspended growth continuously stirred-tank anaerobic reactor. The major components of the treatment system include physical-chemical treatment (pH adjustment and coagulant-polymer addition) and denitrification and selenium removal steps which consist of an anoxic reactor, anaerobic reactor, anaerobic clarifier and a gravity sand filter. The power generating station had a target to reduce the concentration of total selenium from 1.3 mg/L in the untreated FGD blowdown stream down to <0.21 mg/L in the effluent. The authors reported the majority of the selenium treatment took place in the physical-chemical treatment stage (removal of selenite) with only 36% removal in the anaerobic biological treatment stage.

6.5.1.2 Activated sludge systems

EPRI (2019) indicated that there is currently only one commercial activated sludge system known to be operating in the United States that is treating wet flue gas desulfurization wastewater for nitrate and selenium removal.

Mal *et al.* (2017) showed that a sequencing batch reactor (SBR) can remove both selenate and ammonium via, respectively, bioreduction and partial nitrification-denitrification, and thus offers possibilities for treating selenium and ammonium contaminated effluents. These studies were conducted using a bench-scale treatment system.

Interest is growing in the fate of selenium at municipal wastewater treatment facilities, the majority of which use suspended growth systems. This is particularly true in areas with seleniferous soils. Pontarolo *et al.* (2017) conducted a two-year-long study of a biological nutrient removal (BNR) wastewater treatment plant in Colorado with pre-anoxic, anaerobic, anoxic, and aerobic zones. They reported a 49% and 65% reduction in influent recoverable selenium without and with poly-aluminium addition. No attempt was made to optimize selenium removal through the process.

6.5.1.3 Membrane bioreactors

Membrane bioreactors (MBRs) and in particular anaerobic MBRs have been piloted for combined nitrate and selenium reduction. EPRI piloted anaerobic MBR systems for treatment of WFGD blowdown. The results of studies conducted to date are considered by EPRI to be insightful but inconclusive (EPRI, 2019). To this author's knowledge, there are currently no full-scale anaerobic MBR systems in operation for selenium bioremediation.

6.5.2 Granular sludge systems

The upflow anaerobic sludge blanket (UASB) reactor uses self-immobilized small (3 to 5 mm) compact biofilms without carrier support. The environmental conditions created in the reactor result in the development of large, dense, readily settleable particles called granules, which allow for very high biomass concentrations. Wastewater flows from the bottom of the reactor to the top at a high enough velocity (typically 1 m/h) to keep the granules suspended without washout. The upper part of the bioreactor contains a gas/liquid separator to allow produced gases to be vented. UASB reactors are typically employed for high strength wastewaters.

The Adams Avenue Agricultural Drainage Research Centre in California performed a UASB pilot for selenium removal in 2004. The study reported between 58% and more than 90% selenium removal from an influent selenium concentration of approximately 500 μg/L (CH2M Hill, 2010). A number of operational issues were identified with the pilot including short circuiting caused by accumulation of gas within the pilot reactor that became trapped in the sludge, long acclimation period (approximately 6 months), and variability in Se removal efficiency due to temperature sensitivity.

Tan *et al.* (2018a, 2018b) demonstrated that two reactor configurations employing different biomass retention systems (biofilm in biotrickling filter (BTF) and granules in UASB) result in a diverse removal performance when treating synthetic mine wastewater contaminated with SeO_4^{2-}, SO_4^{2-} and Ni^{2+}. The Se removal efficiency of the BTF biofilm was improved by >70% in the presence of SO_4^{2-}, whereas the Se removal performance of the UASB reactor was not affected. Nickel addition initially negatively impacted selenate reduction in both processes, although both processes eventually recovered with the UASB recovering sooner.

6.6 PASSIVE AND SEMI-PASSIVE BIOREACTOR SYSTEMS

The cost of constructing full-scale active selenium treatment facilities like those described in Section 6.4 can be significant. Active biological selenium removal plant projects can cost as much as $3,963 to $7,926 (USD) per cubic metre of capacity to construct (Simm, 2018). The high cost of active treatment has resulted in significant interest in passive treatment technologies for selenium bioremediation particularly in the mining industry.

A passive system can be considered one that does not require a deliberate continuous nutrient feed and can operate with minimal or no electrical equipment and operator attention (Golder, 2020). Passive treatment systems are often referred to as biochemical reactors (BCRs). These systems include an organic medium such as hay, wood chips, or sawdust. Semi-passive systems differ from passive systems in that they include the controlled addition of an electron donor and essential nutrients.

The advantages of passive and semi-passive treatment systems for selenium include low capital and operating cost relative to active treatment. The disadvantages include low hydraulic loading rates and large area requirements, lack of control over organic media degradation, and potential for high levels of residual nutrients. By design, passive treatment lacks precision in process control but can provide acceptable performance in certain applications.

The need to consistently meet ultra-low effluent selenium limits has increased interest in semi-passive treatment systems where electron donor and nutrient addition are more precisely controlled. Gravel bed reactors (GBR™) and submerged rock fills (SRFs) are two examples.

One key issue associated with both passive and semi-passive treatment is the long-term fate of reduced selenium. These systems, unlike the active systems discussed previously, do not include controlled solids wasting to remove reduced selenium from the system. These systems remove selenium from the system for the remainder of the system life. This raises concern with respect to what happens if there is an upset and/or in cases where the conditions develop such that the selenium could/can be remobilized?

6.6.1 Constructed wetlands

Constructed wetlands are designed and constructed to use vegetation, soils, and associated microbial activity to provide treatment. Wetlands create a layer of biological detritus that through decomposition creates an anoxic/anaerobic substrate rich in organic carbon. The aquatic environment supports the growth of bacteria supporting the reduction of selenium oxyanions as an energy source. According to Kadlec and Wallace (2009) selenium is reduced to elemental selenium as well as organic forms such as dimethyl selenide and dimethyl diselenide. The elemental selenium is sequestered in the wetland sediment.

Constructed wetlands include surface flow wetlands, subsurface flow wetlands, and variations on surface flow wetlands such as vertical downflow wetlands. Constructed wetlands are designed to allow inflow rates and water depths to be regulated which influences hydraulic and mass loading of the system as well as the hydraulic residence time. Typical retention times can be several days or more (CH2M Hill, 2010). The area requirements for municipal and industrial wastewaters can be 4 to 405 hectares and more than 4,000 hectares for large agricultural drainage flows (CH2M Hill, 2010).

The treatment effectiveness among surface flow wetlands varies widely. Kadlec and Wallace (2009) reported selenium removal rates of between 0% and 96% for 10 full-scale surface flow wetland sites. Constructed wetlands have been attempted at full scale at several power plants and interest appears to have waned due to performance issues (Kennedy, Personal Communications). Some have witnessed late Winter and early Spring selenium release for instance (Kennedy, Personal Communications).

Constructed wetlands may also pose an ecological risk to wildlife by exposure to accumulated constituents. Monitoring of ecological effects may be required particularly for selenium impacts on fish and other vertebrates. Practices to exclude wildlife, such as covering, can be appropriate.

6.6.2 Biochemical reactors

Biochemical reactors (BCRs) consist of a lined area that has been filled with organic substrate and can be considered as an enhancement to constructed wetlands. These systems are generally operated in a gravity downflow mode. An underdrain, overlain by gravel, is constructed at the bottom of the reactor. Water level control is installed to control the water level in the BCR. Organic substrates in the reactor include hay, manure, and woodchips. The organic media degrade over time and will ultimately need to be replaced. Anoxic/anaerobic conditions are created in the reactor and selenium reducing bacteria convert selenium oxyanions to elemental selenium. The effluent from a BCR will typically have elevated BOD (biological oxygen demand) and will require further aerobic treatment. CH2M Hill (2010) present several BCR case studies with a range of selenium removal efficiencies. As indicated previously, these systems have lower capital and operation and maintenance costs than active treatment systems. However, the ability for these systems to consistently meet ultra-low selenium discharge limits is uncertain.

6.6.3 Gravel bed reactors

Gravel bed reactors (GBR™) are referred to as a semi-passive treatment alternative (Mancini *et al.*, 2019). The GBR™ consists of an engineered bed of gravel/media through which water containing nutrients of concern is passed and treated. Electron donor(s) and nutrients are added to the water at the inlet of the GBR™ to promote the activity of selenium reducing bacteria, sequentially immobilizing the reduced selenium in the gravel bed. The top and bottom of the treatment zone are lined using synthetic membranes to prevent water loss/influx, creating hydraulic isolation from the surrounding environment. A schematic of a typical GBR™ is presented in Figure 6.8.

Mancini *et al.* (2019) provide general performance information for two case studies treating water in an urban stream in California and mine impacted water at a coal mine in West Virginia. The influent selenium concentrations for the California and West Virginia applications were approximately 40 ppb and 30 ppb, respectively. Both systems are reported to reduce selenium to less than 5 ppb. These types of systems offer lower capital, operating and maintenance cost relative to active systems. The one concern with these types of systems is the long-term fate of sequestered selenium given there is no active control of solids wasting.

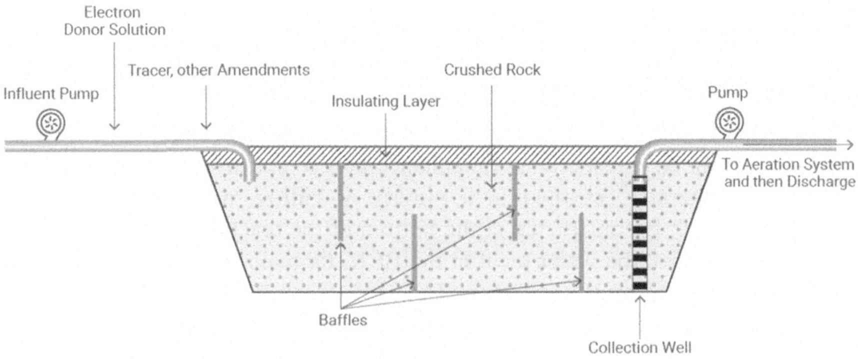

Figure 6.8 GBR™ main system components (figure from Mancini *et al.*, 2019).

6.6.4 Submerged rock fills in mining applications

Submerged Rock Fill (SRF) uses a backfilled mine pit as a bioreactor (Golder, 2020). Mine water and nutrients are injected into the backfill, then flow horizontally through the backfill, and the treated water is pumped out of the backfill for polishing treatment and discharge. Denitrification and selenium reduction occur in the backfill. The long-term fate of sequestered selenium is of concern with SRF as it is with GBRs.

One of the mining companies in British Columbia, Canada, has posted on the company website that they plan to expand their SRF for selenium bioremediation from 10 million to 20 million litres per day (Teck, 2020). The expanded facility will replace a previously planned active treatment system. The rationale for choosing the SRF over a tank based system includes: *"SRFs can treat large volumes of water with less energy and with a smaller environmental footprint compared to tank-based facilities, SRFs are also quicker to build, less complex to operate, and have lower capital and operating costs".*

6.7 OTHER REACTOR TYPES

6.7.1 Fungal based bioreactors

Fungi are heterotrophic, eukaryotic and achlorophyllous organisms, which have the ability to secrete large amounts of enzymes and reductive proteins. Different fungal strains have shown their ability to degrade a wide range of environmental pollutants, from dyes to pharmaceutical compounds, heavy metals, trace organic contaminants and endocrine disrupting contaminants. Some researchers have investigated the potential of fungal based bioreactors for selenium bioremediation (Espinosa-Ortiz *et al.*, 2015, 2016; Negi *et al.*, 2020).

Espinosa-Ortiz *et al.* (2015) investigated the performance of a novel fungal bioreactor system containing pellets of *Phanerochaete chrysosporium*. The bioreactor was initially operated under batch conditions and then switched to continuous operation for 41 days. The reactor was fed with selenite at selenium and glucose loading rates of 10 mg Se L/d and 0.95 g glucose L/d, respectively. The reactor hydraulic retention time was 24 hours. The reported selenium removal at steady state was approximately 70%. The reactor was tested under fluctuating loads and showed significant resilience. Espinosa-Ortiz *et al.* (2017) also studied the simultaneous biomineralization of Se and Te by *P. chrysosporium*.

6.7.2 Electro-biochemical reactor

The electro-biochemical reactor (EBR), produced by INOTEC, is a patented high-efficiency denitrification, metals and inorganics removal technology. The system is based on the concept of providing electrons to the microbial community using electrodes and a low voltage potential (1–3 V) at low milli-Amp levels (Opara *et al.*, 2014).

INOTEC lists several selenium removal EBR projects on its website (website visited August 30, 2020). These include 17 projects (11 bench-scale tests, 5 pilots, and 1 full-scale facility). Selenium was not a target constituent for 2 of the 17 projects. The one full-scale facility on INOTEC's selenium experience list is for a gold mining project where influent selenium is reportedly reduced from 690 ppb to approximately 50 ppb. The influent nitrate at the gold mining site is reportedly approximately 180 mg/L as nitrate-nitrogen. The bioreactor HRT is listed as 24 hours.

6.7.3 Hydrogen based membrane biofilm reactor

The hydrogen-based membrane biofilm reactor (MBfR) is designed to reduce oxidized pollutants using hydrogen gas as an electron donor. Hydrogen or a mixture of hydrogen and carbon dioxide (4:1) is delivered by diffusion through the walls of non-porous hollow fibre membranes. A hydrogen oxidizing biofilm forms on the membrane outer walls and reduces electron acceptors in the bulk liquid including nitrate and selenate.

Several studies (Chung *et al.*, 2006; Lai *et al.*, 2014; Van Ginkel *et al.*, 2011a, 2011b) have reported high SeO_4^{2-} removal rates employing the MBfR technology. Studies with actual WFGD wastewater showed a high selenate reduction efficiency (Van Ginkel *et al.*, 2011a, 2011b).

The MBfR technology has been commercialized by APTwater of Sacramento (California, USA). The APTwater website references the selenium removal technology as ARoSel targeting selenium, naturally occurring and from flue gas desulfurization, oil production and refining. As of August 30, 2020, the APTwater website referenced the ARoSel system as "In Development".

6.8 FUTURE PERSPECTIVES FOR OPTIMIZING BIOLOGICAL SELENIUM REMOVAL TECHNOLOGIES

Implementation of more stringent aquatic water quality guidelines for selenium, particularly in North America, has driven the development and commercialization of a small number of active selenium bioremediation technologies. Several private companies and members of the Electrical Power Research Institute (EPRI) in the United States have conducted numerous pilot studies of selenium bioremediation technologies. Golder's (2020) recently completed report on the state of knowledge of selenium treatment technologies concluded: "*despite numerous installations, selenium treatment technologies have not reached full maturity and should be regarded as developmental.*"

A 2018 presentation (Simm, 2018) presented at the EPRI Selenium Summit in the Netherlands summarized some of the key issues and challenges associated with biological Se removal and identified key research and development needs based upon observations at full-scale facilities. One of the primary drivers for having this summit in Europe was to connect leading European researchers working on the fundamental science of selenium bioremediation with North American practitioners engaged in implementing selenium bioremediation projects to meet current and proposed stringent effluent selenium regulations. It is this author's opinion that the practice of selenium bioremediation is lagging the regulatory implementation. Some of the key issues and research requirements associated with selenium bioremediation are presented in this section.

6.8.1 Selenium measurement and speciation

The primary driver for ultra-low selenium limits is the protection of sensitive aquatic species and waterfowl. The concentration and speciation of selenium affects its bioaccumulation potential. It can be difficult to measure selenium speciation in complex environmental matrices like mine water and wet flue gas desulfurization (WFGD) blowdown. Being able to accurately measure influent and effluent selenium concentrations is critically important.

Based on what is known about the potential for the production of organic Se species in biological treatment systems, monitoring for their presence could be extremely beneficial in predicting the toxicity of the effluent. According to LeBlanc *et al.* (2018), sensitive and robust analytical methods for the measurement of Se speciation in water samples are required for such an endeavour, as environmentally relevant concentrations of organic Se may be only a small fraction of the total Se in a given sample.

6.8.2 Bioavailability of reduced selenium species in treated effluents

Waterborne inorganic selenate and selenite typically bioaccumulate 100 to 4000 times in aquatic food chains, but organic selenoamino acids can produce bioaccumulation factors in excess of 350,000 (Tan *et al.*, 2016). There have been limited studies on effluent selenium speciation and its direct impact upon the aquatic environment. However, the studies that have been conducted suggest reduced selenium species in the effluent from a selenium bioremediation facility are of potential concern.

Amweg *et al.* (2003) studied an algal-bacterial selenium reduction (ABSR) process for treating agricultural drainage water in the San Joaquin Valley of California. These authors used laboratory-based algal bioaccumulation tests and *in situ* microcosms with a variety of invertebrates to measure differences in Se bioavailability before and after ABSR treatment. These authors reported a 2 to 4 times increase in Se concentration in aquatic biota tissue exposed to ABSR system effluent relative to organisms exposed to untreated water. This is despite the fact the ABSR system removed approximately 80% of the total influent Se.

LeBlanc and Wallschläger (2016) measured organo-selenium in natural and industrial waters. These authors showed industrial biological treatment systems designed for remediation of selenium-contaminated waters increased both the concentration of organic selenium species in the effluent, relative to influent water, and the fraction of organic selenium which increased to up to 8.7% of the total selenium in the effluent, from less than 1.1% in the influent.

Golder *et al.* (2020) reported on one biological system where environmental effects monitoring indicated an approximate 50% reduction in selenium concentration in the receiving waters. Despite this reduction, the near-stream selenium concentrations in benthic invertebrates increased approximately 7-fold compared to before treatment. The owner at this particular site attributed the increase to organo-selenium species in the effluent and ultimately implemented an effluent oxidation step to convert reduced selenium species to selenate.

The management of reduced selenium or organic selenium species in the effluent of biological selenium treatment facilities serves as a potential challenge that needs to be addressed. Research needs to be conducted on the reactor conditions leading to organo-selenium production and optimization of post treatment methods for mitigating their impact.

6.8.3 Bioprocess operations

6.8.3.1 Bioreactor sizing and design optimization

Selenium oxyanions are the dominant selenium species in selenium impacted waters requiring treatment. Nitrate and sulfate are typically present in significant concentrations (parts per million) relative to selenium oxyanions (parts per billion). As indicated in Section 6.2, significant selenium oxyanion reduction does not typically take place until the majority of nitrate has been reduced.

Contrary results on the inhibition or stimulation of selenium oxyanion reduction in the presence of NO_3^- and SO_4^{2-} have been reported. In addition, previous studies have shown both sulfate reducing bacteria and denitrifiers can reduce selenium oxyanions (Simm, 2017). Tan *et al.* (2016) have suggested optimizing bioreactor operational parameters for establishing and maintaining a microbial community with coexisting denitrifying as well as sulfate and selenate reducing bacteria is yet to be fully understood and operated at full scale.

The EPRI State of the Knowledge Document (2019) suggested: *"Even though the complex nature of selenium reduction is not fully understood, WFGD wastewater biological treatment systems have operated for more than 10 years and are demonstrating significant selenium reduction. Identifying and understanding the pathways for selenium reduction may lead to new insights into operating systems more efficiently."*

There is a need for a biochemical model and process simulator that takes into account the interactions between nitrate, selenium oxyanions, and sulfate and that can be used to provide operational insights for both pilot and full-scale facilities (Simm, 2017). Such a model could ultimately reduce piloting requirements and be used for system sizing and design optimization.

Boltz (2019) is currently developing a mathematical model that is capable of describing the biochemical transformation of selenate (SeO_4^{2-}), selenite (SeO_3^{2-}), nitrate (NO_3^-), and sulfate (SO_4^{2-}), amongst other relevant compounds, by functional bacterial groups within the system's operational environmental conditions. Model features include a description of the simulated processes, state variables, Gujer matrices, stoichiometric relationships, kinetic expressions, conversion factors, and parameter values (see Chapter 4). Preliminary model results suggest good agreement between model predictions and pilot plant observations. Additional research is required to verify the model's efficacy.

6.8.3.2 Better understanding and optimization of passive treatment designs

Reduced selenium (S^0 and selenides) is not routinely removed in passive systems and will accumulate over time. The long-term fate of these accumulations is not well understood. There is a need for research that identifies the selenium sequestration capacity of various passive systems, potential for selenium re-release and the factors controlling re-release, and best practices for environmentally sound system closure.

6.8.3.3 Optimization of selenium reduction at municipal wastewater treatment plants

Municipal wastewater treatment plants in areas with seleniferous soils are likely to have elevated influent Se and stringent effluent Se discharge limits. This is certainly the case in the western US in states like Colorado and Arizona. The work of Pontarolo *et al.* (2017) is instructive and there is a need for further research focused on optimization of Se removal in municipal activated sludge systems and in suspended growth systems in general.

6.8.3.4 Selenium treatment residuals handling and long-term management

Most selenium bioremediation technologies produce residuals with a high selenium concentration. In many cases, selenium laden residuals from active treatment systems are dewatered and landfilled. The long-term fate of these residuals is a concern especially the potential for re-mobilization and future environmental contamination. There is significant interest in selenium recovery technologies (Hageman, 2015) and/or encapsulation.

REFERENCES

Amweg E. L., Stuart D. L. and Weston D. P. (2003). Comparative bioavailability of selenium to aquatic organisms after biological treatment of agricultural drainage water. *Aquatic Toxicology*, **63**, 13–25.

Boltz J. (2019). SeSANS Mathematical and Process Model. Presented at Electric Power Research Institute (EPRI) Selenium Summit 2019, 10 September 2019.

CH2M Hill (2010). Review of Available Technologies for the Removal of Selenium from Water. Prepared for the North American Metals Council – Selenium Working Group, June 2010.

CH2M Hill (2013). NAMC White Paper Report Addendum. Prepared for the North American Metals Council – Selenium Working Group, 29 March 2013.

Chapman P. M., Adams W. H., Marjorie Brooks M., Charles G., Delos C. G., Samuel N., Luoma S. N., Mahar W. A., Ohlendorf H. M., Presser T. S. and Shaw P. (2010). Ecological Assessment of Selenium in the Environment. CRC Press, Taylor & Francis Group, Boca Raton, London, New York.

Chung J., Nerenberg R. and Rittmann B. E. (2006). Bio-reduction of selenate in a hydrogen-based membrane biofilm reactor. *Environmental Science & Technology*, **40**, 1664–1671.

Electrical Power Research Institute (EPRI) (2009). Evaluation of GE ABMet at Duke Energy's Belews Creek Stream Station. Report 66517.

Electrical Power Research Institute (EPRI) (2019). State of the Knowledge: Biological Treatment for Wet Flue Gas Desulfurization Wastewater.

Espinosa-Ortiz E. J., Rene E. R., Guyot F., van Hullebusch E. D. and Lens P. N. L. (2015). Biomineralization of tellurium and selenium-tellurium nanoparticles by the white-rot fungus *Phanerochaete chrysosporium*. *International Biodeterioration & Biodegradation*, **102**, 361–369.

Espinosa-Ortiz E. J., Pechaud Y., Lauchnor E., Rene E. R., Gerlach R., Peyton B. M. and Lens P. N. (2016). Effect of selenite on the morphology and respiratory activity of Phanerochaete chrysosporium biofilms. *Bioresource Technology*, **210**, 138–145. https://doi.org/10.1016/j.biortech.2016.02.074.

Espinosa-Ortiz E. J., Rene E. R., van Hullebusch E. D. and Lens P. N. L. (2017). Removal of selenite from wastewater in a *Phanerochaete chrysosporium* pellet based fungal bioreactor. *International Biodeterioration & Biodegradation*, **124**, 258–266.

Golder Associates Ltd (2020). State of the Knowledge on Selenium Treatment Technologies, North American Metals Council – Selenium Working Group, April 2020.

Grady C. L., Daigger G. T. and Lim H. C. (1999). Biological Wastewater Treatment. Marcel Dekker, Inc, New York.

Hageman S. P. W. (2015). Bio-induced Solid Selenium for Recovery from Water. PhD thesis, Wageningen University, October 2015.

Janz D. M., DeForest D. K., Brooks M. L., Chapman P. M., Gilron G., Hoff D., Hopkins W. A., McIntyre D. O., Mebane C. A., Palace V. P., Skorupa J. P. and Wayland M. (2010). Selenium toxicity to aquatic organisms. In: Ecological Assessment of Selenium in the Aquatic Environment, P. M. Chapman, W. J. Adams, M. L. Brooks, C. G. Delos, S. N. Luoma, W. A. Maher, H. M. Ohlendorf, T. P. Presser and D. P. Shaw (eds.), CRC Press, Boca Raton, Florida, pp. 141–231.

Kadlec R. H. and Wallace S. D. (2009). Treatment Wetlands, 2nd edn. CRC Press, Boca Raton, FL, pp. 475–480.

Kennedy B. (Ex-Manager, Duke Energy), personal communication.

Kennedy B. and Henderson D. (2020). Selenium Removal from FGD Wastewaters: A Critical Look at Biotreatment Technologies. Presentation Made at North American Metals Council – Selenium Working Group Meeting, 11 June 2020.

Kumkrong P., LeBlanc K. L., Mercier P. H. J. and Mester Z. (2018). Selenium analysis in waters. Part 1: regulations and standard methods. *Science of the Total Environment*, **640–641**, 1611–1634.

Lai C. Y., Yang X., Tang Y., Rittmann B. E. and Zhao H. P. (2014). Nitrate shaped the selenate-reducing microbial community in a hydrogen-based biofilm reactor. *Environmental Science & Technology*, **48**, 3395–3402.

Lau A., Pudvay M., Bradberry J., D'Alesandro R., Frank S., Garaventa A., Wojciechowski S., Stover E., Stover R., Campana C., Bernstein R. and McNamee H. (2012). Design and start-up of a full-scale biological selenium removal system for flue gas desulfurization (FGD) wastewater from a power generating station. Presented at the Annual IWC Conference, 4–8 November 2012, San Antonio, Texas.

LeBlanc K. L. and Wallschläger D. (2016). Production and release of Selenomethionine and related organic selenium species by microorganisms in natural and industrial waters. *Environmental Science & Technology*, **50**(12), 6164–6171.

LeBlanc K. L., Kumkrong P., Mercier P. H. J. and Mester Z. (2018). Selenium analysis in waters, part 2: speciation methods. *Science of the Total Environment*, **640–641**, 1635–1651.

Lemly A. D. (2004). Aquatic selenium pollution is a global environmental safety issue. *Ecotoxicology and Environmental Safety*, **59**, 44–56.

Lemly A. D. (2007). A procedure for NEPA assessment of selenium hazards associated with mining. *Environmental Monitoring and Assessment*, **125**, 361–375. https://doi.org/10.1007/s10661-006-9445-9.

Mal J., Nancharaiah Y. V., van Hullebusch E. D. and Lens P. N. L. (2017). Biological removal of selenate and ammonium by activated sludge in a sequencing batch reactor. *Bioresource Technology*, **229**, 11–19.

Mancini S., Cox E., deVlaming L., Bechard K., James R., Przepiora A. and Risacher F. (2019). Gravel Bed Reactors: A Water Treatment Technology for Industrial & Mining Applications, Published in British Columbia Environmental Industry Guide (page 19).

McKevitt B. (2019). Continuous Improvement at Teck's West Line Creek Water Treatment Plant. 26th Annual BC/MEND Metal Leaching/Acid Rock Drainage Workshop, 4–5 December 2019, Vancouver, BC.

Nancharaiah Y. V. and Lens P. N. L. (2015a). Selenium biomineralization for biotechnological applications. *Trends in Biotechnology*, **33**(6), 323–330.

Nancharaiah Y. V. and Lens P. N. L. (2015b). The ecology and biotechnology of selenium-respiring bacteria. *Microbiology and Molecular Biology Reviews*, **79**(1), 61–80.

Negi B. B., Sinharoy A. and Pakshirajan K. (2020). Selenite removal from wastewater using fungal pelleted airlift bioreactor. *Environmental Science and Pollution Research*, **27**, 992–1003. https://doi.org/10.1007/s11356-019-06946-6

Opara A., Peoples M. J., Adams J. D. and Martin A. S. (2014). Electro-biochemical reactor (EBR) technology for selenium removal from British Columbia's coal-mining wastewaters. https://www.inotec.us/uploads/5/1/2/8/5128573/opara_et_al_ebr_technology_for_selenium_removal_from_bc_coal_mining_wastewaters.pdf (accessed 18 June 2021).

Patterson M. (2017). Selenium Water Treatment Demonstration Facility. Presented to the North American Metals Council – Selenium Working Group, 8 June 2017.

Pickett T. (2020). Evolution of the SeHAWK® Bioreactor System for Selenium Treatment. Presented at the North American Metals Council – Selenium Working Group, 11 June 2020.

Pontarolo D., Sandy T., Keller N., Gearhart N., Patel V. and Jimenez J. (2017). Fate and Forms of Selenium in Biological Nutrient Removal Wastewater Treatment Plants. WEFTEC 2017 Proceedings, pp. 4118–4132.

Seiler R. L., Skorupa J. P., Naftz D. L. and Nolan B. T. (1965). Irrigation-induced Contamination of Water, Sediment, and Biota in the Western United States – Synthesis of Data from the National Irrigation Water Quality Program. U.S. Geological Survey; Professional Paper 1665; United States Department of the Interior, Denver, Colorado.

Shrift A. (1964). A selenium cycle in nature. *Nature*, **201**, 1304–1305.

Simm R. (2017). State of the Art of Selenium Treatment and Process Model Research. Presented at the North American Metals Council – Selenium Working Group Meeting, 8 June 2017.

Simm R. (2018). State of the Art and Vendor Market for Biological Selenium Removal in North America, EPRI (Electrical Power Research Institute), Selenium Summit 2018, Leeuwarden, The Netherlands.

Soda S., Kashiwa M., Kagami T., Kuroda M., Yamashita M. and Ike M. (2011). Laboratory scale bioreactors for soluble selenium removal from selenium refinery wastewater using anaerobic sludge. *Desalination*, **279**(1–3), 433–438.

Tan L. C., Nancharaiah Y. V., van Hullebusch E. E. and Lens P. N. L. (2016). Selenium: environmental significance, pollution, and biological treatment technologies. *Biotechnology Advances*, **34**, 886–907.

Tan L. C., Nancharaiah Y. V., van Hullebusch E. and Lens P. N. L. (2018a). Effect of elevated nitrate and sulfate concentrations on selenate removal by mesophilic anaerobic granular sludge bed reactors. *Environmental Science: Water Research and Technology*, **4**, 303–314. doi: doi:10.1039/c7ew00307b

Tan L. C., Papirio S., Luongo V., Nancharaiah Y. V., Cennamo P., Esposito G., van Hullebusch E. D. and Lens P. N. L. (2018b). Comparative performance of anaerobic attached biofilm and granular sludge reactors for the treatment of model mine drainage wastewater containing selenate, sulfate and nickel. *Chemical Engineering Journal*, **345**, 545–555.

Teck Resources, Water Stewardship (2020). https://www.teck.com/responsibility/approach-to-responsibility/sustainability-report-disclosure-portal/material-topics/water-stewardship/ (accessed 18 June 2021).

Van Ginkel S. W., Yang Z., Kim B. O., Sholin M. and Rittmann B. E. (2011a). Effect of pH on nitrate and selenate reduction in flue gas desulfurization brine using the H_2-based membrane biofilm reactor (MBfR). *Water Science & Technology*, **63**, 2923–2928.

Van Ginkel S. W., Yang Z., Kim B. O., Sholin M. and Rittmann B. E. (2011b). The removal of selenate to low ppb levels from flue gas desulfurization brine using the H_2-based membrane biofilm reactor (MBfR). *Bioresour Technol*, **102**, 6360–6364.

Witt J., Prouty A. and Aulbach J. (2019). Physio-Biological Removal of Selenium from Mining Impacted Waters. Paper Presented at the International Water Conference, November 2019, Orlando, Florida.

Chapter 7

In situ and *ex situ* bioremediation of seleniferous soils and sediments

Shrutika L. Wadgaonkar and Piet N. L. Lens

7.1 INTRODUCTION

The importance of environmental selenium (Se) research has been increasingly recognized during the last decade (Nancharaiah & Lens, 2015a; Tan *et al.*, 2016). The concerns about Se toxicity began in the 1930s, when symptoms for alkali disease and blind staggers were observed in livestock grazing on grass grown on Se-enriched soil in South Dakota (Tinggi, 2003). On the other hand, Se deficiency was brought to the forefront in the 1960s with identification of a peculiar heart muscle disease symptom, called Keshan's disease, in China (Chen, 2012).

In animals and humans, selenium plays an important role in the redox regulation of intracellular signaling, redox homeostasis and thyroid hormone metabolism (Huawei, 2009; Papp *et al.*, 2007). To avoid deficiency and toxicity, the United States National Academy of Sciences Panel on Dietary Oxidants and Related Compounds recommended a dietary allowance of 55 μg Se day^{-1} in humans and set an upper tolerable limit of 400 μg Se day^{-1} (World Health Organization, 2011). The United Kingdom Expert Group on Vitamins and Minerals (2003) recommended a minimum intake of 60 μg Se day^{-1} for women and 70 μg Se day^{-1} for men.

The amount of selenium in the food chain, and thus in the human diet, depends on the selenium concentrations in the soil. Therefore, soil is the most important part of

doi: 10.2166/9781789061055_0217

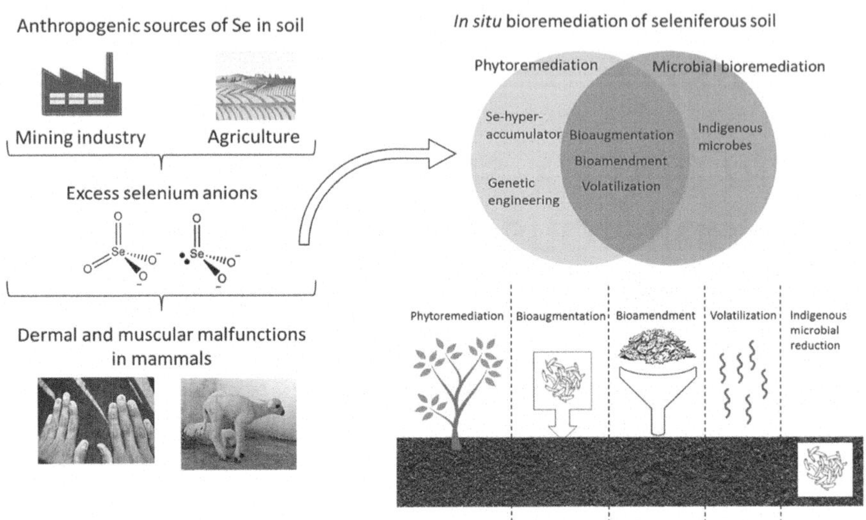

Figure 7.1 General schematic of causes, effects and bioremediation options for seleniferous soils and sediments (Wadgaonkar, 2017).

the environment in selenium cycling (Hagarova *et al.*, 2005; Wadgaonkar *et al.*, 2018a). This chapter overviews research on the selenium cycle in the context of bioremediation of seleniferous soils and sediments. Figure 7.1 provides a general overview of seleniferous soil contamination and *in situ* bioremediation techniques applied. Recent studies on the natural and anthropogenic sources of selenium in soils and the adverse effects of elevated soil-Se content on the environment are overviewed in this chapter. Particular attention is paid to the cause and effect of selenium toxicity on flora and fauna associated with seleniferous soils and sediments as well as technologies for bioremediation of seleniferous soils coupled to selenium recovery.

7.2 METABOLIC ROLE OF SELENIUM

7.2.1 Selenium essentiality

In humans, selenium plays a complex metabolic role in protection of body tissues against oxidative stress, maintenance of the immune system and modulation of growth and development (Figure 7.2). Most of the assimilated selenium in tissues is available in the form of proteins called selenoproteins, in which selenium exists as selenocysteine (SeCys). Table 7.1 lists selenoproteins identified in humans and their functions. Selenium is an essential part of the antioxidant enzyme glutathione peroxidase which protects cell membranes from damage caused by lipid peroxidation. Selenoproteins like thioredoxin reductase and glutathione peroxidase play a role in cancer prevention by preventing intracellular oxidative

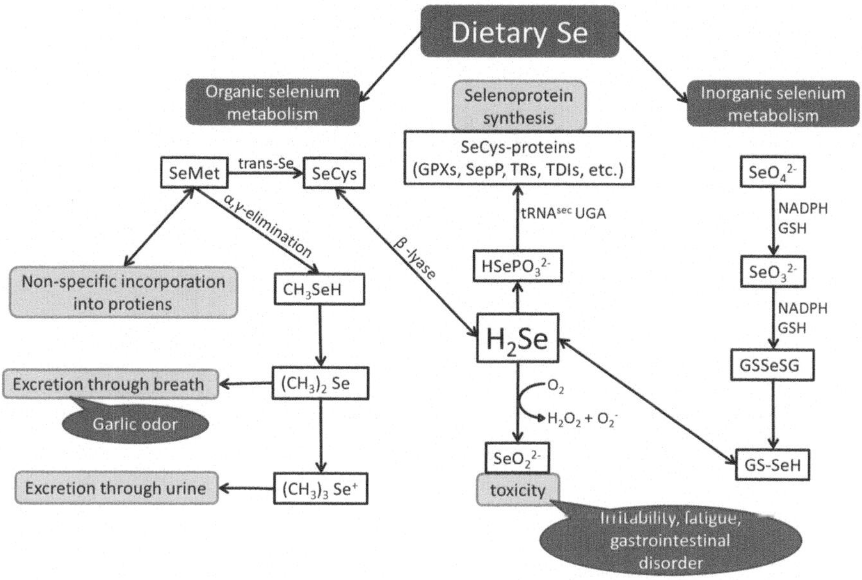

Figure 7.2 Schematic representation of selenium metabolism in mammals. Se, selenium; SeMet, selenomethionine; SeCys, selenocysteine; H_2Se, hydrogen selenide; $HSePO_3^{2-}$, selenophosphate; CH_3SeH, methylselenol; $(CH_3)_2Se$, dimethyl selenide; SeO_2, selenium dioxide; $(CH_3)_3Se^+$, trimethyl selenonium ion; NADPH, nicotinamide adenine dinucleotide phosphate; GSH, glutathione; TRs, thioredoxin reductases; TDIs, thyroixine de-iodinases; GPXs, glutathione peroxidases; SepP, selenoprotein P (Fairweather-tait *et al.*, 2010; Huawei, 2009; Tinggi, 2003).

stress (Kalender *et al.*, 2013; Selenius *et al.*, 2010). Selenoproteins play an important role not only in the regulation of the intracellular redox state and anti-inflammatory functions, but may also play a significant role in glucose metabolism, calcium metabolism and glycoprotein folding (Rayman, 2012). Selenium forms a structural component of specific selenoproteins incorporated in the form of Se-methionine (Se-Met) in plants and Se-cysteine (SeCys) in animals. Active sites of selenoproteins consisting of SeCys have redox functions, such as scavenging free radicals (Figure 7.3), thus preventing oxidative stress and cancer (Misra *et al.*, 2015; Tapiero *et al.*, 2003).

7.2.2 Selenium toxicity

Selenium is a contaminant of potential environmental concern and is hence an important element from an environmental pollution point of view. The element was discovered to be toxic when the livestock grazing on grass grown in

Table 7.1 List of selenoproteins identified in humans and their functions

Selenoprotein	Tissue/Position	Role/ Functions	References
Glutathione peroxidase (GPX1)	Cell cytosol	Antioxidant reducing H_2O_2 and phospholipase A2 cleaved lipid hydroperoxides and storage vehicle for Se	Brown and Arthur (2001), Diamond (2015), Papp *et al.* (2007), Rayman (2012)
Glutathione peroxidase (GPX2)	Gastrointestine	Protection from toxicity of ingested lipid hydroperoxides in mammals; intracellular defense mechanisms against oxidative damage by preventing production of reactive oxygen species in colon	Brown and Arthur (2001), Papp *et al.* (2007), Rayman (2012)
Glutathione peroxidase (GPX4)	Membrane associated phospholipid hydroperoxide	Reductive destruction of lipid hydroperoxides and small soluble hydroperoxides; capable of metabolizing cholesterol and cholesterol ester hydroperoxides in oxidized low density lipoprotein	Brown and Arthur (2001), Diamond (2015), Papp *et al.* (2007), Rayman (2012)
Glutathione peroxidase (GPX3)	Extracellular	Possible antioxidant in renal tubules	Brown and Arthur (2001), Papp *et al.* (2007), Rayman (2012)
Selenium-binding protein 1 (SBP1)	Nucleus and cytoplasm of prostrate tissue	Toxification/detoxification process, cell-growth regulation, intra-Golgi protein transport, aging and lipid metabolisms	Diamond (2015), Papp *et al.* (2007)
Thioredoxin reductases-1 (TR1)	Cell cytosol	Regulation of intracellular redox state	Papp *et al.* (2007), Rayman (2012)
Thioredoxin reductases-2 (TR2)	Mitochondria	Regulation of intracellular redox state	Papp *et al.* (2007), Rayman (2012)
Thioredoxin reductases-3 (TR3)	Testis	Regulation of intracellular redox state	Papp *et al.* (2007), Rayman (2012)
Iodothyronine deiodinases type1 (DI1)	Kidney, liver, thyroid, brown adipose tissue	Inactive thyroxine metabolism to active 3,3'-5'triiodothyronine	Papp *et al.* (2007), Rayman (2012)

Iodothyronine deiodinases type2 (DI2)	Thyroid, central nervous system, pituitary, skeletal muscle, adipose tissue	Activation of thyroid hormones	Papp *et al.* (2007), Rayman (2012)
Iodothyronine deiodinases type3 (DI3)	Placenta, central nervous system, fetus	Inactivation of thyroid hormone	Papp *et al.* (2007), Rayman (2012)
Selenoprotein-P (SelP)	Plasma, brain, liver and testis	Selenium homeostasis; antioxidant activity	Brown and Arthur (2001), Papp *et al.* (2007), Rayman (2012)
Selenoprotein-W (SelW)	Brain, colon, heart, skeletal muscle, prostrate	Antioxidant activity, cardiac and skeletal muscle metabolism	Brown and Arthur (2001), Papp *et al.* (2007);
Selenoprotein-N (SelN)	Endoplasmic reticulum	Unknown	Papp *et al.* (2007)
Selenoprotein-S (SelS)	Endoplasmic reticulum	Role in innate immune response, inflammation, regulation of cytokines	Papp *et al.* (2007)
Selenoprotein-K (SelK)	Endoplasmic reticulum	Unknown	Papp *et al.* (2007)
Selenoprotein-R (SelR)	Cytosol and nucleus	Methionine metabolism, protein repair, antioxidant activity	Papp *et al.* (2007)
Selenoprotein-H (SelH)	Nucleus	DNA binding protein, regulation of glutathione synthesis genes and phase II detoxification	Kurokawa and Berry (2013), Fairweather-tait *et al.* (2010)
Selenoprotein-M (SelM)	Moderate expression on heart, lungs, kidney, uterus and placenta; and high expression in thyroid and brain	Unknown	Papp *et al.* (2007)
Seleno-phosphate synthetase	Cell cytosol	This enzyme regulates selenocysteine incorporation in selenoproteins to prevent toxicity	Papp *et al.* (2007)

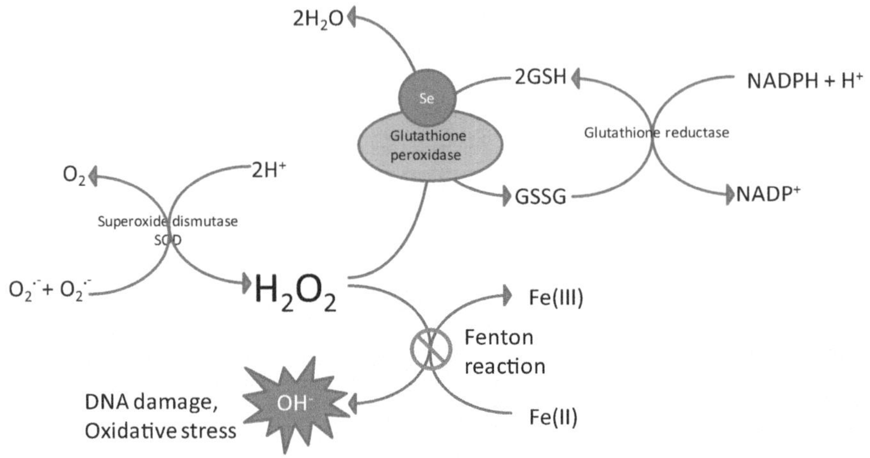

Figure 7.3 Role of selenoprotein glutathione peroxidase in scavenging free radicals (Navarro-Alarcon & Cabrera-Vique, 2008).

selenium-enriched soil developed disorders known as alkali disease and blind staggers in 1930 in South Dakota (Tinggi, 2003). Selenium has a tendency to bioaccumulate in the aquatic environment and becomes toxic to aquatic organisms, such as fish, at elevated concentrations. Selenium becomes toxic to cormorants and other birds that prey on aquatic organisms harboring elevated levels of this element (Miller *et al.*, 2013). In 1984, high incidences of deformities and mortalities were recorded in waterfowl in the Kesterson wildlife reservoir (California, USA), where agricultural drainage water and industrial effluent containing a high selenium content was discharged.

Selenium bioaccumulation has exposed the fish and waterfowl in wetlands and evaporation ponds to a severe threat, which has caused deformities, impaired reproduction and eventually death of fish and birds (Zhang *et al.*, 2006). Transfer of selenium to the higher trophic levels and accumulation in the food chain due to selenium bioaccumulation (De La Riva *et al.*, 2014) caused additional damaging effects such as nausea, vomiting and diarrhea due to selenium poisoning (selenosis) in humans and animals along with dermal and neurological dysfunction associated with deformation of nails and hoofs, unsteady gait or even paralysis and cardiovascular symptoms (Duntas & Benvenga, 2014). Dietary uptake higher than 400 μg day^{-1} was linked to hair and nail loss and disruption of the nervous and digestive systems in humans and animals (Misra *et al.*, 2015). Lemly (2014) calculated the monetary loss due to reduced fish productivity and studied the impact of selenium pollution in water bodies associated with high selenium concentrations on tissues of aquatic organisms (e.g., fish), particularly morphological abnormalities and teratogenic deformities.

7.2.3 Selenium deficiency

The perspective of researchers towards selenium changed in 1960, with the identification of a peculiar heart muscle disease symptom called Keshan's disease in selenium deficient populations in China (Chen, 2012). A selenium concentration in staple food lower than the critical standard of 100 μg kg^{-1} can lead to the development of selenium deficiency diseases (Wang *et al.*, 2016). Selenium deficiency causes reproductive disorders and heart failure in humans, white muscle disease in young animals, fatal diseases such as hepatosis and Mulberry heart disease in pigs and exudative diathesis in poultry (Mehdi *et al.*, 2013). High incidences of white muscle disease and heart necrosis were observed in selenium deficient sheep and cattle in New Zealand and Western Oregon, USA (Tinggi, 2003). These and other studies have contributed to the understanding of physiological functions of selenium in higher animals and humans. The narrow window of 40–400 μg day^{-1} between selenium deficiency and toxicity has led to selenium being appropriately termed as an 'essential toxin' (Lenz & Lens, 2009) and makes environmental selenium research (Winkel *et al.*, 2012) critical to maintain a balance between providing the necessary level and avoiding toxicity.

7.2.4 Selenium bioavailability

Both these Se toxicity and deficiency disorders are the ramifications of the Se bioavailability in the respective soils. Several studies are being carried out to resolve the issue of Se imbalance across the globe, where efforts are being made to remove Se from the seleniferous regions using *in situ* and *ex situ* bioremediation technologies (this Chapter; Bañuelos & Lin, 2005; Lindblom *et al.*, 2014) or fortify Se-deficient soils with organic and inorganic Se compounds (Chapter 9; Bañuelos *et al.*, 2015; Lyons, 2010).

7.3 SELENIUM GEOCHEMISTRY IN SELENIFEROUS SOILS AND SEDIMENTS

The selenium content in soil varies greatly throughout the world. Higher amounts of bioavailable forms of selenium in soils greatly influence the amount of selenium in the food chain. Elevated amounts of water-soluble selenium in soil can lead to contamination of water bodies and groundwaters due to the leaching caused by rainfall and irrigation (Wu, 2004). Effluent discharges from mining industries and coal fired power plants have resulted in large-scale Se deposition in marine sediments (Ellwood *et al.*, 2016). However, quantification of the total selenium does not actually give information about the chemical species, thus giving no information about its bioavailability to plants.

Various forms of selenium are soluble, exchangeable, bound to organic matter, sulfides, carbonates and oxides. Selenium occurs mainly in four oxidation states in soil, viz. selenate (VI), selenite (IV), elemental selenium (0) and selenide (-II).

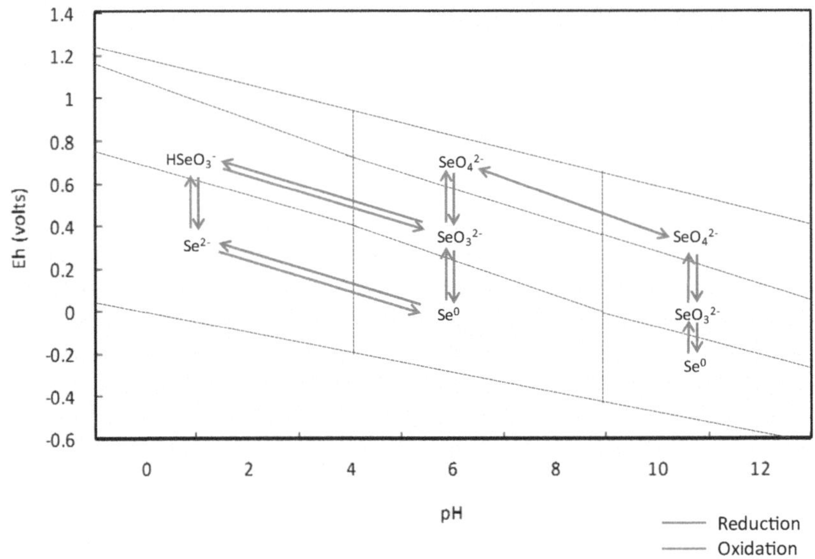

Figure 7.4 Effect of pH and redox potential (E_h) of soil on Se speciation (Mayland *et al.*, 1989).

The chemical forms of selenium and their solubility in soil mainly depend on the prevailing redox conditions and pH (Figure 7.4), salinity and carbonate content of the soil (Hagarova *et al.*, 2005). Other factors that contribute to selenium speciation, and thus adsorption, bioavailability and toxicity, are organic matter content and composition, iron oxide levels as well as amount and type of clay (Table 7.2; Bajaj *et al.*, 2011; Schilling *et al.*, 2015).

In acidic clay soils and soils with a high organic matter content, selenium is mainly present in the form of selenides and selenium sulfides (SeS_2/Se_nS_{8-n}), which have limited solubility and bioavailability (Kabata-Pendias & Pendias, 2001). Selenium binds to proteins, fulvic acids, clay and organic matter in acidic soils (Cuvardic, 2003; Wu, 2004). In well drained neutral soils, selenium is predominantly available in the form of selenite (Mayland *et al.*, 1989). In alkaline and aerated soils, selenium occurs as selenate which is mobile and available to plants. Selenite is oxidized to selenate, the more mobile and bioavailable form of selenium in soil environments (Dhillon & Dhillon, 2014).

Se in seleniferous soil is often of lithogenic origin. For the soils in northwest India, selenium was transported via Se-rich sediments by seasonal rivulets from sub-Himalayan ranges called the Shiwalik hills in the north of Punjab, India (Dhillon & Dhillon, 2009). Intensive irrigation on the local agricultural lands during the last few decades has increased the Se deposition in the soil and has led to accumulation of up to 1.4 kg Se per hectare every year (Dhillon & Dhillon,

Table 7.2 Soil conditions that affect the speciation, mobility and bioavailability of selenium (Kabata-Pendias & Pendias, 2001).

Soil factor	Variables	Major Se form	Mobility
pH	Alkaline	Selenates	High
	Neutral	Selenites	Moderate
	Acidic	Selenides	Low
Redox potential	High	Selenites	High
	Low	Selenides	Low
Organic matter	Undecayed	Adsorbed	Low
	Decayed	Complexed	High
	Enhanced biomethylation	Volatilized	High
Clay content	High	Adsorbed	Low
	Low	Soluble	High

2003), which affects agricultural crop yields (Figure 7.5). In order to avoid further contamination of these agricultural soils via rivulets containing Se-rich sediments, storage and sedimentation of the irrigation water may be practiced prior to irrigation on the agricultural fields (Dhillon & Dhillon, 2014). The Se-rich sediments may then be extracted and treated using a suitable technique. Se is a scarce and critical element (Nancharaiah *et al.*, 2016a) and its recovery from soil may not only assist in the clean-up of seleniferous soil, but the recovered Se may complement as a raw material for its wide range of applications from healthcare to electronic industries.

Figure 7.5 Agricultural soil with high Se content, where wheat has been grown, in northwest India adversely affects the productivity of the crops and leads to bioaccumulation and biotransfer of Se to higher trophic levels (Wadgaonkar, 2017).

7.4 BIOREMEDIATION OF SELENIFEROUS SOILS

7.4.1 *In situ* treatment

Wadgaonkar *et al.* (2019a) set up microcosms to evaluate the effect of the organic amendment and bioaugmentation on reduction of Se in a seleniferous soil from northwest India. The organic amendment and bioaugmentation exhibited a similar Se removal performance as those of the control set-up without any amendment or bioaugmentation. This suggested that under ideal environmental conditions, the soil indigenous microbial population and organic content available in the soil were sufficient to achieve Se reduction *in situ*. Flury *et al.* (1997) also did not find major differences between the control and tests in a field study that attempted volatilization of Se from the Kesterson reservoir (California, USA). The drawback of the *in situ* method is that it only converts the bioavailable Se into insoluble Se forms, where the risk of its re-oxidation to soluble Se forms in soils with high redox potential and alkaline pH cannot be ruled out.

7.4.2 *Ex situ* treatment by soil flushing

Wadgaonkar *et al.* (2018b) characterized the physico-chemical parameters, total Se content and sequential extraction of Se in the soil collected from the seleniferous regions of Punjab (India). These characteristics allowed the further design of experiments for soil flushing and washing. Soil flushing was performed to assess Se migration in a soil column by simulating artificial rainfall or irrigation. Se migration and accumulation from the upper to the lower layers in the soil columns was observed suggesting reduction of soluble Se to insoluble Se forms in the deeper soil layers. Nevertheless, with time, this insoluble Se fraction may be slowly re-oxidized to soluble Se forms *in situ* and further contaminate groundwater.

7.4.3 *Ex situ* treatment by soil washing

Wadgaonkar *et al.* (2018b) optimized the soil washing of seleniferous soil using different washing solutions in order to achieve a maximum removal of bioavailable forms of Se from the seleniferous soil. Oxidizing agents were found to be most efficient in extracting Se from the studied seleniferous soil (Wadgaonkar *et al.*, 2018b). Biological treatment of the Se-rich soil leachate was further evaluated using microbial reduction in a continuous bioreactor (Wadgaonkar *et al.*, 2019a) and by phytoremediation using aquatic plants (Ohlbaum *et al.*, 2018). Figure 7.6 shows a schematic approach of the washing of seleniferous soil and biotreatment of the Se-rich soil leachate.

Prior to reconstitution of washed soil, suitability of the soil to be discharged in the environment should be determined. Unlike soil flushing, continuous agitation during soil washing disturbs the layers present in the soil column naturally, which might affect the soil characteristics. In addition, after soil washing, the soil characteristics and the bioavailability of Se might change in the course of time

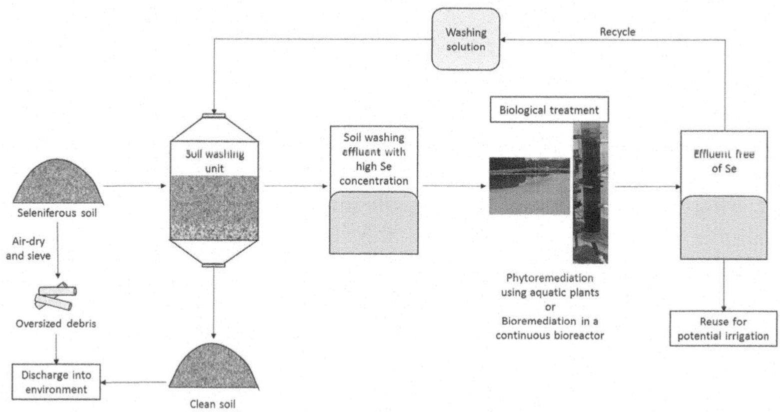

Figure 7.6 Schematic approach of the washing of seleniferous soils and biotreatment of the Se-rich soil washing effluent (Wadgaonkar, 2017).

depending on the soil environmental conditions. Long-term experiments to determine the physico-chemical characteristics of the washed soil, as well as alterations in the Se fractions during sequential extraction must be performed. Fertility of the soil may be checked using germination tests. Pot experiments to compare plant growth and Se uptake in seleniferous soil and washed seleniferous soil may be performed.

Phytoremediation of Se-rich leachate containing residual oxidizing agents ($K_2S_2O_8$ and $KMnO_4$) was evaluated using aquatic plants such as *Lemna minor* and *Eregia densa* (Ohlbaum *et al.*, 2018). *L. minor* was found to be more efficient in removing Se from soil leachate and Se accumulation than *E. densa*. However, the presence of residues of the oxidizing agents in the Se-rich effluent adversely affected not only the Se removal efficiency but also growth of both plant species. Although application of oxidizing agents is promising for effective removal of Se from the seleniferous soils (Wadgaonkar *et al.*, 2018b), the use of chemical agents must be discouraged, as it might aggravate the post-treatment operations for both treated soil and the leachate.

7.5 BIOLOGICAL TREATMENT OF SELENIFEROUS SOIL WASHING WATER AND SELENIUM-CONTAMINATED GROUNDWATER

7.5.1 UASB reactors

7.5.1.1 Treatment of soil leachate

The leachate from seleniferous soils produced via *ex situ* treatment (soil flushing or soil washing) can be treated in an upflow anaerobic sludge bed (UASB) reactor (see Chapter 6; Sinharoy & Lens, 2020). Se removal (up to 90%) and retention in the form of biogenic Se(0) in the granular sludge of the UASB was achieved during biological treatment of seleniferous soil washing water (Wadgaonkar *et al.*, 2019a). Along with the granular sludge, the indigenous soil microorganisms and the organic matter, extracted into the leachate during soil washing, play an important role to achieve efficient Se removal in a UASB reactor. Microbial reduction of soluble selenium oxyanions (selenate and selenite) to insoluble elemental selenium is the best available cost effective and eco-friendly option (Nancharaiah & Lens, 2015b).

Depending on the selenium concentration, external electron donors need to be dosed for complete selenate removal. Most of the research on bioreduction of Se oxyanions using anaerobic granular sludge utilize expensive carbon sources such as lactate (Dessì *et al.*, 2016; Mal *et al.*, 2016), methanol (Eregowda *et al.*, 2020) or glucose (Espinosa-Ortiz *et al.*, 2015). Few studies have explored the possibility of using inexpensive electron donors like methane (Lai *et al.*, 2016) as the sole electron donor for bioreduction of Se oxyanions. Wadgaonkar (2017) investigated the anaerobic bioreduction of selenate to elemental Se by marine lake sediment in the presence of methane as a sole electron donor. Complete bioreduction of selenate was observed in serum bottles under high pressure conditions and in a biotrickling filter (BTF) reactor. However, due to the slow growing nature of the inoculum (Bhattarai *et al.*, 2017), the operation time to achieve complete selenate removal from the medium was significantly higher than that of other studies that provide lactate as the electron donor (Dessì *et al.*, 2016; Mal *et al.*, 2016).

7.5.1.2 Presence of tellurium

Anthropogenic activities such as mining and refinery industries can lead to contaminated soil-water environments containing both Se and Te (Perkins, 2011). Recovery of Se and Te as by-products from copper mining industries focusses on anode slime which contains high concentrations of Se and Te (Jorgenson, 2002). Both Se and Te are considered as critical elements because of their wide range of applications and increasing demand in the electronic industries (Nancharaiah *et al.*, 2016a; Ramos-Ruiz *et al.*, 2016).

Wadgaonkar *et al.* (2018c) investigated the effect of metalloid co-contaminants (Te) on Se removal and simultaneous removal of both Se and Te from wastewater. Simultaneous removal of both selenite and tellurite from synthetic wastewater by anaerobic granular sludge was sustainable in a lab-scale UASB reactor. Microbial transformations converted both oxyanions (selenite and tellurite) to their respective elemental forms (Se(0) and Te(0)), which were entrapped in the extracellular polymeric substances (EPS) matrix of the granular sludge, easing the recovery of these critical elements. Interestingly, characterization using high-end microscopic and spectroscopic techniques revealed the formation of biogenic Se (0)-Te(0) nanostructures, entrapped in the sludge granules, during simultaneous reduction of Se and Te oxyanions (Wadgaonkar *et al.*, 2018c).

7.5.1.3 Presence of other oxyanions

7.5.1.3.1 Granular versus biofilm reactor systems

The practical application of biological treatment for Se-laden wastewater may have important limitations because of the presence of co-contaminants such as nitrate (NO_3^-) and sulfate (SO_4^{2-}) (Tan 2017) and heavy metal ions (e.g. Ni^{2+}) (Mal *et al.*, 2016). Tan *et al.* (2018d) demonstrated that a UASB reactor was capable of removing selenate (SeO_4^{2-}) in the presence of NO_3^- and SO_4^{2-} without prior inoculum adaptation, achieving a 100% NO_3^-, 30% SO_4^{2-} and 80% Total Se (Se_{tot}) removal efficiency after 90 days of reactor operation. Comparing the effect of the oxyanions SO_4^{2-} and NO_3^-, individually, on Se removal in continuous operation, a correlation with different reactor configurations, microbial growth system, and co-oxyanions was observed. SO_4^{2} was shown to be a controlling factor in biofilm systems using a drip flow reactor (DFR) (Tan *et al.*, 2018a) and biotrickling filter (BTF) (Tan *et al.*, 2018c) achieving both higher Se removal and more biofilm growth/formation when SO_4^{2-} was present. In contrast, the UASB reactor did not reveal any changes, either increase or decrease, in Se removal efficiencies, when SO_4^{2-} was removed from (Tan *et al.*, 2018d) or included (Tan *et al.*, 2018c) in the feed solution. In contrast to SO_4^{2-}, it was observed that the removal of NO_3^- from the feed solution possibly caused an increase in Se_{tot} concentration in the UASB effluent, thus negatively impacting Se removal efficiencies by the end of the reactor run (Tan *et al.*, 2018d).

The influence of NO_3^- and SO_4^{2-} on SeO_4^{2-} removal is most likely interlinked with (i) the bioreactor type used, (ii) possible interaction among the bioreduced products of the oxyanions, and (iii) microbial community changes that occurred during the operation. SeO_4^{2-} reduction can be carried out by various microorganisms including denitrifiers and sulfate reducers. The microbial community analysis revealed the presence of high relative abundances of denitrifiers along with a small proportion of sulfate reducers, irrespective of the presence of NO_3^- (Tan *et al.*, 2018a, 2018d). Enzymes such as SeO_4^{2-} reductase, NO_3^- reductase or periplasmic NO_2^- reductase have been reported to catalyze SeO_4^{2-} reduction to Se^0 (DeMoll-Decker & Macy, 1993; Nancharaiah & Lens, 2015a). It was hypothesized that the presence of NO_3^- can promote a higher state of metabolic activity in microorganisms (Oremland *et al.*, 1999). This implies that NO_3^- plays a significant role in shaping the microbial community in the bioreactor and specific metabolic pathways/activities of denitrifiers could be linked to the increase of SeO_4^{2-} removal in the presence of NO_3^- and its by-products (Lai *et al.*, 2014).

7.5.1.3.2 Adsorption coupled to biological selenium removal processes

Tan *et al.* (2018b) and Cálix *et al.* (2019) investigated the possibility of including an ion exchange (IX) process in the treatment scheme for removing both SeO_4^{2-} and SO_4^{2-} from synthetic mine wastewater. Tan *et al.* (2018b) validated the applicability of Amberlite® IRA-900, a strong anionic IX resin, for the simultaneous removal of SeO_4^{2-} and SO_4^{2-} achieving >70% adsorption efficiencies. Modified Langmuir multi-component isotherms using a complete competition model indicated that IRA-900 was not selective towards SeO_4^{2-} and that both oxyanions competed for the active sites. The non-selectivity of IRA-900 gives the advantage of adsorbing both oxyanions saving the need for pre-treatment of SO_4^{2-} to avoid inhibition of SeO_4^{2-} adsorption. Chemical regeneration was optimal at 20 min using 0.5 M HCl and the resin was reusable for up to six adsorption-desorption cycles.

Despite the advantages of biological techniques, there are still issues with reaching the regulatory discharge limit for Se_{tot}. Cálix *et al.* (2019) evaluated the feasibility of combining two unit processes, an IX column and a UASB reactor, for the overall improvement of SO_4^{2-} and Se_{tot} removal (Figure 7.7). The IX process was evaluated as a pre- or post-treatment process for the biological process. IX as a post-treatment was demonstrated to be less effective due to competition with other bioreduced products and the presence of suspended solids. On the other hand, IX as a pre-treatment system allowed the UASB reactor to receive lower concentrations of SO_4^{2-} and SeO_4^{2-} facilitating a better biological removal process and achieving an overall higher removal efficiency of 99% SO_4^{2-} and 97% Se_{tot}. The final treated effluent contained <100 mg L^{-1}

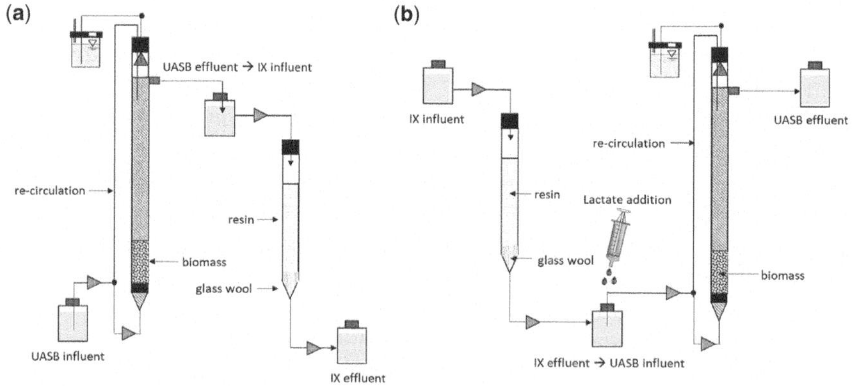

Figure 7.7 Schematic of the integrated treatment system comprising of an IX column and a UASB reactor. The process flow configuration was evaluated as either (a) IX as post treatment of the UASB or (b) IX as pre-treatment of the UASB influent (Cálix *et al.*, 2019).

SO_4^{2-} and <0.3 mg L^{-1} Se_{tot} (Cálix *et al.*, 2019). Though the Se_{tot} concentration still does not reach the regulatory discharge limit, it should be noted that all dissolved Se was removed and only colloidal Se was remaining. Therefore, the addition of a (electro-)coagulation process (Staicu *et al.*, 2015b) as a finishing step could further lower the Se_{tot} concentration and achieve the regulatory limit (Nancharaiah & Lens, 2015b). Additionally, with the lowering of the SO_4^{2-} concentration supplied to the biological process, less COD requirement would be needed, decreasing the operational costs. Finally, bioregeneration of the resin by reduction of the adsorbed SO_4^{2-} and SeO_4^{2-} by biogranules was shown to be feasible and could potentially be used as an alternative to chemical regeneration. Cálix *et al.* (2019) demonstrated that integration of unit processes can allow the full removal of Se and SO_4^{2-} from wastewaters.

7.5.1.4 Presence of heavy metals

Tan *et al.* (2018c) demonstrated the importance of the chemical oxygen demand (COD)/SO_4^{2-} ratio and the impact of Ni^{2+} on SeO_4^{2-} removal in a UASB reactor in comparison to that of a BTF reactor. Poor SO_4^{2-} removal (<30%) was observed at ~1.8 COD/SO_4^{2-}; while increasing the COD/SO_4^{2-} ratio to 2.8 improved the SO_4^{2-} removal to >80% with generation of high sulfide (HS^-) concentrations (up to 250 mg S L^{-1}). Addition of Ni^{2+} to these sulfidogenic conditions allowed simultaneous removal of Ni and sulfide via co-precipitation. Formation of Ni_3S_2 precipitates was confirmed by X-ray diffraction (XRD). However, Ni^{2+} addition negatively impacted both SeO_4^{2-} and SO_4^{2-} removal by 20–50%, possibly due to the sudden metal toxicity to the biofilms/biogranules. Compared to the BTF, the UASB reactor performance recovered quickly. This could be attributed to the robustness of anaerobic sludge granules, allowing for better response to operational changes and stress.

7.5.2 Aerobic reactors

Selenium-containing wastewaters can be treated, aerobically, in activated sludge systems, where the selenium oxyanions are reduced to elemental selenium by the microorganisms present in the activated sludge flocs. Aerobic reduction of selenium oxyanions to either volatilized selenium compounds followed by gas trapping (Kagami *et al.*, 2013) or selenium trapped in the biomass of aerobic microorganisms such as *Bacillus cereus* (Dhanjal & Cameotra, 2010) and *Escherichia coli* (Dobias *et al.*, 2011) as well as biomass from aerobic granular sludge (Nancharaiah *et al.*, 2018), continuous (Jain *et al.*, 2016) or sequencing batch (Mal *et al.*, 2017) reactors, activated sludge provides a one-step process for treatment of selenite-containing wastewaters.

A continuously operated activated sludge reactor operated at neutral pH and 30°C removed 33.98 and 36.65 mg of total selenium per g of total suspended solids (TSS) at TSS concentrations of 1300 and 3000 mg L^{-1}, respectively (Jain *et al.*,

2016). The selenium fed activated sludge showed better settleability and hydrophilicity, but poorer dewaterability at higher TSS concentrations as compared to the control activated sludge. The selenium fed activated sludge also showed a less negative surface charge density as compared to the control activated sludge (Jain *et al.*, 2016). However, the operated activated sludge reactors crashed upon continuous feeding of selenium after 10–20 days at the applied loading rates and operation could not be recovered (Jain *et al.*, 2016), most likely due to irreversible toxicity of selenite to the aerobic bacteria. This selenite toxicity could be overcome by sequencing batch feeding of the activated sludge system (Mal *et al.*, 2017), which further allowed the integration of selenite removal with ammonium removal from the wastewater.

Some aerobic selenite reducing bacteria are capable of tolerating and reducing high selenite concentrations, for example, endophytes present in plants growing on seleniferous soils (*Pseudomonas moraviensis* subsp. stanleyae; Staicu *et al.*, 2015a) or as contaminant in minimal salt medium (*Delftia lacustris;* Wadgaonkar *et al.*, 2019b). *D. lacustris* not only reduces Se oxyanions to elemental selenium, but also produces seleno-ester compounds (organo-Se compound) depending on the initial concentrations of Se oxyanions in the medium (Wadgaonkar *et al.*, 2019b). Chakraborty *et al.* (2019) grew this bacterium in a co-culture with *Phanerochaete chrysosporium* and immobilized the bacterium on the fungal pellets. Thus, the co-culture was able to couple the reduction of selenite to elemental Se(0) (by *D. lacustris*) to the degradation of phenol (by *P. chrysosporium*).

7.5.3 Membrane reactors

Membrane filtration technology using reverse osmosis (RO) and nanofiltration (NF) has been applied for treating Se-laden wastewaters (Chapter 6; Staicu *et al.*, 2017). However, despite the effectiveness of these technologies, the operational cost is high due to the required hydraulic pressure. Membrane bioreactors, on the other hand, combine the biological process and particle separation and are a cheaper option compared to NF or RO (Hu *et al.*, 2017). Although membrane technology still has the major drawback of higher operating cost than conventional processes, this can be overcome by decreasing the downstream unit processes and overall reduction of the reactor size (Hu & Stuckey, 2006).

Figure 7.8(b) suggests a possible reactor configuration for an anaerobic reactor with a submerged membrane. One of the main disadvantages of using membrane technology, particularly in an anaerobic system, is the higher potential for biofouling. This could, however, be mitigated by using a high flow rate and biogas recirculation as a means for membrane scouring. Additionally, it is possible to either fabricate membranes with covalent Se attached to the membrane or utilize the colloidal Se^0 as a means to reduce biofouling. Vercellino *et al.* (2013a, 2013b) reported a reduction of the biofilm formation and thickness by > 5 logs, as well as

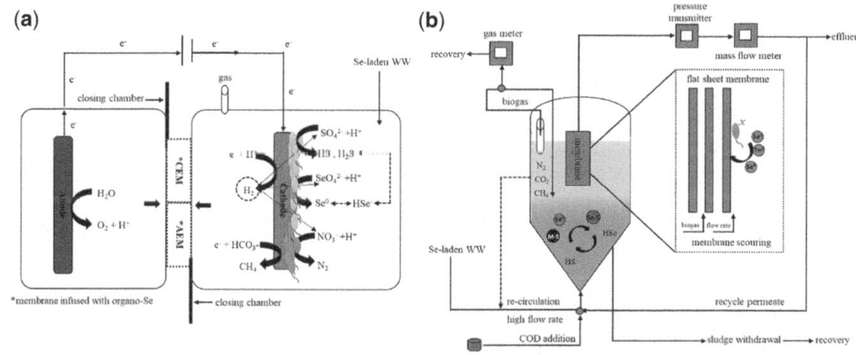

Figure 7.8 Innovative reactor configurations proposed for Se-laden wastewaters with co-contaminants: (a) two-chamber bioelectrochemical system (BES) for autotrophic reduction of oxyanion utilizing hydrogen as electron donor generated electrochemically with an anionic/cationic exchange\membrane AEM/CEM) for ion transfer from one chamber to another and (b) modified upflow anaerobic sludge blanket reactor with submerged membrane module (e.g., flat sheet or hollow fiber membrane) utilizing biogas production and high flow rate for membrane scouring (Tan, 2017).

a flux loss decrease from 55 to 15% due to the biofouling attenuation when organo-Se was covalently attached to the membrane surface.

7.5.4 Bioelectrochemical processes

A bioelectrochemical system (BES) employs microorganisms and electrodes to catalyze redox reactions and this system can be utilized in a two-chamber microbial fuel cell using either the anode, cathode or both as bioelectrodes (Ntagia *et al.*, 2020; Sánchez *et al.*, 2020). Electroactive microorganisms can extract electrons that can be utilized to generate electricity, treat wastewater by redox reactions and also recover nutrients (i.e., P, N), sulfur (Ntagia *et al.*, 2020) and metals (Nancharaiah *et al.*, 2016a). Removal of individual selenate (Sravan *et al.*, 2020), selenite (Catal *et al.*, 2009), NO_3^- (Nancharaiah *et al.*, 2016b) and SO_4^{2-} (Luo *et al.*, 2017) using BESs has been reported in the literature. However, application of a BES to complex Se-laden wastewater with different co-contaminants present has yet to be reported.

Although lactate is an efficient electron donor and carbon source, alternative electron donors/carbon sources are recommended since using lactate adds operational cost considering that most industrial Se-containing wastewaters have a low COD content. Autotrophic reduction of NO_3^-, SO_4^{2-} and SeO_4^{2-} using H_2 as the electron donor was found to be successful in membrane biofilm reactors (Ontiveros-Valencia *et al.*, 2016; van Ginkel *et al.*, 2011). It is, therefore, feasible to carry out autotrophic reduction of NO_3^-, SO_4^{2-} and SeO_4^{2-} at the biocathode side by autotrophs utilizing the generated H_2. Figure 7.8(a) suggests a possible reactor configuration that utilizes the ability of a BES to generate hydrogen gas (H_2) through an electrochemical process. An IX membrane (anionic or cationic) separates the two chambers and allows for charge balancing by moving H^+ and OH^- as well as other possible value-added chemicals produced in each chamber. The IX membrane can be infused with organo-Se to reduce the biofouling that can occur at the side of the bioelectrode (Vercellino *et al.*, 2013b).

7.6 COUPLING SELENIFEROUS SOIL REMEDIATION TO RESOURCE RECOVERY

7.6.1 Biofortification

Extraction of selenium from seleniferous soil by soil washing and its accumulation in anaerobic granular sludge (Dessì *et al.*, 2016; Mal *et al.*, 2016) or aquatic plants (Li *et al.*, 2020a, 2020b; Ohlbaum *et al.*, 2018) helps with easier recovery of Se from soil. The selenium rich granular sludge or aquatic plants may be used for biofortification of selenium in crops (see Chapter 9). In addition, Se-containing aquatic plants can be considered for developing a dietary selenium supplement. Similar application of Se hyperaccumulator plants has been proposed to compensate the effects of selenium deficiency in selenium deficient regions (Moreno *et al.*, 2013; Yasin *et al.*, 2014). Several studies (Curtin *et al.*, 2006; Lyons, 2010) have investigated the effect of the addition of inorganic-Se (selenate and selenite) containing fertilizers for selenium biofortification in food crops in selenium deficient countries such as the UK and New Zealand. In Finland, selenium deficient soils have led to a low selenium status in humans and animals. A nationwide strategy was adopted on application of multimineral fertilizers on agricultural soils to improve selenium dietary intake (Parkman & Hultberg, 2002).

Maintaining accurate and low selenium dosage in agricultural land can be difficult owing to diverse cropping systems and selenium losses due to leaching and volatilization. Also, since selenium is a scarce resource with low recovery potential when applying foliar fertilizer, amendment with organic-Se such as Se-enriched hyperaccumulator plant material to Se-deficient soils may prove as a better fertilizer alternative. Bañuelos *et al.* (2015) concluded that amending soils with organic selenium sources such as biomass of the hyperaccumulator *S. pinnata* is useful for enriching food crops, such as broccoli and carrots, with organic-Se in selenium deficient regions of the world. Bañuelos and Mayland (2000) suggested that Canola (*Brassica napus*) grown as a selected plant species for field phytoremediation of Se-contaminated soils may be harvested and utilized as Se-enriched forage for marginally Se-deficient lambs and cows to help meet their normal selenium intake requirements.

In the case of biogranules, it is possible to simply dry the biomass and use it as a fertilizer for soil augmentation and foliar application for those regions with Se deficiency to increase the Se bioavailability (Wang *et al.*, 2016). In the case of biofilms with carriers, detachment of biofilms through centrifugation/sonication or heating can be considered. However, it should be noted that the biomass produced while treating soil washing liquor or industrial mine wastewater also contains various other contaminants that can be harmful to the environment. Therefore, detailed studies on the toxic effect and health ramification of using Se-rich biomass as fertilizer should be conducted.

7.6.2 Recovery of biologically produced nanomaterials

Resource recovery is an important aspect to consider when applying biological treatment techniques. Recovered and purified Se^0 nanoparticles, along with metal and sulfur complexes, can be reused for various semiconductor (i.e., batteries) and catalyst applications, as an adsorbent for toxic metals (i.e., mercury), and as antimicrobial agents for medical devices and membranes (Chapter 10; Zonaro *et al.*, 2015). Biomass from the bioreactors contains various valuable materials such as metals, sulfur or sulfur complexes, and Se components that have the potential to be recovered (Sinharoy & Lens, 2020). The recovery process of Se and other complexes from biomass has, however, not yet been explored. So far, a method for recovery and purification of Se^0 nanoparticles from biomass was suggested as a series of solid-liquid steps: high-speed centrifugation, and sonication followed by hexane separation (Jain *et al.*, 2015). As such, an eco-friendlier and simple recovery approach should be developed.

Other components in the soil leachate may influence the production and composition of the biogenic Se nanostructures. Along with biogenic Se(0), the possibility of formation of chalcogenide alloys (e.g., Se-Te and Se-S) or metal chalcogenide nanoparticles (e.g., PbSe, CdSe and CuSe) in continuous bioreactor systems cannot be ruled out. The determination of the structure and composition of the biogenic selenium nanoparticles requires detailed analysis of the granular sludge, for example, using high-end techniques such as transmission electron microscopy coupled with energy-dispersive X-ray spectroscopy (TEM-EDS), X-ray photoelectron spectroscopy (XPS) and Raman spectroscopy. These analytical techniques are not routine and require specific equipment.

REFERENCES

Bajaj M., Eiche E., Neumann T., Winter J. and Gallert C. (2011). Hazardous concentrations of selenium in soil and groundwater in North-West India. *Journal of Hazardous Materials*, **189**(3), 640–646.

Bañuelos G. S. and Lin Z.-Q. (2005). Phytoremediation management of selenium-laden drainage sediments in the San Luis Drain: a greenhouse feasibility study. *Ecotoxicology and Environmental Safety*, **62**(3), 309–316.

Bañuelos G. S. and Mayland H. F. (2000). Absorption and distribution of selenium in animals consuming canola grown for selenium phytoremediation. *Ecotoxicology and Environmental Safety*, **46**, 322–328.

Bañuelos G. S., Arroyo I., Pickering I. J., Yang S. I. and Freeman J. L. (2015). Selenium biofortification of broccoli and carrots grown in soil amended with Se-enriched hyperaccumulator *Stanleya pinnata*. *Food Chemistry*, **166**, 603–608.

Bhattarai S., Cassarini C., Naangmenyele Z., Rene E. R. and Lens P. N. L. (2017). Microbial sulfate-reducing activities in anoxic sediment from Marine Lake Grevelingen: screening of electron donors and acceptors. *Limnology*, **19**(1), 31–41, doi: 10.1007/s10201-017-0516-0.

Brown K. M. and Arthur J. R. (2001). Selenium, selenoproteins and human health: a review. *Public Health Nutrition*, **4**(2B), 593–599.

Cálix E. M., Tan L. C., Rene E. R., Nancharaiah Y. V., Van Hullebusch E. D. and Lens P. N. L. (2019). Simultaneous removal of sulfate and selenate from wastewater by process integration of an ion exchange column and upflow anaerobic sludge blanket bioreactor. *Separation Science and Technology*, **54**(8), 1387–1399.

Catal T., Bermek H. and Liu H. (2009). Removal of selenite from wastewater using microbial fuel cells. *Biotechnology Letters*, **31**, 1211–1216.

Chakraborty S., Rene E. R. and Lens P. N. L. (2019). Reduction of selenite to elemental Se(0) with simultaneous degradation of phenol by co-cultures of *Phanerochaete chrysosporium* and *Delftia lacustris*. *Journal of Microbiology (Seoul, Korea)*, **57**(9), 738–747.

Chen J. (2012). An original discovery: selenium deficiency and Keshan disease (an endemic heart disease). *Asia Pacific Journal of Clinical Nutrition*, **21**, 320–326.

Curtin D., Hanson R., Lindley T. N. and Butler R. C. (2006). Selenium concentration in wheat (*Triticum aestivum*) grain as influenced by method, rate, and timing of sodium selenate application. *New Zealand Journal of Crop and Horticultural Science*, **34**, 329–339.

Cuvardic M. S. (2003). Selenium in soil. *Proceedings for Natural Sciences, Matica Srpska Novi Sad*, **104**, 23–37.

De La Riva D. G., Vindiola B. G., Castañeda T. N., Parker D. R. and Trumble J. T. (2014). Impact of selenium on mortality, bioaccumulation and feeding deterrence in the invasive Argentine ant, *Linepithema humile* (Hymenoptera: Formicidae). *Science of the Total Environment*, **481**, 446–452.

DeMoll-Decker H. and Macy J. M. (1993). The periplasmic nitrite reductase of *Thauera selenatis* may catalyze the reduction of selenite to elemental selenium. *Archives of Microbiology*, **160**, 241–247.

Dessì P., Jain R., Singh S., Seder-Colomina M., van Hullebusch E. D., Rene E. R., Ahammad S. Z., Carucci A. and Lens P. N. L. (2016). Effect of temperature on selenium removal from wastewater by UASB reactors. *Water Research*, **94**, 146–154.

Dhanjal S. and Cameotra S. S. (2010). Aerobic biogenesis of selenium nanospheres by *Bacillus cereus* isolated from coalmine soil. *Microbial Cell Factories*, **9**, 52.

Dhillon K. S. and Dhillon S. K. (2003). Quality of underground water and its contribution towards selenium enrichment of the soil-plant system for a seleniferous region of northwest India. *Journal of Hydrology*, 120–130.

Dhillon K. S. and Dhillon S. K. (2014). Development and mapping of seleniferous soils in northwestern India. *Chemosphere*, **99**, 56–63.

Dhillon S. K. and Dhillon K. S. (2009). Phytoremediation of selenium-contaminated soils: the efficiency of different cropping systems. *Soil Use and Management*, **25**, 441–453.

Diamond A. (2015). The subcellular location of selenoproteins and the impact on their function. *Nutrients*, **7**, 3938–3948.

Dobias J., Suvorova E. and Bernier-latmani R. (2011). Role of proteins in controlling selenium nanoparticle size. *Nanotechnology*, **22**, 195605.

Duntas L. H. and Benvenga S. (2014). Selenium: an element for life. *Endocrine*, **48**, 756–775.

Ellwood M. J., Schneider L., Potts J., Batley G. E., Floyd J. and Maher W. A. (2016). Volatile selenium fluxes from selenium-contaminated sediments in an Australian coastal lake. *Environmental Chemistry*, **13**, 68–75.

Eregowda T., Rene E. R., Matanhike L. and Lens P. N. L. (2020). Effect of selenate and thiosulfate on anaerobic methanol degradation using activated sludge. *Environ Sci Pollut Res*, **27**(24), 29804–29811.

Espinosa-Ortiz E., Rene E. R., van Hullebusch E. D. and Lens P. N. L. (2015). Removal of selenite from wastewater in a *Phanerochaete chrysosporium* pellet based fungal bioreactor. *International Biodeterioration & Biodegradation*, **102**, 361–369.

Expert group on vitamins and minerals (2003). Safe Upper Levels of Vitamins and Minerals. Food Safety Agency.

Fairweather-tait S. J., Collings R. and Hurst R. (2010). Selenium bioavailability : current knowledge and future research. *American Journal of Clinical Nutrition*, **91**, 1484S–1491S.

Flury M., Frankenberger W. T., Jr. and Jury W. A. (1997). Long-term depletion of selenium from Kesterson dewatered sediments. *Science of The Total Environment*, **198**, 259–270.

Hagarova I., Zemberyova M. and Bajcan D. (2005). Sequential and single step extraction procedures for fractionation of selenium in soil samples. *Chemical Papers*, **59**(2), 93–98.

Hu A. Y. and Stuckey D. C. (2006). Treatment of dilute wastewaters using a novel submerged anaerobic membrane bioreactor. *Journal of Environmental Engineering*, **132**, 190–198.

Hu Y., Wang X. C., Hao Ngo H., Sun Q. and Yang Y. (2017). Anaerobic dynamic membrane bioreactor (AnDMBR) for wastewater treatment: A review. *Bioresource Technology*, **247**, 1107–1118.

Huawei Z. (2009). Selenium as an essential micronutrient: roles in cell cycle and apoptosis. *Molecules*, **14**(3), 1263–1278.

Jain R., Seder-Colomina M., Jordan N., Dessi P., Cosmidis J., van Hullebusch E. D., Weiss S., Farges F. and Lens P. N. L. (2015). Entrapped elemental selenium nanoparticles affect physicochemical properties of selenium fed activated sludge. *Journal of Hazardous Materials*, **295**, 193–200.

Jain R., Matassa S., Singh S., van Hullebusch E. D., Esposito G. and Lens P. N. L. (2016). Reduction of selenite to elemental selenium nanoparticles by activated sludge. *Environmental Science and Pollution Research*, **23**(2), 1193–1202.

Jorgenson J. D. (2002). Selenium and tellurium, in: U.S. Geological Suvery Minerals Yearbook. pp. 1–7.

Kabata-Pendias A. and Pendias H. (2001). Trace Elements in Soils and Plants Trace Elements in Soils and Plants.

Kagami T., Narita T., Kuroda M., Notaguchi E., Yamashita M., Sei K., Soda S. and Ike M. (2013). Effective selenium volatilization under aerobic conditions and recovery from the aqueous phase by *Pseudomonas stutzeri* NT-I. *Water Research*, **47**, 1361–1368.

Kurokawa S. and Berry M. J. (2013). Selenium. Role of the essential metalloid in health. *Metal Ions in Life Science*, **13**(34), 499–534.

Kalender S., Uzun F. G., Demir F., Uzunhisarcikli M. and Aslanturk A. (2013). Mercuric chloride-induced testicular toxicity in rats and the protective role of sodium selenite and vitamin E. *Food and Chemical Toxicology*, **55**, 456–462.

Lai C.-Y., Yang X., Tang Y., Rittmann B. E. and Zhao H.-P. (2014). Nitrate shaped the selenate-reducing microbial community in a hydrogen-based biofilm reactor. *Environmental Science and Technology*, **48**, 3395–3402.

Lai C.-Y., Wen L.-L., Shi L.-D., Zhao K.-K., Wang Y.-Q., Yang X., Rittmann B. E., Zhou C., Tang Y., Zheng P. and Zhao H.-P. (2016). Selenate and nitrate bioreductions using methane as the electron donor in a membrane biofilm reactor. *Environmental Science and Technology*, **50**, 10179–10186.

Lemly A. D. (2014). Teratogenic effects and monetary cost of selenium poisoning of fish in Lake Sutton, North Carolina. *Ecotoxicology and Environmental Safety*, **104**(1), 160–167.

Lenz M. and Lens P. N. L. (2009). The essential toxin: the changing perception of selenium in environmental sciences. *Science of the Total Environment*, **407**(12), 3620–3633.

Li J., Loi G., Otero-Gonzalez L., Du Laing G., Ferrer I. and Lens P. N. L. (2020a). Selenate and selenite uptake, accumulation and toxicity in *Lemna minuta*. *Water Science and Technology*, **81**(9), 1852–1862.

Li J., Lens P. N. L., Otero-Gonzalez L. and Du Laing G. (2020b). Production of selenium- and zinc-enriched *Lemna* and *Azolla* as potential micronutrient-enriched bioproducts. *Water Research*, **172**, Article 115522.

Lindblom S. D., Fakra S. C., Landon J., Schulz P., Tracy B. and Pilon-Smits E. A. H. (2014). Inoculation of selenium hyperaccumulator *Stanleya pinnata* and related non-accumulator *Stanleya elata* with hyperaccumulator rhizosphere fungi – investigation of effects on Se accumulation and speciation. *Physiologia Plantarum*, **150**(1), 107–118.

Luo H., Teng W., Liu G., Zhang R. and Lu Y. (2017). Sulfate reduction and microbial community of autotrophic biocathode in response to acidity. *Process Biochemistry*, **54**, 120–127.

Lyons G. (2010). Selenium in cereals: improving the efficiency of agronomic biofortification in the UK. *Plant and Soil*, **332**(1), 1–4.

Mal J., Nancharaiah Y. V., van Hullebusch E. D. and Lens P. N. L. (2016). Effect of heavy metal co-contaminants on selenite bioreduction by anaerobic granular sludge. *Bioresource Technology*, **206**, 1–8.

Mal J., Nancharaiah Y. V., van Hullebusch E. D. and Lens P. N. L. (2017). Biological removal of selenate and ammonium by activated sludge in a sequencing batch reactor. *Bioresource Technology*, **229**, 11–19.

Mayland H. F., James L. F., Panter K. E. and Sonderegger J. L. (1989). Selenium in seleniferous environments. *Selenium in Agriculture and the Environment*, **23**, 15–50.

Mehdi Y., Hornick J.-L., Istasse L. and Dufrasne I. (2013). Selenium in the environment, metabolism and involvement in body functions. *Molecules (Basel, Switzerland)*, **18**(3), 3292–3311.

Miller L. L., Rasmussen J. B., Palace V. P., Sterling G. and Hontela A. (2013). Selenium bioaccumulation in stocked fish as an indicator of fishery potential in pit lakes on reclaimed coal mines in Alberta, Canada. *Environmental Management*, **52**(1), 72–84.

Misra S., Boylan M., Selvam A., Spallholz J. and Björnstedt M. (2015). Redox-active selenium compounds – from toxicity and cell death to cancer treatment. *Nutrients*, **7**(5), 3536–3556.

Moreno R. G., Burdock R., Cruz M., Álvarez D. and Crawford J. W. (2013). Managing the selenium content in soils in semiarid environments through the Recycling of Organic Matter. *Applied and Environmental Soil Science*, ArticleID283468, 1–10.

Nancharaiah Y. V. and Lens P. N. L. (2015a). Selenium biomineralization for biotechnological applications. *Trends in Biotechnology*, **33**, 323–330.

Nancharaiah Y. V. and Lens P. N. L. (2015b). The ecology and biotechnology of selenium respiring bacteria. *Microbiology and Molecular Biology Reviews*, **79**, 61–80.

Nancharaiah Y. V., Mohan S. V. and Lens P. N. L. (2016a). Biological and bioelectrochemical recovery of critical and scarce metals. *Trends in Biotechnology*, **34**, 137–155.

Nancharaiah Y. V., Mohan S. V. and Lens P. N. L. (2016b). Recent advances in nutrient removal and recovery in biological and bioelectrochemical systems. *Bioresource Technology*, **215**, 173–185.

Nancharaiah Y. V., Sarvajith M. and Lens P. N. L. (2018). Selenite reduction and ammoniacal nitrogen removal in an aerobic granular sludge sequencing batch reactor. *Water Research*, **131**, 131–141.

Navarro-Alarcon M. and Cabrera-Vique C. (2008). Selenium in food and the human body: A review. *Sci.Tot. Environ.*, **400**(1–3), 115–141.

Ntagia E., Prévoteau A. and Rabaey K. (2020). Electrochemical treatment of sulfur pollution. In: Environmental Technologies to Treat Sulfur Pollution: Principles and Engineering, 2nd edn, P. N. L. Lens (ed.), IWA Publishing, London (UK), pp. 247–276.

Ohlbaum M., Wadgaonkar S. L., van Bruggen J. J. A., Nancharaiah Y. V. and Lens P. N. L. (2018). Phytoremediation of seleniferous soil leachate using the aquatic plants *Lemna minor* and *Egeria densa*. *Ecological Engineering*, **120**, 321–328.

Ontiveros-Valencia A., Penton C. R., Krajmalnik-Brown R. and Rittmann B. E. (2016). Hydrogen-fed biofilm reactors reducing selenate and sulfate: community structure and capture of elemental selenium within the biofilm. *Biotechnology and Bioengineering*, **113**, 1736–1744.

Oremland R. S., Blum J. S., Bindi A. B., Dowdle P. R., Herbel M. and Stolz J. F. (1999). Simultaneous reduction of nitrate and selenate by cell suspensions of selenium-respiring bacteria. *Applied and Environmental Microbiology*, **65**, 4385–4392.

Papp L. V., Lu J., Holmgren A. and Khanna K. K. (2007). From selenium to selenoproteins: synthesis, identity, and their role in human health. *Antioxidants & Redox Signaling*, **9**(7), 775–806.

Parkman H. and Hultberg H. (2002). Occurence and effects of selenium in the environment – a literature review. IVL-rapport B1486, Swedish Environmental Research Institute, Göteborg.

Perkins W. T. (2011). Extreme selenium and tellurium contamination in soils – an eighty year-old industrial legacy surrounding a Ni refinery in the Swansea Valley. *Science of the Total Environment*, **412–413**, 162–169.

Ramos-Ruiz A., Field J. A., Wilkening J. V. and Sierra-Alvarez R. (2016). Recovery of elemental tellurium nanoparticles by the reduction of tellurium oxyanions in a methanogenic microbial consortium. *Environmental Science and Technology*, **50**, 1492–1500.

Rayman M. P. (2012). Selenium and human health. *The Lancet*, **379**(9822), 1256–1268.

Sánchez C., Dessì P., Duffy M. and Lens P. N. L. (2020). Microbial electrochemical technologies: electronic circuitry and characterization tools. *Biosensors & Bioelectronics*, **150**, 111884.

Schilling K., Johnson T. M., Dhillon K. S. and Mason P. R. D. (2015). Fate of selenium in soils at a seleniferous site recorded by high precision Se isotope measurements. *Environmental Science and Technology*, **49**(16), 9690–9698.

Selenius M., Rundlöf A.-K., Olm E., Fernandes A. P. and Björnstedt M. (2010). Selenium and the selenoprotein thioredoxin reductase in the prevention, treatment and diagnostics of cancer. *Antioxid. Redox Signal.*, **12**(7), 867–880.

Sinharoy A. and Lens P. N. L. (2020). Biological removal of selenate and selenite from wastewater: options for selenium recovery as nanoparticles. *Current Pollution Reports*, **6**(3), 230–249.

Sravan J. S., Nancharaiah Y. V., Lens P. N. L. and Mohan S. V. (2020). Cathodic selenium recovery in bioelectrochemical system: regulatory influence on anodic electrogenic activity. *Journal of Hazardous Materials*, **399**, 122843.

Staicu L. C., Ackerson C. J., Cornelis P., Ye L., Berendsen R. L., Hunter W. J., Noblitt S. D., Henry C. S., Cappa J. J., Montenieri R. L., Wong A. O., Musilova L., Sura-de Jong M., van Hullebusch E. D., Lens P. N. L., Reynolds R. J. B. and Pilon-Smits E. A. H. (2015a). *Pseudomonas moraviensis* subsp. stanleyae, a bacterial endophyte of hyperaccumulator Stanleya pinnata, is capable of efficient selenite reduction to elemental selenium under aerobic conditions. *Journal of Applied Microbiology*, **119**(2), 400–410.

Staicu L. C., van Hullebusch E. D., Lens P. N. L., Pilon-Smits E. A. H. and Oturan M. A. (2015b). Electrocoagulation of colloidal biogenic selenium. *Environmental Science and Pollution Research*, **22**(4), 3127–3137.

Staicu L. C., van Hullebusch E. D. and Lens P. N. L. (2017). Industrial selenium pollution: wastewaters and physical-chemical treatment technologies. In: Bioremediation of Selenium Contaminated Wastewater, E. van Hullebusch (ed.), Springer International Publishing, New York (USA), pp. 103–130.

Tan L. C. (2017). Anaerobic treatment of mine wastewater for the removal of selenate and its co-contaminants.

Tan L. C., Nancharaiah Y. V., van Hullebusch E. D. and Lens P. N. L. (2016). Selenium: environmental significance, pollution, and biological treatment technologies. *Biotechnology Advances*, **34**, 886–907.

Tan L. C., Espinosa-Ortiz E. J., Nancharaiah Y. V., van Hullebusch E. D., Gerlach R. and Lens P. N. L. (2018a). Selenate removal in biofilm systems: effect of nitrate and sulfate on selenium removal efficiency, biofilm structure and microbial community. *Journal of Chemical Technology and Biotechnology*, **93**(8), 2380–2389.

Tan L. C., Calix E. M., Rene E. R., Nancharaiah Y. V., Van Hullebusch E. D. and Lens P. N. L. (2018b). Amberlite IRA-900 ion exchange resin for the sorption of selenate and sulfate: equilibrium, kinetic, and regeneration studies. *Journal of Environmental Engineering*, **144**(11), 04018110.

Tan L. C., Papirio S., Luongo V., Nancharaiah Y. V., Cennamo P., Esposito G., van Hullebusch E. D. and Lens P. N. L. (2018c). Comparative performance of anaerobic attached biofilm and granular sludge reactors for the treatment of model mine drainage wastewater containing selenate, sulfate and nickel. *Chemical Engineering Journal*, **345**, 545–555.

Tan L. C., Nancharaiah Y. V., Van Hullebusch E. D. and Lens P. N. L. (2018d). Effect of elevated nitrate and sulfate concentrations on selenate removal by mesophilic anaerobic granular sludge bed reactors. *Environmental Science: Water Research & Technology*, **4**(2), 303–314.

Tapiero H., Townsend D. M. and Tew K. D. (2003). The antioxidant role of selenium and seleno-compounds. *Biomedicine & Pharmacotherapy*, **57**(3), 134–144.

Tinggi U. (2003). Essentiality and toxicity of selenium and its status in Australia: a review. *Toxicology Letters*, **137**, 103–110.

van Ginkel S. W., Yang Z., Kim B., Sholin M. and Rittmann B. E. (2011). The removal of selenate to low ppb levels from flue gas desulfurization brine using the H_2-based membrane biofilm reactor (MBfR). *Bioresource Technology*, **102**, 6360–6364.

Vercellino T., Morse A., Tran P., Hamood A., Reid T., Song L. and Moseley T. (2013a). The use of covalently attached organo-selenium to inhibit *S. aureus* and *E. coli* biofilms on RO membranes and feed spacers. *Desalination*, **317**, 142–151.

Vercellino T., Morse A., Tran P., Song L., Hamood A., Reid T. and Moseley T. (2013b). Attachment of organo-selenium to polyamide composite reverse osmosis membranes to inhibit biofilm formation of *S. aureus* and *E. coli*. *Desalination*, **309**, 291–295.

Wadgaonkar S. L. (2017). Novel Bioremediation Processes for Treatment of Seleniferous Soils and Sediment. PhD thesis, IHE, Delft, The Netherlands.

Wadgaonkar S. L., Nancharaiah Y. V., Esposito G. and Lens P. N. L. (2018a). Environmental impact and bioremediation of seleniferous soils and sediments. *Critical Reviews in Biotechnology*, **38**(6), 941–956.

Wadgaonkar S. L., Ferraro A., Race M., Nancharaiah Y. V., Dhillon K. S., Fabbricino M., Esposito G. and Lens P. N. L. (2018b). Optimization of soil washing to reduce the selenium levels of seleniferous soil from Punjab, Northwestern India. *Journal of Environmental Quality*, **47**(6), 1530–1537.

Wadgaonkar S. L., Mal J., Nancharaiah Y. V., Maheshwari N. O., Esposito G. and Lens P. N. L. (2018c). Formation of Se(0), Te(0), and Se(0)–Te(0) nanostructures during simultaneous bioreduction of selenite and tellurite in a UASB reactor. *Applied Microbiology and Biotechnology*, **102**(6), 2899–2911.

Wadgaonkar S. L., Ferraro A., Nancharaiah Y. V., Dhillon K. S., Fabbricino M., Esposito G. and Lens P. N. L. (2019a). In situ and ex situ bioremediation of seleniferous soils from northwestern India. *Journal of Soils and Sediments*, **19**(2), 762–773.

Wadgaonkar S. L., Nancharaiah Y. V., Jacob C., Esposito G. and Lens P. N. L. (2019b). Microbial transformation of Se oxyanions in cultures of *Delftia lacustris* grown under aerobic conditions. *Journal of Microbiology (Seoul, Korea)*, **57**(5), 362–371.

Wang J., Li H., Yang L., Li Y., Wei B., Yu J. and Feng F. (2016). Distribution and translocation of selenium from soil to highland barley in the Tibetan plateau Kashin-Beck disease area. *Environmental Geochemistry and Health*, 1–12.

Winkel L. H. E., Johnson C. A., Lenz M., Grundl T., Leupin O. X., Amini M. and Charlet L. (2012). Environmental selenium research: from microscopic processes to global understanding. *Environmental Science and Technology*, **46**(2), 571–579.

World Health Organization (2011). Selenium in Drinking-Water. World Health Organization, Geneva.

Wu L. (2004). Review of 15 years of research on ecotoxicology and remediation of land contaminated by agricultural drainage sediment rich in selenium. *Ecotoxicology and Environmental Safety*, **57**(3), 257–269.

Yasin M., El Mehdawi A. F., Jahn C. E., Anwar A., Turner M. F. S., Faisal M. and Pilon-Smits E. A. H. (2014). Seleniferous soils as a source for production of selenium-enriched foods and potential of bacteria to enhance plant selenium uptake. *Plant and Soil*, **386**, 385–394.

Zhang Y., Romero H., Salinas G. and Gladyshev V. N. (2006). Dynamic evolution of selenocysteine utilization in bacteria: a balance between selenoprotein loss and evolution of selenocysteine from redox active cysteine residues. *Genome Biology*, **7** (10), R94.

Zonaro E., Lampis S., Turner R. J., Junaid S. and Vallini G. (2015). Biogenic selenium and tellurium nanoparticles synthesized by environmental microbial isolates efficaciously inhibit bacterial planktonic cultures and biofilms. *Frontiers in Microbiology*, **6**, 658.

Part III

Selenium Biofortification

Chapter 8

Selenium hyperaccumulation in plants

Leonardo Warzea Lima and Michela Schiavon

8.1 INTRODUCTION

Selenium (Se) is a metal(loid) that functions as a structural component of essential selenoproteins serving a variety of metabolic functions in many living organisms, including humans and animals (Kieliszek, 2019). However, its essentiality to vascular plants has not yet been established (Schiavon & Pilon-Smits, 2017a). Edible plants represent a source of Se for human diet and their capacity to accumulate Se largely varies depending on phytoavailable soil Se, the plant species, and the existence of selective uptake mechanisms (White, 2016).

Tissue Se concentrations higher than 100 mg Se kg^{-1} DW typically cause toxicity to most plants because Se amino acids formed during Se assimilation might be incorporated in proteins in place of their sulfur (S) analogs, hence leading to protein folding disturbance and function disruption (Van Hoewyk, 2013). Nevertheless, current literature supports a role for Se as an antioxidant in plants at low concentrations (1–10 mg Se kg^{-1} DW) owing to its capacity to foster, either directly or indirectly, the activity of radical-scavenging enzymes and the synthesis of non-enzymatic antioxidant compounds (Chauhan *et al.*, 2019; Kamran *et al.*, 2019). Doing this, Se contrasts the generation of reactive oxygen species (ROS) in cells, which can injure a variety of biomolecules, especially DNA and membrane components (Natasha *et al.*, 2018). Other positive outcomes of low Se concentrations in plants include growth stimulation, reduction of toxic metal accumulation, retardation of leaf senescence, and protection from

doi: 10.2166/9781789061055_0245

pathogens and predators (Pilon-Smits, 2019; Schiavon & Pilon-Smits, 2017a; Valdez Barillas *et al.*, 2011).

Beneficial effects of Se have been described in several plant species, whether non-Se-hyperaccumulators or Se-hyperaccumulators. The latter, in particular, are native to seleniferous soils and can be potentially used in certain phytotechnologies, such as phytoremediation, biofortification, and agromining, by virtue of their extremely high capacity to accumulate Se without experiencing toxicity (Reeves *et al.*, 2018; Schiavon & Pilon-Smits, 2017b). Recent studies have highlighted several mechanisms that might explain why some plants developed the Se hyperaccumulation and hypertolerance traits, such as the existence of specific tissue Se sequestration patterns and the constitutive elevate expression of genes coding for Se/S transporters and Se/S assimilation enzymes, antioxidant enzymes, and proteins involved in hormone biosynthesis, signaling and interactions with ecological partners (Lima *et al.*, 2018; Wang *et al.*, 2018). It is believed that Se hyperaccumulation is the result of convergent evolution of Se transporters and specific or alternative biochemical pathways in different angiosperm clades, which occurred during geological periods when seleniferous soils were widely distributed (White, 2016). In support of this hypothesis are several pieces of evidence. Compared to non-hyperaccumulators, Se-hyperaccumulators have higher shoot/root Se ratios, Se/S ratios (indicative of more selective mechanisms for Se uptake and translocation over S), and higher Se translocation from sources to sinks (El Mehdawi *et al.*, 2018; Schiavon *et al.*, 2015). Also, they contain more organic Se, especially in the form of methylselenocysteine (SeMetCys), which suggests intense Se assimilation and the existence of biochemical mechanisms aimed at preventing misincorporation of Se-amino acids in proteins (White, 2018). SeMetCys primarily accumulates in epidermal cells and reproductive organs of these Se-hyperaccumulator plants (Pilon-Smits, 2019).

In this chapter, we aim to describe (i) differences in uptake, accumulation, and metabolic fate of Se between the non-Se accumulator and Se hyperaccumulator species, (ii) key factors responsible for the evolution of the Se hyperaccumulation and hypertolerance traits, and (iii) potential uses of Se-hyperaccumulators in agro- and phytotechnologies.

8.2 VARIATION IN Se ACCUMULATION BETWEEN HYPERACCUMULATORS AND NON-HYPERACCUMULATORS

8.2.1 Se uptake in plants

Despite Se accumulation by plants tightly relating to soil Se, angiosperm species can differ in their capacity to take up Se when growing in the same environment. This suggests that Se accumulation in plants also depends on the selectivity of the transport mechanisms used to acquire Se.

Although plants do not have an essential demand for Se, they can take up different inorganic and organic forms of this element, whose oxidation state in soils ranges between −2 and +6 (Hartikainen, 2005). Among inorganic Se species, selenate (SeO_4^{2-}) is the most available to plants in oxic soils of high redox potential (Hartikainen, 2005). Selenate is the chemical (toxic) analog to sulfate, and thus plants make use of sulfate transporters to allow selenate to enter the root cells (El-Mehdawi *et al.*, 2018). In anoxic soils characterized by low redox potential and pH from neutral to acidic, like paddy soils, selenate tends to be converted to selenite (SeO_3^{2-}), which is acquired by plants using either phosphate transporters, silicon transporters, or aquaporins, depending on the selenite species (selenite anion (SeO_3^{2-}), hydrogenselenite ion ($HSeO_3^-$), and selenous acid (H_2SeO_3)) (Shahid *et al.*, 2018). Plants can also take up elemental Se, but not selenide or colloidal elemental Se (Winkel *et al.*, 2015).

8.2.2 Se accumulators and hyperaccumulating plants

There is broad variation in the capacity of plants to accumulate Se in their organs, which is largely affected by soil Se content and phytoavailability (White, 2016, 2018). Plants thriving in soils either naturally rich in available Se or contaminated with Se due to anthropogenic activities or dust depositions in coal-burning areas are inclined to accumulate more Se than plants colonizing low Se areas (Schiavon and Pilon-Smits, 2017a, b; White, 2016). The Se content in soils is commonly below 2 μg g^{-1} but can reach several hundred μg g^{-1} (up to 1.2 mg g^{-1}) in soils derived from sedimentary rocks, especially Cretaceous sediments rich in selenites and selenides associated with sulfide minerals (Winkel *et al.*, 2015). These soils are termed seleniferous and are located in the Great Plains of the USA, Canada, Brazil, Australia, India, China, and Russia, and usually support a distinctive pattern of vegetation (Oldfield, 2002; Winkel *et al.*, 2015). Plants growing in seleniferous areas can exclude or actively remove Se from their tissues, or tolerate high internal Se concentrations (White, 2018). Despite such a different behavior towards Se, all these plants have a high Se concentration in their tissues and therefore must exhibit a minimal ability to tolerate elevated Se concentrations.

Based on the maximum Se concentration in their shoot, higher plants can be sorted into two major ecological groups, that is, Se non-hyperaccumulators and Se-hyperaccumulators. Se non-hyperaccumulators include non-accumulators and accumulator species. The majority of plants fall in the category of non-accumulators and are unable to accumulate and tolerate more than 100 μg Se g^{-1} DW. Therefore, these plants cannot colonize seleniferous soils and, in most cases, their Se concentration is below 1 μg g^{-1} DW (White, 2016, 2018). In contrast, Se-accumulators can tolerate tissue Se concentrations approaching 1 mg Se g^{-1} DW and can populate both non-seleniferous and seleniferous soils.

Furthermore, the Se concentration in their shoot tissues is often indicative of soil phytoavailable Se (Pilon-Smits, 2019).

8.2.3 Se-hyperaccumulating plant species

Se-hyperaccumulating species are rare and include plants with an extremely high ability to accumulate and tolerate Se (Pilon-Smits, 2019; White, 2016). They can contain more than 1 mg Se g^{-1} DW in their shoots while growing in their native environment, which is generally restricted to seleniferous soils. The bioconcentration factor (BF) for these species is generally very high (e.g., over 800:1 in *Astragalus bisulcatus*), and it may perform better than the absolute concentration threshold in discriminating hyperaccumulators from non-accumulators (Statwick *et al.*, 2016). The BF, as well as the translocation factor (TF), often negatively correlates with the plant biomass in non-hyperaccumulators because of Se toxicity. Conversely, in hyperaccumulators, despite the apparent costs required to actively concentrate high Se in tissues through energy-dependent mechanisms, Se seems to stimulate plant growth (Statwick *et al.*, 2016).

The discovery of Se-hyperaccumulators was made by Orville and coworkers in the 1930s in the western United States (Rosenfeld & Beath, 1964), and Se-hyperaccumulators that populate natural seleniferous soils in these areas accumulate up to 15 mg Se g^{-1} DW (Pilon-Smits, 2019). Since then, Se hyperaccumulation has been recorded in about 45 taxa (Cappa & Pilon-Smits, 2014; Pilon-Smits, 2019; White, 2016). Figure 8.1 depicts representative Se hyperaccumulator plants. The genus *Astragalus* of the Fabaceae family comprises 25 Se hyperaccumulator species, while the Asteraceae genera *Xylorhiza* and *Symphyotrichum* contain 3 Se hyperaccumulator species each (Table 8.1). In other cases, there are only one or two species per genus, such as *Stanleya pinnata* and *Stanleya bipinnata* from the genus *Stanleya*, *Cardamine violifolia* from the genus *Cardamine* of the Brassicaceae family, and *Neptunia amplexicaulis* from the genus *Neptunia* of the Fabaceae family (El Mehdawi and Pilon-Smits, 2012; Harvey *et al.*, 2020; Rosenfeld & Beath, 1964; White, 2016). Except for *Cardamine violifolia*, which is native to seleniferous soils in the Yutangba region in China, and *Neptunia amplexicaulis*, which grows in seleniferous soils in Queensland (Australia), Se-hyperaccumulators are commonly native to seleniferous soils of the Western USA (Both *et al.*, 2020; Harvey *et al.*, 2020; Pilon-Smits, 2017).

Species of the genera *Astragalus* and *Stanleya* are likely the strongest Se-hyperaccumulators, being able to accumulate up to 10 mg Se g^{-1} when growing in soils with 2–10 μg Se g^{-1} (Schiavon & Pilon-Smits, 2017a). However, it must be noted that within an Se-hyperaccumulating species there are populations that largely differ in their capacity to accumulate Se, as well as individuals within the same population (Pilon-Smits, 2017). Such differences are mainly due to genetic variability, local Se availability, and perhaps associated rhizosphere and endophytic microorganisms (Pilon-Smits, 2017).

Figure 8.1 Representative Se-hyperaccumulators within different families. Photos available at: https://naturalcommunities.net/products/symphyotrichum-ericoides-heath-aster; http://www.easterncoloradowildflowers.com/Astragalus_bisulcatus.htm; https://www.shutterstock.com/it/search/stanleya+pinnata; https://en.wikipedia.org/wiki/Atriplex_confertifolia; https://en.wikipedia.org/wiki/Gutierrezia_microcephala; https://www.prairiemoon.com/astragalus-racemosus-creamy-milk-vetch-prairie-moon-nursery.html; http://www.canbr.gov.au/cpbr/cd-keys/RFK7/key/RFK7/Media/Html/entities/Coelospermum_decipiens.htm; https://www.fs.fed.us/wildflowers/plant-of-the-week/Castilleja-coccinea.shtml (photo by Christopher David Benda).

8.2.4 Se uptake in Se-hyperaccumulators

Se-hyperaccumulators possess a high expression of specific transporters mediating Se uptake and possibly delivering to the shoot, which might justify their extremely high capacity to accumulate Se. Studies so far indicate that these plants can discriminate between Se and its analog S, while non-hyperaccumulators cannot. This assumption is based on the observation that tissue Se/S ratios displayed by these plants are higher compared to non-hyperaccumulators (Schiavon *et al.*, 2015). The mechanism of this hypothesized preference for Se over S by Se-hyperaccumulators has not been clarified yet, but the existence of a sulfate transporter with a higher affinity for Se over S has been recently proposed in *S. pinnata*. This Se-hyperaccumulator has elevated and constitutive expression of the gene encoding in the root for high affinity sulfate transporter *Sultr1;2*. This transporter in the non-hyperaccumulators usually plays a major role in sulfur primary uptake and is regulated by both sulfate availability and the S status of the plant, typically down-regulated by high sulfate concentration in the external

Table 8.1 Species, distribution and maximum Se shoot concentration of several Se hyperaccumulators as retrieved by White (2016).

Species	Plant Distribution	Se Concentration (mg Se kg^{-1} DW)
Asteraceae (Asterales)		
Dieteria canescens (Pursh) Nutt.	Midwest USA	1600
Grindelia squarrosa (Pursh) Dunal	Lower Brule Reservation, SD, USA	930
Gutierrezia microcephala (DC.) A.Gray	Thompson, UT, USA	1287
Oonopsis foliosa Greene	Lascar, CO, USA	3630
Oonopsis wardii (A.Gray) Greene	Albany County, WY, USA	9120
Symphyotrichum ascendens (Lindl.) G.L.Nesom	Soda Springs, ID, USA	4455
Symphyotrichum ericoides (L.) G.L. Nesom	Pine Ridge, Fort Collins, CO, USA	1378
Symphyotrichum lateriflorum (L.) Á. Löve & D.Löve	SD, USA	1800
Xylorhiza glabriuscula Nutt.	Huerfano County, CO, USA	1750
Xylorhiza parryi Greene	Albany County, WY, USA	5390
Xylorhiza venusta (M.E.Jones) A. Heller	Midwest USA	3486
Fabaceae		
Acacia cana Maiden	NW Queensland, Australia	1121
Astragalus albulus Wooton & Standl.	La Ventana, NM, USA	530
Astragalus beckwithii var. *purpureus* M.E.Jones	Cameron, AZ, USA	3135
Astragalus bisulcatus (Hook.) A.Gray	Pine Ridge, Fort Collins, CO, USA	13,685
Astragalus bisulcatus var. *haydenianus* (A. Grey) Barneby	Cuba, NM, USA	2377
Astragalus canadensis L.	Las Vegas, NE, USA	1110
Astragalus crotalariae A.Gray	Truckhaven, CA, USA	2175
Astragalus eastwoodiae M.E.Jones	Utah, USA	1664
Astragalus flavus Torr. & A.Gray	Aztec, NM, USA	1361
Astragalus flavus var. *argillosus* (M.E.Jones) Barneby	Greenriver, UT, USA	631
Astragalus flavus var. *candicans* A.Gray	Thompson, UT, USA	1322
Astragalus grayi S.Watson	Carbon County, WY, USA	4450
Astragalus osterhoutii M.E.Jones	Kremmling, CO, USA	2678
Astragalus pattersonii A.Gray	Thompson, UT, USA	8512

(*Continued*)

Table 8.1 Species, distribution and maximum Se shoot concentration of several Se hyperaccumulators as retrieved by White (2016) (*Continued*).

Species	Plant Distribution	Se Concentration (mg Se kg^{-1} DW)
Astragalus pectinatus (Hook.) G.Don	Teton County, MT, USA	5170
Astragalus praelongus E.Sheld.	Leupp, AZ, USA	4835
Astragalus praelongus var. *ellisiae* (Rydb.) B.L.Turner	Valmont, NM, USA	656
Astragalus preussii A.Gray	Thompson, UT, USA	4188
Astragalus racemosus Pursh.	WY, USA	14,920
Astragalus rafaelensis M.E.Jones	Jensen, TX, USA	716
Astragalus sabulosus M.E.Jones	Thompson, UT, USA	2210
Astragalus toanus M.E.Jones	ID, USA	990
Neptunia amplexicaulis Domin	Richmond, Queensland, Australia	4334
Brassicaceae (Brassicales)		
Cardamine hupingshanensis	Yutangba, Enshi, China	1965
Cardamine violofolia	Yutangba, China	2700
Stanleya bipinnata Greene	Laramie, WY, USA	2490
Stanleya pinnata (Pursh) Britton	Pine Ridge, Fort Collins, CO, USA	>4000
Stanleya pinnata var. *integrifolia* (E. James) Rollins	Vernal, UT, USA	977
Amaranthaceae (Caryophyllales)		
Atriplex confertifolia (Torr. & Frém.) S. Watson	Thompson, UT, USA	1734
Atriplex nutallii S.Watson	WY, USA	930
Rubiaceae (Gentianales)		
Coelospermum decipiens Baill.	Cape York Peninsula, Queensland, Australia	1141
Orobanchaceae (Lamiales)		
Castilleja angustifolia var. dubia	Lysite, WY, USA	3460

Se hyperaccumulators whose Se shoot concentration and distribution are unknown are not reported.

medium and in the plant (El Kassis *et al.*, 2007). *Sultr1;2* from *S. pinnata* (*SpSultr1;2*) is not subjected to such a feedback inhibition by high sulfate concentrations (El Mehdawi *et al.*, 2018). Furthermore, increased expression of *Sultr1;2* is associated with high Se uptake rates in this hyperaccumulator. Consistent with already described metal hyperaccumulation mechanisms (Craciun *et al.*, 2012; Hanikenne *et al.*, 2008; Lochlainn *et al.*, 2011), the overexpression of *Sultr1;2* may be attributed in part to gene duplication events, which would account for the enhanced Se transport capacity and likely allowed at least one of the gene copies to evolve toward greater specificity for Se over S. Also, both the constitutively high expression and concomitant unresponsiveness of *SpSultr1;2* to S might have been the result of either mutation in cis-regulatory sequences of the transporter or changes in its coding sequence that may have altered the transporter affinity and specificity for the substrate (El Mehdawi *et al.*, 2018).

In addition to selenate, plants can take up organic Se species very efficiently. In a transcriptome study, the gene *LHT1* was substantially more expressed in *S. pinnata* than in the non-hyperaccumulator *Stanleya elata* (Wang *et al.*, 2018). *LHT1* encodes a high affinity transporter for cellular amino acid uptake from the soil into the mesophyll cells (Hirner *et al.*, 2006). Owing to their broad substrate specificity, LHT1 proteins may mediate the uptake of Se amino acids, either methylated or not, which can be released to the soil once the plant residues from Se-hyperaccumulators are decomposed by soil microorganisms. Se-hyperaccumulators are generally rich in organic Se and their deposition may affect soil Se speciation (Pilon-Smits, 2019). Therefore, the upregulation of *LHT1* in these species might represent a strategy to get more highly available organic Se in the surrounding environment.

Once inside roots cells, inorganic Se can be delivered to the plastids to access the S-assimilation pathway, or loaded into the xylem and be transferred to the shoot for assimilation into the chloroplasts of mesophyll cells (Sors *et al.*, 2005). Organic Se produced in the roots or accumulated after uptake can be also conveyed to the shoot through the xylem flow. *S. pinnata* plants show high expression of the low-affinity sulfate transporter *Sultr2;1*, which may be, at least in part, responsible for the enhanced accumulation of Se in the shoot (El Mehdawi *et al.*, 2018).

8.3 METABOLIC PATHWAYS SUPPORTING Se HYPERACCUMULATION

8.3.1 Se metabolism in non-hyperaccumulating plants

Selenium (Se) is not considered a nutrient for plants, but it can be beneficial to them in lower concentrations, and in contrast, it is toxic to most species at higher tissue levels. This antagonistic characteristic is puzzling and fascinating at the same time and establishes Se as a unique element when its interaction with plants is considered. Non-Se-hyperaccumulator plants take up and translocate Se inadvertently using sulfur transporters. The inorganic Se is then nonspecifically assimilated and reduced via sulfur metabolism to the amino acids SeCys (selenocysteine) and SeMet (selenomethionine), resulting in protein malfunction and systemic stress.

8.3.2 Organo-Se synthesis in hyperaccumulating plants

This scenario is, however, more complex for the Se-hyperaccumulator species. These plants evolved to cope with the high concentrations of Se in their tissues, therefore, they can accumulate thousands of mg Se kg^{-1} DW without showing–any symptoms of stress. The main mechanism used by these plants to avoid the misincorporation of Se into proteins is elegant and efficient: the accumulation of organic forms of Se. This is accomplished by some different processes including methylation of the amino acid SeCys, and exclusion of Se by volatilization. The main forms of Se found in these plants are selenocystathionine and methylselenocystheine (MeSeCys), the latter is formed by a unique isoform of the enzyme SeCys methyltransferase (SMT) found in Se-hyperaccumulators (Freeman *et al.*, 2010). The methylated form of cysteine cannot be incorporated into proteins. Thus, it can be accumulated at high levels without inducing any stress symptoms (Figure 8.2). This enzyme is constitutively expressed in Se-hyperaccumulators, and previous studies showed that SMT from Se-hyperaccumulator species *Astragalus bisulcatus* can induce the accumulation of the organic and less toxic MeSeCys when expressed in the non-Se-hyperaccumulator plants *Arabidopsis thaliana* and *Brassica juncea*, resulting in tolerance to Se, confirming its important role in Se-hyperaccumulator species (LeDuc *et al.*, 2004).

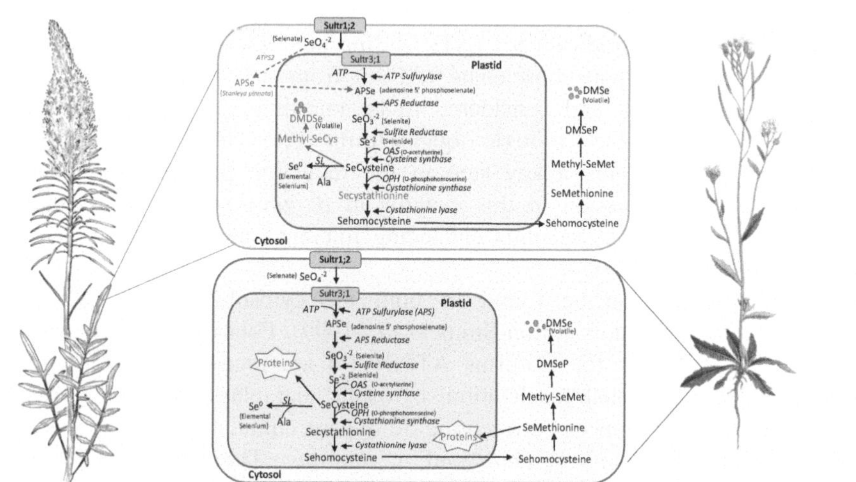

Figure 8.2 Selenium assimilation in hyperaccumulators and non Se-hyperaccumulators. Biochemical routes specific to hyperaccumulators are highlighted in red. ATPS2 = ATP sulfurylase isoform 2, SL = selenocysteine lyase, Ala = alanine, DMSe = dimethyl selenide, DMDSe = dimethyldiselenide. Adapted from Schiavon and Pilon-Smits (2017a).

Another important and unique metabolic pathway found in the Se-hyperaccumulator plants is the ability to volatilize Se from the leaves. This process starts with the oxidation of the MeSeCys, forming methylselenocysteineselenideoxide (MeSeCysSeO) (Chin & Lindsay, 1994). MeSeCysSeO is then converted to methaneselenol by the enzyme cysteine sulfoxide lyase, which is then converted to dimethyldiselenide (DMDSe), a volatile form of selenium that can be excluded from the leaves (Valdez Barillas *et al.*, 2011). This process can protect the plant from the possible Se toxicity, but it is also hypothesized to help the plant to avoid herbivores through deterrence (Schiavon & Pilon-Smits, 2017a).

Se is stored in vacuoles of epidermal leaf cells, and it can be strategically stored in high concentrations in cells at the edge of the leaf blade in *S. pinnata*, possibly for herbivory protection (Freeman *et al.*, 2010). There is a seasonal variation in selenium accumulation in the *S. pinnata* leaves, where higher Se concentrations can be found in early spring and lower concentrations in the fall, while sulfur levels spike in midsummer (El Mehdawi & Pilon-Smits, 2012).

8.3.3 Enzymology of organo-Se formation

Most of the enzymes in the sulfur assimilation pathways are constitutively overexpressed in the Se-hyperaccumulator species when compared to non-Se-hyperaccumulators. The first enzyme in the selenate/sulfate assimilation is the enzyme adenosine triphosphate (ATP)-sulfurylase (ATPS), responsible for forming adenosine 5'phosphoselenate (APSe) from ATP and selenate/sulfate. This step is important and considered to be rate-limiting in the reduction of inorganic selenium to the less toxic organic forms (Pilon-Smits *et al.*, 1999). The ATPS enzyme represents a key step in the reduction process of selenate to selenite, the overexpression of this enzyme in *B. juncea* resulted in enhanced accumulation of organic Se, thus enhancing tolerance and accumulation of Se (Pilon-Smits *et al.*, 1999).

ATP-sulfurylase can be found in both the cytosol and the plastids in non-Se-hyperaccumulators (Pilon-Smits *et al.*, 1999). Four different isoforms are known in *A. thaliana*: the isoforms ATPS1-3-4 are found in plastids and the isoform ATPS2 has dual localization, cytosolic and plastidial. There are also some qualitative differences between the Se-hyperaccumulator *S. pinnata* and the non-Se-hyperaccumulators *S. elata* and *A. thaliana*. The Se-hyperaccumulator ATP sulfurylase isoform 2 (ATPS2) enzyme transcription is higher in roots, and it has a unique cytosolic localization (Figure 8.2), due to a stop codon in the DNA sequence coding for the protein, resulting in an interrupted chloroplast transit peptide in the amino acid sequence (Jiang *et al.*, 2018). This would indicate the selenium could be reduced in the root cells, and subsequently transported to the shoot in its organic forms in *S. pinnata*.

In the past 20 years, several different genes from the sulfur assimilation pathway have been manipulated to study their effect on plant Se tolerance and accumulation. The overexpression of the enzyme cystathionine synthase, responsible for the formation of Se cystathionine from SeCys (Figure 8.2), in *Brassica juncea* resulted in three-fold high Se volatilization from either selenate or selenite, indicating expression of this enzyme as a limiting factor for Se volatilization in plants (Van Huysen *et al.*, 2003). As stated before (see 8.3.2), the conversion of inorganic Se to its less toxic organic forms is the main step towards Se tolerance in Se-hyperaccumulator plants, and the enzyme SMT plays a crucial role in this process. LeDuc *et al.* (2004) demonstrated that overexpressing the SMT from the hyperaccumulator *A. bisulcatus* in the non-Se-hyperaccumulators *A. thaliana and B. juncea*, enhanced the Se volatilization, tolerance, and the accumulation of MeSeCys. The effect was more pronounced when the transgenic plants were supplied with selenite, which indicates the reduction of selenate to selenide is a limiting step in the assimilation pathway.

Another enzyme that was extensively studied is the selenocysteine lyase (SL), which breaks down SeCys into elemental Se and alanine (this process is widely used by non-Se-hyperaccumulators to detoxify the excess of Se). Some studies showed the overexpression of SL can reduce the incorporation of Se into proteins, enhance volatilization and overall tolerance in *A. thaliana*, and improve Se accumulation by two-fold in *B. juncea*, a promising trait for Se phytoremediation and fortification in crop fields (Bañuelos *et al.*, 2002, 2015).

The Se-hyperaccumulator plants also show tissue-specific patterns of Se accumulation. The Se levels are usually high in young leaves and reproductive tissues like the flower, silique, seeds, pollen, and ovules due to its remobilization from aging leaves (Cappa & Pilon-Smits, 2014).

8.4 EVOLUTION OF THE Se HYPERACCUMULATION TRAIT

8.4.1 Main driving-factors

The Se hyperaccumulation is a derived trait that possibly evolved independently in 45 different taxa, in 14 genera from 6 dicot plant families, including the Brassicaceae, Fabaceae, and Asteraceae families (White, 2016). Because the occurrence of Se-hyperaccumulators correlates with the Se content in the soil, usually in seleniferous areas, it can be hypothesized that the Se bioavailability, concentration, and distribution in the soil were possibly the most important driving factors for the evolution of this trait. However, these might not be the only aspects to consider when analyzing the evolution of Se hyperaccumulation, because just a small number of plant species found in seleniferous areas can tolerate extreme concentrations of Se. Therefore, the ecological benefits from the high selenium concentration, the upregulation of genes related to the antioxidant activity and hormonal stress resistance were possibly strong components, in combination with seleniferous soils, driving the evolution of this trait (Schiavon & Pilon-Smits, 2017a).

8.4.2 Metabolic defense mechanisms

The Se-hyperaccumulator species *S. pinnata* demonstrates a higher expression of genes involved in the biosynthesis of different hormones related to biotic and abiotic stress resistance, including jasmonic acid, ethylene, and salicylic acid, as well as an upregulation of different antioxidant enzymes, when compared to the non-Se-hyperaccumulator *S. elata* (Wang *et al.*, 2018). The cellular antioxidant defense mechanisms correspond to the enzymatic and non-enzymatic responses to directly quench the reactive oxygen species (ROS), such as the superoxide radical (O_2^-) or hydrogen peroxide (H_2O_2), thus neutralizing the deleterious effects of the different reactive processes. In other words, Se-hyperaccumulator plants are more prepared to deal with different stresses and are selected in evolution to tolerate and accumulate Se. It was found, for example, that genes involved in the synthesis of glutathione were upregulated in the Se-hyperaccumulator *S. pinnata* (Wang *et al.*, 2018). Glutathione is a tripeptide formed from the amino acids glutamate, cysteine and glycine, and is a major metabolite in the defense system against ROS and its degenerative effects during the antioxidant stress, which could be caused by high concentrations of Se.

8.4.3 Plant ecology

The ecological aspects of hyperaccumulation also played an important role in the evolution of the trait. It is known that the extreme Se concentrations found in the hyperaccumulator species *A. bisulcatus* and *S. pinnata* can promote protection against different herbivores due to deterrence (Quinn *et al.*, 2010), and pathogenic fungi in *B. Juncea* (Hanson *et al.*, 2003). There is evidence that hyperaccumulator plants can also benefit from the higher Se concentration when competing with neighbor plants, by increasing the Se concentration in the soil up to 10 times, due to litter deposition and exudation, which affects the germination and growth of Se-sensitive plants (El Mehdawi *et al.*, 2012).

Endophytic bacterial communities harbored by Se hyperaccumulators might also have played a role in the evolution of the Se hyperaccumulation and hypertolerance traits in these species and are different in composition compared to endophytic bacterial communities harbored by non-Se-accumulators (Cochran *et al.*, 2018). In a recent study, endophytic bacteria highly resistant to selenate and selenite (up to 200 mM) and able to reduce selenite to elemental Se, as well as to reduce nitrite and produce siderophores, have been isolated and identified in *S. pinnata* and *A. bisulcatus* plants growing in their natural environments (Sura de-Jong *et al.*, 2015). These isolates may increase the tolerance of plants to Se by reducing its potential toxicity in tissues, and provide them with additional

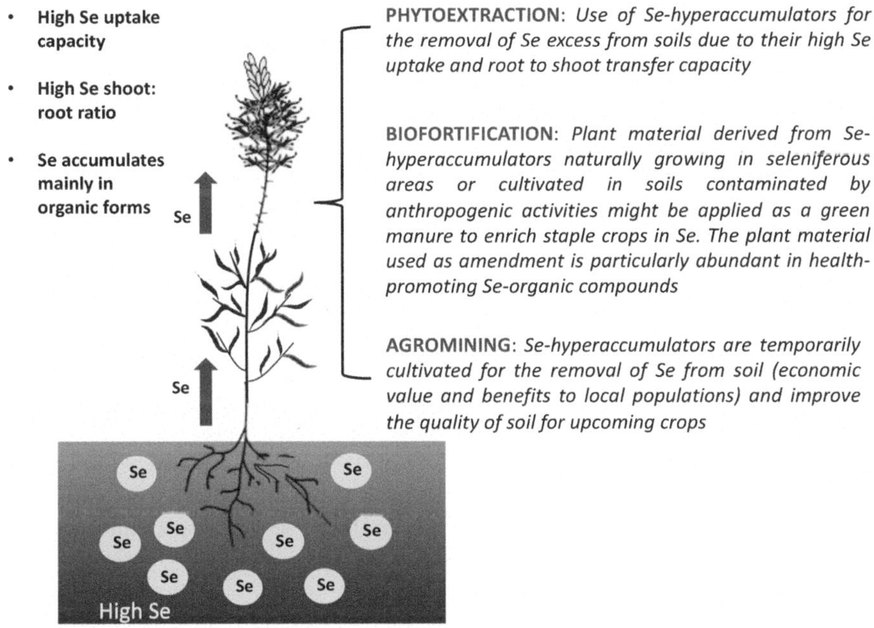

Figure 8.3 Potential application of Se-hyperaccumulators (Schiavon & Pilon-Smits, 2017b).

benefits. Indeed, some of the isolates showed plant growth-promoting properties when inoculated in non-hyperaccumulators. In turn, organic forms of Se released by hyperaccumulators in their habitats may alter microbial species composition, favoring those taxa that utilize the essential nutrient Se more efficiently.

8.5 POTENTIAL USES OF Se-HYPERACCUMULATORS IN PHYTOTECHNOLOGIES

Se-hyperaccumulators find potential practical application in different types of phytotechnologies, such as (i) phytoextraction of Se from mining-wastes and reclamation of Se-polluted soils, (ii) biofortification of staple crops to combat Se deficiency in vulnerable populations living in low Se areas, and (iii) agromining (Figure 8.3) (Schiavon & Pilon-Smits, 2017b).

8.5.1 Phytoremediation

Selenium, similar to other metalloids (e.g., arsenic) and metals (e.g., thallium, cobalt, cadmium, and rhenium), is a by-product occurring in mining wastes that may be at concentrations suitable for economic phytoextraction (phytomining). Se phytoextraction could be carried out in those areas where wastes from Se-rich coal represent a potential threat to the environment (Reeves *et al.*, 2018). Such an approach has been adopted in the USA, Canada, the UK, and Australia (Remigio *et al.*, 2020), and has also been applied to soils that are contaminated with Se due to extensive use of Se-rich irrigation waters. Se-hyperaccumulators might be ideally used in Se phytoextraction of Se contaminated substrates in areas where they are naturally widespread. Thus, *A. bisulcatus* or *S. pinnata* could be used in the USA, whereas *C. violifolia* and *N. amplexicaulis* could be used in China and Australia, respectively.

The great capacity of Se-hyperaccumulators to concentrate Se at extremely high concentrations in their shoot and convert it into relatively less-toxic, human health beneficial organic forms, make them useful for phytoextraction, as harvesting could be restricted only to the above-ground parts of the plant. Furthermore, the Se-hyperaccumulator *S. pinnata* has been shown to volatilize Se at high rates when used for the remediation of Se-rich irrigation wastewater under high sulfate conditions (Parker *et al.*, 2003). Because of this efficient Se metabolism, *S. pinnata* has been proposed as a promising species to use for Se phytoremediation. Similarly, the hyperaccumulator *A. bisulcatus* substantially enhanced the emission rate of Se from soil (Pletsch, 2003). Freeman and Bañuelos (2011) found that hyperaccumulation and volatilization of Se by *S. pinnata* genotypes and their associated microbes could remove approximately 30% of the total soil Se in 0–30 cm Se-laden agricultural drainage sediment containing 9.0 ($\pm$3.8) µg Se g^{-1} DW. However, despite these promising results and the advantages offered by Se-hyperaccumulators, so far only a few large field-scale studies have been conducted using these plants, mainly because of their non-agronomic status, coupled with their low biomass yield and slow growth (Zambrano *et al.*, 2018). Also, significant differences in Se accumulation might exist between different ecotypes of Se hyperaccumulators, as evidenced for *S. pinnata* by Feist and Parker (2001). Rather, Se accumulators (e.g., *Brassica juncea* L. Czern) producing high biomass yields and of moderate capacity to take up and translocate Se to the above-ground tissues have been widely employed (Bañuelos *et al.*, 2002; Dhillon & Bañuelos, 2017).

8.5.2 Biofortification

Se-hyperaccumulators growing in natural seleniferous areas or used for the reclamation of Se-contaminated soils can be a source of Se-laden plant material to be employed as green manure to enhance the content of Se in agricultural soils and food crops (Bañuelos *et al.*, 2015, 2016; Schiavon & Pilon-Smits, 2017b; Stonehouse *et al.*, 2020). Bañuelos *et al.* (2015, 2016) showed that Se-enriched plant material from *S. pinnata* plants, initially grown to remove Se by phytoextraction from Se-laden agricultural drainage sediment, is valuable as a soil amendment for enriching broccoli and carrots with healthful organic-Se compounds, especially SeMet.

The plant material derived from Se-hyperaccumulators can otherwise be recycled as forage for livestock (Bañuelos *et al.*, 2009). In this way, Se phytoremediation and Se biofortification technologies can be proficiently joined. However, before being launched into the food chain, Se-laden material from Se-hyperaccumulators must be subjected to rigorous controls to verify the absence of potential contaminants (e.g., heavy metals and metalloids) that might be toxic to direct consumers (livestock), crops, and food crop consumers (humans).

Se-hyperaccumulators are particularly rich in Se organic forms (e.g., SeMet and MetSeCys). Therefore, crops to be amended with green manure obtained from Se-hyperaccumulators could be selected via breeding or otherwise engineered for elevated uptake rates and accumulation of organic Se in the edible produce. Food crops enriched with organic Se compounds might possess higher nutritional values compared to crops that principally accumulate inorganic Se, as organic Se compounds are more readily available for essential Se metabolism in humans and animals (Davis, 2012). Alternatively, Se-hyperaccumulators could be used in co-cropping or intercropping (Schiavon & Pilon-Smits, 2017b). In this case, their residues will deposit on the soil at the end of the growing season, and their decomposition by Se-resistant soil microorganisms will lead to the release of Se organic compounds that might be easily acquired by the neighboring or upcoming crops. The microbiome of Se-hyperaccumulators has been recently found to be different from that of non-hyperaccumulators, with a higher average relative abundance of *Pedobacter* and *Deviosa* (Cochran *et al.*, 2018).

8.5.3 Agromining

Compared to other Se phytotechnologies, Se agromining as an alternative type of agriculture conducted on low productive agricultural (ultramafic) lands is poorly widespread. Agromining is a variation of phytomining, which foresees the cultivation of hyperaccumulators and aims to harvest their biomass to retrieve target metal/loids. The recovery of metal/loids (bio-ores) involves the drying, ashing, and managing of the plant biomass. This is an important process in view of the fact that ore grades are progressively declining, resulting in increased waste products from the mining industry. Agromining is economically viable only for certain metal/loids, including Se, nickel (Ni), and tellurium (Te), as it depends on the element market price, annual yield per unit area, and the existence of regions enriched in the target element. Till now, prices per metric ton are high for Ni (US$14,000), Se (US$44,000), and Te (US$65,000) (USGS, 2020).

The idea of growing hyperaccumulators on infertile soils raises from the consideration that most of them are highly nutrient-efficient, and thus have low fertilization requirements. Agromining might provide better economic incomes to local populations and should be intended as temporary farming for metal/loids and not as a steady replacement of food crops. Also, after hyperaccumulators are harvested for metal/loid recovery, the soil fertility is supposed to have improved enough by their cultivation to support the growth of upcoming crops (van der Ent *et al.*, 2015). Despite the benefits of using hyperaccumulators for agromining, the adoption of Se-hyperaccumulators for economic extraction of elemental Se is still a very limited practice.

8.6 CONCLUSION

Se-hyperaccumulators are intriguing species characterized by an extraordinary capacity to accumulate Se in all their tissues and a higher transfer capacity of Se from the roots to their shoot. Research performed so far has highlighted the potential evolutionary drivers and selective pressures responsible for the development of the hyperaccumulation trait, and the ecological benefits associated with high Se concentrations in tissues. Pieces of evidence are suggesting that Se-hyperaccumulators possess specific mechanisms for Se uptake and root to shoot translocation, while the specific metabolic routes through which they prevent Se amino acid misincorporation in proteins have been widely described. Ongoing studies aimed to overexpress genes from hyperaccumulators into non-hyperaccumulator model species may help to elucidate the process of hypertolerance and hyperaccumulation in plants. These genes could be further transferred into plant species suitable for phytoremediation, biofortification, and agromining. Indeed, although Se-hyperaccumulators may be employed in these technologies, their use in field application is unfortunately still limited primarily because of their limited biomass production, which is generally smaller compared to that produced by non-Se accumulators.

REFERENCES

Bañuelos G. S., Lin Z. Q., Wu L. and Terry N. (2002). Phytoremediation of selenium-contaminated soils and waters: fundamentals and future prospects. *Reviews on Environmental Health*, **17**(4), 291–306.

Bañuelos G. S., Robinson J. and Da Roche J. (2009). Developing selenium-enriched animal feed and biofuel from canola planted for managing Se-laden drainage waters in the Westside of Central California. *International Journal of Phytoremediation*, **12**, 243–254.

Bañuelos G. S., Arroyo I., Pickering I. J., Yang S. I. and Freeman J. L. (2015). Selenium biofortification of broccoli and carrots grown in soil amended with Se-enriched hyperaccumulator *Stanleya pinnata*. *Food Chemistry*, **166**, 603–608.

Bañuelos G. S., Arroyo I. S., Dangi S. R. and Zambrano M. C. (2016). Continued selenium biofortification of carrots and broccoli grown in soils once amended with Se-enriched *S. pinnata*. *Frontiers in Plant Science*, **23**, 7–1251.

Both E. B., Stonehouse G. C., Lima L. W., Fakra S. C., Aguirre B., Wangeline A. L., Xiang J., Yin H., Jókai Z., Soós Á., Dernovics M. and Pilon-Smits E. A. H. (2020). Selenium tolerance, accumulation, localization and speciation in a *Cardamine* hyperaccumulator and a non-hyperaccumulator. *Science of the Total Environment*, **703**, 135041.

Cappa J. J. and Pilon-Smits E. A. H. (2014). Evolutionary aspects of elemental hyperaccumulation. *Planta*, **239**, 267–275.

Chauhan R., Awasthi S., Srivastava S., Dwivedi S., Pilon-Smits E. A. H., Dhankher O. P. and Rudra D. (2019). Understanding selenium metabolism in plants and its role as a beneficial element. *Critical Reviews in Environmental Science and Technology*, **49**, 1937–1958.

Chin H. W. and Lindsay R. C. (1994). Mechanisms of formation of volatile sulfur compounds following the action of cysteine sulfoxide lyases. *Journal of Agricultural and Food Chemistry*, **42**, 1529–1536.

Cochran A. T., Bauer J., Metcalf J. L., Lovecka P., Sura de Jong M., Warris S., Mooijman P. J. W., van der Meer I., Knight R. and Pilon-Smits E. A. H. (2018). Plant selenium hyperaccumulation affects rhizosphere: enhanced species richness and altered species composition. *Phytobiomes*, **2**, 82–91.

Craciun A. R., Meyer C.-L., Chen J., Roosens N., De Groodt R., Hilson P. and Verbruggen N. (2012). Variation in HMA4 gene copy number and expression among *Noccaea caerulescens* populations presenting different levels of Cd tolerance and accumulation. *Journal of Experimental Botany*, **63**, 4179–4189.

Davis C. D. (2012). Selenium supplementation and cancer prevention. *Current Nutrition Reports*, **1**, 16–23.

Dhillon K. S. and Bañuelos G. S. (2017). Overview and prospects of selenium phytoremediation approaches. In: Selenium in Plants: Molecular, Physiological, Ecological and Evolutionary Aspects, E. A. H. Pilon-Smits, L. H. E. Winkel and Z.-Q. Lin (eds.), Springer International Publishing, Switzerland, pp. 277–321.

El Kassis E., Cathala N., Rouached H., Fourcroy P., Berthomieu P., Terry N. and Davidian J. C. (2007). Characterization of a selenate-resistant *Arabidopsis* mutant. Root growth as a potential target for selenate toxicity. *Plant Physiology*, **3**, 1231–1241.

El Mehdawi A. F., Cappa J. J., Fakra S. C., Self J. and Pilon-Smits E. (2012). Interactions of selenium hyperaccumulators and nonaccumulators during cocultivation on seleniferous or nonseleniferous soil – the importance of having good neighbors. *The New Phytologist*, **194**(1), 264–277.

El Mehdawi A. F. and Pilon-Smits E. A. H. (2012). Ecological aspects of plant selenium hyperaccumulation. *Plant Biology*, **14**, 1–10.

El Mehdawi A. F., Jiang Y., Guignardi Z. S., Esmat A., Pilon M., Pilon-Smits E. A. H. and Schiavon M. (2018). Influence of sulfate supply on selenium uptake dynamics and expression of sulfate/selenate transporters in selenium hyperaccumulator and nonhyperaccumulator Brassicaceae. *New Phytologist*, **217**, 194–205.

Feist L. J. and Parker D. R. (2001). Ecotypic variation in selenium accumulation among populations of *Stanleya pinnata*. *New Phytologist*, **149**, 61–69.

Freeman J. L. and Bañuelos G. S. (2011). Selection of salt and boron tolerant selenium hyperaccumulator *Stanleya pinnata* genotypes and characterization of Se phytoremediation from agricultural drainage sediments. *Environmental Science & Technology*, **22**, 9703–9710.

Freeman J. L., Tamaoki M., Stushnoff C., Quinn C. F., Cappa J. J., Devonshire J., Fakra S. C., Marcus M. A., McGrath S. P., Van Hoewyk D. and Pilon-Smits E. A. H. (2010).

Molecular mechanisms of selenium tolerance and hyperaccumulation in *Stanleya pinnata*. *Plant Physiology*, **153**(4), 1630–1652.

Hanikenne M., Talke I. N., Haydon M. J., Lanz C., Nolte A., Motte P., Kroymann J., Weigel D. and Kraemer U. (2008). Evolution of metal hyperaccumulation required cis-regulatory changes and triplication of HMA4. *Nature*, **453**, 391–395.

Hanson B. R., Garifullina G. F., Lindblom S. D., Wangeline A., Ackley A., Kramer K., Norton A. P., Lawrence C. B. and Pilon-Smits E. A. H. (2003). Selenium accumulation protects *Brassica juncea* from invertebrate herbivory and fungal infection. *New Phytologist*, **159**, 461–469.

Hartikainen H. (2005). Biogeochemistry of selenium and its impact on food chain quality and human health. *Journal of Trace Elements in Medicine and Biology*, **18**, 309–318.

Harvey M. A., Erskine P. D., Harris H. H., Brown G. K., Pilon-Smits E. A. H., Casey L. W., Echevarria G. and van der Ent A. (2020). Distribution and chemical form of selenium in *Neptunia amplexicaulis* from Central Queensland, Australia. *Metallomics*, **12**, 514–527.

Hirner A., Ladwig F., Stransky H., Okumoto S., Keinath M., Harms A., Frommer W. B. and Koch W. (2006). Arabidopsis LHT1 is a high-affinity transporter for cellular amino acid uptake in both root epidermis and leaf mesophyll. *Plant Cell*, **18**(8), 1931–1946.

Jiang Y., Schiavon M., Lima L. W., Tripti Jones R. R., El Mehdawi A. F., Royer S., Zeng Z., Hu Y., Pilon-Smits E. A. H. and Pilon M. (2018). Comparison of ATP sulfurylase 2 from selenium hyperaccumulator *Stanleya pinnata* and non-accumulator *Stanleya elata* reveals differential intracellular localization and enzyme activity levels. *Biochimica et Biophysica Acta – General Subjects*, **1862**(11), 2363–2371.

Kamran M., Parveen A., Ahmar S., Malik Z., Hussain S., Chattha M. S., Saleem M. H., Adil M., Heidari P. and Chen J. T. (2019). An Overview of hazardous impacts of soil salinity in crops, tolerance mechanisms, and amelioration through selenium supplementation. *International Journal of Molecular Sciences*, **21**(1), 148.

Kieliszek M. (2019). Selenium-fascinating microelement, properties and sources in food. *Molecules*, **24**(7), 1298.

LeDuc D. L., Tarun A. S., Montes-Bayon M., Meija J., Malit M. F., Wu C. P., AbdelSamie M., Chiang C. Y., Tagmount A., deSouza M., Neuhierl B., Böck A., Caruso J. and Terry N. (2004). Overexpression of selenocysteine methyltransferase in *Arabidopsis* and Indian mustard increases selenium tolerance and accumulation. *Plant Physiology*, **135**, 377–383.

Lima L. W., Pilon-Smits E. A. H. and Schiavon M. (2018). Mechanisms of selenium hyperaccumulation in plants: a survey of molecular, biochemical and ecological cues. *Biochimica et Biophysica Acta – General Subjects*, **1862**(11), 2343–2353.

Lochlainn S. Ó., Bowen H. C., Fray R. G., Hammond J. P., King G. J., White P. J., Graham N. S. and Broadley M. R. (2011). Tandem quadruplication of HMA4 in the zinc (Zn) and cadmium (Cd) hyperaccumulator *Noccaea caerulescens*. *PLoS One*, **6**, e17814.

Natasha Shahid M., Niazi N. K., Khalid S., Murtaza B., Bibi I. and Rashid M. I. (2018). A critical review of selenium biogeochemical behavior in soil-plant system with an inference to human health. *Environmental Pollution*, **234**, 915–934.

Oldfield J. E. (2002). Selenium world atlas: Updated Edition; Selenium-Tellurium Dev. Assoc.: Grimbergen, Belgium, pp. 1–59.

Parker D., Feist L., Varvel T., Thomason D. and Zhang Y. (2003). Selenium phytoremediation potential of *Stanleya pinnata*. *Plant Soil*, **249**(1), 157–165.

Pilon-Smits E. A. H. (2017). Mechanisms of plant selenium hyperaccumulation. In: Selenium in Plants: Molecular, Physiological, Ecological and Evolutionary Aspects, E. A. H. Pilon-Smits, L. H. E. Winkel and Z.-Q. Lin (eds.), Springer International Publishing, Switzerland, pp. 53–66.

Pilon-Smits E. A. H. (2019). On the ecology of selenium accumulation in plants. *Plants (Basel)*, **30**, 197.

Pilon-Smits E. A. H., Hwang S. B., Lytle C. M., Zhu Y. L., Tai J. C., Bravo R. C., Leustek T. and Terry N. (1999). Overexpression of ATP sulfurylase in *Brassica juncea* leads to increased selenate uptake, reduction and tolerance. *Plant Physiology*, **119**, 123–132.

Pletsch M. (2003). Plants and the environment. In: Phytoremediation, B. Thomas (ed.), Encyclopedia of Applied Plant Sciences, Elsevier, pp. 781–786.

Quinn C. F., Freeman J. L., Reynolds R. J. B., Cappa J. J., Fakra S. C., Marcus M. A., Lindblom S. D., Quinn E. K., Bennett L. E. and Pilon-Smits E. A. H. (2010). Selenium hyperaccumulation offers protection from cell disruptor herbivores. *Plant Physiology*, **153**, 1630–1652.

Reeves R. D., van der Ent A. and Baker A. J. M. (2018). Global distribution and ecology of hyperaccumulator plants. In: Agromining: Farming for Metals, Mineral Resource Reviews, A. van der Ent, R. D. Reeves and A. J. M. Baker (eds.), Springer International Publishing AG, pp. 75–92.

Remigio A., Chaney R. L., Baker A. J. M., Edraki M., Erskine P. D., Echevarria G. and van der Ent A. (2020). Phytoextraction of high value elements and contaminants from mining and mineral wastes: opportunities and limitations. *Plant Soil*, **449**, 11–37.

Rosenfeld I. and Beath O. A. (1964). Selenium: Geobotany, Biochemistry, Toxicity, and Nutrition. Academic Press, New York.

Schiavon M. and Pilon-Smits E. A. (2017a). The fascinating facets of plant selenium accumulation – biochemistry, physiology, evolution and ecology. *New Phytologist*, **213**(4), 1582–1596.

Schiavon M. and Pilon-Smits E. A. H. (2017b). Selenium biofortification and phytoremediation phytotechnologies: a review. *Journal of Environmental Quality*, **46**, 10–19.

Schiavon M., Pilon M., Malagoli M. and Pilon-Smits E. A. H. (2015). Exploring the importance of sulfate transporters and ATP sulphurylases for selenium hyperaccumulation—a comparison of *Stanleya pinnata* and *Brassica juncea* (Brassicaceae). *Frontiers in Plant Science*, **6**, 2.

Shahid M., Niazi N. K., Khalid S., Murtaza B., Bibi I. and Rashid M. I. (2018) A critical review of selenium biogeochemical behavior in soil-plant system with an inference to human health. *Environmental Pollution*, **234**, 915–934.

Sors T. G., Ellis D. R. and Salt D. E. (2005). Selenium uptake, translocation, assimilation and metabolic fate in plants. *Photosynthesis Research*, **86**(3), 373–389.

Statwick J., Majestic B. J. and Sher A. A. (2016). Characterization and benefits of selenium uptake by an *Astragalus hyperaccumulator* and a non-accumulator. *Plant Soil*, **404**, 345–359.

Stonehouse G. C., McCarron B. J., Guignardi Z. S., El Mehdawi A. F., Lima L. W., Fakra S. C. and Pilon-Smits E. A. H. (2020). Selenium Metabolism in Hemp (*Cannabis sativa* L.). Potential for Phytoremediation and Biofortification. *Environmental Science & Technology*, **54**(7), 4221–4230.

Sura-de Jong M., Reynolds R. J., Richterova K., Musilova L., Staicu L. C., Chocholata I., Cappa J. J., Taghavi S., van der Lelie D., Frantik T., Dolinova I., Strejcek M., Cochran A. T., Lovecka P. and Pilon-Smits E. A. (2015). Selenium hyperaccumulators harbor a diverse endophytic bacterial community characterized by high selenium resistance and plant growth promoting properties. *Frontiers in Plant Science*, **6**, 113.

USGS (2020). National Minerals Information Center. pubs.usgs.gov (accessed 8 October 2020).

Valdez Barillas J. R., Quinn C. F. and Pilon-Smits E. A. (2011). Selenium accumulation in plants–phytotechnological applications and ecological implications. *International Journal of Phytoremediation*, **13**(Suppl 1), 166–178.

van der Ent A., Baker A. J. M., Reeves R. D., Pollard A. J. and Schat H. (2015). Commentary: Toward a more physiologically and evolutionarily relevant definition of metal hyperaccumulation in plants. *Frontiers in Plant Science*, **6**, 554.

Van Hoewyk D. (2013). A tale of two toxicities: malformed selenoproteins and oxidative stress both contribute to selenium stress in plants. *Annals of Botany*, **112**(6), 965–972.

Van Huysen T., Abdel-Ghany S., Hale K. L., LeDuc D., Terry N. and Pilon-Smits E. A. H. (2003). Overexpression of cystathionine-γ-synthase enhances selenium volatilisation in Brassica juncea. *Planta*, **218**, 71–78.

Wang J., Cappa J. J., Harris J. P., Edger P. P., Zhou W., Pires J. C., Adair M., Unruh S. A., Simmons M. P., Schiavon M. and Pilon-Smits E. A. H. (2018). Transcriptome-wide comparison of selenium hyperaccumulator and nonaccumulator *Stanleya* species provides new insight into key processes mediating the hyperaccumulation syndrome. *Plant Biotechnology Journal*, **16**, 1582–1594.

White P. J. (2016). Selenium accumulation by plants. *Annals of Botony*, **117**(2), 217–235.

White P. J. (2018). Selenium metabolism in plants. *Biochimica et Biophysica Acta – General Subjects*, **1862**(11), 2333–2342.

Winkel L. H., Vriens B., Jones G. D., Schneider L. S., Pilon-Smits E. and Bañuelos G. S. (2015). Selenium cycling across soil-plant-atmosphere interfaces: a critical review. *Nutrients*, **7**(6), 4199–4239.

Zambrano M. C., Yuan L., Yin X. and Bañuelos G. (2018). Element case studies: selenium. In: Agromining: Farming for Metals Extracting Unconventional Resources Using Plants, A. van der Ent, G. Echevarria, A. J. M. Baker and J. L. Morel (eds.), Springer International Publishing, Cham, pp. 241–251.

Chapter 9

Selenium biofortification for human and animal nutrition

Jun Li, Gijs Du Laing, Ivet Ferrer and Piet N. L. Lens

9.1 INTRODUCTION

Selenium (Se) is an essential trace element, playing a crucial role in the functioning of enzymes in humans and animals and protecting cells from damage by free radicals (Hatfield *et al.*, 2014). Selenoproteins, that is, proteins containing selenium, are best known as antioxidants and catalysts for the production of active thyroid hormone (Rayman, 2012). Although the essential role of Se for the growth and survival of plants has not been confirmed yet, it is a beneficial element for plants, which can enhance resistance to stress (see Chapter 8).

Despite the importance of this trace element, intake of Se by animals and humans in a wide range of countries, including several countries in Western Europe and East and Central Africa, is still low, resulting in Se deficiency and causing negative health effects, including increased risk of mortality, poor immune function, and cognitive decline (Broadley *et al.*, 2006; Rayman, 2012; Roekens *et al.*, 1986). An estimated one billion people around the world are affected by selenium deficiency, because of low Se intake (Poblaciones & Rengel, 2017; Rayman, 2004). The recommended daily Se intake in an adult human diet is 0.04–0.4 mg per person per day (Food and Agriculture Organization of the United Nations/World Health Organization [FAO/WHO], 2001). Besides, farm animals (Dermauw *et al.*, 2013) and pets (van Zelst *et al.*, 2016) can be affected by Se deficiencies, leading to economic losses. Therefore, the Se content in the human

doi: 10.2166/9781789061055_0265

and animal diet is a topic of interest to public health systems around the world (Lavu *et al.*, 2012).

Biofortification, that is, the dietary supply of Se through its enrichment in food and feed crops, is being explored as a possible solution for Se deficiency (Lavu *et al.*, 2013; Li *et al.*, 2010; Thavarajah *et al.*, 2008). This chapter gives insights into factors affecting Se toxicity and deficiency for humans and animals, and meanwhile summarizes the different phytotechnologies used for Se biofortification, including conventional plant breeding and genetic engineering, and soil and foliar application of Se-based fertilizers (agronomic biofortification), with specific attention to the use of Se-enriched organic materials and nano-sized selenium (SeNPs), and the addition of beneficial microorganisms into soil for enhancement of Se accumulation in the crops. The factors influencing Se biofortification strategies are also discussed.

9.2 SELENIUM TOXICITY AND DEFICIENCY FOR HUMANS AND ANIMALS

Selenium exists in inorganic forms as selenate (SeO_4^{2-}), selenite (SeO_3^{2-}), selenide (Se^{2-}), elemental Se (Se^0), and in organic forms such as selenocysteine (SeCys) and selenomethionine (SeMet). Due to this diversity in form of occurrence, Se is found in all natural materials on Earth: soil, rocks, waters, air, plants and animals (Fordyce, 2007). For a long time, Se has been identified as a dangerous substance because of its toxicity (Fordyce, 2007). More recently, it has also been recognized as an essential trace element due to its crucial role in the functioning of enzymes of humans and animals (Fordyce, 2013; Rayman, 2000). The range between beneficial and harmful Se concentrations is relatively narrow for animals and humans (Li *et al.*, 2015a). Thus, both toxic and deficient incidences of Se dietary uptake have been reported over the world (Li *et al.*, 2015a).

9.2.1 Se toxicity

Se intoxication events for animals and humans, such as selenosis in America, Canada, China, and Mexico, have occurred occasionally where Se has entered the food chain in excessive amounts (Li *et al.*, 2015a). These events were caused by excessive Se in soil and water. For instance, the Se toxicity for humans and animals discovered in the Enshi District, Hubei Province, and in Ziyang County, Shanxi Province in China was related to the extremely high Se concentrations in the local food and environment (Fordyce *et al.*, 2000). For humans, Se toxicity (selenosis) can result in garlic breath, hair and nail loss, nervous system disorder, poor dental health, and paralysis (Rayman, 2012). For animals, excess Se can cause alkali disease and blind staggers for grazing animals, and hoof loss in hooved animals (Fordyce, 2007; Tan *et al.*, 2002). Alkali disease is characterized by dullness, lack of vitality, emaciation, rough coat, sloughing of the hooves, erosion of the joints and bones, anemia, lameness, liver cirrhosis, and reduced reproductive performance (Fordyce, 2007). Blind staggers results in impaired vision and blindness, anorexia, weakened legs, paralyzed tongue, labored respiration, abdominal pain, emaciation, and death (Fordyce, 2007). Hair loss and other abnormalities in farm animals have also been observed in areas of Columbia, as a result of Se toxicity (Johnson *et al.*, 2009).

9.2.2 Se deficiency

On the contrary, Se deficiency is also frequently observed worldwide and is even more widespread than Se toxicity. It is estimated that 0.5–1 billion people are directly affected by Se deficiency on a global scale due to low dietary Se intake (Haug *et al.*, 2007; Stonehouse *et al.*, 2020). It has been demonstrated that Se deficiency can cause Keshan disease and Kashin-Beck disease (endemic disease) with low Se supplies in the food system, that is, weakening of the heart and also atrophy and necrosis of cartilage tissue in the joints, which has been reported in the middle of China (Stone, 2009), Saudi Arabia, the Czech Republic, Burundi, New Guinea, Nepal, Croatia, and Egypt (Wu *et al.*, 2015). Low Se status has also been associated with a significantly increased risk of cancer incidence and mortality, cardiovascular risk, poor immune function, male infertility and lower reproduction (Fordyce, 2007; Haug *et al.*, 2007). In addition, Se deficiency may also be a factor in some other diseases. For instance, studies have found that the prevalence of iodine deficiency diseases was greater among populations with lower Se status than among those with higher Se status in Africa (Combs, 2001). This is probably attributed to the fact that Se is essential for the metabolic production of thyroid hormone.

Se deficiency adversely affects livestock health around the world, including south and north America, Africa, Australia, UK and New Zealand (Reilly, 1996). Selenium deficiency causes reproductive and immune response impairment of animals, growth depression (ill-thrift), and white-muscle disease, a myopathy of heart and skeletal muscle principally affecting cattle, sheep, poultry and horses (Rayman, 2000).

Se deficiency in humans and animals is attributed to a low Se daily dietary intake, varying considerably between regions. As mentioned previously, Se deficiency has been identified in parts of the world which have a notably low content of Se in soil or water, as Se enters the food chain from the environment through crop and plant uptake, mainly from local water or soil (Haug *et al.*, 2007). Therefore, the Se concentration in foods is determined by geological and geographical factors. Globally, the total Se concentration in soils ranges from 0.01 to 2.0 mg/kg, with a mean of 0.4 mg/kg (He *et al.*, 2010; Rayman, 2008). Some parts of the world have relatively low Se contents in their soils such as Denmark, Finland, New Zealand, eastern and central Siberia and a long belt extending from northeast to southwest China including parts of Heilongjiang, Jilin, Liaoning, Hebei,

Shanxi, Shaanxi, Sichuan and Zhejiang provinces and Inner Mongolia. Therefore, these regions are characteristic for low amounts of Se in their food chains (Combs, 2001).

9.2.3 Se in nutrition

In humans, chronic Se toxicity is observed above levels of 400 μg/day and Se deficiency occurs when the dietary intake of Se is below 40 μg/day (Gupta & Gupta, 2017; Winkel *et al.*, 2012). More specifically, the tolerable upper intake levels are 90 μg/day for children of 1–3 years, 150 μg/day for children of 4–8 years, 280 μg/day for children of 9–13 years, and 400 μg/day for children >14 years and adults (Ngigi, 2019; Sciences, 2000). For livestock, the toxic Se concentration in animal feed is approximately 2–5 mg/kg dry forage, while the minimal requirement of Se is defined as 0.05–0.10 mg/kg (Gupta & Gupta, 2017). The National Research Council (NRC, 2005) has published the following maximum tolerable levels (MTL) for animals: 5 mg Se/kg feed dry matter (DM) for cattle and sheep, 4 mg Se/kg feed DM for pigs, and 3 mg Se/kg feed DM for poultry. The MTL for horses and fish were derived from interspecies extrapolation and amount to 5 and 2 mg Se/kg DM feed, respectively (NRC, 2005).

Table 9.1 summarizes the recommended daily Se intake and Table 9.2 overviews the status of daily Se intake in some countries. The two tables show that the

Table 9.1 Recommended daily Se intake for adults (μg/d).

Countries	Males	Females
Australia (1990)	85	70
Belgium (2000)	70	70
France (2001)	60	50
FAO/WHO (2001)	40	40
Germany, Austria, Switzerland (2013)	30–70	30–70
Italy (1996)	55	55
Japan (1999)	55–60	45
Netherlands (2000)	50–150	50–150
Nordic countries (2014)	60	50
Ireland (1999)	55	55
Scientific Committee Food (2003)	55	55
USA and Canada (2000)	55	55
United Kingdom (1991)	75	60

Adapted from: EC Scientific Committee on Food (2003); Thomson (2004); Rayman (2004) and European Food Safety Authority (EFSA, 2014).

Table 9.2 Estimated selenium intake status of adults in several countries (μg/person per d).

Countries	SE Intake
Australia	57–87
Austria	48
Belgium	28*–61
Canada	98–224
China	
Keshan disease area (e.g., a wide belt from northeast China to southwest China)	7–11*
Moderate Se area (e.g., Guangzhou)	40–120
Selenosis area (e.g., Hubei and Shaanxi provinces)	750–4990
Czech Republic	10–25*
Denmark	38*–47
Finland	
Before 1984	25*
After 1984 (Se biofortification)	67–110
France	29*–43
Germany	35*
Ireland	44–59
Italy	35*–42
Japan	104–199
Latvia	50
New Zealand	55–80
Serbia	30*
Slovakia	27*–43
Sweden	38*
Switzerland	70
UK	29*–39*
USA	60–220

Table adapted from: Combs (2001); Rayman (2004) and EFSA (2014).
* indicates that this level does not meet the WHO recommended requirement (FAO/WHO, 2001).

recommended daily Se intake in some countries has not yet been achieved, such as in some European countries and parts of China. This demonstrates that the food systems of these countries do not provide sufficient Se for consumption. It may thus be assumed that many individuals have a potential risk of Se deficiency, which can increase their risks to various diseases, including those of the heart and lungs, as well as cancer, and make them more vulnerable to infectious diseases due to poor functioning of their immune system. There is a clear need to enhance Se in food systems of these countries to remediate Se deficiency.

9.3 SELENIUM BIOFORTIFICATION STRATEGIES FOR ADDRESSING Se DEFICIENCY

Addressing micronutrient deficiencies to reduce health-related issues can be achieved through various types of interventions, such as through food supplements, dietary diversification, biofortification, or increasing the digestibility of trace elements in foods and products (Lavu *et al.*, 2012; Li *et al.*, 2020). For instance, sodium selenite has been supplemented in feeds in some areas with selenium deficiency in livestock in order to achieve optimal Se intake (EFSA, 2016). Biofortification is one of the most promising, widespread and accepted strategies, aimed at improving the lack of Se in a diet through enrichment of food and feed crops, in particular the edible parts of plants using different phytotechnologies (Snchez *et al.*, 2017). The different strategies are summarized in Figure 9.1.

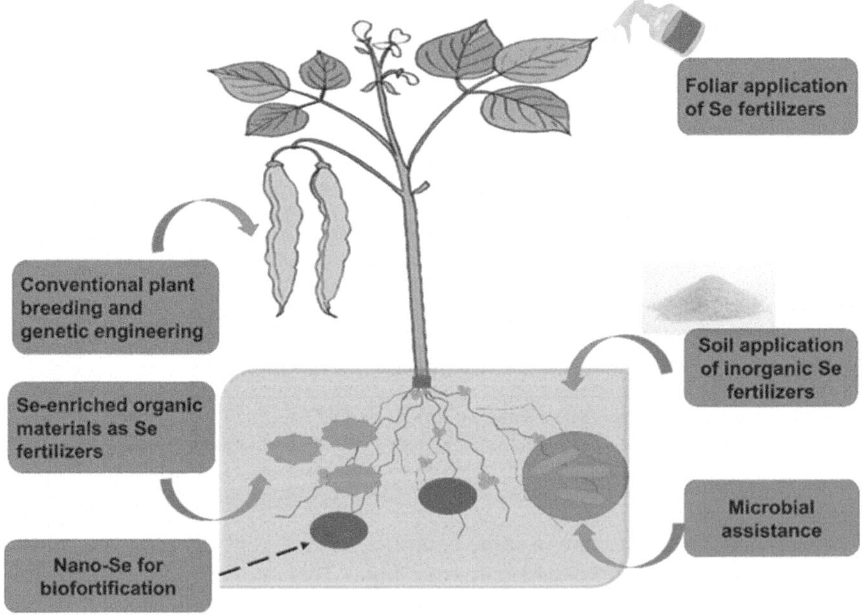

Figure 9.1 Overview of Se biofortification strategies.

9.3.1 Conventional plant breeding and genetic engineering

Breeding of crops aims to screen plant varieties with specific traits, such as the ability for elevated Se uptake or accumulation, to transform Se from inorganic to organic species or to translocate Se quickly from the roots to edible parts. This approach has been explored as a practice for the enhancement of the Se content in edible plants, because there is a huge interspecies and intraspecies genetic variation in plants (Schiavon *et al.*, 2020). Previous studies have demonstrated genetic variation in grain Se concentration of cereal crops, such as wheat (Sharma *et al.*, 2016; Wang *et al.*, 2021; White, 2016), rice (White, 2016; Zhang *et al.*, 2006), oat and barley (White, 2016). Moreover, significant genetic variation effects on seed Se concentration of leguminous crops and edible parts of vegetables have also been observed. Leguminous crops include common bean, mung bean, field pea, lentil and chickpea, and vegetable species include onion, broccoli, Brassicaceae *spp.*, lettuce, tomato, pepper, Chinese cabbage, Indian mustard, pepper, cauliflower and potato (see the review of White (2016) for specific references).

Besides conventional breeding, genetic engineering, as an emerging cutting-edge technology, has shown promise to improve the Se biofortification efficiency (Schiavon *et al.*, 2020). It is aimed at enhancing Se accumulation, preferentially as beneficial selenoamino acids, in the edible part of plants via transgenic methods (White, 2016). So far, genetic engineering for improvement of biofortification has focused on genetic manipulation for (1) reduction of selenate in plants through overexpression of sulfate transporters or adenosine triphosphate (ATP) sulfurylase (APS1 or APS2), which can catalyze the rate-limiting steps of Se assimilation in plants resulting in higher Se accumulation, (2) conversion of Se-cystine (SeCys) to Se-methionine (SeMet) and dimethyl selenide (DMSe) by Cystathionine-γ-synthase (CSeGS) enzyme, leading to more Se volatilization, and (3) avoidance of SeCys misincorporation into proteins by mouse selenocysteine lyase or SeCys methyltransferase (SMT) enzyme (Sarwar *et al.*, 2020; Schiavon *et al.*, 2020; Zhu *et al.*, 2009).

Transgenic lines of Indian mustard (*Brassica juncea* L. Czern.) overexpressing genes encoding the enzymes APS, g-glutamyl-cysteine synthetase (ECS) and glutathione synthetase (GS) were tested under field conditions. The APS, ECS, and GS transgenic plants accumulated 4.3, 2.8, and 2.3 fold more Se in their leaves than the wild type, respectively (Bañuelos *et al.*, 2005). Furthermore, the overexpression of the SMT gene (identified from the Se accumulator *Brassica juncea* L.) in tobacco plants could substantially enhance the tolerance to selenite stress, as shown by the significantly higher fresh weight, plant height, and chlorophyll content than control plants (Chen *et al.*, 2019). More importantly, transgenic plants accumulated a high level of Se and the selenoamino acid Se-methyl-selenocysteine (MeSeCys) (Chen *et al.*, 2019). Genetic engineering can thus improve Se accumulation and give a higher yield with better nutritional quality for biofortification purposes. However, the limitation of plant breeding and genetic engineering is that these have to be applied combined with agronomic Se biofortification, particularly for plants grown in Se deficient regions.

9.3.2 Agronomic biofortification

9.3.2.1 Soil inorganic Se fertilizer application

The agronomic approach of applying a fertilizer on the soil can improve the nutritional quality of the plant without genetic modifications (Storksdieck & Hurrell, 2009). It has been developed as a food-based method to help decrease widespread deficiencies of Se. Selenite and selenate-based fertilizers are typically applied (as granular/blended forms or liquid drenches) into soil to improve their total and bioavailable Se, subsequently resulting in a higher Se concentration in the crop. Although Se is not an essential trace element for plants, it presents chemical similarity to S, and both elements have the same carrier membranes and biochemical pathways for assimilation in the plant (see Chapter 8). Soil application of Se fertilizers can therefore ensure its sufficient concentration in the edible parts of plants (Prado *et al.*, 2017; Sarwar *et al.*, 2020).

Se biofortification of food crops is already successfully practiced in some countries (Se-deficient regions) to increase the Se concentration in staple grains, and subsequent dietary Se intake, by adding inorganic Se fertilizer to soils (Bañuelos *et al.*, 2016; Broadley *et al.*, 2006). For instance, in Finland, a three-fold increase of mean Se intake was observed after agronomic Se biofortification in the form of selenate within 2 years, and the concomitant human serum Se concentration was increased by 70% (Aro *et al.*, 1995). Several plants have been successfully biofortified with Se, such as wheat (Ali *et al.*, 2017; Mao *et al.*, 2014), maize (D'Amato *et al.*, 2019; Mao *et al.*, 2014), rice (Gong *et al.*, 2018; Pandey & Gupta, 2015), soybean (Mao *et al.*, 2014), cabbage (Seo *et al.*, 2008), canola (Mao *et al.*, 2014), potato (de Oliveira *et al.*, 2019), lettuce (Munier-Lamy *et al.*, 2007), pak choi (Li *et al.*, 2015a), and tomato (Carvalho *et al.*, 2003). It should be noted that selenate is superior over selenite for soil application, as selenate is highly soluble and bioavailable in soil, while selenite is less mobile and easily absorbed on oxide surfaces, resulting in less bioavailable Se for plants (Schiavon *et al.*, 2020).

9.3.2.2 Foliar Se fertilizer application

Being an alternative to soil application, foliar spraying of Se fertilizers can efficiently and economically improve Se concentrations in crops. With this method, Se-based fertilizers are homogenously sprayed on plants. Studies showed that the efficiency of foliar Se application is on average eight times higher than soil application, suggesting that foliar application is preferred over soil application (Ros *et al.*, 2016). This is attributed to (1) liquid Se-containing fertilizer being directly applied onto plants via spraying, which avoids the retention of Se by the soil (e.g., binding by soil organic matter), thus improving Se utilization by plants and avoiding Se losses; and (2) the translocation of Se from root to shoot or edible parts of plants not being required through the foliar application, resulting in fast assimilation of Se in plant tissues (Schiavon *et al.*, 2020). Foliar application of Se-based fertilizers has successfully enhanced the Se concentration in many plants, including cereal crops: rice (Farooq *et al.*, 2019), wheat (Wang *et al.*, 2020b), maize (Ngigi, 2019) and beans (Ngigi *et al.*, 2019) as well as vegetable crops: tomato (Zhu *et al.*, 2016), potato (Zhang *et al.*, 2019), cabbage, radish, onion and garlic (Slekovec & Goessler, 2015).

Some practical aspects should be carefully considered when implementing foliar application, such as the applied Se dose, the timing of the foliar Se application, and the plant type. For instance, phytotoxicity could be caused by an unsuitable Se concentration sprayed directly on plant leaves. The plants that received Se-based fertilizer should have sufficient leaf area for maintaining and absorbing Se during the biofortification process, and suitable weather during application should be considered in order to avoid Se losses on rainy and windy days. Besides, the application timing is another item that should be addressed, as an application at different growth stages can result in different Se accumulation by the plants. Wang *et al.* (2020b) demonstrated that foliar application of selenate or selenite at the pre-filling stage was superior in improving the Se concentration of wheat grains than that at the pre-flowering stage. The foliar application of selenite during the potato tuber bulking stage resulted in the greatest Se accumulation in the tubers, compared to the application during the tuber initiation and maturation stages (Zhang *et al.*, 2019).

9.3.2.3 Novel Se fertilizers

9.3.2.3.1 Se-enriched organic materials as Se fertilizers

Biomaterials, such as plant residues, sludge and manures, that come from seleniferous areas potentially contain high levels of Se. These micronutrient-enriched materials may serve as potential micronutrient sources and can thus be utilized for Se biofortification of agricultural crops. If Se-enriched organic biomaterials are used to amend agricultural soils, their decomposition will gradually lead to micronutrient release into the soil solutions, which will become bioavailable for uptake by the crop (Bañuelos *et al.*, 2015). Biofortification using these Se-enriched biomaterials can thus be achieved, particularly of crops growing on micronutrient-deficient soils (Bañuelos *et al.*, 2016; Li *et al.*, 2017).

Some studies have investigated the possibility of using Se-enriched biomaterials as fertilizer to improve the Se concentration in crops for biofortification purposes. The accumulation of Se in canola, grown on soil amended with 1.5 mg/kg seleniferous *Astragalus praelongus* E. and *Medicago saliva* L. tissues, was increased as the application dose of these materials increased (Ajwa *et al.*, 1998). Se-enriched wheat and Raya plant straw were used to biofortify sorghum, maize and berseem (Dhillon *et al.*, 2007), and results showed that the Se concentrations in the plants were consistent with the trend of soluble Se in the soil. Similarly, Se-enriched duckweed, Se-enriched anaerobic granular sludge (Li *et al.*, 2021a) and Se-enriched microalgae (Li *et al.*, 2021b) generated in wastewater treatment systems have been evaluated as potential Se fertilizers for improvement of the Se concentration in green beans (*Phaseolus vulgaris*). These biomaterials produced during wastewater treatment released Se, which was efficiently taken up by the beans without negatively affecting their yield. Application of 0.45 g Se-enriched microalgae biomass into 0.5 kg sandy soil even stimulated the beans growth, resulting in a 43% higher yield (Li *et al.*, 2021b). Se-enriched sludge was found to be the preferred slow-release Se biofertilizer for Se-deficient areas, in comparison to Se-enriched duckweed because Se contained in the Se-enriched sludge was released slowly and was more bioavailable for plant uptake than Se contained in the duckweed (Li *et al.*, 2021a).

The supplementation of soils with Se-enriched organic materials as biofertilizer does not only improve the Se concentration in the plants, but also results in value-added plant-based products, as plants can transform the Se taken up during growth into valuable organic Se species (e.g., SeMet, SeCys and MeSeCys), which have important assets in the nutrition of animals and humans. Bañuelos *et al.* (2015) reported that the Se concentration in the edible parts of broccoli and carrots was directly correlated with the amount of Se-enriched *Stanleya pinnata* applied into coarse-loamy soil ($R^2 = 0.94$) and that MeSeCys was the main accumulating Se species. Likewise, the application of Se-enriched duckweed and sludge into loamy and sandy soil as Se biofertilizer significantly improved the proportion of health-beneficial selenoamino acids (e.g., Se-methionine, 76–89%) in the seeds of beans (*Phaseolus vulgaris*) (Li *et al.*, 2021a).

One of the main advantages of micronutrient-enriched organic materials is that they provide a long-lasting micronutrient source, slowly releasing the micronutrients along with the decomposition of the organic materials in the soil (Ajwa *et al.*, 1998). However, the disadvantage is that the application of these materials can introduce additional organic matter into the soil, which can lead to the immobilization of other nutrients in the soil, eventually decreasing the bioavailability for the plant (Stavridou *et al.*, 2011).

It should be noted that Se biofortification via the application of Se-enriched organic materials may not be feasible in all Se-deficient areas. For instance, the Se-deficient region in northeast China, characterized by a high content of organic matter in soil, is not suitable for supplementation with micronutrient-enriched organic materials, as the presence of too much organic matter in the soil will increase the retention of the released Se, reducing the bioavailability of Se in the soil. In contrast, some regional soils with strong leaching potential (i.e., high precipitation (rainfall) and humid climates) and low Se content can benefit from the addition of micronutrient-enriched organic materials since the added organic matter can act as a micronutrient reservoir to avoid leaching of nutrients and their mobilization to the deeper soil layers (Wang & Gao, 2001).

9.3.2.3.2 Nano-Se for biofortification

In recent years, the application of Se nanoparticles (SeNPs) has been proposed for Se biofortification, as SeNPs can slowly release Se for plant uptake, thus minimizing Se losses in comparison with the fast leaching of inorganic Se (El-Ramady *et al.*, 2020). SeNPs can be synthesized from oxidized Se forms (i.e., selenite and selenate) via chemical or biological reduction using chemical reducing agents (El-Ramady *et al.*, 2020) or bacteria (Staicu *et al.*, 2015), fungi (Mosallam *et al.*, 2018) and plants (Ikram *et al.*, 2021; Schiavon *et al.*, 2020). The effects of SeNPs on plants highly rely on the particle size and synthesis method (chemosynthesized or biosynthesized) of the SeNPs (Hu *et al.*, 2018). Previous studies have identified the potential of SeNPs to promote plant growth, increase Se uptake and improve plant quality (Domokos-Szabolcsy *et al.*, 2012; Hussein *et al.*, 2019). The beneficial effects of SeNPs have been shown for several plants, including tomato (Hernandez-Hernandez *et al.*, 2019; Morales-Espinoza *et al.*, 2019), pomegranate (Zahedi *et al.*, 2019), wheat (Hu *et al.*, 2018), rice (Wang *et al.*, 2020a), garlic (Li *et al.*, 2020) and tobacco (Domokos-Szabolcsy *et al.*, 2012). Besides, SeNPs have lower toxicity for plants compared to selenite and selenate (El-Ramady *et al.*, 2020; Li *et al.*, 2020). SeNPs taken up by plants were quickly oxidized to selenite and transformed to organic Se species (e.g., SeCys, SeMet and MeSeCys) in the plant root (Hu *et al.*, 2018; Wang *et al.*, 2020a). However, further investigations are still required to understand the phytotoxicity of SeNPs on various plants and the potential risks to the environment of using SeNPs.

9.3.2.4 Microbial assistance of biofortification

A novel approach in biofortification studies is to make use of plant-microbe interactions for improvement of Se uptake by the plant. Plant growth promoting rhizobacteria (PGPR) not only promote plant growth via different plant-driven mechanisms, for example, the production of phytohormones, nitrogen fixation and stress mitigation, but also affect the Se mobility, speciation and bioavailability in soils (Sarwar *et al.*, 2020; Yasin *et al.*, 2015a).

Early studies have reported that the addition of beneficial microorganisms to soil or inoculation of plants with microbes could enhance Se accumulation in crops. For instance, Yasin *et al.* (2015a) demonstrated that inoculation of wheat plants with both selenium-tolerant bacterial strains *Bacillus cereus*-YAP6 and *Bacillus licheniformis*-YAP7 not only significantly enhanced wheat growth, but also increased the uptake of Se and other nutrients, for example, S, Ca and Fe. Yasin *et al.* (2015b) further reported that the inoculation of the bacterial consortium G1 stimulated the growth of the Se accumulator Indian mustard (*Brassica juncea*) grown on seleniferous soil for Se-enriched plant material production, which resulted in a higher Se accumulation. Similarly, inoculation of the arbuscular mycorrhizal fungi increased the Se content in the shallot bulb (*Allium cepa* L. *Aggregatum* group) by 530% (Golubkina *et al.*, 2019).

9.4 FACTORS AFFECTING Se BIOFORTIFICATION EFFICIENCY

Since low concentrations of plant Se can decrease the dietary intake of Se, it is vital to increase Se uptake by plants and to produce plants with higher Se concentrations and bioavailability in their edible tissues (Bañuelos *et al.*, 2017). This is the key issue for effectively developing a biofortification strategy. The Se biofortification efficiency depends on a number of factors associated with the Se concentration in plants (also called bioavailability) during biofortification, such as plant species, Se species and source (chemical Se fertilizer, natural source of Se or organic Se), soil pH and redox conditions, soil organic matter, and the presence of competitive ions (Fordyce, 2007).

Plant species: Plants have been classified as hyperaccumulators (>1000 mg/kg, such as *Stanleya*), secondary accumulators (100–1000 mg/kg, such as Brassica species, broccoli), and non-accumulators depending upon Se accumulation inside their cells (Gupta & Gupta, 2017). Vegetables (e.g., Brassica species: pak choi and cabbage) normally accumulate more Se than legumes (beans), followed by cereals (wheat and rice). The Se concentration accumulated in fruits is generally low, whereas high concentrations (ranging from 0.03–512 mg/kg) have been reported in Brazil nuts as a result of natural biofortification (Prado *et al.*, 2017).

Se application methods: Different application methods of Se-based fertilizer affect Se accumulation and transformation in plants. Foliar application is generally more efficient in enhancing the Se concentration in plants in comparison with soil application (see Section 9.3). Studies showed that the efficiency of foliar Se application is on average eight times higher than soil application (Ros *et al.*, 2016). Besides, application of Se fertilizers at different plant growth stages can also result in a different biofortification efficiency.

Se species and source: The uptake rates and mechanisms of selenite, selenate and organic Se are different. Some studies showed that selenite is adsorbed and taken up in a faster passive way and readily reduced to organic compounds in plants, while selenate is taken up in an active way and easily distributed from roots to shoots (Arvy, 1993; Gupta & Gupta, 2017). Selenate reduction occurs via substitution for sulfate in the ATP sulfurylase reductase system, which is an ATP-consuming process and rate-limiting step, resulting in lower selenate accumulation in plants compared to selenite (Van Hoewyk, 2013). However, Ros *et al.* (2016) showed that biofortification using selenate-based fertilizers has a high potential to increase Se uptake by crops and subsequent Se intake by animals and humans. This is attributed to the fact that selenate is not easily adsorbed into the soil matrix in comparison with selenite, resulting in higher bioavailable Se concentrations in the soil, while selenite is readily adsorbed in the soil environment (Ros *et al.*, 2016).

Soil pH and redox conditions: Soil pH and redox conditions have an important effect on Se availability since a combination of these factors determines the Se species present in a given soil environment. For instance, selenate is the predominant Se species in near-neutral pH environments under aerobic conditions, whereas selenite predominates at lower pH and redox potential. Selenate is much more mobile, and thus plant-available in soils than selenite which is tightly bound to positively charged binding sites in soils (Eich-Greatorex *et al.*, 2007). Besides, soil pH negatively correlates with the amount of Se adsorbed by soil (Li *et al.*, 2015b). Most studies have demonstrated that relatively high pH values in soil solutions lead to a higher Se accumulation by plants in comparison with low soil pH (Li *et al.*, 2016, 2017). This is attributed to the fact that soil with low pH contains a high amount of H^+, which do not compete for positively charged binding sites with selenite/selenate in soil, thus leading to a relatively high bioavailable Se in the soil solution.

Soil organic matter: Organic matter influences Se availability in different ways. On the one hand, organic matter has a significant capacity to remove Se from the soil solution, and immobilize Se by both biotic and abiotic mechanisms, thus reducing Se bioavailability. On the other hand, organic matter can improve the soil structure and stimulate oxidizing conditions, thus enhancing Se bioavailability (Li *et al.*, 2017). The release of organic matter-immobilized Se through mineralization will increase the bioavailable Se concentration in a soil.

Competitive ions: The Se accumulation in plants can also be influenced by the presence of other ions, especially phosphate (PO_4^{3-}) and sulfate (SO_4^{2-}). Interactions between Se and those ions may occur in the soil or in the plant (Bingham, 1989). Li *et al.* (2008) studied the Se uptake in wheat under phosphorus and sulfur-starved conditions and demonstrated that selenite uptake is an active process mediated partly by phosphorus transporters. Likewise, the Se uptake can be negatively influenced by the addition of sulfur due to the chemical similarity between these two elements. Studies have demonstrated that selenate is taken up by sulfate transporters, thus the competition for the same transporters could inhibit Se uptake by plants when sulfur is applied (Li *et al.*, 2008). For instance, a decrease in the Se concentration in the shoots and roots of corn (*Zea mays*) was observed when the sulfur concentration in solution increased (Huang *et al.*, 2008). Supplementation of sulfur in the calcareous alluvial and yellow-brown soil reduces the Se contents in soybean (*Glycine max* L.) seeds (Deng *et al.*, 2021).

9.5 FUTURE PERSPECTIVES

Different biofortification strategies have been documented for improving human and animal dietary Se intake. However, an adequate comparison of these approaches from economy, health, environment and social acceptance perspectives is still needed. The most cost-effective biofortification method with enhanced Se accumulation and crop yield as well as farmers' inclination as a function of crop growth conditions and the socio-economic environment should be selected. Microbial-assisted biofortification can play an important role in optimization of these biofortification strategies when aiming to simultaneously improve crop Se accumulation and crop yield. However, each biofortification practice must be carefully evaluated to prevent plant-derived food products from having toxic Se levels that may be harmful to the organisms feeding on them.

Additionally, specific issues related to each biofortification strategy should be further addressed. For instance, applying high quantities of Se fertilizers as soil or foliar application may not always be the most sustainable strategy, as this application can result in the leaching of excessive Se, thus requiring regular applications, which can make this approach more costly. Besides, the widespread use of Se for biofortification might cause Se contamination in the environment (e. g., water, soil and plant), which in turn poses potential threats for human and animal health. Selecting crops with a high ability to accumulate greater Se concentrations is needed in the conventional breeding approach, however the crop biodiversity and dietary diversity should also not be neglected. Genetic engineering is the most controversial method because of the fear of disturbing natural gene functions in food crops and potentially causing hazardous effects on humans and animals. Further research should, on the one hand, explore the specific genes contributing to higher Se accumulation and, on the other hand, assess the safety issues and tackle ethical barriers.

In terms of the application of Se-enriched organic materials into the soil, risk assessments should be carefully conducted to avoid other contaminants also ending up in the soil and edible products. Besides, more studies are still needed on the application of SeNPs as an emerging technology. The mechanism of SeNPs uptake from soil, as well as their translocation and transformation in higher plants need to be further unraveled. Understanding the effects of the SeNPs application on the soil microbial ecology is also necessary. Any unpredictable health effect arising from this strategy should be systematically evaluated, also involving the chemical modification and transformation of SeNPs during biofortification and food processing.

REFERENCES

Ajwa H. A., Bañuelos G. S. and Mayland H. F. (1998). Selenium uptake by plants from soils amended with inorganic and organic materials. *Journal of Environment Quality*, **27**(5), 1218–1227.

Ali F., Peng Q., Wang D., Cui Z. W., Huang J., Fu D. D. and Liang D. L. (2017). Effects of selenite and selenate application on distribution and transformation of selenium fractions in soil and its bioavailability for wheat (*Triticum aestivum* L.). *Environmental Science and Pollution Research*, **24**(9), 8315–8325.

Aro A., Alfthan G. and Varo P. (1995). Effects of supplementation of fertilizers on human selenium status in Finland. *The Analyst*, **120**(3), 841–843.

Arvy M. P. (1993). Selenate and selenite uptake and translocation in bean-plants (*Phaseolus vulgaris*). *Journal of Experimental Botany*, **44**(6), 1083–1087.

Bañuelos G., Terry N., LeDuc D. L., Pilon-Smits E. A. H. and Mackey B. (2005). Field trial of transgenic Indian mustard plants shows enhanced phytoremediation of selenium-contaminated sediment. *Environmental Science & Technology*, **39**(6), 1771–1777.

Bañuelos G. S., Arroyo I., Pickering I. J., Yang S. I. and Freeman J. L. (2015). Selenium biofortification of broccoli and carrots grown in soil amended with Se-enriched hyperaccumulator *Stanleya pinnata*. *Food Chemistry*, **166**, 603–608.

Bañuelos G. S., Arroyo I. S., Dangi S. R. and Zambrano M. C. (2016). Continued selenium biofortification of carrots and broccoli grown in soils once amended with Se-enriched S. *pinnata*. *Frontiers in Plant Science*, **7**, 1251.

Bañuelos G. S., Lin Z.-Q. and Broadley M. (2017). Selenium biofortification. In: Selenium in Plants: Molecular, Physiological, Ecological and Evolutionary Aspects, E. A. H. Pilon-Smits, L. H. E. Winkel and Z.-Q. Lin (eds.), Springer International Publishing, Cham, pp. 231–255.

Bingham R. L. M. A. L. P. F. T. (1989). Factors affecting selenium accumulation by agricultural crops. In: Selenium in Agriculture and the Environment, L. W. Jacobs (ed.), Elsevier, The Netherlands, pp. 65–94.

Broadley M. R., White P. J., Bryson R. J., Meacham M. C., Bowen H. C., Johnson S. E., Hawkesford M. J., McGrath S. P., Zhao F. J., Breward N., Harriman M. and Tucker M. (2006). Biofortification of UK food crops with selenium. *Proceedings of the Nutrition Society*, **65**(2), 169–181.

Carvalho K. M., Gallardo-Williams M. T., Benson R. F. and Martin D. F. (2003). Effects of selenium supplementation on four agricultural crops. *Journal of Agricultural and Food Chemistry*, **51**(3), 704–709.

Chen M., Zeng L., Luo X., Mehboob M. Z., Ao T. and Lang M. (2019). Identification and functional characterization of a novel selenocysteine methyltransferase from *Brassica juncea* L. *Journal of Experimental Botany*, **70**(21), 6401–6416.

Combs G. F., Jr. (2001). Selenium in global food systems. *British Journal of Nutrition*, **85**(5), 517–547.

D'Amato R., De Feudis M., Guiducci M. and Businelli D. (2019). Zea mays L. Grain: increase in nutraceutical and antioxidant properties due to Se fortification in low and high water regimes. *Journal of Agricultural and Food Chemistry*, **67**(25), 7050–7059.

de Oliveira V. C., Faquin V., Andrade F. R., Carneiro J. P., da Silva E. C., de Souza K. R. D., Pereira J. and Guilherme L. R. G. (2019). Physiological and physicochemical responses of potato to selenium biofortification in tropical soil. *Potato Research*, **62**(3), 315–331.

Deng X., Zhao Z., Lv C., Zhang Z., Yuan L. and Liu X. (2021). Effects of sulfur application on selenium uptake and seed selenium speciation in soybean (*Glycine max* L.) grown in different soil types. *Ecotoxicology and Environmental Safety*, **209**, 111790.

Dermauw V., Yisehak K., Belay D., Van Hecke T., Du Laing G., Duchateau L. and Janssens G. P. J. (2013). Mineral deficiency status of ranging zebu (*Bos indicus*) cattle around the

Gilgel Gibe catchment, Ethiopia. *Tropical Animal Health and Production*, **45**(5), 1139–1147.

Dhillon S. K., Hundal B. K. and Dhillon K. S. (2007). Bioavailability of selenium to forage crops in a sandy loam soil amended with Se-rich plant materials. *Chemosphere*, **66**(9), 1734–4173.

Domokos-Szabolcsy E., Marton L., Sztrik A., Babka B., Prokisch J. and Fari M. (2012). Accumulation of red elemental selenium nanoparticles and their biological effects in *Nicotinia tabacum*. *Plant Growth Regulation*, **68**(3), 525–531.

EFSA (2014). Scientific opinion on dietary reference values for selenium. *EFSA Journal*, **12** (10), 3846.

EFSA (2016). Safety and efficacy of selenium compounds (E8) as feed additives for all animal species: Sodium selenite, based on a dossier submitted by Todini and Co SpA. *EFSA Journal*, **14**(3).

Eich-Greatorex S., Sogn T. A., Øgaard A. F. and Aasen I. (2007). Plant availability of inorganic and organic selenium fertiliser as influenced by soil organic matter content and pH. *Nutrient Cycling in Agroecosystems*, **79**(3), 221–231.

El-Ramady H., Faizy S. E. D., Abdalla N., Taha H., Domokos-Szabolcsy É., Fari M., Elsakhawy T., Omara A. E.-D., Shalaby T., Bayoumi Y., Shehata S., Geilfus C.-M. and Brevik E. C. (2020). Selenium and nano-selenium biofortification for human health: Opportunities and challenges. *Soil Systems*, **4**(3), 57.

FAO/WHO (2001). Human Vitamin and Mineral Requirements. Report of a Joint FAO/WHO Expert Consultation, Bangkok, Thailand.

Farooq M. U., Tang Z. C., Zeng R., Liang Y. K., Zhang Y. J., Zheng T. D., Ei H. H., Ye X. Y., Jia X. M. and Zhu J. (2019). Accumulation, mobilization, and transformation of selenium in rice grain provided with foliar sodium selenite. *Journal of the Science of Food and Agriculture*, **99**(6), 2892–2900.

Food E. S. C.o. (2003). Opinion of the Scientific Committee on Food on the Revision of Reference Values for Nutrition Labelling, (Ed.) C.o.t.E. Communities. Brussels.

Fordyce F. (2007). Selenium geochemistry and health. *AMBIO: A Journal of the Human Environment*, **36**(1), 94–97.

Fordyce F. M. (2013). Selenium deficiency and toxicity in the environment. In: Essentials of Medical Geology: Revised Edition, O. Selinus (ed.), Springer Netherlands, Dordrecht, pp. 375–416.

Fordyce F. M., Guangdi Z., Green K. and Xinping L. (2000). Soil, grain and water chemistry in relation to human selenium-responsive diseases in Enshi District, China. *Applied Geochemistry*, **15**(1), 117–132.

Golubkina N., Zamana S., Seredin T., Poluboyarinov P., Sokolov S., Baranova H., Krivenkov L., Pietrantonio L. and Caruso G. (2019). Effect of selenium biofortification and beneficial microorganism inoculation on yield, quality and antioxidant properties of shallot bulbs. *Plants (Basel)*, **8**(4), 102.

Gong R. Y., Ai C. Y., Zhang B. J. and Cheng X. L. (2018). Effect of selenite on organic selenium speciation and selenium bioaccessibility in rice grains of two Se-enriched rice cultivars. *Food Chemistry*, **264**, 443–448.

Gupta M. and Gupta S. (2017). An overview of selenium uptake, metabolism, and toxicity in plants. *Frontiers in Plant Science*, **7**, 2074.

Hatfield D. L., Tsuji P. A., Carlson B. A. and Gladyshev V. N. (2014). Selenium and selenocysteine: roles in cancer, health, and development. *Trends in Biochemical Sciences*, **39**(3), 112–120.

Haug A., Graham R. D., Christophersen O. A. and Lyons G. H. (2007). How to use the world's scarce selenium resources efficiently to increase the selenium concentration in food. *Microbial Ecology in Health and Disease*, **19**(4), 209–228.

He Z. L., Shentu J. and Yang X. E. (2010). Manganese and selenium. In: Trace Elements in Soils, P. S. Hooda (ed.), Blackwell, Chichester, pp. 481–495.

Hernandez-Hernandez H., Quiterio-Gutierrez T., Cadenas-Pliego G., Ortega-Ortiz H., Hernandez-Fuentes A. D., Cabrera de la Fuente M., Valdes-Reyna J. and Juarez-Maldonado A. (2019). Impact of selenium and copper nanoparticles on yield, antioxidant system, and fruit quality of tomato plants. *Plants (Basel)*, **8**(10), 355.

Hu T., Li H., Li J., Zhao G., Wu W., Liu L., Wang Q. and Guo Y. (2018). Absorption and bio-transformation of selenium nanoparticles by wheat seedlings (*Triticum aestivum* L.). *Frontiers in Plant Science*, **9**, 597.

Huang Y. Z., Hu Y. and Liu Y. X. (2008). Interactions between sulfur and selenium uptake by corn in solution culture. *Journal of Plant Nutrition*, **31**(1), 43–54.

Hussein H.-A. A., Darwesh O. M. and Mekki B. B. (2019). Environmentally friendly nano-selenium to improve antioxidant system and growth of groundnut cultivars under sandy soil conditions. *Biocatalysis and Agricultural Biotechnology*, **18**, 101080.

Ikram M., Javed B., Raja N. I. and Mashwani Z.-U.-R. (2021). Biomedical potential of plant-based selenium nanoparticles: a comprehensive review on therapeutic and mechanistic aspects. *International Journal of Nanomedicine*, **16**, 249–268.

Johnson C. C., Fordyce F. M. and Rayman M. P. (2009). Symposium on 'Geographical and geological influences on nutrition' Factors controlling the distribution of selenium in the environment and their impact on health and nutrition. conference on 'Over- and undernutrition: challenges and approaches'. *Proceedings of the Nutrition Society*, **69** (1), 119–132.

Lavu R. V. S., Du Laing G., Van de Wiele T., Pratti V. L., Willekens K., Vandecasteele B. and Tack F. 2012. Fertilizing soil with selenium fertilizers: impact on concentration, speciation, and bioaccessibility of selenium in leek (*Allium ampeloprasum*). *Journal of Agricultural and Food Chemistry*, **60**(44), 10930–10935.

Lavu R. V. S., De Schepper V., Steppe K., Majeti P. N. V., Tack F. and Du Laing G. (2013). Use of selenium fertilizers for production of Se-enriched Kenaf (*Hibiscus cannabinus*): effect on Se concentration and plant productivity. *Journal of Plant Nutrition and Soil Science*, **176**(4), 634–639.

Li H. F., McGrath S. P. and Zhao F. J. (2008). Selenium uptake, translocation and speciation in wheat supplied with selenate or selenite. *New Phytolgist*, **178**(1), 92–102.

Li H. F., Lombi E., Stroud J. L., McGrath S. P. and Zhao F. J. (2010). Selenium speciation in soil and rice: influence of water management and Se fertilization. *Journal of Agricultural and Food Chemistry*, **58**(22), 11837–11843.

Li J., Liang D., Qin S., Feng P. and Wu X. (2015a). Effects of selenite and selenate application on growth and shoot selenium accumulation of pak choi (*Brassica chinensis* L.) during successive planting conditions. *Environment Science and Pollution Research*, **22**(14), 11076–11086.

Li Z., Man N., Wang S., Liang D. and Liu J. (2015b). Selenite adsorption and desorption in main Chinese soils with their characteristics and physicochemical properties. *Journal of Soils and Sediments*, **15**(5), 1150–1158.

Li J., Peng Q., Liang D., Liang S., Chen J., Sun H., Li S. and Lei P. (2016). Effects of aging on the fraction distribution and bioavailability of selenium in three different soils. *Chemosphere*, **144**, 2351–2359.

Li Z., Liang D., Peng Q., Cui Z., Huang J. and Lin Z. (2017). Interaction between selenium and soil organic matter and its impact on soil selenium bioavailability: a review. *Geoderma*, **295**, 69–79.

Li Y., Zhu N., Liang X., Zheng L., Zhang C., Li Y. F., Zhang Z., Gao Y. and Zhao J. (2020). A comparative study on the accumulation, translocation and transformation of selenite, selenate, and SeNPs in a hydroponic-plant system. *Ecotoxicology and Environmental Safety*, **189**, 109955.

Li J., Otero-Gonzalez L., Parao A., Tack P., Folens K., Ferrer I., Lens P. N. L. and Du Laing G. (2021a). Valorization of Se-enriched sludge and duckweed generated from wastewater as micronutrient biofertilizer. *Chemosphere*, **281**, 130767.

Li J., Ferrer I., Lens P. N. L. and Du Laing G. (2021b). Evaluation of selenium-enriched microalgae biomass and extract produced from domestic wastewater in raceway pond as biostimulant for plant growth and Se accumulation. *Journal of Applied Phycology*, (In Press).

Mao H., Wang J., Wang Z., Zan Y., Lyons G. and Zou C. (2014). Using agronomic biofortification to boost zinc, selenium, and iodine concentrations of food crops grown on the loess plateau in China. *Journal of Soil Science and Plant Nutrition*, **14**(2), 459–470.

Morales-Espinoza M. C., Cadenas-Pliego G., Perez-Alvarez M., Hernandez-Fuentes A. D., Cabrera de la Fuente M., Benavides-Mendoza A., Valdes-Reyna J. and Juarez-Maldonado A. (2019). Se nanoparticles induce changes in the growth, antioxidant responses, and fruit quality of tomato developed under NaCl stress. *Molecules*, **24**(17), 3030.

Mosallam F. M., El-Sayyad G. S., Fathy R. M. and El-Batal A. I. (2018). Biomolecules-mediated synthesis of selenium nanoparticles using *Aspergillus oryzae* fermented Lupin extract and gamma radiation for hindering the growth of some multidrug-resistant bacteria and pathogenic fungi. *Microbial Pathogenesis*, **122**, 108–116.

Munier-Lamy C., Deneux-Mustin S., Mustin C., Merlet D., Berthelin J. and Leyval C. (2007). Selenium bioavailability and uptake as affected by four different plants in a loamy clay soil with particular attention to mycorrhizae inoculated ryegrass. *Journal of Environmental Radioactivity*, **97**(2–3), 148–158.

Ngigi P. B. (2019). Assessment and Optimization of Dietary Selenium Intake in Kenya: Exploration of Biofortification as a Solution to the Hidden Hunger. PhD dissertation, Faculty of Bioscience Engineering, Ghent University, Ghent, Belgium.

Ngigi P. B., Lachat C., Masinde P. W. and Du Laing G. (2019). Agronomic biofortification of maize and beans in Kenya through selenium fertilization. *Environmental Geochemistry and Health*, **41**(6), 2577–2591.

Pandey C. and Gupta M. (2015). Selenium and auxin mitigates arsenic stress in rice (*Oryza sativa* L.) by combining the role of stress indicators, modulators and genotoxicity assay. *Journal of Hazardous Materials*, **287**, 384–391.

Poblaciones M. J. and Rengel Z. (2017). Combined foliar selenium and zinc biofortification in field pea (*Pisum sativum*): accumulation and bioavailability in raw and cooked grains. *Crop and Pasture Science*, **68**(3), 265.

Prado R. d. M., Cruz F. J. R. and Ferreira R. L. d. C. (2017). Selenium biofortification and the problem of its safety. In: Superfood and Functional Food – an Overview of Their Processing and Utilization, V. Y. Waisundara (ed.), INTECH, London, pp. 229–238.

Rayman M. P. (2000). The importance of selenium to human health. *The Lancet*, **356**(9225), 233–241.

Rayman M. P. (2004). The use of high-selenium yeast to raise selenium status: how does it measure up? *British Journal of Nutrition*, **92**(4), 557–573.

Rayman M. P. (2008). Food-chain selenium and human health: emphasis on intake. *British Journal of Nutrition*, **100**(2), 254–268.

Rayman M. P. (2012). Selenium and human health. *The Lancet*, **379**(9822), 1256–1268.

Reilly C. (1996). Selenium in Food and Health. Blackie Academic and Professional, London.

Roekens E. J., Robberecht H. J. and Deelstra H. A. (1986). Dietary selenium intake in Belgium for different population groups at risk for deficiency. *Diätetische Selen-Aufnahme in Belgien bei verschiedenen Bevölkerungsgruppen mit den Risiken eines Defizits*, **182**(1), 8–13.

Ros G. H., van Rotterdam A. M. D., Bussink D. W. and Bindraban P. S. (2016). Selenium fertilization strategies for bio-fortification of food: an agro-ecosystem approach. *Plant and Soil*, **404**(1), 99–112.

Sarwar N., Akhtar M., Kamran M. A., Imran M., Riaz M. A., Kamran K. and Hussain S. (2020). Selenium biofortification in food crops: Key mechanisms and future perspectives. *Journal of Food Composition and Analysis*, **93**, 103615.

Schiavon M., Nardi S., Dalla Vecchia F. and Ertani A. (2020). Selenium biofortification in the 21(st) century: status and challenges for healthy human nutrition. *Plant and Soil*, **453**, 245–270.

Sciences N. A. O. (2000). Dietary Reference Intakes for Vitamin C, Vitamin E, Selenium, and Carotenoids, N.A.O. Sciences. The National Academies Press, Washington, DC, p. 528.

Seo T. C., Spallholz J. E., Yun H. K. and Kim S. W. (2008). Selenium-enriched garlic and cabbage as a dietary selenium source for broilers. *Journal of Medicinal Food*, **11**(4), 687–692.

Sharma S., Gupta R. and Singh D. (2016). Variation in selenium tolerance, accumulation, and growth parameters of different wheat cultivars. *Communications in Soil Science and Plant Analysis*, **47**(2), 203–212.

Slekovec M. and Goessler W. (2015). Accumulation of selenium in natural plants and selenium supplemented vegetable and selenium speciation by HPLC-ICPMS. *Chemical Speciation & Bioavailability*, **17**(2), 63–73.

Snchez E., SidaArreola J., vilaQuezada G., OjedaBarrios D., SotoParra J. and PreciadoRangel P. 2017. Can biofortification of zinc improve the antioxidant capacity and nutritional quality of beans? *Emirates Journal of Food and Agriculture*, **29**(3), 237.

Staicu L. C., van Hullebusch E. D., Oturan M. A., Ackerson C. J. and Lens P. N. (2015). Removal of colloidal biogenic selenium from wastewater. *Chemosphere*, **125**, 130–138.

Stavridou E., Thorup-Kristensen K. and Young S. D. (2011). Assessment of selenium mineralization and availability from catch crops. *Soil Use and Management*, **27**(3), 305–311.

Stone R. (2009). Diseases. A medical mystery in middle China. *Science*, **324**(5933), 1378–1381.

Stonehouse G. C., McCarron B. J., Guignardi Z. S., El Mehdawi A. F., Lima L. W., Fakra S. C. and Pilon-Smits E. A. H. (2020). Selenium metabolism in Hemp (*Cannabis sativa* L.)-potential for phytoremediation and biofortification. *Environment Science & Technology*, **54**(7), 4221–4230.

Storksdieck S. and Hurrell R. F. (2009). The impacts of trace elements from plants on human nutrition: A case for biofortification. In: Biofortified Agricultural Products, G. S. Buñuelos and Z. Q. Lin (eds.), CRC Press, Boca Raton, FL.

Tan J. A., Zhu W., Wang W., Li R., Hou S., Wang D. and Yang L. (2002). Selenium in soil and endemic diseases in China. *Science of the Total Environment*, **284**(1), 227–235.

Thavarajah D., Ruszkowski J. and Vandenberg A. (2008). High potential for selenium biofortification of lentils (*Lens culinaris* L.). *Journal of Agricultural and Food Chemistry*, **56**(22), 10747–10753.

The National Research Council (2005). Mineral Tolerance of Animals: Second Revised Edition, 2005. The National Academies Press, Washington, DC.

Thomson C. D. (2004). Selenium and iodine intakes and status in New Zealand and Australia. *British Journal of Nutrition*, **91**(5), 661–672.

Van Hoewyk D. (2013). A tale of two toxicities: malformed selenoproteins and oxidative stress both contribute to selenium stress in plants. *Annals of Botany*, **112**(6), 965–972.

van Zelst M., Hesta M., Gray K., Staunton R., Du Laing G. and Janssens G. P. J. (2016). Biomarkers of selenium status in dogs. *BMC Veterinary Research*, **12**, 15.

Wang Z. and Gao Y. (2001). Biogeochemical cycling of selenium in Chinese environments. *Applied Geochemistry*, **16**(11), 1345–1351.

Wang K., Wang Y., Li K., Wan Y., Wang Q., Zhuang Z., Guo Y. and Li H. (2020a). Uptake, translocation and biotransformation of selenium nanoparticles in rice seedlings (Oryza sativa L.). *Journal of Nanobiotechnology*, **18**(1), 103.

Wang M., Ali F., Wang M., Dinh Q. T., Zhou F., Bañuelos G. S. and Liang D. (2020b). Understanding boosting selenium accumulation in Wheat (*Triticum aestivum* L.) following foliar selenium application at different stages, forms, and doses. *Environmental Science and Pollution Research*, **27**(1), 717–728.

Wang M., Ali F., Qi M., Peng Q., Wang M., Bañuelos G. S., Miao S., Li Z., Dinh Q. T. and Liang D. (2021). Insights into uptake, accumulation, and subcellular distribution of selenium among eight wheat (Triticum aestivum L.) cultivars supplied with selenite and selenate. *Ecotoxicology and Environmental Safety*, **207**, 111544.

White P. J. (2016). Selenium accumulation by plants. *Annals of Botany*, **117**(2), 217–235.

Winkel L. H., Johnson C. A., Lenz M., Grundl T., Leupin O. X., Amini M. and Charlet L. (2012). Environmental selenium research: from microscopic processes to global understanding. *Environmental Science & Technology*, **46**(2), 571–579.

Wu Z., Bañuelos G. S., Lin Z.-Q., Liu Y., Yuan L., Yin X. and Li M. (2015). Biofortification and phytoremediation of selenium in China. *Frontiers in Plant Science*, **6**, 136.

Yasin M., El-Mehdawi A. F., Pilon-Smits E. A. and Faisal M. (2015a). Selenium-fortified wheat: potential of microbes for biofortification of selenium and other essential nutrients. *International Journal of Phytoremediation*, **17**(8), 777–786.

Yasin M., El Mehdawi A. F., Jahn C. E., Anwar A., Turner M. F. S., Faisal M. and Pilon-Smits E. A. H. (2015b). Seleniferous soils as a source for production of selenium-enriched foods and potential of bacteria to enhance plant selenium uptake. *Plant and Soil*, **386**(1), 385–394.

Zahedi S. M., Hosseini M. S., Daneshvar Hakimi Meybodi N. and Teixeira da Silva J. A. (2019). Foliar application of selenium and nano-selenium affects pomegranate (*Punica granatum* cv. *Malase Saveh*) fruit yield and quality. *South African Journal of Botany*, **124**, 350–358.

Zhang L., Shi W., Wang X. and Zhou X. (2006). Genotypic differences in selenium accumulation in rice seedlings at early growth stage and analysis of dominant factors influencing selenium content in rice seeds. *Journal of Plant Nutrition*, **29**(9), 1601–1618.

Zhang H., Zhao Z., Zhang X., Zhang W., Huang L., Zhang Z., Yuan L. and Liu X. (2019). Effects of foliar application of selenate and selenite at different growth stages on selenium accumulation and speciation in potato (*Solanum tuberosum* L.). *Food Chemistry*, **286**, 550–556.

Zhu Y. G., Pilon-Smits E. A., Zhao F. J., Williams P. N. and Meharg A. A. (2009). Selenium in higher plants: understanding mechanisms for biofortification and phytoremediation. *Trends in Plant Science*, **14**(8), 436–442.

Zhu Z., Chen Y., Zhang X. and Li M. (2016). Effect of foliar treatment of sodium selenate on postharvest decay and quality of tomato fruits. *Scientia Horticulturae*, **198**, 304–310.

Part IV

Selenium Nanoparticles and Quantum Dots

Chapter 10

Se nanoparticle manufacturing for medical applications

Alessandro Presentato, Elena Piacenza and Raymond J. Turner

10.1 INTRODUCTION

The use of biological material or various life forms to produce nanomaterials is routed in the idea that their use will be more eco-friendly than chemically synthesized materials. The present chapter focuses on the current knowledge of how biological organisms and their associated biomolecules or biomass reduce selenium (Se) oxyanions to Se^0 atoms (Figure 10.1). The Se atoms then subsequently assemble on the nanoscale, thus producing 'biogenic' Se nanoparticles (BioSeNPs).

The world of 'very small materials' implies the manipulation of matter at the molecular or atomic level (Horikoshi & Serpone, 2013), a field known as nanotechnology. Materials scaled down to the nano range (1–100 nm) are defined as nanomaterials (NMs), and possess unique physical-chemical properties arising from their high surface-to-volume ratio, large surface energy, and high spatial confinement (Cao, 2004; Yuwen & Wang, 2013). NMs exist in various sizes and shapes including nanoparticles (NPs), nanorods (NRs), quantum dots (QDs), nanowires (NWs), and nanotubes (NTs) (Rao *et al.*, 2004). These materials have enhanced chemical, catalytic, mechanical, electrical and opto-magnetic properties (Appenzeller, 1991; Yuwen & Wang, 2013). Hence, NMs can be applied in a vast range of applications such as: biomedicine and biotechnologies, energy production, food and chemical industries, environmental engineering, mechanics,

doi: 10.2166/9781789061055_0289

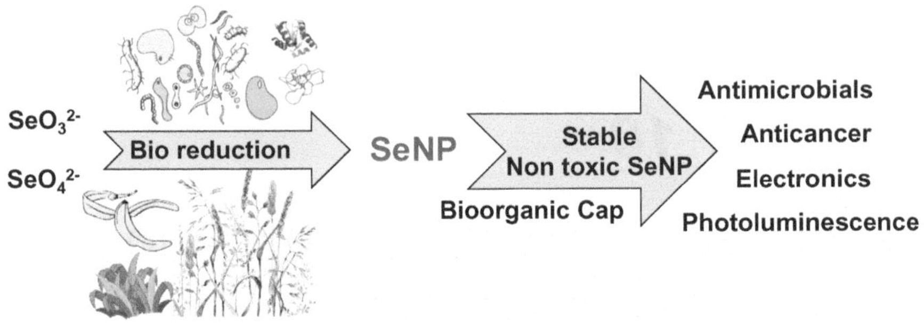

Figure 10.1 Illustration of catalytic sources for the biological applications in SeNP manufacturing.

sensors, optics, and material science (Ahmed *et al.*, 2016; Cao, 2004; Horikoshi & Serpone, 2013; Sharma *et al.*, 2015). The research interest around this field has grown exponentially in the past decade, reflected in the high number of scientific reports for the fabrication of novel nanomaterials through chemical (chemogenic) or biological (biogenic) means.

SeNPs can be produced through a variety of physical (laser ablation, UV radiation, and hydrothermal techniques), and/or chemical (precipitation, acid decomposition, catalytic reduction using a variety of reducing agents) methods. However, these approaches require high temperatures and/or harsh/hazardous chemicals and pH (Dwivedi *et al.*, 2011; Quintana *et al.*, 2002; Zhang *et al.*, 2010a). Thus, biosynthesis approaches have been explored that turn out to be safer, less expensive and utilize more eco-friendly materials and operational conditions.

Selenium is an essential trace element in humans and many microorganisms as an important cofactor in biochemical processes (see Chapter 3). Indeed, the 21st amino acid, selenocysteine, is an important active center of at least 25 different selenoproteins (Rayman, 2012), and its deficiency can cause diverse diseases in humans (Brigelius-Flochè & Maiorino, 2013; Morenoreyes *et al.*, 1998; Zhang *et al.*, 2010b). Despite this, high concentrations of selenium compounds (i.e., the highly soluble oxyanions selenate [SeO_4^{2-}] and selenite [SeO_3^{2-}]) in the environment can be toxic at relatively low concentrations (Presser & Ohlendorf, 1987; Weres *et al.*, 1989). This is due to the oxyanions' mobility through the trophic chain and its tendency to bioaccumulate. Yet, there are now many organisms identified that are highly tolerant to excessive selenium loads. Bacteria resistant to and/or able to respire selenium oxyanions started to be identified in the late 1980s and through the 1990s (Stolz & Oremland, 1999; Stolz *et al.*, 2006). However, it was not recognized until the late 2000s that this process led to potentially valuable technologies in bioremediation and nanomaterial manufacturing (Gadd, 2010; Nancharaiah & Lens, 2015).

This chapter overviews biological sources as reducing agents of Se oxyanions, either as whole organisms/cells or biomass components. This chapter does not go into the engineering of the reaction process, which would in most cases be batch reactors and fermenter systems scaled to match reagent availability and production needs (see Chapter 6). Since 2015, there have been a number of reviews published in the area of Se nanomaterial production and their subsequent (bio)technological uses, both potential and realized. For general information on SeNPs see: Khurana *et al.* (2019); Hosnedlova *et al.* (2018); Menon *et al.* (2018); Kielczykowska *et al.* (2018); Guan *et al.* (2018); Sharma *et al.* (2017); and Verma (2015). For reviews specifically on biogenic SeNPs, see Tugarova and Kamnev (2017); Wadhwani *et al.* (2016); and Tan *et al.* (2016).

10.2 BIOLOGICAL SYNTHESIS OF SELENIUM NANOPARTICLES

Research on Se polluted sites has led the exploration of diverse organisms that can tolerate high Se loads. Some plant species have been found growing in soils and aquatic systems that were highly impacted by Se pollutants, and several Se oxyanion resistant bacteria were isolated associated with these plants (Wu, 2004). Utilizing the microorganisms capable of transforming Se toxicants became a strategy to attenuate the presence of these compounds in the environment. Their study led to the observation of an orange-red coloration, that is now known to be the result of the formation of biogenic SeNPs. It was also found that certain anaerobic bacteria could respire selenium oxyanions, which often resulted in the accumulation of elemental selenium (Se^0).

The spectral properties of BioSeNPs differ considerably from those of amorphous Se^0, formed by chemical oxidation of hydrogen selenide (H_2Se), and of black, vitreous Se^0 formed chemically by reduction of selenite with ascorbate (Pettine *et al.*, 2013). The microbial synthesis of Se^0 nanospheres results in a unique, complex, and compact nano structural arrangement of Se atoms. Furthermore, it turns out that the natural organic material supplied by the organism that is found to 'cap' the SeNPs leads to a more stable SeNP than that found through chemical/physical methods (Piacenza *et al.*, 2018a; Wadhwani *et al.*, 2016). Yet, their shape and size vary considerably depending on the biogenic system used. This probably reflects the wide diversity of enzymes involved in the dissimilatory reduction that differ in various microorganisms. To date, these conditions cannot be achieved by current chemical/physical synthesis methods (Hosnedlova *et al.*, 2018; Oremland *et al.*, 2004). It is now recognized that using biological systems to produce SeNPs is an eco-friendly approach to manufacture high quality and stable SeNPs with unique properties (Ingale & Chaudhari, 2013; Wadhwani *et al.*, 2016).

10.2.1 Bacteria mediated selenium nanomaterial production

For quite some time, it was known that microorganisms had an important role in the transformation of selenium compounds from one oxidation state to another (Favre-Bonte *et al.*, 2005; Herbel *et al.*, 2003; Stolz *et al.*, 2006). These transformations are key to the selenium biogeochemical cycle. For example, studies reported on the ability of various soil bacteria to reduce SeO_4^{2-} and SeO_3^{2-} (Dowdle & Oremland 1998; Sarathchandra & Watkinson, 1981), while some other microbes generate volatile organo-selenium species such as hydrogen selenide (H_2S) or methylated selenium species (e.g., $HSeCH_3$; $Se(CH_3)_2$) (Favre-Bonte *et al.*, 2005; McCarty *et al.*, 1993; Stolz *et al.*, 2006). The biotic process behind the reduction of selenium oxyanions elicited by bacteria either under oxic (Antonioli *et al.*, 2007; Bajaj *et al.*, 2012; Hunter & Kuykendall, 2007; Hunter & Manter, 2009; Piacenza *et al.*, 2018b, 2019; Presentato *et al.*, 2018a) or anoxic (DeMoll-Decker & Macy, 1993; Hunter & Kuykendall, 2006; Kessi, 2006) conditions most often results in the formation of red-colored selenium deposits scaled at the nano range in the form of spheres with a diameter ranging from 50 to 500 nm (Jain *et al.*, 2014; Oremland *et al.*, 2004). An example of cell and culture coloration is shown in Figure 10.2. There is now a

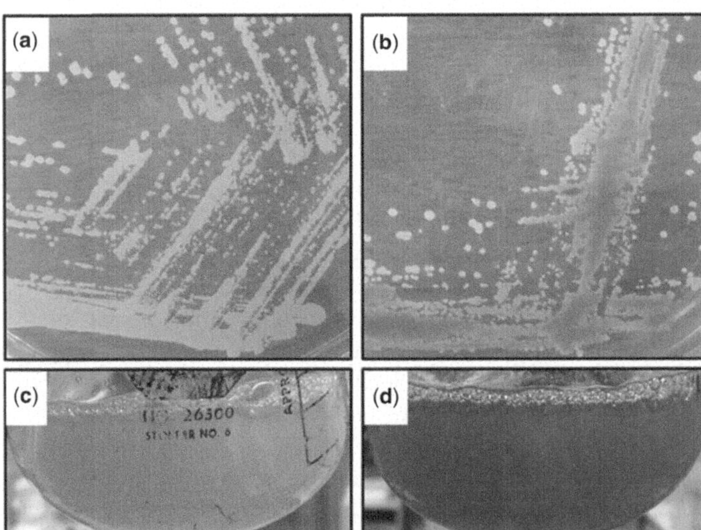

Figure 10.2 *Rhodococcus aetherivorans* BCP1 grown onto solid (A and B) and liquid (C and D) Luria–Bertani rich medium. B and D highlight the typical color change – from yellow to dark red – of the bacterial culture upon addition of 2 mM sodium selenite to the cultivation medium, due to the formation of Se^0 nanomaterial.

wide diversity of microorganisms that have been identified to respire different species of Se (Maltman *et al.*, 2016; Nancharaiah & Lens, 2015). However, the biochemical mechanism(s) of reduction of the selenium oxyanions have not yet been fully elucidated in all systems.

It is difficult to define a common mechanism for the assembly of biogenic SeNPs in selenium tolerant microorganisms, as their accumulation has been observed to be intracellular, membrane-bound, and extracellular. Some aspects of Se biochemistry are common between bacteria, likely related to the metabolic pathways laid out in *Escherichia coli* (Turner *et al.*, 1998). However, differences in genomics and physiology would add up to unique mechanisms in distant strains. This aspect is of interest, as unrelated bacteria, from a phylogenetic perspective, can give rise to diverse SeNMs in terms of structure and properties. Therefore, most biogenic SeNP synthesis outcomes are not predictable, nor reproducible, by the measure of conventional chemical procedures (Oremland *et al.*, 2004). This is because there is a collective knowledge gap regarding the mechanism exploited by bacteria to synthesize SeNMs; particularly, it is not clear how a bacterial cell factory controls the parameters of size, shape, and polydispersity of the final nanomaterial product. By deciphering the biochemical route used for NM synthesis, the fabrication of novel and unique nanomaterials may be honed using either living biological catalysts or specific macromolecules derived from them. Though, at this time, the best production conditions are still empirically determined, leading to biotechnological challenges towards scale-up.

Thauera selenatis, formerly defined as a *Pseudomonas* sp. strain (Macy *et al.*, 1989), was the first bacterial strain cultured in axenic conditions where its ability to respire selenate for energetic purposes was established (DeMoll-Decker & Macy, 1993; Macy *et al.*, 1993; Schröder *et al.*, 1997; Rech & Macy, 1992). Whatever the source of selenium precursor, the main phenotype was always represented by the color of the *T. selenatis* culture turning red, highlighting how the biotic reduction process led to the accumulation of selenium deposits, which occurred as extracellular SeNPs in this strain. It was speculated that *T. selenatis* could utilize an intracellular reductant to further detoxify selenite to Se^0 during selenate respiration, leading inevitably to the accumulation of Se^0 atoms within the cells. This observation raised critical questions: (i) How do elemental selenium atoms assemble forming NPs? (ii) How is the latter secreted to the extracellular environment? (iii) Is the process reversed where atoms are exported and then extracellularly assembled? (iv) Do NPs form both intracellularly and extracellularly with no Se or SeNP transport? The evidence is still ambiguous as experimental observations do not resolve the mechanism. *T. selenatis* cells entering the stationary phase during anaerobic growth in the presence of acetate as a carbon source and selenate as a terminal electron acceptor showed both intra- and extracellular selenium particles (Macy *et al.*, 1993). However, there was no evidence of cell lysis or distortion of the membrane suggesting two independent processes.

The use of microorganisms to produce SeNMs has other advantages over the chemical procedures, beyond being eco-friendly. The parameters of choice of the cell factory to perform a biogenic synthesis including growth temperature (Wang *et al.*, 2010), pH, type of media (Piacenza *et al.*, 2019), bacterial cell physiology (i.e., actively growing or resting cells (Presentato *et al.*, 2016, 2018b; Srivastava & Mukhopadhyay, 2013), culture incubation time (Ahmad *et al.*, 2015; Zhang *et al.*, 2011), and concentration of the metal(loid) precursor supplied (Piacenza *et al.*, 2019) can all be optimized to drive the production of a certain NM type(s). Above all, it is important to consider the metabolic potential of a given bacterial strain investigated. Strains belonging to either *Rhodococcus* or *Ochrobactrum* genera have high metabolic versatility (Martínková *et al.*, 2009) that allow them to survive harsh conditions. For example, *Rhodococcus aetherivorans* BCP1 is capable of simultaneously producing SeNPs and SeNRs when cultured in the presence of selenite under oxic conditions (Presentato *et al.*, 2018a). Examples of these nanomaterials are shown in Figure 10.3. This was also reported for *Ochrobactrum* sp. MPV1 cultivated under metabolically controlled growth conditions or in complex medium amended with very high concentrations (10 mM) of SeO_3^{2-}, producing high yields of SeNPs (Piacenza *et al.*, 2018b, 2019).

The stress on bacteria from Se oxyanions can represent a trigger to produce energetically valuable hydrophobic storage compounds to keep them thriving when the cells are experiencing an oligotrophic environment (Alvarez *et al.*, 1996). Surfactant-like molecules (i.e., biosurfactants) are produced to absorb and attenuate the stress derived from the carbon source exploited for energy production (Kumar *et al.*, 2014; Piacenza *et al.*, 2018b) or the toxicity exerted by a high metal(loid) load (Piacenza *et al.*, 2019). The presence of amphiphilic compounds, that act like surfactants, was identified in the biogenic SeNM extract. This was determined through fluorescence spectroscopy. A hydrophobic

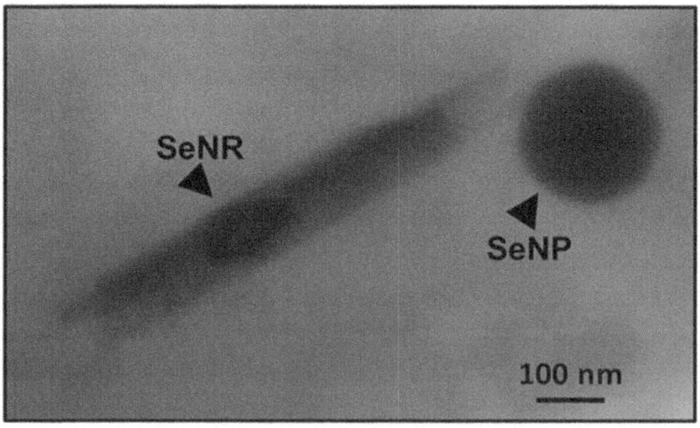

Figure 10.3 Transmission electron micrographs of extracted selenium nanomaterials in the form of nanorod (SeNR) and particle (SeNP) produced by *Rhodococcus aetherivorans* BCP1.

fluorescent dye [$DiOC_{18}(3)$] was used to label either biogenic NM extract or a sample containing liposomes, unveiling that the fluorescence signals were very much alike (Piacenza *et al.*, 2018b; Presentato *et al.*, 2018b). The presence of these amphiphilic molecules in the biogenic NM extract can have multiple roles, acting either as a stabilizing agent of the NM, as a driving force for the nucleation, or for the one-dimensional growth of nanomaterials (i.e., NRs), or even as a mediating component of the NM functional properties. This is inferred from chemical synthetic approaches, where the addition of surfactant compounds to the reaction mixture leads to NRs as products (Eastoe & Tabor, 2014; Evans & Wennerstrom, 1994). Finally, the nanostructures can exist in either amorphous or crystalline configurations, where the latter gives higher thermodynamic stability in suspension, which is a trait typically found for NRs (Jeong *et al.*, 2006). It are the biosurfactants in nanorod synthesis that lead to the stable crystalline rods (Presentato *et al.*, 2018b).

10.2.2 Fungi mediated selenium nanomaterial production

The high sensitivity of most fungi to Se compounds (Hanson *et al.*, 2004) has led to their limited exploitation for SeNM production as compared to bacteria. Regardless, fungal species have several advantages over other microorganisms, such as their ease and cost-effectiveness in culturing, scaling up and downstream processing (issues reviewed in – Piacenza *et al.*, 2018a), as well as their ability to adsorb and accumulate metals (Thakkar *et al.*, 2010). To date, the tolerance of fungi towards Se seems to be linked to long and/or constant exposure of these microorganisms to Se-containing compounds in their natural habitats, such as the rhizosphere of Se-hyperaccumulator plants (Shoeibi *et al.*, 2017). In this respect, fungal strains belonging to *Alternaria*, *Curvularia*, *Cladosporium*, *Pleurotus*, *Aspergillus*, *Chaetomium*, *Trichoderma*, *Aureobasidium*, *Mortierella*, *Phoma*, and *Agaricarius* feature higher tolerance towards Se (Liang *et al.*, 2019; Sarkar *et al.*, 2011; Vetchinkina *et al.*, 2019; Wangeline *et al.*, 2011).

Sarkar and coworkers (2011) were among the first groups to report the exploitation of fungi, using filtered fungal-free spent media of *Alternaria alternata*, for SeNP biosynthesis. These biogenic SeNPs showed high polydispersity, ranging in size between 30 and 150 nm, and an amorphous nature (Sarkar *et al.*, 2011). They appeared to be coated by a protein layer, which in turn made the SeNPs stable over several months (Sarkar *et al.*, 2011). Similarly, *Aspergillus terreus* displayed a good proficiency in bioconverting SeO_3^{2-}, producing relatively stable extracellular SeNPs with an average diameter of 48 nm (Zare *et al.*, 2013). Selenite was also the precursor for the extracellular SeNP formation by the basidiomycete *Lentinula edodes*. This species was also capable of biotically transforming the organo-Se compound 1,5-diphenyl-3-selenopentanedione (DAPS-25), forming nanosized Se products with an average size of 180 nm (Vetchinkina *et al.*, 2013).

The effect of using actively growing fungal cultures or extracts for the biosynthesis of SeNPs was investigated by Vetchinkina *et al.* (2019). Strains of *L. edodes* and *Pleurotus ostreatus* produced polydisperse (50–150 nm) SeNPs with uniform shape, while those obtained from the mycelium extracts of the same species were less regular in shape. Similar results were obtained by exploiting *Grifola frondosa*, *Ganoderma lucidum*, *Agaricarius bisporus*, and *A. arvensis*: small (20–50 nm) and homogeneous SeNPs were detected using the intracellular extracts of their mycelium (Vetchinkina *et al.*, 2019). *Aureobasidium pullulans*, *Mortierella humilis*, and *Phoma glomerata* generated extracellular, granular and amorphous SeNPs between 40 and 60 nm. Se oxide (SeO_2) NPs that were heterogeneous in shape and size have been detected on the *Trichoderma harzianum* cells' outer surface (Liang *et al.*, 2019).

Besides myceliating fungi, the single-celled yeasts have also been explored in SeNM production. Yeasts accumulate high amounts of Se during their growth (Marinescu *et al.*, 2011). This has been observed for *Saccharomyces cerevisiae*, whose bioaccumulation capacity was found to be regulated by several external factors, such as temperature, fermentation time, pH, shaking speed, and Se concentration (Esmaeili *et al.*, 2012). Similarly, the aerobic bioconversion of SeO_3^{2-} oxyanions carried out by the same yeast strain led to the biosynthesis of 30–100 nm SeNPs, which demonstrated good antimicrobial properties against Gram-positive and Gram-negative pathogens (Hariharan *et al.*, 2012).

At this time, the synthetic manufacturing of SeNPs by fungi is still in its infancy. The mechanistic information is still sparce. Studies only suggest a potential role played by glutathione, as well as cysteine residues present in the hypha cell walls, for the bioconversion of Se precursors to Se^0 on the mycelium through Painter-type reactions (Poluboyarinov *et al.*, 2009).

10.2.3 Plants and selenium nanomaterials production

Besides bacteria and fungi, other biological systems that have been investigated for SeNM production include plants (and plant material) in part due to their resistance towards a wide range of toxic compounds. Although a broad spectrum of plants has been explored for a breadth of metal NM generation, particularly silver (Shoeibi *et al.*, 2017), their use in SeNM production is currently limited. Trials with plant leaves (Alam *et al.*, 2019; Fardasadegh *et al.*, 2018; Li *et al.*, 2007; Sownadarya *et al.*, 2017), fruit, and their plant wastes/extracts have been explored.

Capsicum annuum leaf extract was used to biotically reduce SeO_3^{2-}, simultaneously forming amorphous SeNPs surrounded by a protein layer, which controlled the nucleation and growth of the NPs (Li *et al.*, 2007). Additionally, the SeNP size and structure was affected by the concentration of the leaf extract as well as the pH of the medium (Li *et al.*, 2007). Selenious acid was used as a Se-precursor for the generation of SeNPs by *Trigonellafoeum graecum* seed extract, by exploiting ascorbic acid as an initiator of the reduction process.

The resulting SeNPs featured high polydispersity, between 50 and 150 nm in size, and a good crystalline structure, which seemed to be mediated by the presence of bioactive compounds of the plant extracts containing C=C, NH_2, COOH, and C=O functional groups (Ramamurthy *et al.*, 2013). These associated biomolecules were considered responsible for the slight cytotoxicity observed under a prolonged exposure of human breast cancer cells to increasing concentrations of the biogenic SeNPs (Ramamurthy *et al.*, 2013). On the contrary, SeNPs recovered from *Allium sativum* extract were not cytotoxic towards human kidney epithelial Vero cells, a feature that the authors ascribed to the presence of biocompatible molecules containing N=O as stabilizing agents (Anu *et al.*, 2017). The size of these SeNPs ranged from 40 to 100 nm, being homogeneous in shape and relatively monodispersed in solution (Anu *et al.*, 2017).

Trigonal SeNPs of ca. 8 nm were synthesized by the dried fruit extract of *Vitis vinifera*, whose lignin molecules appeared to be associated with the NPs in solution. These lignin molecules conferred these SeNPs' high thermodynamic stability (Sharma *et al.*, 2015). Larger SeNPs (46–78 nm) were obtained from *Clausena dentata* leaf extract with proteins found as NP capping agents. These SeNPs had good insecticidal activity towards mosquito larvae (Sownadarya *et al.*, 2017). Microwave irradiation was exploited by Fardasadegh *et al.* (2018) to obtain 50 nm SeNPs from *Pelargonium zonale* leaf extract through SeO_3^{2-} bioreduction. This approach was considered to be mediated by tannins, flavonoids and other metabolites containing highly reactive -OH groups, while proteins were responsible for the stabilization of the formed SeNPs. Furthermore, these biogenic SeNPs showed good antimicrobial properties towards both bacterial and fungal pathogens, having a higher efficacy against the bacteria (Fardasadegh *et al.*, 2018). Similar results were reported for SeNPs of ca. 10–20 nm in size obtained from *Psidium guajava* leaf extract, where ascorbic acid residues and phenolic compounds were responsible for the bioconversion of SeO_3^{2-} (Alam *et al.*, 2019).

10.3 ROLE OF BIOMOLECULES IN THE SYNTHESIS OF SELENIUM NANOPARTICLES

Several biochemical reactions can catalyze the reduction of selenate or selenite generating the seed atoms of the nanomaterials, Se^0. The most widely reported reaction is the Painter-like reaction with thiols. However, one can have other reductive biochemical reactions with a variety of metabolites (such as plant flavonoids and lignin). Key biochemical electron mediators (cytochromes and quinones) can also catalyze the reduction reaction as demonstrated with the quinone analogue lawsone to produce SeNPs with *E. coli* (Wang *et al.*, 2011) or with *Rhodobacter capsulatus* (Borghese *et al.*, 2014). There is also the potential for direct enzymatic reactions with the oxyanion as the substrate. In this context, some early studies reported evidence that the catalytic enzyme may in fact nucleate the crystal growth of SeNPs (DeMoll-Decker & Macy, 1993; Rech & Macy, 1992).

10.3.1 Proteins involved in selenium nanoparticle synthesis

Tugarova and Kamnev reviewed proteins involved in SeNP synthesis in 2017. This review highlights that although some enzymes have been implicated in Se oxyanion reduction and NP formation, their role is found mostly in NP size modulation and capping, where they provide thermodynamic stability to the NP. Below follows a brief description of enzymes involved in the reduction reactions followed by key studies finding proteins associated with the caps of the SeNPs.

The research in this field mostly recognizes that the biomolecular reduction of selenite occurs through Painter-like reactions (Painter, 1941) with glutathione (GSH) molecules or similar cysteine peptides (cysteine, bacillithiol, and mycothiol). The reactions (10.1) to (10.3) follow the process of reduction with glutathione and subsequent reduction to Se^0 mediated by a NADP(H) dependent reductase:

$$3GSH + 6SeO_3{}^{2-} + 4H^+ \rightarrow 3GSSeSG + 2O_2^- + 5H_2O \quad (10.1)$$

$$GSSeSG + NADP(H) \rightarrow GSSe^- + GSH + NAD(P)^+ \quad (10.2)$$

$$GSSe^- + H^+ \rightarrow GSH + Se^0 \quad (10.3)$$

The reaction with GSHs leads to a seleno diglutathione (GS-Se-SG) intermediate. This GSSeSG intermediate can be further acted on by the enzyme glutathione reductase to generate GSH and Se^0. Alternatively, the GSSeSG can be acted on by thioredoxin with further reductive cycling by the enzyme thioredoxin reductase leading to glutathione seleno persulfide ($GS\text{-}Se^-$), which then rapidly dismutates into GSH and Se^0. Both of these enzyme-catalyzed reactions utilize NADPH as a source of electrons (Ganther, 1971; Turner *et al.*, 1998).

Within the group of selenate respiring bacteria, several complex iron-sulfur molybdoenzymes (CISM) have been demonstrated to use selenium oxyanions as substrates, leading to red selenium particles (reviewed by Nancharaiah & Lens, 2015; Zannoni *et al.*, 2008). Early reports observed the CISM nitrate reductases of *E. coli* to have this catalytic activity, using selenate as a substrate (Avazeri *et al.*, 1997). This NarG or NarZ nitrate reductase dependent reduction was later confirmed in other bacteria (Sabaty *et al.*, 2001), with shared mechanisms and thus demonstrating cross-reactivities between microbial selenate and nitrate enzymatic reduction (Watts *et al.*, 2005). Another CISM enzyme encoded by the *ynfEFGH* operon in *Salmonella enterica* was also shown to be responsible for the respiration of selenate to selenite, and subsequently converting the latter to SeNMs (Connelly *et al.*, 2016). A cartoon description of a CISM enzyme process

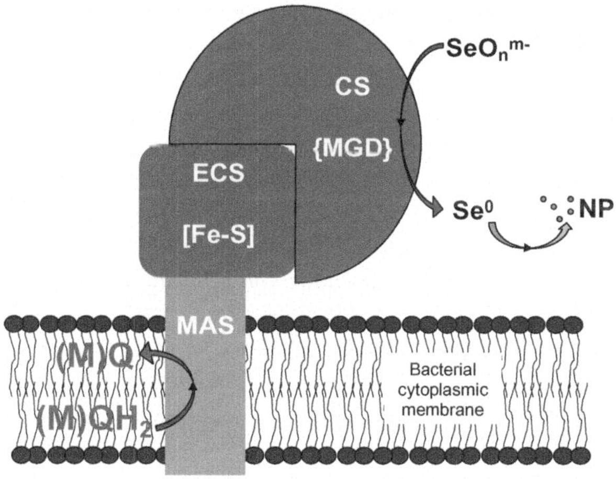

Figure 10.4 Description of a complex iron-sulfur molybdoenzymes (CISM) involvement in generating SeNPs. CS = catalytic subunit; ECS – electron conduit subunit; MAS = membrane anchor subunit, may or may not contain cytochromes; (M)Q = (menaquinone) or quinone; [Fe-S] = iron sulfur clusters, different atom number clusters are present from [4Fe-4S], [3Fe-4S] to [2Fe-2S]; the CS may or may not contain an Fe-S or cytochrome as well; MGD = Bis molybdopterin guanine dinucleotide; in some organisms, Mo may be replaced with W; NP = nanoparticles.

is shown in Figure 10.4. Unfortunately, purification of pure CISMs protein complexes is extremely difficult, and thus, they would be difficult to use directly as a catalyst in SeNP manufacturing. However, the knowledge of these enzyme systems allows research to focus on the genetic regulation of these systems for application in biotechnological processes for selenium bioremediation and nanomaterial production.

T. selenatis is one of the most studied selenate/selenite-respiring species (Butler *et al.*, 2012). Beyond the CISM related SerABC enzyme for its respiration of selenate (Schroder *et al.*, 1997), *T. selenatis* anaerobic respiration of selenite finds a single protein associated with SeNPs. Cell-free spent medium from cultures of *T. selenatis* containing SeNPs was analyzed for its protein content, and a single 95 kDa protein, named selenium factor A (SefA), was found. This protein was detected when *T. selenatis* cells were grown in the presence of nitrate (electron acceptor) during anaerobic growth in complex media and in the presence of selenate or selenite. The absence of either selenate or selenite led to undetectable levels of SefA, highlighting that the selenium oxyanions likely induced the expression of *sefA*. Bioinformatic analysis revealed that, although this protein was somehow associated with extracellular SeNPs, it does not possess a signal peptide responsible for its secretion. Moreover, the heterologous expression of the *sefA* gene in *E. coli* revealed the presence of the corresponding protein in both the soluble fraction and in the extracellular environment. *E. coli* cells expressing the *sefA* gene accumulated large SeNPs in the cytoplasm, while these NPs were not observed outside the cells, suggesting that (i) SefA might be a substrate for protein export, (ii) its secretion was not dependent on the binding with Se^0 atoms, and (iii) there may exist a specific export system for SeNPs in *T. selenatis,* since *E. coli* cells expressing *sefA* did not reveal any trace of extracellular SeNPs.

SefA presents a possible tool for SeNP manufacturing. When SefA was added to the reaction mixture containing glutathione and selenite in a 4:1 molar ratio, stable SeNPs were generated. However, in the absence of SefA only vitreous Se deposits were detected, underlining that SefA functions to bind and stabilize Se^0. This process occurs naturally in the cytoplasm of *T. selenatis* cells, and the SeNPs are subsequently exported to the extracellular environment once the spheres reach a particular size (ca. 150 nm in diameter). The whole mechanism by which *T. selenatis* can secrete these NPs through the inner membrane, periplasm, and outer membrane remains unknown (Dobieux *et al.*, 2011). Although this evidence suggests SefA has a role in the assembly of biogenic SeNPs under anaerobic conditions, it is not easy to find equivalent genetic determinants for a similar function in aerobic bacterial strains (Gonzalez-Gil *et al.*, 2016; Lenz *et al.*, 2011). It is also speculated that protein binding to SeNPs leads to a gain in thermodynamic stability. Hence, the SefA helps the NPs avoid aggregation. Such stabilization was also displayed by Bovine Serum Albumin bound to SeNPs *in vitro*, thus acting as a capping agent likely generating a sort of protein shell (Bücking *et al.*, 2010).

Dobias and colleagues (2011) tried to unravel the role of proteins in biomineralized selenium, with the aim of gaining the advantage of avoiding the maintenance of live cultures for NPs synthesis. *E. coli* derived SeNPs and associated proteins were purified by performing a sucrose gradient fractionation, whose fractions were analyzed by sodium dodecyl sulfate-polyacrylamide gel electrophoresis (SDS-PAGE). Among the proteins associated with SeNPs, some remained bound to NPs despite washing steps of the NM extract with increasing concentrations of denaturing solutions or after boiling in 10% w/v SDS solution. A similar observation was made for biochemicals associated with SeNPs from a variety of Gram negative and positive environmental isolates (Bulgarini *et al.*, 2021). This highlights the strong interaction of certain proteins to SeNPs providing the hypothesis that proteins involved in NP formation could have covalent bonds to the metal atoms of the NP. A similar finding was reported for SeNPs produced by *Lysinibacillus* sp. ZYM-1, where hexane treatments used in extracting and purifying the NPs did not successfully remove all the biomolecules present in the biogenic colloidal suspension, and thermogravimetric analysis revealed the presence of proteins in the coating (Che *et al.*, 2017). The *in vitro* synthesis of biogenic and chemically synthesized SeNPs in the presence of *E. coli* cell-free extract revealed that none of the proteins associated with SeNPs are known to be involved in selenium metabolism. Instead, the proteins found were involved in carbohydrate or fatty acid metabolism. Only two proteins (elongation factor Tu and 3-oxoacyl synthase) were present in all *in vitro* conditions tested, which suggested a non-specific mode of binding.

Comamonas testosteroni S44 grown under aerobic conditions in the presence of selenite had several proteins bound to SeNPs resulting from physical-chemical interactions (Xu *et al.*, 2018). Dobias *et al.* (2011) defined four proteins of varying size and isoelectric point (isocitrate lyase [AceA], isocitrate dehydrogenase [Idh], outer membrane protein C precursor [OmpC], and alcohol dehydrogenase [AdhP]) were found bound to the colloidal Se. Among these proteins, AdhP was selected for further studies aimed towards the formation of SeNPs by *in vitro* synthesis or using *E. coli* cell-free extract. These studies revealed that AdhP binding to SeNPs gave rise to larger NPs (122 nm average diameter) as compared to those produced in the presence of the cell-free extract (ranging from 10–90 nm). Yet, no effect of different proteins on the structure and crystallinity of the SeNPs was found (Dobias *et al.*, 2011). Other proteins like Mop and CysB from *C. testosteroni* S44, as well as PhoB1 and PhoB2 from *Agrobacterium* strain GW4, utilized to synthesize SeNPs *in vitro*, were also found to bind to the final SeNP product.

An explanation put forward for the protein-SeNP interaction focuses on the charged amino acids (i.e., Asp, Glu, Arg, and Lys) content of the associated proteins. Several proteins that are of low cell abundance and have a high charged amino acid content are found adsorbed onto the BioSeNPs surface; as reported in the case of SeNPs synthesized by an actinobacteril (Ramya *et al.*, 2019). However, the zinc-dependent alcohol dehydrogenase (AdH), which contains a low charged amino acid content, is also frequently found on the surface of SeNPs (Dobias *et al.*, 2011; Kessi *et al.*, 1999; Lampis *et al.*, 2017; Lenz *et al.*, 2011; Xu *et al.*, 2018). Thus, the diversity of proteins associated with the SeNPs to date suggests a non-specific mode of interaction of proteins with NPs, yet their robustness of association to the particles suggest a covalent bonding style (Xu *et al.*, 2018).

Having proteins, particularly large bulky proteins like AdH, likely provides stability to the SeNPs through steric hindrance inhibiting the Se cores of the NPs from interacting with each other, thus preventing their aggregation. This supports the observations that proteins are important in governing the size of NMs and contributing to generate a more monodisperse size population. This provides stability from a thermodynamic perspective, as this biogenic nanomaterial has a low tendency to form aggregates. This is likely because of repulsive interactions derived from charged proteins (Srivastava & Mukhopadhyay, 2013) and the steric hindrance effect exerted by proteins (Piacenza *et al.*, 2018a). Indeed, the thermodynamic stability of a colloidal suspension is a feature that is as fundamental and important as the eco-friendliness of the synthetic procedure adopted, when it comes to a potential application of nanomaterials.

10.3.2 Other biomolecules involved in selenium nanoparticle synthesis

Besides proteins, it has been hypothesized that bacterially derived extracellular polymeric substance (EPS) might act as a potential capping agent driving the assembly of SeNPs. This is thought to occur mostly in those bacterial strains for which the synthesis localization is either near the membrane or in the extracellular environment. The complex macromolecular composition of EPS, which contains proteins, polysaccharides, metabolites, lipid vesicles, humic-like molecules, and extracellular DNA (eDNA) (More *et al.*, 2014; Sheng *et al.*, 2010) offers many chemical functional groups capable of interacting with SeNMs. Moreover, EPS has been found to influence the surface properties, size, and shape of SeNPs (Jain *et al.*, 2014). Since many bacterial species have demonstrated either extracellular or periplasmic located reduction of selenite to elemental selenium (Jiang *et al.*, 2012; Li *et al.*, 2014), it is reasonable to suggest a possible involvement of EPS in the nanomaterial production extracellularly. The *in vitro* reduction of selenite with glutathione in the presence or absence of extracted EPS material, respectively, led to the generation of either stable spheres (NPs) or wires (NWs). This indicates that EPS might act as a template and/or capping agent and can tune the NM morphology (Zhang *et al.*, 2010a). This is similar to the polymers polyvinyl alcohol or polyethylene glycol used in chemogenic synthesis protocols (Shah *et al.*, 2010; Zheng *et al.*, 2012).

The complexity of the EPS associated with BioSeNP extracts was investigated by Jain and colleagues (2015) through Fourier transformed infrared spectroscopy (FTIR), which unveiled the presence of proteins, carbohydrates, and humic-like compounds. Since it is expected that sugar residues show more hydroxyl moieties as compared to proteins, it is reasonable to hypothesize interactions occur between these hydroxyl moieties and Se^0. Since only very small amounts of DNA were found in the SeNP extracts, its role as a capping agent derived during cell lysis events as a mechanism for SeNP release, was dismissed. Bulgarini *et al.* (2021) quantified the carbohydrate, protein and lipid levels of SeNPs from five different environmental isolates including *B. mycoides* SeITE01 and *S. maltophilia* SeITE02. The amount and content ratios varied remarkably between strains, yet the lipids were dominant for all isolates. It is presently unknown if these biochemicals are involved in the reduction or just dictating size and structure. Clearly the bioorganic capping layer is typically very complex, yet it is responsible for the SeNPs surface charge, which laid within the theoretical stability ($\zeta > |25 \text{ mV}|$) range for colloidal suspensions, as a function of the pH and ionic strength of the dispersing medium (Jain *et al.*, 2015).

The capability of several plants and plant extracts to bioconvert a broad spectrum of Se-containing compounds into Se^0 is considered to originate mostly by Painter-type thiol reactions (Jablonski & Anderson, 1982). Nevertheless, secondary metabolites of plant extracts, such as alkaloids, flavonoids, proteins, polysaccharides, glycosides, and saponins were reported to mediate *in vitro* reduction of SeO_4^{2-} (Li *et al.*, 2007; Ramamurthy *et al.*, 2013). Gallic acid (GA), a common plant polyphenol, was described for its ability to interact with selenite forming antioxidant GA-Se nanofibers, which could potentially be used in anticancer treatments (Barnaby *et al.*, 2011).

Overall, the *in vitro* production of BioSeNPs using various extracted biomolecules is still in its infancy. Although the use of plant and microbial extracts has been successful, there are still issues of reproducibility. Thus, producing BioSeNM with live microbial cultures of selected strains with tuned physiology presently has the most promise.

10.3.3 Approaches to monitor microbial manufactured selenium nanoparticles

Although the chemical and biophysical characterization of SeNPs is beyond the context of this chapter, a brief commentary on the methods used to characterize the nanomaterial is worthwhile. Beyond the microbiology and plant biology of BioSeNP manufacturing, to understand the biological process, information acquired from various analytical methods needs to be employed in BioSeNM characterization and quality control, because a serious limitation of biogenic synthesis is the relatively poor reproducibility. Adequate characterization of the bioNMs is an overlooked crucial step prior to evaluating the BioSeNM functional properties. A variety of well-established physical-chemical methods routinely used for the characterization of macromolecular structures and biomolecules are available to the study of BioSeNPs.

Dynamic light scattering (DLS) for size and dispersion as well as Zeta surface potential measurements are standard to the nanotechnology field for an impression of the material stability and quality of NMs. Scanning and transmission electron microscopy (SEM or TEM) coupled with energy dispersive X-ray (EDX) spectroscopy provide accurate information regarding the actual size, shape and elemental compositional of the NMs under study. X-ray diffraction (XRD) pattern analysis is also becoming a standard in the nanomaterial field.

Evaluating the composition of the biomolecular organic material associated with the biogenic nanomaterial is still a challenging task due to biology's intrinsic complexity. Nevertheless, techniques such as UV-visible absorption and vibrational (Fourier Transmission Infrared (FTIR) or Raman) spectroscopies can be used to identify the biomolecular classes present within the organic material. Nuclear magnetic resonance (NMR) and routine protein chemistry approaches are also available. Of course, the modern system biology approaches of genomics, metabolomics, proteomics and lipidomics can also be applied to evaluate the presence of specific nucleic acids, metabolites, proteins and lipids, respectively. The metal composition can be determined by atomic absorption or inductively coupled plasma (ICP) methods. The oxidation state and coordination can be inferred through X-ray synchrotron techniques such as X-ray absorption near edge spectroscopy (XANES). The overall quality of the Se core can also be inferred from its fluorescence, whereas electron diffraction or XRD patterns will reveal the crystalline nature/configuration of the Se within the nanomaterials.

A good first step towards repeatable characterization for the BioSeNP area is to investigate the types and ratios of various biomolecules associated with BioSeNPs. Bulgarini *et al.* (2021) suggest a facile approach to establish trends from various biogenic processes. Next, a set of key physical-chemical approaches should be applied in order to effectively compare SeNPs produced by different biological processes. This has been suggested for the antimicrobial activity of biogenic silver nanoparticles giving a standard characterization protocol that includes DLS, Zeta, XRD, absorption spectra, and TEM analysis (Duran *et al.*, 2016). However, such a mandate has not yet been proposed for the SeNM field.

10.4 TOXICITY OF SELENIUM-BASED NANOMATERIALS

As with any metal(loid), the toxicity of Se depends on its concentration, speciation, and other compounds that may influence toxicity through antagonistic or synergistic mechanisms. Thus, fundamentally, the toxicity of SeNMs will come from three components: (i) the selenium and any composite heteroatoms; (ii) properties of the NP itself, such as size, thermodynamic stability, and its amorphous/crystalline nature; and (iii) the capping material, which can influence the SeNMs bioavailability and could also have toxicity properties of its own. Differences between chemogenic and biogenically synthesized NMs lie in subtle differences in crystallinity and size distributions, but more so in the capping material. The stabilizing and capping agents of chemogenic SeNMs have defined and well-known physical-chemical properties, which could make them biologically active, whereas the heterogeneous composition and chemical behavior of the organic material associated with biogenic SeNMs are still not elucidated. Nevertheless, since the organic material features biomolecules, it can play an active role in influencing the bioavailability and toxicity of BioSeNMs.

10.4.1 Toxicity of selenium

Before considering the toxicity of SeNPs, a brief assessment of the biochemistry and toxicity of selenium is worthwhile. Since selenium is an essential trace element for most lifeforms, its biochemistry and physiology has similar issues as other trace elements: that is, how does one get enough, yet not too much? For most of these elements, carefully regulated import and export transporters as well as specific binding proteins that act as holding reservoirs exist in cells. Under physiological conditions, the most toxic species of Se is the highly reduced selenide (S^{2-}, II-), which is highly reactive, yet readily oxidized to the less bioavailable Se^0. Of the oxyanion forms, selenite (SeO_3^{2-}, IV) is far more toxic than selenate (SeO_4^{2-}, VI). These oxyanions typically get into cells through either phosphate or sulfate uptake systems, where they are used to form either selenomethionine or selenocysteine via specific biochemistry. Some of this chemistry is mediated by thiol (RSH) redox homeostasis peptides and proteins (e.g., thioredoxin and glutaredoxin) through Painter-like reactions (Turner *et al.*, 1998). In eukaryotes and Gram-negative bacteria, the most common peptide exploited for this bioprocess is glutathione, while for Gram-positives this peptide is in carbohydrate-modified version either as mycothiols (MSHs; typical of actinobacteria) or bacillithiols (BSHs; typical of *Bacillus* spp.), which show a greater redox stability as compared to GSH molecules (Presentato *et al.*, 2016). These peptides are found at concentrations in the order of 10 mM. The product of the RSH-mediated reduction of selenite is oxidized thiols and reduced selenium either as a Se-amino acid or as Se^0 atoms which may assemble into SeNMs.

The mechanism(s) of how metal nanomaterials enter a cell is unclear. Yet once inside the cell, the metal NP may decompose providing a localized high concentration pulse of the antimicrobial metal. Any membrane disruption mediated by the NP during passage could uncouple the electron transfer chain which would produce reactive oxygen species (ROS) (Wang *et al.*, 2017). Additionally, ROS are produced if the metal is redox active. The toxicity of selenite/selenate is founded on the basis of cellular thiol redox homeostasis in that some biochemical reactions produce a lot of reactive oxygen species (ROS) and there must be enough metabolic reducing potential to recover the oxidized peptides and proteins from damage. With this in mind, the Se oxyanions react with RSH containing biomolecules in the cell. In this way, the Se oxyanions consume the reducing equivalents that would normally be used to buffer naturally produced ROS. Thus, one observes increased ROS from selenite exposure because the redox buffering has been unbalanced. The ROS measured is from the electron transport chain and not directly catalyzed by the Se atoms. However, some of the Se metabolism steps can lead to ROS species due to imbalanced redox reactions (Turner *et al.*, 1998).

A consideration of SeNPs toxicity is that they can often be synthesized with other metals present (heteroatoms). An example of this is Se-based quantum dots (QDs) which contain cadmium (Cd). Although they are primarily synthesized chemically, approaches towards their biogenic production are underway (Siresj, 2014; Suresh, 2014). SeCd QDs have not been found to have toxicity *per se* as they are capped for high stability. Indeed, no signs of necropsy in mice or in tissue cultures were found if there was no QD decomposition. However, if decomposition of SeCd QDs occurs, a cellular response with a dose-dependent pattern like that of molecular Cd is observed (see review Farre *et al.*, 2009). Thus, the heteroatom toxicity can overwhelm any effect of an increased Se load. The nature of the polymer coating on the SeCd QDs thus has a large influence on toxicity via governing their decomposition rate in the environment (Sharma *et al.*, 2017).

It has been well discussed how organic molecules involved in the capping of the Se atom core lead to size, shape, and thermodynamic stability characteristics

(Piacenza *et al.*, 2018a). However, it is important to acknowledge the source of this biogenic material and biomolecules from the organism that manufactured it. The examples in this chapter provide evidence that the biogenic material has different and enhanced activities compared to chemically synthesized nanomaterials. These properties are in part due to the biomolecules making up the organic material. Thus, to properly investigate the toxicity of BioSeNMs, it is imperative to consider the content of the biogenic extracts, that is, both the organic material and the Se^0 atoms.

10.4.2 Selenium nanoparticle toxicity to Prokaryotes

We are now in a time often referred to as the Antibiotic Resistance Era, where antimicrobial resistance (AMR) is rampant in most pathogens. In response, there has been increased exploration towards novel antimicrobial agents and antibiotic stewardship. One response to this call is metal and metalloid based antimicrobials (MBAs) (Turner, 2017). The application of metal salts used alone or mixed with other antimicrobials is used for wound treatment, in device coatings and for high touch surfaces to limit contamination (Monych *et al.*, 2019). In the last decade, metal and antimicrobial conjugated nanomaterials have been explored with significant promise to become the next generation of antimicrobial agents (see reviews: Huh & Kwon, 2011; Rudramurthy *et al.*, 2016; Sanchez-Lopez *et al.*, 2020; Vimbela *et al.*, 2017; Want *et al.*, 2017; Yah & Simate, 2015). At this time, silver nanoparticles AgNP are leading with many formulations and applications now out to market. Nevertheless, nanomaterials of gold (Au), zinc (Zn), nickel (Ni), titanium (Ti), copper (Cu), and iron (Fe) are all promising. Alternative technologies include antimicrobial-conjugated silica NPs (Bernardos *et al.*, 2019), other organic nanomaterials (reviewed in Raghunath & Perumal 2017), and even carbon nanomaterials either alone or as antibiotic carriers (Maas, 2016). The strong movement towards the exploitation of metal(loid) NMs as antimicrobials derives from the assumption that NPs will be able to overcome the common mechanisms of resistance towards traditional antibiotics. This is based on observations that MBAs have broad biochemical targets mediating their antimicrobial effects (Lemire *et al.*, 2013). Thus, spontaneous mutations can give rise to resistance towards an antibiotic that has a single cell target. However, it is highly unlikely that multiple mutations will occur to protect against all antimicrobial mechanisms of MBAs (Wang *et al.*, 2017). A challenge is, however, that the toxicity of a metal(loid) is correlated to its speciation.

To date, a number of general reviews have been published evaluating the antimicrobial mechanisms of metal nanomaterials and biogenic nanoparticles, however, few mention SeNMs. Yet, a few studies are finding that biogenically produced SeNPs have good antimicrobial activities. Unfortunately, what is often overlooked when evaluating efficacy of metal-based antimicrobials is that growth media and bacterial strain differences can lead to a large range of antimicrobial inhibition effects, which makes it difficult to compare efficacies between studies. Furthermore, most studies miss the difference between the chemogenic and biogenic NM capping material, and this cap may play a significant role in the NMs antimicrobial efficacy.

To date, the mechanisms of metal NP antimicrobial activities include: (i) photocatalytic production of ROS or direct Fenton style reactions; (ii) cell membrane and wall damage with or without direct release of the metal(loid) ions; (iii) disruption of energy production (uncoupling the proton motive force or inhibiting specific bioenergetics enzymes); and (iv) direct inhibition of specific enzyme activity and central biology processes (Bernardos *et al.*, 2019; Huh & Kwon, 2011; Rudramurthy *et al.*, 2016; Wang *et al.*, 2017). It is likely that SeNPs could affect prokaryotes in similar ways.

The bacterial response to selenium exposure has been reviewed previously (Zannoni *et al.*, 2008), and would be applicable to SeNPs if at some point the SeNPs would decompose (naturally or via biological actions) and expose a bacterium to a local dose of oxidized selenium. In order to make the Se toxic, the biochemical processing from the microbe would need to change the Se(0) species by reducing the Se to the minus II state or oxidizing it to plus IV. From here the selenite ions could cause cell damage in two major ways within bacterial cells. 1. Reduction: The toxicity originates from the oxidation of cellular components or stealing electrons from the bioenergetic enzymes and electron transfer chain. This oxidation can lead to reactive oxygen species production, which would lead to further biomolecule damage. 2. Specific chemistry: Methylation reactions seem to be quite common (Chasteen & Bentley, 2003), where a variety of methylated species can be produced, including but not limited to: CH_3SeH, CH_3SeCH_3, CH_3SeSCH_3, and $CH_3SeSeCH_3$. If released these compounds are volatile and lost, however, if they are trapped within the cell, particularly through thiol based biochemical reactions, they can be damaging by inhibiting enzymes. These end point reactions could be the alkyl capping of active site functional -SH groups, such as in cysteines and cofactors such as coenzyme-A and lipoamide. These reactions result in the irreversible inhibition of the subsequent biochemistry and related metabolism.

The AMR issue is amplified by the capability and preference of most bacteria to grow as a surface attached community, which is referred to as a biofilm, showing remarkable antimicrobial tolerance. Human infections see 65–80% caused by the formation and proliferation of pathogens as biofilms (Bryers, 2008; NIH, 2002). BioSeNPs produced by *Stenotrophomona smaltophilia* SeITE02 and *Ochrobactrum* sp. MPV1 showed excellent efficacy against *E. coli*, *P. aeruginosa,* and *S. aureus* both planktonically and as a biofilm (Zonaro *et al.*, 2015). Similar results were observed by Piacenza *et al.* (2017), who used *Bacillus mycoides* SelTE01 to produce BioSeNPs with good efficacy against planktonic growth of *P. aeruginosa* and *S. aureus*. In these studies, the BioSeNPs had greater antimicrobial efficacy than the chemogenic SeNPs. Incorporating the

BioSeNPs on hydroxyapatite coated surfaces led to excelled antibiofilm activity. In these studies, it appears that some of the antimicrobial activity originates from the biochemical organic capping of their SeNMs, as efficacy appears to be lost when this organic cap is extracted (Cremonini *et al.*, 2018).

10.4.3 Selenium nanoparticles toxicity towards Eukaryotes

The cytotoxic effects of nanomaterials on various body tissues and their interaction with various cell types is still poorly understood. The characteristics of the NMs including size, shape, quantity, charge and surface structure along with the target cell type and incubation conditions influence their effect on a vast array of biological properties (Beyth *et al.*, 2015). One sees positive effects on cell growth and proliferation as well as vasodilation, whereas negative effects can be genotoxicity, carcinogenesis, apoptosis, and inhibition of cell proliferation. As noted below, some of these negative effects are exploited therapeutically.

There does not seem to be much cytotoxicity associated with BioSeNMs produced by bacteria towards normal eukaryotic cells. Antibacterial BioSeNMs showed only minor effects on human dendritic cells and fibroblasts (Cremonini *et al.*, 2016). Their lack of cytotoxicity is an exciting aspect of BioSeNMs as antimicrobials. Fluorescence microscopy performed on both Gram-positive and Gram-negative pathogens revealed the ability of BioSeNMs produced from plants to disrupt the bacterial cell structure, yet no cytotoxic effect was detected on human cell lines, indicating their biocompatibility for future applications (reviewed by Alam *et al.*, 2019). Overall, chemogenic SeNMs show more toxicity than Se oxyanions in aquatic toxicity studies (fish), whereas for mammals the oxyanions tend to be more toxic than the chemogenic SeNMs.

Beyond the toxicity towards humans, there are eukaryotic pathogens where BioSeNMs may be useful for their control, even though this aspect has been far less explored. Indeed, BioSeNMs have been explored for topical treatment of lesion infections from the protozoan *Leishmania* with an IC_{50} of 25 μg/ml (Soflaei *et al.*, 2014). Similarly, selenium supplementation helps control Trypanosomal infections (da Silva *et al.*, 2014), which provides another possible application of BioSeNMs.

10.5 MEDICAL APPLICATIONS OF SeNP FOR HUMAN HEALTH

10.5.1 Benefits of selenium for human health

Selenium is not only an important essential trace element, but over the past 30 years its role in protection of oxidative stress-induced DNA damage and oncogenic-induced DNA adduct formation has been elucidated. This protection occurs primarily through modulating glutathione peroxidases and thioredoxin reductase. Selenium can also induce apoptosis in transformed and cancer cells. The mechanism for this apoptosis is considered to be through p53, mitogen-activated protein kinases (MAPK), and other cell signaling pathways, triggering redox-dependent apoptosis. Selenium has more recently shown promise as an anti-diabetic and anti-inflammatory agent, mostly through its antioxidant ROS scavenging abilities (Khurana *et al.*, 2019). In another context, selenium supplementation has been shown to ameliorate the toxicity of other heavy metals such as mercury (Watanabe, 2002) and arsenics (Gailer *et al.*, 2002). Such observations endorse selenium as a key nutritional element for humans, primarily for its antioxidant properties (Kielczykowska *et al.*, 2018), but is also noted for its antitumor, endocrine regulation, enhancing immune responses, and cognitive abilities (Guan *et al.*, 2018).

The efficacy of Se use for human applications for the above functions was noted to be enhanced and far more effective in the NM form (Guan *et al.*, 2018). There is considerable interest in the use of BioSeNMs, which have higher bioavailability and lower toxicity than inorganic or organoselenium nutritional supplements (Hosnedlova *et al.*, 2018). The inorganic ions are more quickly cleared by the body and nonspecific, whereas SeNPs are less toxic, provide a more even release as a nutritional supplement, and are more specific to tumors, thus leading to significant medical application potential (Guan *et al.*, 2018). The SeNPs for nutritional delivery are often capped with chitosan. In the past decade, Se delivered in the form of SeNPs has seen increased interest for medical and therapeutic use (Soumya *et al.*, 2018).

Although it is recognized that biogenically produced SeNPs are eco-friendly (Ingale & Chaudhari, 2013), few defined studies have compared the therapeutic differences between chemogenic and biogenically produced SeNMs. One can only project that due to the nature of the biochemical capping material, BioSeNMs would behave differently as compared to their chemogenic counter parts. Regardless, few studies to date exist where the cytotoxicity of bacterial produced SeNMs were evaluated (Cremonini *et al.*, 2016). From these studies one can only presume that the BioSeNMs can be used in the same biomedical applications as the chemogenic SeNMs. However, significant caution must be taken as there may be allergic and immunogenic responses to the biological derived molecules of the NP caps. As indicated above, these biochemical caps are composed mostly of uncharacterized biomolecules. Therefore, more research is required into BioSeNMs cytotoxicity before they are used in nutrition supplements in agriculture or for human health.

10.5.2 Biological synthesis of selenium nanoparticles for medical applications

Most of the synthetic procedures to produce metal(loid) nanomaterials rely on dangerous operational conditions and the use of toxic reagents, posing serious concerns regarding the disposal of the generated waste (Zhang *et al.*, 2006). This aspect also represents a disadvantage in terms of the clinical applications of synthetic NMs (Jeevanandam *et al.*, 2018). On the other hand, the important role played by biological catalysts (e.g., bacteria and fungi) in reducing various sources of bulk metal(loid) precursors into the less bioavailable and less toxic nanoscale forms is now widely accepted and recognized as a valid, eco-friendly, and cost-effective strategy to produce NMs (Bhainsa & D'Souza, 2006; Song & Kim, 2009; Suresh *et al.*, 2004; Vetchinkina *et al.*, 2019). This has led the biotechnology search for improved processes, but these have problems as well.

10.5.3 Limitations of biological SeNM synthesis

As promising as the biosynthesis of SeNMs can be, there are still major drawbacks for the implementation of these procedures, such as the need to elucidate details of the biological processes behind NM production and the physical-chemical characteristics of the biogenic extracts. Indeed, the parameters influencing SeNM production, the biochemical processes governing their biogenic synthesis, their extensive physical-chemical characterization and their potential biotechnological applications are still the black holes of this emerging scientific field. All these slow down the implementation of biological systems as cell factories for SeNM synthesis. Making a parallel with the workflow behind chemical synthetic procedures, key aspects influencing the biogenic production of SeNMs are the choice of reducing agents, precursor-to-reducing agent ratio, reaction time, temperature and pH of the system. Additionally, the choice of microorganisms to exploit for SeNM production and their physiological states are among the most important parameters to consider, directly implying the need to deeply understand the bacterial genetic and metabolic backgrounds. For instance, the use of microorganisms highly tolerant towards Se oxyanions assure the use of higher concentrations of Se oxyanions which should provide higher yields. With more tolerant organisms, a greater concentration range can be used to control the biosynthesized nanomorphologies. Se precursor concentration and incubation time are fundamental parameters observed in all studies that need to be taken into account to optimize NM biosynthesis. This appears to be as complex as the biological system itself, bringing to light the necessity to further investigate the bacterial response to selenium species.

A major drawback of the biogenic SeNMs synthesis is represented by their recovery from the biomass used as a catalyst. Anecdotal evidence suggests only a small fraction of NMs are extracted out of the total amount produced, although this is hard to ascertain as in most studies yields and mass balances are not reported. Poor SeNM recovery from the biomass strongly discourages the implementation of large-scale manufacturing of NMs, making it imperative to develop defined recovery strategies. In this regard, the localization of biogenic SeNMs plays a fundamental role in terms of the feasibility and ease of their recovery. Extracellular production is preferred, as cell disruption by means of expensive or non-*eco-friendly* procedures (e.g., sonication or chemical lysis) can limit the amount of product obtained. On the other hand, stable and crystalline SeNRs are generally obtained through bacterial intracellular processes, likely due to the high local concentrations of Se^0 atoms available for their deposition and growth as well as the availability of reductive processes (Presentato *et al.*, 2018a). Thus, depending on the desired final product, different cell factories could be used for SeNM biosynthesis, leading to the necessity to optimize multiple procedures for their recovery. Nevertheless, the identification and purification of bacterial enzymes responsible for Se precursor bioconversion may lead to the development of *in vitro* synthesis approaches that allow manufacturers to overcome issues encountered when working with live microorganisms.

Although biogenic SeNMs are highly valuable from a physical-chemical point of view (Piacenza *et al.*, 2018a), their polydispersity in size represents another disadvantage for applications, as several properties of material confined in the nano range directly depend on a uniform size and shape. Thus, focusing the research on obtaining the most monodisperse NM population possible is of importance for many uses. This will be directly linked to increasing the knowledge around the microbial cell factory exploited and the consequences of varying their growth or exposure conditions to the metalloid precursors. In this regard, it has been suggested that the use of bacterial cells under resting mode or metabolically controlled conditions can limit the number of NS nucleation events occurring within the cells, decreasing the chances to obtain polydisperse populations (Piacenza *et al.*, 2018b; Presentato *et al.*, 2018b). Yet, using a low bacterial cell density results in decreased oxyanion bioconversion, leading to the necessity to adjust the precursor concentration to obtain a high amount of NMs produced (Piacenza *et al.*, 2019).

The complex organic material generally present within the BioSeNM extracts also requires more investigations aimed at elucidating the composition and concentration of the biomolecules present. Issues remain in characterizing the associated organic material, as the limit of detection of most techniques used to date to identify or quantify these biomolecules is too low for the amount of material normally recovered in lab bench-scale experiments. As indicated in Section 10.3.3, various techniques are used to identify the biomolecules associated with the BioSeNPs, although reports systematically evaluating the identification, concentration and ratios of these molecules are still missing in the literature. Additionally, the inherent complexity of biological systems of active growing cells in complex media makes it unlikely to always obtain the same concentration of the different biomolecules in the organic capping material. A solution could be found in using bacteria under defined growth conditions, which may represent more controllable work systems to standardize production procedures.

10.6 PERSPECTIVES OF THE BIOGENIC MANUFACTURING OF SELENIUM NANOMATERIALS

10.6.1 Medical applications

The field of BioSeNPs is progressing forward from its adolescence and it is time to move from the general discovery that SeNMs can be made with various microorganisms towards focused studies to control biogenic synthesis of SeNMs with specific sizes and/or shapes. Additionally, biogenic production research should also be tuning the BioSeNPs towards specific applications. Here we propose directions for the field:

- A recognized problem in the field of using BioSeNPs therapeutically, particularly in the use as an anticancer agent, is that studies have been poorly designed (as noted in reviews by Hosnedlova *et al.*, 2018; Khurana *et al.*, 2019). The issues are highlighted around missing appropriate comparisons to other Se sources, and from our point of view, missing thought toward the influences from the subtle differences in NP properties (size and shape) or consideration of the role of the biochemical capping material in their application and roles in toxicity or lack thereof.
- There is significant potential in the use of BioSeNPs to fight AMR, both as an antiseptic in pathogen control in topical treatments or for touch surfaces in pathogen transfer. However, their use as an oral antibiotic is yet to be properly explored and it will take longer for such a product to pass to the market. Additionally, when groups are evaluating their BioSeNPs for antimicrobial efficacy, defined pathogen indicator strains from strain depositories should be used, with universal robust antimicrobial testing protocols in order to provide effective and useful comparisons.
- There are still gaps to be filled in the microbial nanotechnology field. There is still a lot of fundamental research to be done. A complete understanding of the biochemical and biological processes behind selenate/selenite conversion to elemental selenium (Se^0), as well as the route exploited by microorganisms towards the assembly of these Se^0 atoms in nanostructures (NSs) is still lacking. Directed systematic studies are required to enhance our understanding of these processes.
- Characterization of the BioSeNPs needs to be standardized. The full chemical and biophysical characterization of the BioSeNP is not always carried out and also the assessment of the cap's biomolecule character and composition needs to be more routine. Defined purification and quantifying yields need to be developed and implemented to industrial scales if we are to see biogenic NMs used routinely.

10.6.2 Non-medical applications

Given the similarities between chemogenic and biogenic SeNMs in terms of physical-chemical features, the investigation regarding BioSeNMs potential applications is still in its infancy as compared to those chemically synthesized. At this time, we see applications of BioSeNPs being almost entirely focused on biomedical applications. However, the particular features of Se at the nanoscale as well as the necessity to develop new 'low-carbon technologies' must push us to explore new properties of BioSeNMs beyond biomedical ones. Making a parallel with applications already evaluated for chemogenic NMs can support the development of innovative applications for BioSeNMs. Applications to consider are novel catalysts, nanobiosensors, environmental pollutant capture, photo and electrical devices, and others suggested in the review by Wadhwani *et al.* (2016). Here, we suggest the strength of the strategy to use microbes in selenium pollution remediation is not only to detoxify an environment, but also for the bioconversion of Se into useful SeNMs, allowing remediation biotechnologies to link to bioconversions for a value added outcome.

REFERENCES

Ahmad M. S., Yasser M. M., Sholkamy E. N., Ali A. M. and Mehanni M. M. (2015). Anticancer activity of biostabilized selenium nanorods synthesized by *Streptomyces bikiniensis* strain Ess_amA-1. *International Journal Nanomedicine*, **10**, 3389–3401.

Ahmed S., Annu Ikram A. S. and Yudha S. S. (2016). Biosynthesis of gold nanoparticles: A green approach. Journal of Photochemistry and Photobiology, B, **161**, 141–153.

Alam H., Khatoon N., Raza M., Ghosh P. C. and Sardar M. (2019). Synthesis and characterization of nano Selenium using plant biomolecules and their potential applications. *BioNanoScience*, **9**, 96–104.

Alvarez H. M., Mayer F., Fabritius D. and Steinbüchel A. (1996). Formation of intracytoplasmic lipid inclusions by *Rhodococcus opacus* strain PD630. *Archives of Microbiology*, **165**, 377–386.

Antonioli P., Lampis S., Chesini I., Vallini G., Rinalducci S., Zolla L. and Righetti P. G. (2007). Stenotrophomonas maltophilia SeITE02, a new bacterial strain suitable for bioremediation of selenite-contaminated environmental matrices. *Applied Environmental Microbiology*, **73**(21), 6854–6863.

Anu K., Singaravelu G., Murugan K. and Benelli G. (2017). Green-synthesis of selenium nanoparticles using garlic cloves (*Allium sativum*): biophysical characterization and cytotoxicity on Vero cells. *Journal of Cluster Science*, **28**, 551–563.

Appenzeller T. (1991). The Man Who Dared to Think Small. *Science*, **254**(5036), 1300.

Avazeri C., Turner R. J., Pommier J., Weiner J. H., Giordano G. and Vermeglio A. (1997). Tellurite and Selenate Reductase Activity of Nitrate Reductases from *Escherichia coli:* Correlation with Tellurite Resistance. *Microbiology*, **143**, 1181–1189.

Bajaj M., Schmidt S. and Winter J. (2012). Formation of Se (0) Nanoparticles by *Duganella* sp. and *Agrobacterium* sp. Isolated from Se-laden soil of North-East Punjab, India. *Microbial Cell Factories*, **11**(1), 64.

Barnaby S., Sarker N., Dowdell A. and Banerjee I. (2011). The spontaneous formation of selenium nanoparticles on gallic acid assemblies and their antioxidant properties. *Fordham Undergraduate Research Journal*, **1**, 41–46.

Bernardos A., Piacenza E., Sancenon F., Mehrdad H., Maleki A., Turner R. J. and Martinez-Manez R. (2019). Mesoporous silica-based materials with bactericidal properties. *Small*, article ID 1900669.

Beyth N., Houri-Haddad Y., Domb A., Khan W. and Hazan R. (2015). Alternative antimicrobial approach: nano-antimicrobial materials. *Evidence-based Complementary and Alternative Medicine*. article ID246012.

Bhainsa K. C. and D'Souza S. F. (2006). Extracellular biosynthesis of silver nanoparticles using the fungus *Aspergillus fumigatus*. *Colloids and Surface B: Biointerfaces*, **47**(2), 160–164.

Borghese R., Baccolini C., Francia F., Sabatino P., Turner R. J. and Zannoni D. (2014). Reduction of Chalcogen Oxyanions and Generation of Nano-precipitates by the Photosynthetic Bacterium *Rhodobacter capsulatus*. *Journal of Hazardous Materials*. **269**, 24–30.

Brigelius-Flohé R. and Maiorino M. (2013). Glutathione peroxidases. *Biochimica et Biophysica Acta (BBA) – General Subjects*, **1830**(5), 3289–3303.

Bryers J. D. (2008) Medical biofilms. *Biotechnology Bioengineering*, **100**, 1–18.

Bücking W., Massadeh S., Merkulov A., Xu S. and Nann T. (2010). Electrophoretic properties of BSA-coated quantum dots. *Analytical Bioanalytical Chemistry*, **396**(3), 1087–1094.

Bulgarini A., Lampis S., Turner R. J. and Vallini G. (2021). Biomolecular composition of capping layer and stability of biogenic selenium nanoparticles synthesized by five bacterial species. *Microbial Biotechnology*, **14**(1), 198–212.

Butler C. S., Debieux C. M., Dridge E. J., Splatt P. and Wright M. (2012). Biomineralization of selenium by the selenite-respiring bacterium *Thauera selenatis*. *Biochemical Society Transactions*, **40**, 1239–1243.

Cao G. (2004). Introduction. In: Nanostructures & Nanomaterials, Synthesis, Properties and Applications, G. Cao (ed.), Imperial College Press, London, United Kingdom, pp. 1–14.

Chasteen T. G. and Bentley R. (2003) Biomethylation of selenium and tellurium microorganisms and plants. *Chemical Reviews*, **103**, 1–25.

Che L., Dong Y., Wu M., Zhao Y., Liu L. and Zhou H. (2017). Characterization of Selenite Reduction by Lysinibacillus sp. ZYM 1 and Photocatalytic Performance of Biogenic Selenium Nanospheres. *ACS Sustainable Chemistry & Engineering*, **5**(3), 2535–2543.

Connelly K. R. S., Stevenson C., Kneuper H. and Sargent F. (2016). Biosynthesis of selenate reductase in Salmonella enterica: critical roles for the signal peptide and DmsD. *Microbiology*, **162**(12), 2136–2146.

Cremonini E., Boarett M. N., Vandecandelaere I., Zonaro E., Coenye T., Lleo M. M., Lampis S. and Vallini G. (2018). Biogenic selenium nanoparticles synthesized by Stenotrophomonas maltophilia SelTE02 antibacterial and antibiofilm efficacy as a result of the progressive alteration of their organic coating layer. *Microbial Biotechnology*, **11**, 1037–1047.

Cremonini E., Zonaro E., Donini M., Lampis S., Boaretti M., Dusi S., Melotti P., Lleo M. M. and Vallini G. (2016). Biogenic selenium nanoparticles: characterization, antimicrobial activity and effects on human dendritic cells and fibroblasts. *Microbial Biotechnology*, **9**(6), 758–771.

da Silva M. T. A., Silva-Jardim I. and Thiemann O. H. (2014) Biological implications of selenium and its role in trypanosomiasis treatment. *Current Medicinal Chemistry*, **21**, 1772–1780.

DeMoll-Decker H. and Macy J. M. (1993). The periplasmic nitrite reductase of *Thauera selenatis* may catalyze the reduction of selenite to elemental selenium. *Archives Microbiology*, **160**, 241–247.

Dobias J., Suvorova E. I. and Bernier-latmani R. (2011). Role of Proteins in controlling selenium nanoparticle size. *Nanotechnology*, **22**(19), 195605.

Dobieux C. M., Dridge E. J., Mueller C. M., Splatt P., Paszkiewicz K., Knight I., Florance H., Love J., Titball R. W., Lewis R. J., Richardson D. J. and Butler C. S. (2011). A bacterial process for selenium nanosphere assembly. *PNAS*, **108**(33), 13480–13485.

Dowdle P. R. and Oremland R. S. (1998). Microbial oxidation of elemental selenium in soil slurries and bacterial cultures. *Environmental Science and Technology*, **32**(23), 3749–3755.

Duran N., Nakazato G. and Seabra A. B. (2016). Antimicrobial activity of biogenic silver nanoparticles, and silver chloride nanoparticles: an overview and comments. *Applied Microbiology and Biotechnology*, **100**, 6555–6570.

Dwivedi C., Shah C., Singh K., Kumar M. and Bajaj P. (2011). An organic acid-induced synthesis and characterization of selenium. *Journal of Nanoparticles and Nanotechnology*, Article ID 651971.

Eastoe J. and Tabor R. F. (2014). Surfactants and nanoscience. In: Colloidal Foundation of Nanoscience, D. Berti and G. Palazzo (eds.), Elsevier B.V., London, United Kingdom, pp. 135–157.

Esmaeili S., Khosravi-Darani K., Pourahmad R. and Komeili R. (2012). An experimental design for production of selenium-enriched yeast. *World Applied Sciences Journal*, **19**, 31–37.

Evans D. F. and Wennerstrom H. (1994). Colloidal stability. In: The Colloidal Domain: Where Physics, Chemistry, Biology and Technology Meet, D. F. Evans and H. Wennerstrom (eds.), Wiley-VCH, New York, USA.

Fardasadegh B., Vaghari H., Mohammad-Jafari R., Najian Y. and Jafarizadeh-Malmiri H. (2018). Biosynthesis, characterization and antimicrobial activities assessment of fabricated selenium nanoparticles using *Pelargonium zonale* leaf extract. *Green Process Synth*, **8**, 191–198.

Farre M., Gajda-Schrantz K., Kantiani L. and Barcelo D. (2009). Ecotoxicity and analysis of nanomaterials in the aquatic environment. *Analytical Bioanalytical Chemistry*, **393**, 81–95.

Favre-Bonte S., Ranjard L., Colinon C., Prigent-Combaret C., Nazaret S. and Cournoyer B. (2005). Freshwater selenium-methylating bacterial thiopurine methyltransferases: diversity and molecular phylogeny. *Environmental Microbiology*, **7**(2), 153–164.

Gadd G. M. (2010). Metals, minerals and microbes: geomicrobiology and bioremediation. *Microbiology*, **156**, 609–643.

Gailer J., George G. N., Pickering I. J., Prince R. C., Yonis H. S. and Winzerling J. J. (2002). Biliary excretion of [GS(2)AsSe] after intravenous injection of rabbits with arsenite and selenate. *Chemical Research Toxicology*, **15**, 1466–1471.

Ganther H. E. (1971). Reduction of the selenotrisulfide derivative of glutathione to a persulfide analog by a gluthatione reductase. *Biochemistry*, **10**, 4089–4098.

Gonzalez-Gil G., Lens P. N. L. and Saikaly P. E. (2016). Selenite reduction by anaerobic microbial aggregates: microbial community structure, and proteins associated to the produced selenium spheres. *Frontiers in Microbiology*, **7**, 571.

Guan B., Yan R., Li R. and Zhang X. (2018). Selenium as a pleotropic agent for medical discovery and drug delivery. *International Journal Nanomedicine*, **13**, 7473–7490.

Hanson B. R., Lindblom S. D., Loeffler M. L. and Pilon-Smits E. A. H. (2004). Selenium protects plants from phloem-feeding aphids due to both deterrence and toxicity. *New Phytology*, **162**, 655–662.

Hariharan H., Al-Harbi N., Karuppiah P. and Rajaram S. (2012). Microbial synthesis of selenium nanocomposite using saccharomyces cerevisiae and its antimicrobial activity against pathogens causing nosocomial infection. *Chalcogenide Letters*, **9**, 509–515.

Herbel M. J., Switzer B. J., Oremland R. S. and Borglin S. E. (2003). Reduction of elemental selenium to selenide: experiments with anoxic sediments and bacteria that respire Se-oxyanions. *Geomicrobiology Journal*, **20**(6), 587–602.

Horikoshi S. and Serpone N. (2013). General introduction to nanoparticles. In: Microwaves in Nanoparticle Synthesis: Fundamentals and Applications, S. Horikoshi and N. Serpone (eds.), Wiley-VCH Verlag, Weinheim, Germany, pp. 1–24.

Hosnedlova B., Kepinska M., Skalickova S., Fernandez C., Ruttkay-Nedecky B., Peng A., Baron M., Melcova M., Opatrilova R., Zidkova J., Bjorklund G., Sochor J. and Kizek R. (2018). Nano-selenium and its nanomedicine applications, a critical review. *International Journal Nanomedicine*, **13**, 2107–2128.

Huh A. J. and Kwon Y. J. (2011). 'Nanoantibiotics': a new paradigm for treating infectious diseases using nanomaterials in the antibiotics resistant era. *Journal of Control Release*, **156**, 128–145.

Hunter W. J. and Kuykendall L. D. (2006). Identification and characterization of an *Aeromonas salmonicida* (syn *Haemophilus piscium*) strain that reduces selenite to elemental red selenium. *Current Microbiology*, **52**(4), 305–309.

Hunter W. J. and Kuykendall L. D. (2007). Reduction of selenite to elemental red selenium by *Rhizobium* sp. strain B1. *Current Microbiology*, **55**(4), 344–349.

Hunter W. J. and Manter D. K. (2009). Reduction of selenite to elemental red selenium by *Pseudomonas* sp. strain CA5. *Current Microbiology*, **58**(5), 493–498.

Ingale A. G. and Chaudhari A. N. (2013). Biogenic synthesis of nanoparticles and potential applications: an eco-friendly approach. *Journal of Nanomedicine & Nanotechnology*, **4**, 1–7.

Jablonski P. P. and Anderson J. V. (1982). Light-dependent reduction of selenite by sonicated pea chloroplasts. *Phytochemistry* **21**, 2179–2184.

Jain R., Gonzalez-Gil G., Singh V., van Hullebusch E. D., Farges F. and Lens P. N. L. (2014). Biogenic selenium nanoparticles, production, characterization and challenges. In: Nanobiotechnology, A. Kumar and J. N. Govil (eds.), Studium Press LLC, USA, pp. 361–390.

Jain R., Jordan N., Weiss S., Foerstendorf H., Heim K., Kacker R., Hübner R., Kramer H., van Hullebusch E. D., Farges F. and Lens P. N. L. (2015). Extracellular Polymeric Substances Govern the Surface Charge of Biogenic Elemental Selenium Nanoparticles. *Environmental Science & Technology*, **49**(3), 1713–1720.

Jeevanandam J., Barhoum A., Chan Y. S., Dufresne A. and Danquah M. K. (2018). Review on nanoparticles and nanostructured materials: history, sources, toxicity and regulations. *Beilstein Journal of Nanotechnology*, **9**, 1050–1074.

Jeong U., Camargo P. H. C., Lee Y. H. and Xia Y. (2006). Chemical transformation: a powerful route to metal chalcogenide nanowires. *Journal of Material Chemistry*, **16** (40), 3893–3897.

Jiang S., Ho C. T., Lee J. H., van Duong H., Han S. and Hur H. G. (2012). Mercury capture into biogenic amorphous selenium nanospheres produced by mercury resistant *Shewanella putrefaciens* 200. *Chemosphere*, **87**(6), 621–624.

Kessi J. (2006). Enzymic systems proposed to be involved in the dissimilatory reduction of selenite in the purple non- sulfur bacteria *Rhodospirillum rubrum* and *Rhodobacter capsulatus*. *Microbiology*, **152**(3), 731–743.

Kessi J., Ramuz M., Wehrli E., Spycher M. and Bachofen R. (1999). Reduction of selenite and detoxification of elemental selenium by the phototrophic bacterium *Rhodospirillum rubrum*. *Applied and Environmental Microbiology*, **65**(11), 4734–4740.

Khurana A., Tekula S., Saifi M. A., Venkatesh P. and Godugu C. (2019) Therapeutic applications of selenium nanoparticles. *Biomedicine and Pharmocotherapy*, **111**, 802–812.

Kielczykowska M., Kocot J., Pazdzior M. and Musik I (2018) Selenium – a fascinating antioxidant of protective properties. *Advances in Clinical and Experimental Medicine*, **27**, 245–255.

Kumar C., Sujhitha P., Mamidyala S., Usharani P., Das B. and Reddy C. (2014). Ochrosin, a new biosurfactant produced by halophilic *Ochrobactrum* sp. strain BS-206 (MTCC 5720): purification, characterization and its biological evaluation. *Process Biochemistry*, **49**(10), 1708–1717.

Lampis S., Zonaro E., Bertolini C., Cecconi D., Monti F., Micaroni M., Turner R. J., Butler C. S. and Vallini G. (2017). Selenite biotransformation and detoxification by *Stenotrophomonas maltophilia* SeITE02: novel clues on the route to bacterial biogenesis of selenium nanoparticles. *Journal of Hazardous Materials*, **324**, 3–14.

Lemire J., Harrison J. J. and Turner R. J. (2013) Antimicrobial activity of metals: Mechanisms, molecular targets and applications. *Nature Reviews Microbiology*, **11**, 371–384.

Lenz M., Kolvenbach B., Gygax B., Moes S. and Corvini P. F. X. (2011). Shedding light on selenium biomineralization: proteins associated with bionanominerals. *Applied Environmental Microbiology*, **77**(13), 4676–4680.

Li D. B., Cheng Y. Y., Wu C., Li W. W., Li N., Yang Z. C., Tong Z. H. and Yu H. Q. (2014). Selenite reduction by *Shewanella oneidensis* MR-1 is mediated by fumarate reductase in periplasm. *Scientific Reports*, **4**, 3735.

Li S., Shen Y., Xie A., Yu X., Zhang X., Yang L. and Li C. (2007). Rapid, room temperature synthesis of amorphous selenium/protein composites using *Capsicum annuum* L extract. *Nanotechnology*, **18**, 405101.

Liang X., Perez M. A. M. J., Nwoko K. C., Egbers P., Feldmann J., Csetenyi L. and Gadd G. M. (2019). Fungal formation of selenium and tellurium nanoparticles. *Applied Microbiology and Biotechnology*, **103**, 7241–7259.

Maas M. (2016). Carbon nanomaterials as antibacterial colloids. *Materials*, **9**, 617.

Macy J. M., Michel T. A. and Kirsch D. G. (1989). Selenate reduction by a *Pseudomonas* species: a new mode of anaerobic respiration. *FEMS Microbiology Letters*, **61**(12), 195–198.

Macy J. M., Rech S. A., Dorsch M., Stackebrandt E. and Sly L. I. (1993). Thauera selenatis gen. nov., sp. nov., a Member of the Beta Subclass of Proteobacteria with a Novel Type of Anaerobic Respiration. International Journal of Systematic Bacteriology, **43**(1), 135–142.

Maltman C., Walter G. and Yurkov V (2016). A diverse community of metal(loid) oxide respiring bacteria is associated with tube worms in the vicinity of the Juan de Fuca Ridge Black Smoker Field. *PLoS One*, **11**, e0149812.

Marinescu G., Stoicescu A. G. and Teodorof L. (2011). Industrial nutrient medium use for yeast selenium preparation. *Food & Technology*, **35**, 45–53.

Martínková L., Uhnáková B., Pátek M., Nesvera J. and Kren V. (2009). Biodegradation potential of the genus *Rhodococcus*. *Environmental International*, **35**(1), 162–177.

McCarty S., Chasteen T., Marshall M., Fall R. and Bachofen R. (1993). Phototrophic bacteria produce volatile, methylated sulfur and selenium compounds. *FEMS Microbiology Letters*, **112**(1), 93–98.

Menon S., Devi Ks S., Santhiya R., Rajeshkumar S. and Venkat Kumar S. O. (2018). Selenium nanoparticles: A potent chemotherapeutic agent and an elucidation of its mechanism. *Colloids Surfaces B: Biointerfaces*, **170**, 280–292.

Monych N., Gugala N. and Turner R. J. (2019). Metal based antimicrobials. In: Biomaterials Science series No. 5. Antimicrobial Materials for Biomedical Applications, A. J. Domb, K. R. Kunduru and F. Shady (eds.), Royal Society of Chemistry Press. Chapter 9, pp. 252–276. ISBN: 978-1-78801-188-4.

More T. T., Yadav J. S. S., Yan S., Tyagi R. D. and Surampalli R. Y. (2014). Extracellular polymeric substances of bacteria and their potential environmental applications. *Journal of Environmental Management*, **144**, 1–25.

Morenoreyes R., Suentes C., Mathieu F., Bagaux F., Zhu D., Rivera M. T., Boelaert M., Nève J., Perlmutter N. and Vabderpas J. (1998). Kashin-beck osteoarthropathy in rural tibet in relation to selenium and iodine status. *The New England Journal of Medicine*, **339**(16), 1112–1120.

Nancharaiah Y. V. and Lens P. N. L. (2015). Ecology and Biotechnology of Selenium-respiring bacteria. *Microbiology and Molecular Biology Reviews*, **79**, 61–80.

NIH. (2002). Research on microbial biofilms National Heart, Lung and Blood Institute. PA-03-047.

Oremland R. S., Herbel M. J., Blum J. S., Langley S., Beveridge T. J., Ajayan P. M., Sutto T., Ellis A. V. and Curran S. (2004). Structural and spectral features of selenium nanospheres produced by Se-respiring bacteria. *Applied Environmental Microbiology*, **70**(1), 52–60.

Painter E. P. (1941) The chemistry and toxicity of selenium compounds, with special reference to the selenium problem. *Chemical Reviews*, **28**, 179–213.

Pettine M., Gennari F. and Campanella L. (2013). The reaction of selenium (IV) with ascorbic acid: Its relevance in aqueous and soil systems. *Chemosphere*, **90**, 245–250.

Piacenza E., Presentato A., Zonaro E., Lemire J., Demeter M., Vallini G., Turner R. J. and Lampis S. (2017) Antimicrobial activity of biogenically produced spherical Se-nanomaterials embedded in organic material against *Pseudomonas aeruginosa* and *Staphylococcus aureus* strains on hydroxyapatite-coated surfaces. *Microbial Biotechnology*, **10**, 804–818.

Piacenza E., Presentato A. and Turner R. J. (2018a). Stability of biogenic metal(loid) nanomaterials related to the colloidal stabilization theory of chemical nanostructures. *Critical Review in Biotechnology*, **38**(8), 1137–1156.

Piacenza E., Presentato A., Ambrosi E., Speghini A., Turner R. J., Vallini G. and Lampis S. (2018b). Physical-chemical properties of biogenic selenium nanostructures produced by *Stenotrophomonas maltophilia* SeITE02 and *Ochrobactrum* sp. MPV1. *Frontiers in Microbiology*, **9**, 3178.

Piacenza E., Presentato A., Bardelli M., Lampis S., Vallini G. and Turner R. J. (2019). Influence of bacterial physiology on the processing of selenite, biogenesis of nanomaterials and their thermodynamic stability. *Molecules*, **24**, 2532.

Poluboyarinov P. A., Vikhreva V. A., Leshchenko P. P., Aripovskii A. V. and Likhachev A. N. (2009). Elemental selenium formation upon destruction of the organoselenium compound DAFS-25 molecule by growing fungal mycelium. *Moscow University Biological Sciences Bulletin*, **64**, 164–168.

Presentato A., Piacenza E., Anikovskiy M., Cappelletti M., Zannoni D. and Turner R. J. (2016). Rhodococcus aetherivorans BCP1 as cell factory for the production of intracellular tellurium nanorods under aerobic conditions. *Microbial Cell Factories*, **15**, 204.

Presentato A., Piacenza E., Anikovskiy M., Cappelletti M., Zannoni D. and Turner R. J. (2018a). Biosynthesis of selenium-nanoparticles and -nanorods as a product of selenite bioconversion by the aerobic bacterium *Rhodococcus aetherivorans* BCP1. *New Biotechnology*, **41**, 1–8.

Presentato A., Piacenza E., Darbandi A., Anikovskiy M., Cappelletti M., Zannoni D. and Turner R. J. (2018b). Assembly growth and conductive properties of tellurium nanorods produced by *Rhodococcus aetherivorans* BCP1. *Scientific Reports*, **8**, 3923.

Presser T. S. and Ohlendorf H. M. (1987). Biogeochemical cycling of selenium in the San Joaquin Valley, *California*. Environonmental Management, **11**(6), 805–821.

Quintana M., Haro-Poniatowski E., Morales J. and Batina N. (2002). Synthesis of selenium nanoparticles by pulsed laser ablation. *Applied Surface Science*, **195**, 175–186.

Raghunath A. and Perumal E. (2017). Metal oxide nanoparticles as antimicrobial agents: a promise for the future. *International Journal of Antimicrobial Agents*, **49**, 137–152.

Ramamurthy C. H., Sampath K. S., Arunkumar P., Suresh Kumar M., Sujatha V., Premkumar K. and Thirunavukkarasu C. (2013). Green Synthesis and characterization of Selenium nanoparticles and its augmented cytotoxicity with doxorubicin on cancer cells. *Bioprocess and Biosystems Engineering*, **36**, 1131–1139.

Ramya S., Shanmugasundaram T. and Balagurunathan R. (2019). Actinobacterial enzyme mediated synthesis of selenium nanoparticles for antibacterial, mosquito larvicidal and anthelminthic applications. *Particulate Science and Technology*, **38**, 63–72.

Rao C. N. R., Muller A. and Cheetham A. K. (eds.) (2004). The Chemistry of Nanomaterials: Synthesis, Properties and Applications. Wiley-VCH Verlag GmbH & Co. KGaA, Weinheim.

Rayman M. P. (2012). Selenium and human health. *The Lancet*, **379**, 1256–1268.

Rech S. A. and Macy J. M. (1992). The Terminal Reductases for Selenate and Nitrate Respiration in *Thauera selenatis* are two distinct enzymes. *Journal of Bacteriology*, **174**, 7316–7320.

Rudramurthy G. R., Swamy M. K., Sinniah U. R. and Ghasemzadeh A. (2016). Nanoparticles: Alternatives against drug-resistant pathogenic microbes. *Molecules*, **21**, 836.

Sabaty M., Avazeri C., Pignol D. and Vermeglio A. (2001). Characterization of the reduction of selenite and tellurite by nitrate reductases. *Applied Environmental Microbiology*, **67**, 5122–5126.

Sanchez-Lopez E., Gomes D., Esteruelas G., Bonilla Lopez-Machado A. L., Galindo R., Cano A., Espina M., Ettcheto M., Camins A., Silva A. M., Durazzo A., Santini A., Garcia M. L. and Souto E. B. (2020). zMetal-based nanoparticles as antimicrobial agents: an overview. *Nanomaterials*, **10**, 292.

Sarathchandra S. U. and Watkinson J. H. (1981). Oxidation of elemental selenium to selenite by *Bacillus megaterium*. *Science*, **211**(4482), 600–601.

Sarkar J., Dey P., Saha S. and Acharya K. (2011). Mycosynthesis of selenium nanoparticles. *Micro & Nano Letters*, **6**, 599–602.

Schröder I., Rech S., Kraffy T. and Macy J. M. (1997). Purification and Characterization of the Selenate Reductase from *Thauera selenatis*. *Journal Biological Chemistry*, **272**(38), 23765–23768.

Shah C. P., Singh K. K., Kumar M. and Bajaj P. N. (2010). Vinyl monomers-induced synthesis of polyvinyl alcohol-stabilized selenium nanoparticles. *Materials Research Bulletin*, **45**(1), 56–62.

Sharma D., Kanchi S. and Bisetty K. (2015). Biogenic synthesis of nanoparticles: A review. *Arabian Journal of Chemistry*, **12**(8), 3576–3600.

Sharma V. K., McDonald T. J., Sohn M., Anquandah G. A. K., Pettine M. and Zboril R. (2017). Assessment of toxicity of selenium and cadmium selenium quantum dots: a review. *Chemosphere*, **188**, 403–413.

Sheng G. P., Yu H. Q. and Li X. Y. (2010). Extracellular polymeric substances (EPS) of microbial aggregates in biological wastewater treatment systems: a review. *Biotechnology Advances*, **28**(6), 882–894.

Shoeibi S., Mozdziak P. and Golkar-Narenji A. (2017). Biogenesis of selenium nanoparticles using green chemistry. *Top Curr Chem (Z)*, **375**, 88.

Siresj A. K. (2014). Extracellular bio-production and characterization of small monodispersed CdSe quantum dot nanocrystallites. *Spectrochimica Acta PartA: Molecular and Biomolecular Spectroscopy*, **130**, 334–349.

Soflaei S., Dalimi A., Abdoli A., Kamali M., Nasiri V., Shakibaie M. and Tat M. (2014). Anti-leishmanial activities of selenium nanoparticles and selenium dioxide on *Leishmania infantum*. *Comparative Clinical Pathology*, **23**, 15–20.

Song J. Y. and Kim B. S. (2009). Rapid biological synthesis of silver nanoparticles using plant leaf extracts. *Bioprocess and Biosystems Engineering*, **32**(1), 79–84.

Sownadarya P., Ramkumar G. and Shivakumar M. S. (2017). Green synthesis of selenium nanoparticles conjugated *Clausena dentata* plant leaf extract and their insecticidal potential against mosquito vectors. *Artif Cells Nanomedicine Biotechnology*, **45**, 1490–1495.

Srivastava N. and Mukhopadhyay M. (2013). Biosynthesis and structural characterization of selenium nanoparticles mediated by *Zooglea ramigera*. *Powder Technology*, **244**, 26–29.

Stolz J. F., Basu P., Santini J. M. and Oremland R. S. (2006). Arsenic and selenium in microbial metabolism. *Annual Review of Microbiology*, **60**(1), 107–130.

Stolz J. F. and Oremland R. S. (1999). Bacterial respiration of arsenic and selenium. *FEMS Microbiology Reviews*, **23**, 615–627.

Suresh A. K. (2014). Extracellular bio-production and characterization of small monodispersed CdSe quantum dot nanocrystallites. *Spectrochimica Acta. Part A, Molecular and Biomolecular Spectroscopy*, **130**, 344–349.

Suresh K., Prabagaran S. R., Sengupta S. and Shivaji S. (2004). Bacillus indicus sp. nov., an arsenic-resistant bacterium isolated from an aquifer in West Bengal, India. *International Journal of Systematic and Evolutionary Microbiology*, **54**(4), 1369–1375.

Tan L. C., Nancharaiah Y. V., van Hullebusch E. D. and Lens P. N. L. (2016). Selenium: environmental significance, pollution, and biological treatment technologies. *Biotechnology Advances*, **34**, 886–907.

Thakkar K. N., Mhatre S. S. and Parikh R. Y. (2010). Biological synthesis of metallic nanoparticles. *Nanomedicine*, **6**, 257–262.

Tugarova A. V. and Kamnev A. A. (2017) Proteins in microbial synthesis of selenium nanoparticles. *Talanta*, **174**, 539–547.

Turner R. J. (2017). Metal-based antimicrobial strategies. *Microbial Biotechnology*, **10**, 1062–1065.

Turner R. J., Weiner J. H. and Taylor D. E. (1998). Selenium Metabolism in *Escherichia coli. BioMetals*, **11**, 223–227.

Verma P. (2015). A review on synthesis and their antibacterial activity of silver and selenium nanoparticles against biofilm forming Staphylococcus aureus. *World Journal of Pharmacy and Pharmaceutical Sciences*, **4**, 652–677.

Vetchinkina E., Loshchinima E., Kupryashina M., Burov A. and Nikitina V. (2019). Shape and size diversity of gold, silver, selenium, and silica nanoparticles prepared by green synthesis using fungi and bacteria. *Industrial & Engineering Chemistry Reaserch*, **58** (37), 17207–17218.

Vetchinkina E., Loshchinina E., Kursky V. and Nikitina V. (2013). Reduction of organic and inorganic selenium compounds by the edible medicinal basidiomycete *Lentinula edodes* and the accumulation of elemental selenium nanoparticles in its mycelium. *Journal of Microbiology*, **51**, 829–835.

Vimbela G. V., Ngo S. M., Fraze C., Yang L. and Stout D. A. (2017). Antibacterial properties and toxicity from metallic nanomaterials. *International Journal of Nanomedicine*, **12**, 3941–3965.

Wadhwani S. A., Shedbalkar U. U., Singh R. and Chopade B. A. (2016). Biogenic selenium nanoparticles: current status and future prospects. *Applied Microbiology and Biotechnology*, **100**, 2555–2566.

Wang T., Yang L., Zhang B. and Liu J. (2010). Extracellular biosynthesis and transformation of selenium nanoparticles and application in H_2O_2 biosensor. *Colloids Surface B: Biointerfaces*, **80**(1), 94–102.

Wang X., Liu G., Zhou J., Wang J., Jin R. and Lv H. (2011). Quinone-mediated reduction of selenite and tellurite by *Escherichia coli. Bioresource Technology*, **102**, 3268–3271.

Wang L., Hu C. and Shao L. (2017). The antimicrobial activity of nanoparticles: present situation and prospects for the future. *International Journal of Nanomedicine*, **12**, 1227–1249.

Wangeline A. L., Valdez J. R., Lindblom S. D., Bowling K. L., Reeves F. B. and Pilon-Smits E. A. (2011). Characterization of rhizosphere fungi from selenium hyper accumulator and non hyper accumulator plants along the eastern Rocky Mountain Front Range. *American Journal of Botany*, **98**, 1139–1147.

Watanabe C. (2002). Modification of mercury toxicity by selenium: practical importance? *Tohoku Journal Experimental Medicine*, **196**, 71–77.

Watts C. A., Ridley H., Dridge E. J., Leaver J. T., Reilly A. J., Richardson D. J. and Butler C. S. (2005). Microbial reduction of selenate and nitrate: common themes and variations. *Biochemical Society Transactions*, **33**, 173–175.

Weres O., Jaouni A. R. and Tsao L. (1989). The distribution, speciation and geochemical cycling of selenium in a sedimentary environment, Kesterson Reservoir, California, U.S.A. *Applied Geochemistry*, **4**(6), 543–563.

Wu L. (2004). Review of 15 years of research on ecotoxicology and remediation of land contaminated by agricultural drainage sediment rich in selenium. *Ecotoxicology Environmental Safety*, **57**, 257–269.

Xu D., Yang L., Wang Y., Wang G., Rensing C. and Zheng S. (2018). Proteins enriched in charged amino acids control the formation and stabilization of selenium nanoparticles in *Comamonas testosteroni* S44. *Scientific Reports*, **8**(1), 4766.

Yah C. S. and Simate G. S. (2015). Nanoparticles as potential new generation broad spectrum antimicrobial agents. *DARU Journal Pharmaceutical Sciences*, **23**, 43.

Yuwen L. and Wang L. (2013). Nanoparticles and Quantum Dots. In: Handbook of Chalcogen Chemistry: New Perspectives in Sulfur, Selenium and Tellurium, 2nd edn, F. Devillanova and W. W. Du Mont (eds.), The Royal Society of Chemistry, Cambridge, United Kingdom, pp. 232–260.

Zannoni D., Borsetti F., Harrison J. J. and Turner R. J. (2008). The bacterial response to the chalcogen metalloids Se and Te. *Advances in Microbial Physiology*, **53**, 1–71.

Zare B., Babaie S., Setayesh N. and Shahverdi A. R. (2013). Isolation and characterization of a fungus for extracellular synthesis of small selenium nanoparticles. *Nanomedicine*, **1**, 13–19.

Zhang B., Ye X., Dai W., Hou W., Zuo F. and Xie Y. (2006). Biomolecule-assisted synthesis of single-crystalline selenium nanowires and nanoribbons via a novel flake-cracking mechanism. *Nanotechnology*, **17**(2), 385–390.

Zhang S., Rocourt C. and Cheng W. H. (2010a). Selenoproteins and the aging brain. *Mechanism of Ageing and Development*, **131**(4), 253–260.

Zhang Y., Wang J. and Zhang L. (2010b). Creation of highly stable selenium nanoparticles capped with hyperbranched polysaccharide in water. *Langmuir*, **26**(22), 17617–17623.

Zhang W., Chen Z., Liu H., Zhang L., Gao P. and Li D. (2011). Biosynthesis and structural characteristics of selenium nanoparticles by *Pseudomonas alcaliphila*. *Colloids Surface B: Biointerfaces*, **88**(1):196–201.

Zheng S., Li X., Zhang Y., Xie Q., Wong Y. S., Zheng W. and Chen T. (2012). PEG-nanolized ultrasmall selenium nanoparticles overcome drug resistance in hepatocellular carcinoma HepG2 cells through induction of mitochondria dysfunction. *International Journal of Nanomedicine*, **7**, 3939–3949.

Zonaro E., Lampis S., Qazi S. J. S., Turner R. J. and Vallini G. (2015). Biogenic selenium and tellurium nanoparticles synthesized by environmental microbial isolates efficaciously inhibit bacterial planktonic cultures and biofilms. *Frontiers Microbiology*, **6**, 584.

Chapter 11

Biological synthesis of metal selenide nanoparticles and their applications

Raj Mohan Balakrishnan and Vrushali Vinayak Kadam

11.1 INTRODUCTION

Semiconductors are characterized by an electronic band structure in which the highest-occupied energy band, called the valence band (VB), and the lowest-occupied energy band, called the conduction band (CB), are separated by a bandgap, that is, a region of forbidden energies (Liqiang *et al.*, 2006). In recent years, semiconductor nanostructures have gained interest due to their applications in various fields including solar cells, laser technology, light-emitting diodes, nanoscale electronic devices, components of armor, parts of automobiles, superabsorbents, packaging films, catalysis, and biosensors. Semiconductor nanocrystals such as silicon and germanium belong to group IV; gallium nitride (GaN), gallium phosphide (GaP), gallium arsenide (GaAs), indium phosphide (InP), and indium arsenide (InAs) belong to group III–V, while zinc oxide (ZnO), zinc sulfide (ZnS), cadmium sulfide (CdS), cadmium selenide (CdSe) and cadmium telluride (CdTe) belong to group II–VI of the periodic table (Suresh, 2013).

Among the semiconductors, selenium (Se) is an attractive material with intriguing physical and chemical properties. It has photoconductive features with

doi: 10.2166/9781789061055_0323

a high refractive index, which are useful for application in photodetectors and sensors. Selenite or Se amino acids enhance the reducing ability of glutathione (GSH) in model systems, and in the presence of Se, the SH groups in GSH peroxidase change their redox state during the catalytic process, thus inducing the antioxidant potential in the biological systems (Rotruck *et al.*, 1973). Its high reactivity with other chemicals enables Se to be a versatile template for various monodisperse (metal-selenide) colloids (Park *et al.*, 2015). The metal selenides encompass an important class of chalcogenide semiconductor nanostructures. Metal selenide semiconductors have a narrower bandgap than the corresponding metal (oxy)sulfides due to shallower valence bands being formed by Se 4p orbitals than those formed by the O 2p and S 3p orbitals. Besides their narrow bandgap, their other features of stability and flexible band structure together make them extensively investigated in the solar energy conversion field, such as photovoltaic and photoelectrochemical cells (Chen *et al.*, 2019).

Among the metal selenides, CdSe, zinc selenide (ZnSe), lead selenide (PbSe), copper selenide (CuSe), mercury selenide (HgSe), and indium selenide (In_2Se_3) are the most studied classes of materials having application in the field of light-emitting diodes (LEDs), photovoltaics, biological imaging, thermoelectric devices or single-electron transistors (Figure 11.1). Current research on applications of metal selenide nanostructures in the material science and biomedical fields (Saikia *et al.*, 2013; Zhao *et al.*, 2018) has gained enormous importance in recent years.

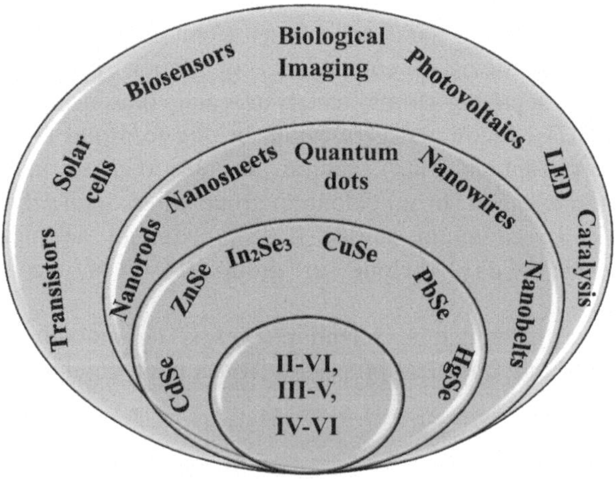

Figure 11.1 Schematic diagram of metal selenide nanostructures and their applications.

Various chemical methods are used for the synthesis of metal selenides, such as solvothermal (Xie *et al.*, 2000), chemical bath deposition (Bakiyaraj & Dhanasekaran, 2013), hydrothermal (Terra *et al.*, 2010), sonochemical (Esmaeili-zare *et al.*, 2012), microwave-assisted (Luo *et al.*, 2014) and physical vapor deposition (Zhou *et al.*, 2015). Physical and chemical methods lead to the synthesis of stable, size-tunable nanostructures with defined properties. However, the use of extensive reaction conditions such as high temperature and pressure, high energy equipment, as well as the use of toxic chemicals and reagents limits their practical applicability. In contrast, the biological mediated route of nanoparticle synthesis is simple, fast, environmentally benign, and cost-effective (Ingale & Chaudhari, 2013; Pantidos & Horsfall, 2014). To date, numerous biological entities such as fungi, bacteria, yeasts and plant extracts have been employed for the synthesis of metal nanoparticles (Figure 11.2).

Several reviews on the chemical and biological synthesis of metallic nanoparticles have been published. On the other hand, the information about the biogenic route for solely metal selenide nanostructure synthesis and its application is scattered. Several reviews of the biological synthesis of chalcogenide semiconductor nanoparticles have been published (Jacob *et al.*, 2015; Lloyd *et al.*, 2011; Mal *et al.*, 2016). The present chapter deals explicitly with the advancement in the biological synthesis of metal selenide nanostructures and their applications.

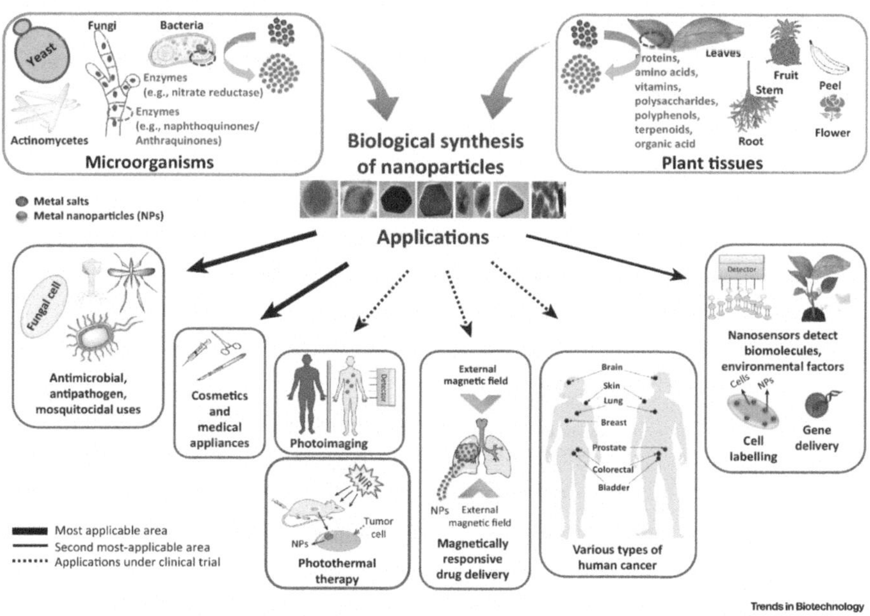

Figure 11.2 Biological synthesis and applications of metal nanoparticles in biomedical and environmental fields (Singh *et al.*, 2016).

11.2 BIOLOGICAL SYNTHESIS OF SELENIUM NANOSTRUCTURES

11.2.1 Selenium geochemistry

Se is not only a strategic element in high-tech electronics and an essential trace element in living organisms, it is also a potential toxin with low threshold concentrations, as represented in Figure 11.3.

Se is present in the environment in organic and inorganic forms (Figure 11.4). The Se distribution in the environment is mainly accounted from (i) utilization of fossil fuels via coal mining, burning, oil refining, thermal power plant effluents and fly ash; (ii) industrialization including metal smelting, production of glass, pigments, varnishes, semiconductors and photocopy machines and (iii) biogeochemical sources such as volcanic activity, weathering and precipitation of minerals and irrigated agriculture on seleniferous soils (Kora & Rastogi, 2016a). Environmental biotechnological applications using bacterial biomineralization have the potential not only to remove Se from contaminated waters but also to sequester it in a reusable form (Nancharaiah & Lens 2015a).

Selenate is the most oxidized and bioavailable form of Se. Reduction of selenate is one way to remove Se from polluted water since selenite adsorbs more readily to surfaces and can be reduced further to elemental Se^0 or selenides, which are less biologically available (Shamberger, 1981; Subedi *et al.*, 2017). Moreover, selenides can form metal precipitates. Biosynthesis of Se^0 or metal selenide can lead to metal deposit formation and is one of the biotechnological bases of bioremediation strategies to alleviate Se pollution (Mal *et al.*, 2016; Nancharaiah & Lens, 2015b; Wang *et al.*, 2018).

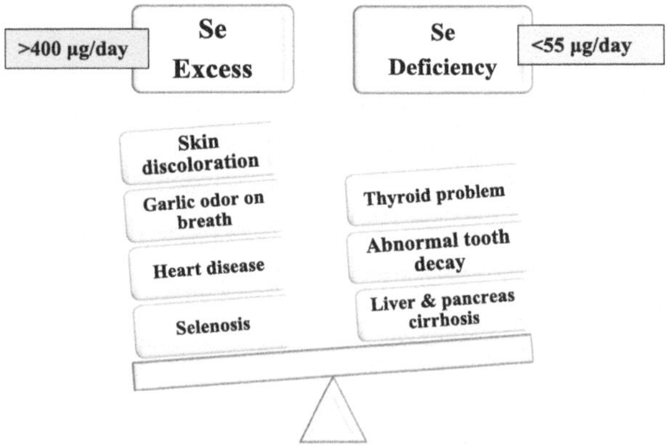

Figure 11.3 Effects of selenium deficiency and excess to human health.

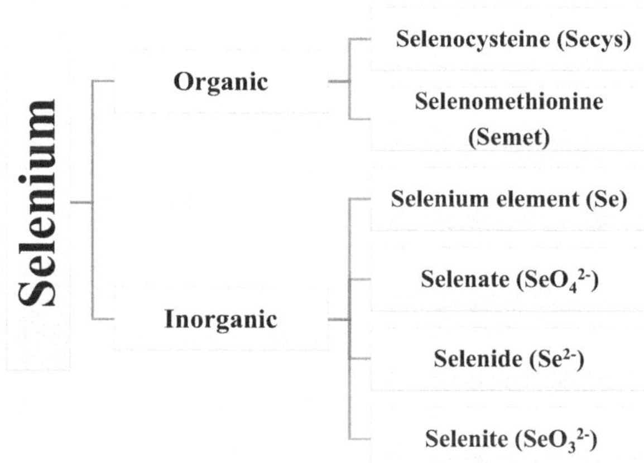

Figure 11.4 Organic and inorganic forms of Se.

11.2.2 Nanoscale elemental selenium formation

11.2.2.1 Bacteria

Table 11.1 presents bacterial species as possible nanofactories using Na_2SeO_3 as a precursor salt. *P. aeruginosa* can reduce the toxic, biologically available, soluble, colorless selenite into non-toxic, biologically unavailable, insoluble red elemental selenium (Kora & Rastogi, 2016b). Elemental selenium particles are stabilized by their natural protein coating extracted from cultures of *Staphylococcus carnosus*, which exhibit considerable activity against the nematode *Steinernemafeltiae, Escherichia coli* and *Saccaromyces cerevisiae* (Estevam *et al.*, 2017).

Srivastava and Mukhopadhyay (2015a) reported the reduction of SeO_3^{2-} to Se^0 during 48 h incubation using the bacterium *Ralstonia eutropha.* The reduction of selenium ions into elemental selenium with a size range of 20–150 nm was observed with lactic acid bacteria within 48 h of incubation. Lactic acid bacteria, namely *Lactobacillus acidophilus, Lactobacillus plantarum* and *Lactobacillus rhamnosus* exhibit significant antifungal activity against pathogenic fungi (Radhika & Gayathri 2015). Various cell factories for chalcogen-based nanomaterial synthesis are *Rhodococcus aetherivorans* BCP1 (Presentato *et al.*, 2018), *Bacillus subtilis* (Chandramohan *et al.*, 2018), *Stenotrophomonas maltophilia* (Zonaro *et al.*, 2015), *Escherichia coli* ATCC 35218 (Kora & Rastogi, 2016a), *B. mycoides* SeITE01 and *S. maltophilia* SeITE02 (Cremonini *et al.*, 2016).

Nguyen *et al.* (2016) isolated a new strain of the *Cronobacter* genus from domestic wastewater, which belongs to the γ-*Proteobacteria* class. The isolated strain exhibits heterotrophic and electrotrophic selenite reduction under microaerobic and anaerobic conditions, respectively (Nguyen *et al.*, 2016). A humic substance analog, anthraquinone-2,6-disulfonate, enhances selenite

Table 11.1 Synthesis of nanoscale elemental selenium by bacteria.

Microorganism	Size (nm)	Site of Biosynthesis	Application	References
P. aeruginosa	47–165	Extracellular	–	Kora and Rastogi (2016b)
Staphylococcus carnosus	439	Intracellular	Antinematode and antimicrobial	Estevam *et al.* (2017)
Ralstonia eutropha	40–120	Extracellular	Antimicrobial	Srivastava and Mukhopadhyay (2015a)
Azospirillum brasilense	400	Extracellular	–	Vogel *et al.* (2018)
Azoarcus sp. CIB	141 ± 37	Extracellular	–	Fernández-Llamosas *et al.* (2016)
Proteus mirabilis YC801	178.3 ± 11.5	Extracellular	–	Wang *et al.* (2018)
Streptomyces sp. M10A65	20–150	Extracellular	Antibacterial, synergistic, larvicidal and anthelminthic	Ramya *et al.* (2020)

biotransformation into Se^0 using the bacterium *Anaeromyxobacter dehalogenans*. The abundance of humic substances in the natural environment can then be exploited to design control strategies for Se pollution (He & Yao, 2011). *Burkholderia fungorum* 95 has been enriched from inner tissues of hybrid poplars grown in soil contaminated by polycyclic aromatic hydrocarbons and *Burkholderia fungorum* DBT1 has been isolated from an oil refinery drainage. The isolated strains showed effective biotransformation of toxic selenite into selenium nanoparticles. Formation of elemental selenium nanoparticles by this bacterium under aerobic conditions was attributed to cytoplasmic enzymatic activation mediated by electron donors (Khoei *et al.*, 2017). *Alcaligenes sp.* CKCr-6A cell-free filtrate exhibits nicotinamide adenine dinucleotide hydrogen (NADH)-dependent reduction of selenate and selenite to elemental Se^0 (Mesbahi-nowrouzi & Mollania, 2018).

11.2.2.2 Fungi

Fungal species such as *Aspergillus oryzae* (Mosallam *et al.*, 2018), *Gliocladium roseum* (Srivastava & Mukhopadhyay, 2015b), *Pichia pastoris* (Elahian *et al.*, 2017), *Mariannaea sp.* HJ (Zhang *et al.*, 2019), and *Trichoderma atroviride* (Joshi *et al.*, 2019) have been exploited for Se remediation by forming nanoselenium. *P. chrysosporium* is able to biotransform selenite but not the selenate to nano-Se. At pH 4.5, *P. chrysosporium* exhibited 40% and 10% of Se removal when incubated with selenite and selenate, respectively (Espinosa-ortiz *et al.*, 2015). Rosenfeld *et al.* (2017) studied the fungal growth, aerobic selenium (IV–VI) reduction, selenium immobilization, and volatilization using six ascomycetes fungi, viz. *Pyrenochaeta sp., Acremonium strictum, Plectosphaerella cucumerina, Stagonospora sp., Alternaria alternate* and *Paraconiothyrium sporulosum*. All the fungi showed 85–93% removal of dissolved Se(IV) within 10 d in the presence of 0.01 mM Se(IV), while only about 20–30% Se(VI) was removed when grown with 0.01 mM Se(VI). All the fungal species produced spherical Se^0 nanoparticles typically of 50 to 300 nm diameter (Rosenfeld *et al.*, 2017).

11.2.3 Nanoscale metal selenide nanoparticles

Researchers have exploited bacteria, fungi, yeast and algae to synthesize metal selenide nanoparticles (Table 11.2). Fungi and bacteria possess various biomolecules that act as reducing agents, having the ability to reduce Se oxyanions to elemental Se^0 and finally to Se^{2-}. Further, the microbially produced Se^{2-} reacts with dissolved metal ions (e.g., Zn, Cd, Pb, Hg, Cu, and In) to form metal selenide nanoparticles. Thus, surface interactions between the biologically reducing agents and precursors exhibit enhanced selectivity and improved stability compared with other physical and chemical methods.

Table 11.2 Microbial mediated synthesis of metal selenides

Microorganism	Precursor	Metal Selenide	References
Bacteria			
Pseudomonas aeruginosa strain RB	$CdCl_2$, Na_2SeO_3	CdSe	Ayano *et al.* (2014)
Shewanella putrefaciens	selenite, $HgCl_2$, pyruvate	HgSe	Ho *et al.* (2015)
Providencia vermicola BGRW	SeO_2, $CdCl_2$	CdSe	Abou-assy *et al.* (2019)
Pantoea agglomerans	$CuSO_4{\cdot}5H_2O$	Cu_2Se	Yue *et al.* (2016)
Escherichia coli	$EuCl_3{\cdot}6H_2O$, SeO_2	EuSe	Kim *et al.* (2016)
Escherichia coli	$CdCl_2$, Na_2SeO_3	CdSe	Yan *et al.* (2014)
Escherichia coli	$CdCl_2$, Na_2SeO_3	CdSe	Xu *et al.* (2019)
Bacillus cereus	Selenite, $PbNO_3$	PbSe	Che *et al.* (2019)
	$AgNO_3$	Ag_2Se	
	$Bi(NO_3)_3$	Bi_2Se_3	
Pseudomonas stutzeri NT-I, Pseudomonas sp. RB, Stenotrophomonas maltophilia TI-1, Ochrobactrum anthropi TI-2, O. anthropi TI-3	Na_2SeO_3, $BiCl_3$	Bi_2Se_3	Kuroda *et al.* (2019)
Pseudomonas stutzeri TS44	$CdCl_2$, Na_2SeO_3	CdSe	Wang *et al.* (2019)
	$CuSO_4$, Na_2SeO_3	CuSe	Cak *et al.* (2019)
	$ZnSO_4$, Na_2SeO_3	ZnSe	

Fungi			
Fusarium Oxysporum	$CdCl_2$, $SeCl_4$	CdSe	Kumar *et al.* (2007)
Helminthosporum solani	$CdCl_2$, $SeCl_4$	CdSe	Suresh (2014)
A. terreus	$PbNO_3$, Na_2SeO_4	PbSe	Jacob *et al.* (2014a, b)
Yeast			
Saccharomyces cerevisiae	Na_2SeO_3, $CdCl_2$	CdSe	Brooks and Lefebvre (2017)
Rhodotorula mucilaginosa PA-1	Na_2SeO_3, $CdCl_2$	CdSe	Cao *et al.* (2020)
Saccharomyces cerevisiae BY4742	Na_2SeO_3, $CdCl_2$	CdSe	Luo *et al.* (2014)
Saccharomyces cerevisiae	Na_2SeO_3, $CdCl_2$	CdSe	Shao *et al.* (2018)
Saccharomyces cerevisiae	Na_2SeO_3, $CdCl_2$	CdSe	Wu *et al.* (2015)
Candida utilis WSH02-08	Na_2SeO_3, $CdCl_2$	CdSe	Tian *et al.* (2017)
Algae			
C. pyrenoidosa and S. obliquus	Na_2SeO_3, $Cd\ (NO_3)_2$	CdSe	Zhang *et al.* (2018)

11.2.3.1 Bacterial mediated metal selenide nanoparticles production

Bacterial mediated synthesis of nanoparticles is an advanced alternative to conventional chemical and physical methods and emerged as a rapidly developing research area in bionanotechnology across the world. Compared to other microbial entities, bacteria confer additional advantages in terms of genetic manipulation for overexpression of specific enzymes responsible for the biomineralization of metal ions (Gahlawat & Choudhury, 2019). *Bacillus cereus* CC-1 was isolated from marine sediments and quantitative real-time polymerase chain reaction (qPCR) was carried out to identify the putative genes involved in Se oxyanion reduction. The induction of these genes was determined by supplementing selenite in the medium. They found that selenite reduction preferably takes place in weakly acidic and neutral conditions. Several reduction mechanisms or a specific Se nanoparticle efflux pump exists in strain CC-1, leading to both extracellular and intracellular synthesis of Se nanoparticles. Additionally, the isolated strain CC-1 can produce PbSe, Ag_2Se, and Bi_2Se_3 nanoparticles when supplemented with these heavy metals along with selenite in the culture medium (Che *et al.*, 2019).

Abou-assy *et al.* (2019) reported synthesis of cubic CdSe quantum dots (QDs) of 2 to 4 nm using *Providencia vermicola* BGRW under the optimized conditions of 0.1 mM SeO_2:0.9 mM $CdCl_2$, 37°C, pH 9 within 24 h (Abou-assy *et al.*, 2019). The effect of reaction variables such as growth period, precursor concentration, reaction time, and exposure to the biological synthesis of CdSe nanoparticles was investigated using fluorescence intensity, as shown in Figure 11.5.

Fellowes *et al.* (2013) reported the biotic and abiotic synthesis of Se and metal selenides. For biotic Se QDs synthesis, an anaerobic culture of *Veillonella atypica* was used, while borohydrate reduction was used for abiotic synthesis. Biogenic Se(II-) was produced by the reduction of Se(IV) by *Veillonella atypica*

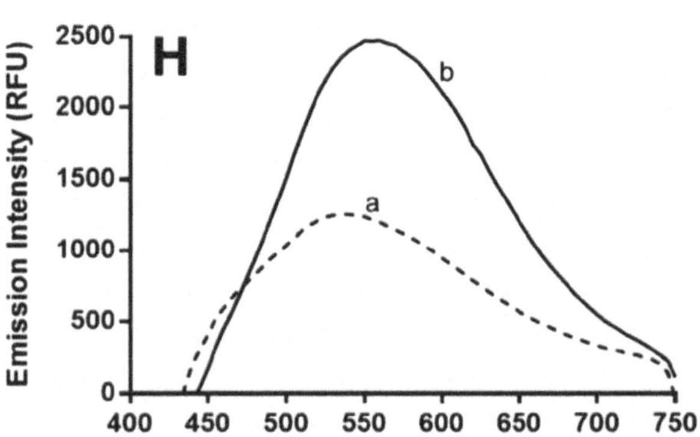

Figure 11.5 Optimized synthesis of CdSe QD in *S. cerevisiae* (a) basic protocol after 60 h and (b) optimized protocol after 84 h (Brooks & Lefebvre, 2017).

and compared directly against borohydride-reduced Se(IV) for the production of glutathione-stabilized CdSe and β-mercaptoethanol-stabilized ZnSe nanoparticles by aqueous synthesis. Biotic QDs exhibited a decreased initial reaction rate and resulted in the narrow size distribution of metal selenides, as biotic Se QDs regulate the rate of nucleation and growth of metal selenides. X-ray absorption spectroscopy (XAS) analysis was carried out to understand the stability of the Se QDs synthesized by biotic and abiotic routes. Biogenic Se QDs retained their reduced state by inhibiting oxidation, whereas abiotic Se QDs had undergone the oxidation. The protein part, the alpha-subunit of methylmalonyl-CoA decarboxylase, involved in the QDs synthesis and stabilization limits the oxidation of the biologically produced Se QDs (Fellowes *et al.*, 2013).

CdSe QDs were synthesized extracellularly using *E. coli*, exhibiting strong fluorescence emission with the addition of 80 mg of mercaptosuccinic acid. The bioimaging and bio-labeling potential of biosynthesized CdSe QDs was explored by incorporating them into the yeast cells (Xu *et al.*, 2019). HgSe QDs were produced and located on the membrane surface of *Shewanella* cells by incubating *Shewanella putrefaciens* 200 with Hg(II) and Se(IV) under anaerobic conditions. Liquid mercury reacts directly with Se nanowires or stellated polyhedral structures in water for coating the Se nanostructures, thereby forming HgSe core-shell structures after 12 hours of incubation at ambient temperature (Ho *et al.*, 2015). *In vivo* synthesis of europium selenide (EuSe) nanoparticles was performed using recombinant *E. coli* cells expressing heavy-metal binding proteins, phytochelatin synthase and metallothionein. The synthesized EuSe nanoparticles exhibited high fluorescence intensities as well as strong magnetic properties (Kim *et al.*, 2016).

Cak *et al.* (2019) reported the bacterial mediated synthesis of ZnSe and CuSe nanostructures and further fabricated Au/ZnSe/p-Si/Al and Au/CuSe/p-Si/Al diode devices as interfacial thin films on p-Si substrates. X-ray diffraction (XRD) analysis of ZnSe/p-Si and CuSe/p-Si thin film revealed cubic and monoclinic hexagonal crystal structures, respectively (Cak *et al.*, 2019).

11.2.3.2 Fungi mediated metal selenide nanoparticles production

Fungi have distinct advantages over other biological entities in terms of high metal tolerance, ease to scale-up and biomass handling (Rosenfeld *et al.*, 2017). Moreover, they are excellent secretors of metabolic enzymes and extracellular proteins, which act as a reducing and capping agent in nanoparticle production (Jacob *et al.*, 2015; Kadam *et al.*, 2019). Monodispersed, hydrophilic, highly stable with broad photoluminescence, and 1% quantum yield CdSe nanoparticles were synthesized using the plant pathogenic fungus *Helminthosporium solani* upon incubation with an aqueous solution of $CdCl_2$ and $SeCl_4$ under ambient conditions (Suresh, 2014).

Jacob *et al.* (2014b) synthesized PbSe QDs using the marine fungus *Aspergillus terreus* isolated from lead- and selenium-contaminated seawater near industrial

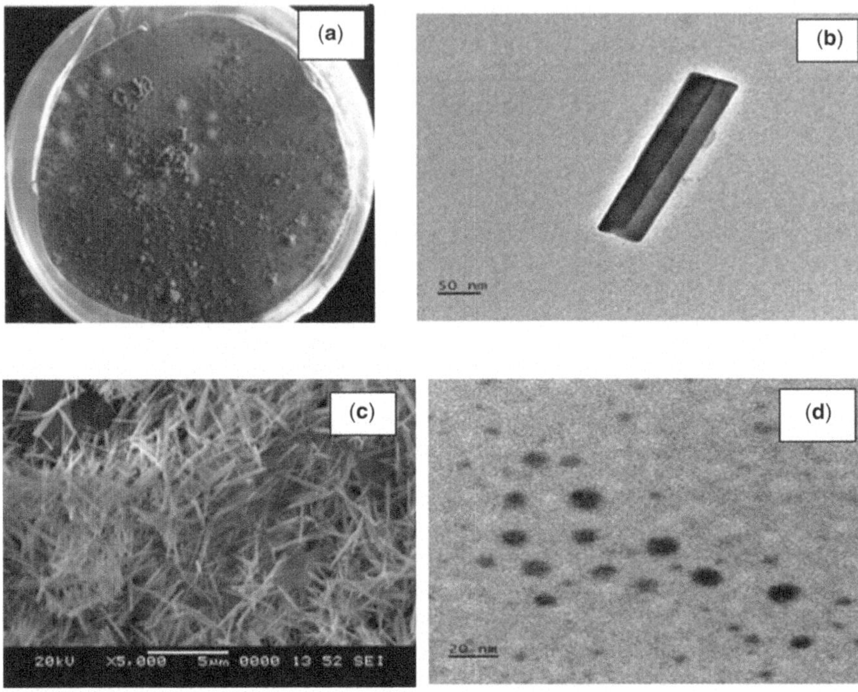

Figure 11.6 Biosynthesis of PbSe QDs (a) *Aspergillus terreus* culture on PDA amended with 20 ppm Pb and Se, (b) TEM and (c) SEM micrograph (Jacob *et al.*, 2014b) and (d) high-resolution TEM (HRTEM) of PbSe QDs (Balakrishnan *et al.*, 2019).

areas in Mangalore. Figure 11.6a shows brown color, compact, columnar and biseriate conidia of *Aspergillus terreus* on potato dextrose agar (PDA) plate supplemented with 20 ppm Pb and Se. Scanning electron microscopy (SEM) analysis confirmed the formation of PbSe nanorods (Figure 11.6b). The biosynthesized nanoparticles aspect ratio ranged between 10 and 70 with an average diameter of 95 nm determined by transmission electron microscopy (TEM) analysis (Figure 11.6c) (Jacob *et al.*, 2014a). Under optimized reaction conditions, spherical PbSe nanoparticles with a diameter range from 10 to 30 nm were obtained, as shown in Figure 11.6d.

Kumar *et al.* (2007) synthesized polydispersed, highly stable semiconductor CdSe QDs of ~11 nm, with a broad photoluminescence spectrum and ~6–7 ns fluorescence half-life using *Fusarium oxysporum*. The components of *F. oxysporum* mycelial cells before and after treatment with the Cd^{2+} and Se(IV) solution were analyzed by sodium dodecyl sulphate-polyacrylamide gel electrophoresis (SDS-PAGE) to assess the optimal conditions for the synthesis of CdSe QDs. TEM analysis showed that the CdSe QDs size was less than 20 nm, and these dots were purified using a gel filtration column. Cu–Zn superoxide dismutase in the cytoplasm decreased following Cd^{2+} and $Se(IV)^{2-}$ treatment. Simultaneously, the level of superoxide increased in the cytoplasm, which may be related to CdSe QDs formation in *F. oxysporum* (Yamaguchi *et al.*, 2016). Diko *et al.* (2020) fabricated 10–30 nm cubic-face centered PbSe nanoparticles using *Trichoderma sp.* WL-Go under optimum reaction conditions of pH 8 with 0.5 g biomass and 1:1 mM of SeO_2:$Pb(NO_3)_2$ and further studied their antioxidant and photocatalytic activity.

11.2.3.3 Metal selenide nanoparticle production by yeast

Yeast strains can be cultivated on simple nutrients with easy downstream processing and gene manipulation making them excellent scaffolds for nanoparticle synthesis. Cao *et al.* (2020) reported *Rhodotorula mucilaginosa* PA-1 could potentially bioremediate Cd and organic pollutants. Further, it aids in synthesizing CdSe QDs through a Se and Cd detoxification mechanism. Shao *et al.* (2018) reported gene modification of *Saccharomyces cerevisiae* through gene regulation to improve the cells ability to synthesize CdSe nanoparticles. A *S. cerevisiae* culture harvested at 12 h, that is, in the stationary phase, was subjected to 1 mM Na_2SeO_3 with a 6 h incubation period followed by the addition of 3 mM $CdCl_2$ in fresh medium for 84 h of incubation. This resulted in the synthesis of CdSe nanoparticles in the cytoplasm and displayed a fluorescence emission at 540 nm (Brooks & Lefebvre, 2017). The fluorescence intensity was 78% higher in the case of an optimized protocol comparable to the basic synthesis protocol.

Tian *et al.* (2017) reported the synthesis of bio-QDs using *C. utilis* WSH02-08 and observed their fluorescence dynamics *in vivo*. Fluorescence intensity, photostable lifetime, photoactivation and photobleaching decay time were obtained and quantified as critical parameters in the synthesis of bio-QDs. Moreover, higher Cd contents were found to favor the formation of the bio-QDs with higher fluorescence intensity and photoactivation index in *C. utilis* WSH02-08, implying the possibility of using fluorescence dynamics parameters as a screening index (Tian *et al.*, 2017). The time point addition of the precursors (i.e., selenium and cadmium) and their concentration, as well as the incubation duration determined the CdSe nanoparticle synthesis efficiency using *Saccharomyces cerevisiae* as a possible nanofactory (Wu *et al.*, 2015).

11.2.4 Mechanism of metal selenide nanoparticle formation

Biological entities such as fungi, bacteria, and yeasts produce nanoparticles as part of their defense mechanisms against environmental pollutants, particularly metals (see Chapter 3). They reduce metal toxicity by coupling them to biological molecules by metal-dependent enzymes, transporters, metal ion regulators, ligands or other binding proteins (Diko *et al.*, 2020; Iravani, 2014; Shao *et al.*, 2018). Further, this reduction of metal ions can result in metal precipitation as a nanomaterial in the microorganism's intra or extracellular spaces. Taking advantage of the microbial defense mechanism towards toxic metals resulted in synthesis of nanomaterials via bioremediation processes. The crucial step in the biosynthesis of NPs by microbial entities is exchanging electrons from a donor molecule to the metal ion that results in precipitation as a nanoparticle. The electron exchange can be done via biological molecules (Mal *et al.*, 2016).

11.2.4.1 Mechanism governing bacterial mediated synthesis of metal selenides

The ability of bacteria to grow and survive under metal stress conditions may be attributed to specific mechanisms of resistance, which include efflux pumps, metal efflux systems, inactivation and complexation of metals, impermeability to metals and the lack of specific metal transport systems, alteration of solubility and toxicity by changes in the redox state of the metal ions, extracellular precipitation of metals, and volatilization of toxic metals by an enzymatic reaction (Gahlawat & Choudhury, 2019). Moreover, microorganisms are capable of mobilization and immobilization of metals and in some cases, the bacteria which could reduce metal ions showed the ability to precipitate metals at a nanometer scale (Iravani, 2014).

Kim *et al.* (2016) used the recombinant *Escherichia coli* expressing the heavy-metal binding protein *Arabidopsis thaliana* phytochelatin synthase (AtPCS) and *Pseudomonas putida* metallothionein (PpMT) for the synthesis of EuSe nanoparticles. The heavy metal binding protein expression in the cells was induced by the addition of 0.5 mM isopropyl-β-d-thiogalactopyranoside (IPTG) in the presence of the precursors, i.e. europium and selenium. The expression of these recombinant heavy-metal binding proteins leads to a reduction of the selenium cations, thus leading to EuSe nanoparticle formation. It was observed that the precursor acts as a catalyst for the recombinant protein expression and aids in providing stability to EuSe nanoparticles through electrostatic attraction (Kim *et al.*, 2016).

Yan *et al.* (2014) proposed a reaction scheme for the synthesis of CdSe QDs as follows:

$$6GSH + 4H^+ + 3SeO_3{}^{2-} \rightarrow 3GS{-}Se{-}SG + 2O^{2-} + 5H_2O \quad (11.1)$$

$$GS{-}Se{-}SG + NADPH \rightarrow GSH + GS{-}Se^- + NADP^+ \quad (11.2)$$

$$GS{-}Cd{-}SG + GS{-}Se^- \rightarrow CdSe\ QDs \quad (11.3)$$

When Na_2SeO_3 was added, it spontaneously reacted with GSH to form GS-Se-SG which was further reduced to low-valence organo-selenium compounds such as seleno-cystine (Cys-Se) by GSH-related enzymes. When

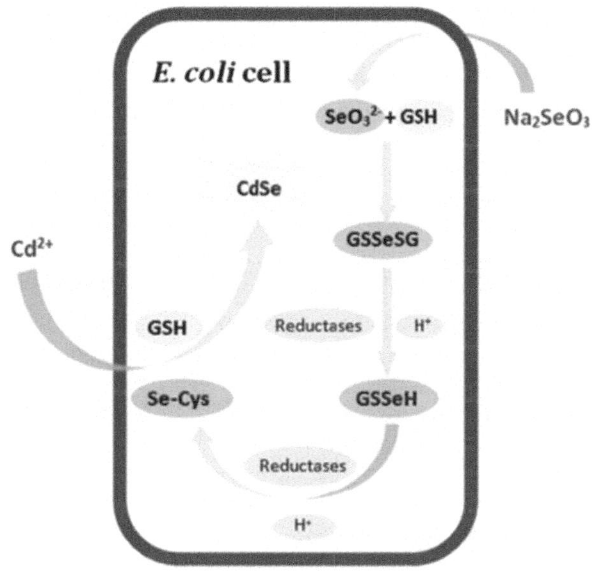

Figure 11.7 Scheme of the biosynthesis of fluorescent CdSe quantum dots by live *E. coli* cells (Yan *et al.*, 2014).

$CdCl_2$ was subsequently added, it was transported into the cytoplasm and transformed into $Cd(SG)_2$, thus leading to the formation of CdSe QDs in *E. coli* cells as shown in Figure 11.7 (Yan *et al.*, 2014).

11.2.4.2 Mechanism governing fungal mediated synthesis of metal selenides

Under physiological stress, fungi synthesize nanomaterials either intracellularly or extracellularly or both using complex biomolecular machinery such as enzymes and proteins, which act as reducing and stabilizing agents for the biosynthesis of nanoparticles (Dorcheh & Vahabi, 2016). Heavy metals bind to the fungal cell wall via electrostatic interactions and are further reduced into nanomaterials leading to intracellular synthesis. Fungal enzyme and metal ion interactions can also direct the extracellular synthesis of nanomaterials, possibly due to secretion of reductase enzymes. However, extracellular synthesis of nanoparticles is advantageous compared to intracellular synthesis in terms of purification and recovery (Roy *et al.*, 2018).

A governing mechanism in the PbSe synthesis is represented in Figure 11.8. The presence of precursor salts in the media exerts metal stress to the fungus, which stimulates detoxification mechanisms by (a) activating surface functional groups, (b) phytochelatin synthase (PS) converting glutathione to phytochelatins (PC),

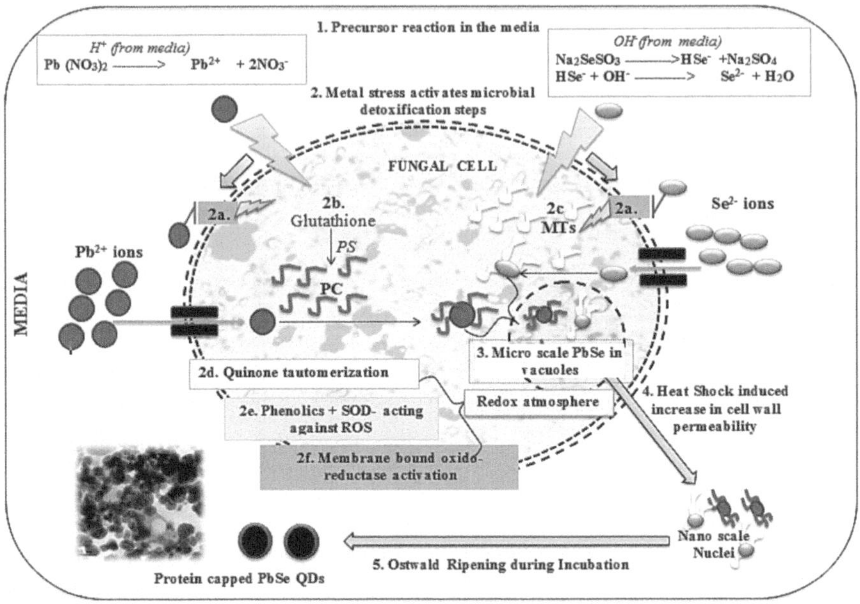

Figure 11.8 Schematic representation of the proposed mechanism for the biosynthesis of PbSe QDs by *Aspergillus terreus* (Jacob *et al.*, 2017).

which bind the metal ions, (c) metallothioneins binding the metal/metalloid ions, (d) quinone tautomerization, (e) superoxide dismutase (SOD) activity and (f) other oxidoreductases. Thus, the involvement of metal-binding peptides, namely metallothioneins and other antioxidant enzymes like superoxide dismutase, plays a prominent role in the microbial metal detoxification system for the biosynthesis of PbSe QDs (Jacob *et al.*, 2017).

11.2.4.3 Mechanism governing yeast mediated synthesis of metal selenides

Yeast has an intrinsic ability to absorb and accumulate high concentrations of toxic metal ions from its surroundings. Yeast cells adapt themselves under metal toxicity conditions using various detoxification mechanisms, viz. bio-precipitation, chelation, and intracellular sequestration (Gahlawat & Choudhury, 2019). Various researchers have exploited this property of yeast cells. For example, Shao *et al.* (2018) used *Saccharomyces cerevisiae* as a possible nanofactory for CdSe QDs synthesis using a Se precursor (Na_2SeO_3) and the Se metabolic flux. The addition of $CdCl_2$ then induced the methionine-to-cysteine pathway to synthesize cysteine, the precursor of GSH, to detoxify Cd^{2+}. The conversion from selenomethionine (SeMet) to selenocysteine (SeCys) is then triggered to biosynthesize CdSe QDs. Overexpression of MET6 gene encodes methionine

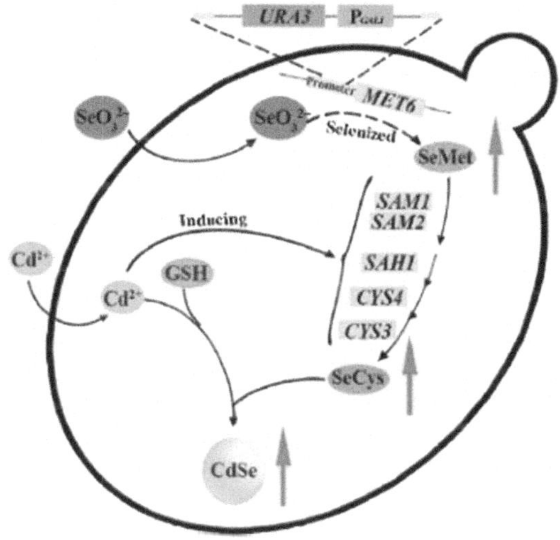

Figure 11.9 Mechanism of CdSe QDs biosynthesis in yeast (Shao *et al.*, 2018).

synthase and thus increased the concentration of SeMet, which enhanced CdSe QD biosynthesis (Figure 11.9).

A yeast cell contains higher levels of GSH in the stationary phase than in the exponential phase. Non-specific incorporation of selenium into organic species results in selenocysteine formation, which, upon further interaction with Cd(II), leads to CdSe biosynthesis. Selenocysteine plays a vital role in the biosynthesis of CdSe nanoparticles in yeast cells (Brooks & Lefebvre, 2017), which supports the hypothesis that selenol-containing biomolecules are involved in the biosynthesis of chalcogenide metal particles.

11.3 APPLICATION OF METAL SELENIDE NANOPARTICLES

Biosynthesized metal selenide nanoparticles have certain advantages such as biocompatibility, stability, reduced toxicity and economic viability (Ingale & Chaudhari 2013). The existence of biomolecules on the biogenic nanoparticles' surface eliminates the additional step of functionalization for certain biomedical applications. Hence, metal selenide nanoparticles find application in specific fields, especially bioimaging and biolabeling, therapeutic agents (Mal *et al.*, 2016), antimicrobials (Jacob *et al.*, 2016) and catalytic applications for removing pollutants from the environment (Diko *et al.*, 2020).

11.3.1 Photocatalysis

Qi *et al.* (2019) reported $Cu_{2-x}Se$ nanospheres as a photocatalyst for the degradation of methylene blue (MB) in the presence of sunlight. Under dark conditions, $Cu_{2-x}Se$

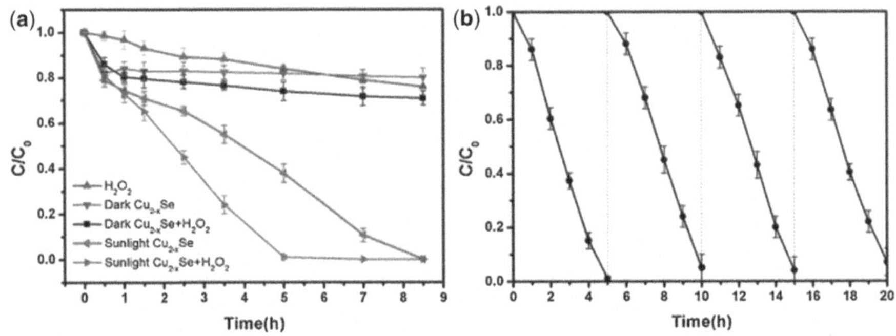

Figure 11.10 Experimental evaluation of the photocatalytic activity of as-biosynthesized $Cu_{2\text{-}x}Se$ nanospheres. (a) The performance of MB degradation photocatalyzed by the $Cu_{2\text{-}x}Se$ nanospheres under different conditions and (b) Repeated photo-degradation of MB catalyzed by the $Cu_{2\text{-}x}Se$ nanospheres for four cycles (Qi *et al.*, 2019).

nanospheres gave a 20.3% removal efficiency after 8.5 h due to mere adsorption of MB by the catalyst. In the presence of H_2O_2 and sunlight the removal efficiency increased to 24.1% and complete removal of MB was achieved within 5 h with the coexistence of H_2O_2 and $Cu_{2\text{-}x}Se$ nanospheres in the presence of light as shown in Figure 11.10.

The photo-generated electron-holes by $Cu_{2\text{-}x}Se$ combining with the free radicals generated from H_2O_2 led to the formation of superoxide radicals, and hydroxyl radicals have subsequently reacted with MB for complete degradation as reported by Qi *et al.* (2019).

Regenerated catalyst under optimized reaction conditions led to 93.4% of MB degradation even at the end of the fourth catalytical cycle showing high catalytic activity and stability of biosynthesized $Cu_{2\text{-}x}Se$ nanospheres.

Yeast mediated synthesized CdSe QDs effectively catalyzed the degradation of malachite green (MG) dye. Under UV-light irradiation for 60 min with a rate constant of 0.0327 min^{-1}, 86.5% degradation of MG dye was observed (Figure 11.11a and b) and 94.28% degradation was observed while exposing to visible light for 60 min with a rate constant of 0.0397 min^{-1} (Figure 11.11c and d). However, complete degradation of MG was achieved under UV light exposure for 120 min. The degradation products were identified as 4-(dimethylamino) phenol and 4-(dimethylamino) benzophenol by gas chromatography-mass spectrometry (GC-MS) analysis (Cao *et al.*, 2020).

Diko *et al.* (2020) reported *Trichoderma sp.* mediated synthesis of PbSe nanoparticles, which exhibited a significant antioxidant activity of 88.60% and photocatalytic activity of 82% in 30 min during the degradation of the rhodamine B dye (10 mg/L, 50 mL). Further, they proposed a mechanism for the degradation of rhodamine B. An incident photon energy $\geq$ intrinsic bandgap

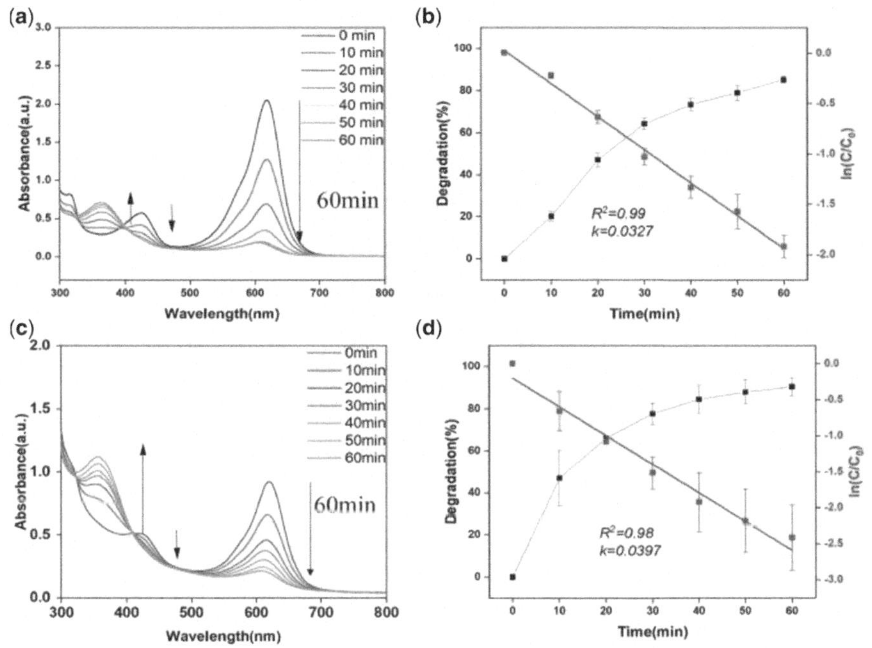

Figure 11.11 The photocatalytic degradation of MG by synthesized CdSe QDs. The MG degradation catalyzed by CdSe QDs under the irradiation of UV-light (a) and Vis-light (c). The MG degradation catalyzed by CdSe QDs under the irradiation of UV-light (b) and Vis-light (d) was quantified based on the standard curve of the MG solution and fitted by first-order kinetics (Cao *et al.*, 2020).

energy of the nanoparticles leads to electron movement from the valance band (VB) to the conduction band (CB), leaving the same number of charged holes in the valance band. These holes in the valence band react with water molecules (H_2O) and hydroxyl ions (OH^-) to yield hydroxyl radicals (OH•), which act as an oxidizing agent, further leading to the degradation of rhodamine B dye (Figure 11.12).

11.3.2 Bioimaging and biolabeling

Bioimaging refers to probes' development to visualize the cellular function, characterization, and measurement of molecular processes in living organisms.

Biosynthesized CdSe QDs were incorporated into yeast cells as illustrated by laser confocal scanning microscopic images, showing great potential in bioimaging and biolabeling applications (Figure 11.13). The confocal images indicate that the biosynthesized CdSe QDs have good biocompatibility for their potential application in *in vitro* cell imaging, which may act as an efficient

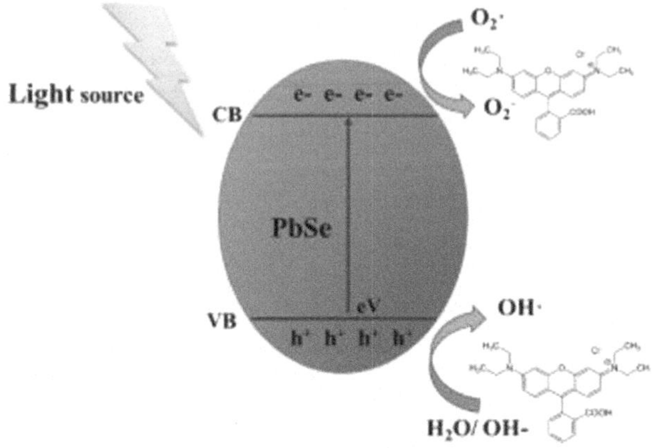

Figure 11.12 Schematic photocatalytic mechanism of PbSe NPs in rhodamine B degradation (Diko *et al.*, 2020).

alternative or complementary tool to conventional organic dyes. Furthermore, it is speculated that the good biocompatibility of CdSe QDs may facilitate the uptake of QDs into cells when the CdSe QDs are extracellularly biosynthesized by using *E. coli* bacteria (Xu *et al.*, 2019).

11.3.3 Therapeutic properties

A variety of nanoparticle systems are being explored for their potential in molecular imaging, with many applications aimed at diagnosing or treating cancer (Pandey & Bodas, 2020).

In vitro studies performed on HeLa, SKOV-3 and 293 T cells showed cytotoxic effects against cancer cell lines using EuSe nanoparticles synthesized from recombinant *E. Coli*. EuSe nanoparticle-treated cells look almost round and wrinkled with shrunken nuclei (Figure 11.14). The severity of damage to cell shape and nuclei increased with an increase in exposure time. Kim *et al.* (2016) showed the anti-cancer properties of EuSe nanoparticles, as small nanoparticles have a large surface area per unit mass, making nanoparticles very reactive in the cellular environment. EuSe nanoparticles can thus be utilized as promising drug-carrying agents in targeted drug delivery against cancer cells (Kim *et al.*, 2016).

Hydrothermally synthesized QDs exhibit noteworthy toxic effects on MCF-7 cells, whereas biologically synthesized QDs can retain cell viability (Figure 11.15). Nanotoxicity of the chemically synthesized QDs limits their biomedical application, whereas the presence of biocompatible capping agents on biogenic QDs aids their application in the biomedical field as a fluorescent dye for broad bio-imaging and labeling (Yan *et al.*, 2014).

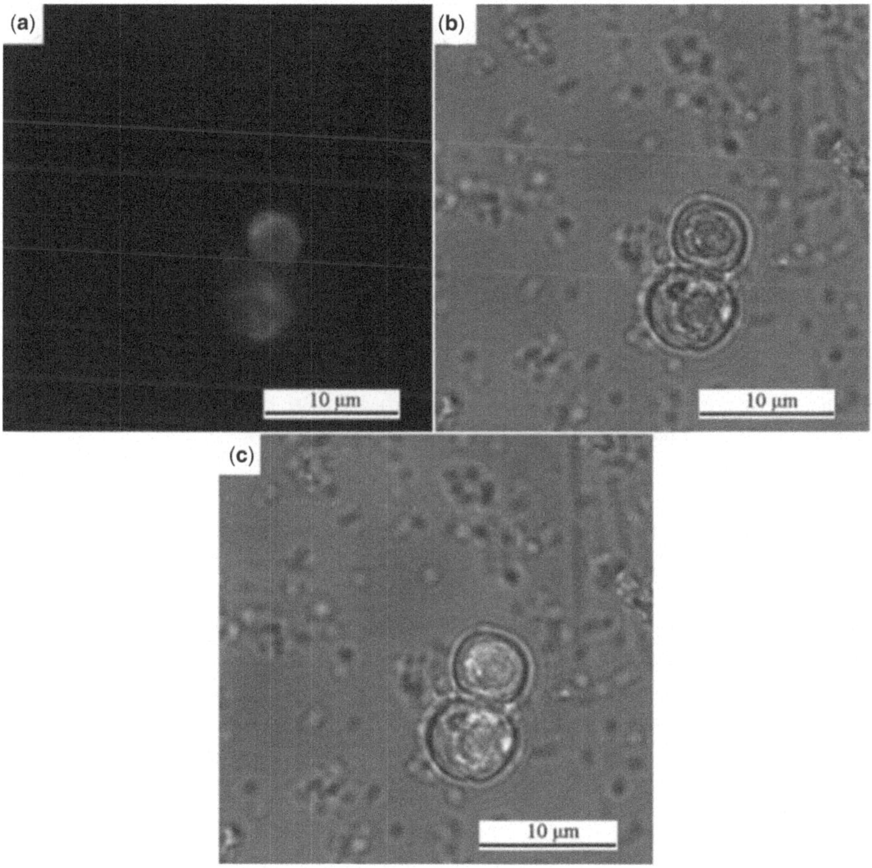

Figure 11.13 Confocal images of yeast cells incorporated with CdSe QDs at 37°C for 2 days. (a) Recorded under excitation by an Ar/Kr laser giving green emission, (b) Bright-field image and (c) Overlaid image (Xu *et al.*, 2019).

11.3.4 Antimicrobial activity

Widespread use of antibiotics has led to the emergence of multidrug-resistant bacterial strains. Conventional antibiotics target the cell wall synthesis, translational machinery, and DNA replication machinery of bacteria. However, a nanoparticle can enter into a bacterial cell by (i) binding to bacterial surface and rupture of a cell wall, (ii) penetration through the bacterial cell wall and interfering in the cellular biochemical pathway of bacteria and (iii) production of reactive oxygen species that rupture their cell wall which leads to cell death (Singh *et al.*, 2020). Therefore, the development of novel nanoparticle-based bactericidal agents is of great clinical importance (see Chapter 10).

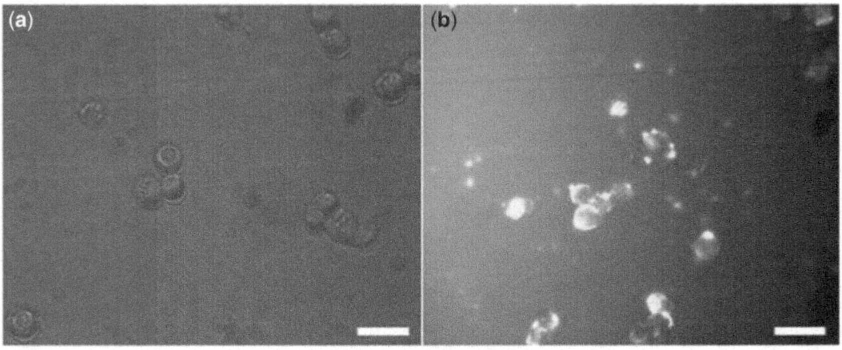

Figure 11.14 Differential interference contrast (DIC) image (a) and fluorescence image (b) of HeLa cells incubated with 2 mM EuSe nanoparticles. Scale bars represent 1 μm (Kim *et al.*, 2016).

Jacob *et al.* (2016) reported a filter paper bioassay for the anti-bacterial activity of the biogenic PbSe quantum rods dispensed on filter paper discs, based on the formation of distinct zones of inhibition by Gram-positive and Gram-negative bacterial pathogens. It was observed that the biogenic PbSe nanorods had an anti-bacterial activity comparable to that of standard antibiotics against pathogens like *B. cereus* and *E. coli* (Figure 11.16). Thus, it can be inferred that the biogenic PbSe quantum rods can serve as a broad-spectrum antibacterial agent.

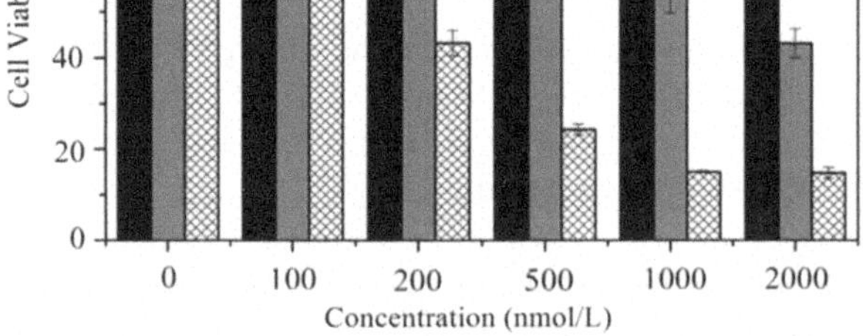

Figure 11.15 Cytotoxicity of biosynthesized QDs compared with two kinds of hydrothermally synthesized QDs (Yan *et al.*, 2014).

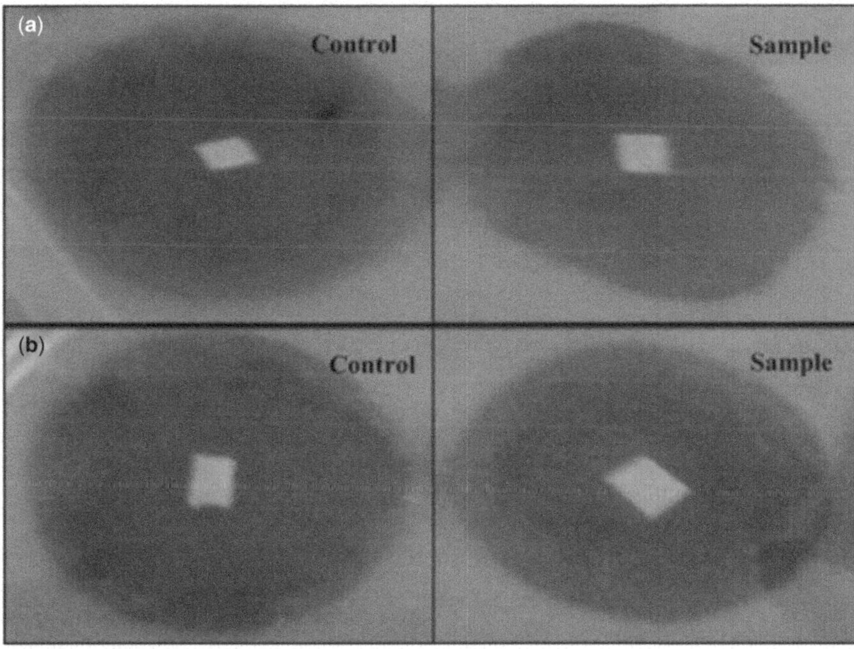

Figure 11.16 Comparison of the zone of inhibition between (a) Gram-positive *Bacillus cercus* and (b) Gram-negative *Escherichia coli* for control and biogenic PbSe quantum rods (sample) (Jacob *et al.*, 2016).

11.4 SCALE-UP OF METAL SELENIDE QDs BIOSYNTHESIS

Microbial entities such as bacteria, fungi and yeasts are efficient biofactories for the synthesis of metal selenide nanoparticles. As discussed in this chapter, microbial mediated synthesis of metal selenide nanoparticles provides an eco-friendly, rapid, green, and efficient route for the fabrication of biocompatible nanostructures with extraordinary physical-chemical and optoelectronic properties. This chapter highlighted the recent developments in the biological synthesis of metal selenide nanoparticles and identified bottlenecks hindering their development. The potential application of biosynthesized metal selenides for cell imaging, bioimaging, as a drug carrier, and as a photocatalyst were overviewed.

Despite this, metal selenide nanoparticles are far from large scale synthesis and application in the field of medicine due to the lack of commercialization of the products and the development of appropriate regulations. Further improvements in metal selenides biosynthesis are required. Understanding the underlying cellular mechanisms is crucial for designing a tailor-made, well-optimized process for the synthesis of nanostructures with desired size, shape, and properties.

REFERENCES

Abou-Assy R. S., El-Deeb B. A., Al-Talhi A. D. and Mostafa N. Y. (2019). Biosynthesis of cadmium selenide quantum dots by *Providencia vermicola*. *African Journal of Microbiology Research*, **13**(6), 106–121.

Ayano H., Kuroda M., Soda S. and Ike M. (2014). Effects of culture conditions of *Pseudomonas aeruginosa* strain RB on the synthesis of CdSe nanoparticles. *Journal of Bioscience and Bioengineering*, **119**(4), 440–445.

Bakiyaraj G. and Dhanasekaran R. (2013). Synthesis and characterization of flower-like ZnSe nanostructured thin films by chemical bath deposition (CBD) method. *Applied Nanoscience*, **3**(2), 125–131.

Balakrishnan R. M., Uddandarao P., Raval K. and Raval R. (2019). A perspective of advanced biosensors for environmental monitoring. In: Tools, Techniques and Protocols for Monitoring Environmental Contaminants, S. K. Brar, K. Hegde and V. Pachapur (eds), Elsevier, pp. 19–51.

Brooks J. and Lefebvre D. D. (2017). Optimization of conditions for cadmium selenide quantum dot biosynthesis in *Saccharomyces cerevisiae*. *Applied Microbiology and Biotechnology*, **101**(7), 2735–2745.

Cak T., Ozdal M., Kundakc M. and Kayal R. (2019). ZnSe and CuSe NP's by microbial green synthesis method and comparison of I–V characteristics of Au/ZnSe/p-Si/Al and Au/CuSe/p-Si/Al structures. *Material Science in Semiconductor Processing*, **103**, 104610.

Cao K., Chen M. M., Chang F. Y., Cheng Y. Y., Tian L. J., Li F., Deng G. Z. and Wu C. (2020). The biosynthesis of cadmium selenide quantum dots by *Rhodotorula mucilaginosa* PA-1 for photocatalysis. *Biochemical Engineering Journal*, **156**, 107497.

Chandramohan S., Sundar K. and Muthukumaran A. (2018). Monodispersed spherical shaped selenium nanoparticles (SeNPs) synthesized by *Bacillus subtilis* and its toxicity evaluation in zebrafish embryos. *Materials Research Express*, **5**(2), 025020.

Che L., Xu W., Zhan J., Zhang L., Liu L. and Zhou H. (2019). Complete genome sequence of *Bacillus cereus* CC-1, a novel marine selenate/selenite reducing bacterium producing metallic selenides nanomaterials. *Current Microbiology*, **76**(1), 78–85.

Chen S., Ma G., Wang Q., Sun S., Hisatomi T., Higashi T., Wang Z., Nakabayashi M., Shibata N., Pan Z. and Hayashi T. (2019). Metal selenide photocatalysts for visible-light-driven Z-scheme pure water splitting. *Journal of Materials Chemistry A*, **7**(13), 7415–7422.

Cremonini E., Zonaro E., Donini M., Lampis S., Boaretti M., Dusi S., Melotti P., Lleo M. M. and Vallini G. (2016). Biogenic selenium nanoparticles: characterization, antimicrobial activity and effects on human dendritic cells and fibroblasts. *Microbial Biotechnology* **9**, 758–771.

Diko C. S., Qu Y., Henglin Z., Li Z., Nahyoon N. A. and Fan S. (2020). Biosynthesis and characterization of lead selenide semiconductor nanoparticles (PbSe NPs) and its antioxidant and photocatalytic activity. *Arabian Journal of Chemistry*, **13**(11), 8411–8423.

Dorcheh S. K. and Vahabi K. (2016). Biosynthesis of Nanoparticles by Fungi: Large-scale Production. Fungal Metabolites. Springer International Publishing, Switzerland, pp. 1–20.

Elahian F., Reiisi S., Shahidi A. and Mirzaei S. A. (2017). High-throughput bioaccumulation, biotransformation, and production of silver and selenium nanoparticles using genetically engineered *Pichia pastoris*. *Nanomedicine: Nanotechnology, Biology and Medicine*, **13**(3), 853–861.

Esmaeili-zare M., Salavati-niasari M. and Sobhani A. (2012). Ultrasonics Sonochemistry Simple sonochemical synthesis and characterization of HgSe nanoparticles. *Ultrasonics – Sonochemistry*, **19**(5), 1079–1086.

Espinosa-ortiz E. J., Gonzalez-gil G. and Saikaly P. E. (2015). Effects of selenium oxyanions on the white-rot fungus *Phanerochaete chrysosporium*. *Applied Microbiology and Biotechnology*, **99**(5), 2405–2418.

Estevam E. C., Griffin S., Nasim M. J., Denezhkin P., Schneider R., Lilischkis R., Dominguez-Alvarez E., Witek K., Latacz G., Keck C. and Schäfer K. H. (2017). Natural selenium particles from *Staphylococcus carnosus*: Hazards or particles with particular promise? *Journal of Hazardous Materials*, **324**, 22–30.

Fernández-Llamosas H., Castro L., Blázquez M. L., Díaz E. and Carmona M. (2016). Biosynthesis of selenium nanoparticles by *Azoarcus sp.* CIB. *Microbial Cell Factories*, **15**(1), 1–10.

Fellowes J. W., Pattrick R. A., Lloyd J. R., Charnock J. M., Coker V. S., Mosselmans J. F., Weng T. C. and Pearce C. I. (2013). Ex situ formation of metal selenide quantum dots using bacterially derived selenide precursors. *Nanotechnology*, **24** (14), 145603.

Gahlawat G. and Choudhury A. R. (2019). A review on the biosynthesis of metal and metal salt nanoparticles by microbes. *RSC Advances*, **9**(23), 12944–12967.

He Q. and Yao K. (2011). Impact of alternative electron acceptors on selenium (IV) reduction by *Anaeromyxobacter dehalogenans*. *Bioresource Technology*, **102**(3), 3578–3580.

Ho C. T., Nguyen A. T., Duong T. T., Dang D. K., Tang T. C. and Hur H. G. (2015). Biologically based method for the synthesis of Hg–Se nanostructures by *Shewanella spp*. *RS Advances*, **5**(27), 20764–20768.

Ingale A. G. and Chaudhari A. N. (2013). Biogenic synthesis of nanoparticles and potential applications: an eco-friendly approach. *Journal of Nanomedicine & Nanotechnology*, **4** (165), 1–7.

Iravani S. (2014). Bacteria in nanoparticle synthesis: current status and future prospects. *International Scholarly Research Notices*, **2014**, 1–18.

Jacob J. M., Raj Mohan B., Bardhan S. K., Jagadeeshbabu P. E. and Aruna M. (2014a). Selenium and lead tolerance in marine *Aspergillus Terreus* for biosynthesis of nano particles – quantum dots/rods. *International Journal of Advances in Chemical Engineering & Biological Sciences*, **1**(1), 6–11.

Jacob J. M., Balakrishnan R. M. and Kumar U. B. (2014b). Biosynthesis of lead selenide quantum rods in marine *Aspergillus terreus*. *Materials Letters*, **124**, 279–281.

Jacob J. M., Lens P. N. L. and Mohan R. (2015). Minireview Microbial synthesis of chalcogenide semiconductor nanoparticles : a review. *Microbial Biotechnology*, **9**(1), 11–21.

Jacob J. M., Mohan B. R. and Gowda K. A. (2016). Insights into the optical and anti-bacterial properties of biogenic PbSe quantum rods. *Journal of Saudi Chemical Society*, **20**(4), 480–485.

Jacob J. M., Sharma S. and Balakrishnan R. M. (2017). Exploring the fungal protein cadre in the biosynthesis of PbSe quantum dots. *Journal of Hazardous Materials*, **324**, 54–61.

Joshi S. M., De Britto S., Jogaiah S. and Ito S. I. (2019). Mycogenic selenium nanoparticles as potential new generation broad spectrum antifungal molecules. *Biomolecules*, **9**(9), 419.

Kadam V. V., Ettiyappan J. P. and Balakrishnan R. M. (2019). Mechanistic insight into the endophytic fungus mediated synthesis of protein capped ZnO nanoparticles. *Materials Science and Engineering: B*, **243**, 214–221.

Khoei N. S., Lampis S., Zonaro E., Yrjälä K., Bernardi P. and Vallini G. (2017). Insights into selenite reduction and biogenesis of elemental selenium nanoparticles by two environmental isolates of *Burkholderia fungorum*. *New Biotechnology*, **34**, 1–11.

Kim E. B., Seo J. M., Kim G. W., Lee S. Y. and Park T. J. (2016). In vivo synthesis of europium selenide nanoparticles and related cytotoxicity evaluation of human cells. *Enzyme and Microbial Technology*, **95**, 201–208.

Kora A. J. and Rastogi L. (2016a). Bacteriogenic synthesis of selenium nanoparticles by *Escherichia coli* ATCC 35218 and its structural characterisation. *IET Nanobiotechnology*, **11**(2), 179–184.

Kora A. J. and Rastogi L. (2016b). Biomimetic synthesis of selenium nanoparticles by *Pseudomonas aeruginosa* ATCC 27853: an approach for conversion of selenite. *Journal of Environmental Management*, **181**, 231–236.

Kumar S. A., Ansary A. A., Ahmad A. and Khan M. I. (2007). Extracellular biosynthesis of CdSe quantum dots by the fungus, *Fusarium oxysporum*. *Journal of Biomedical Nanotechnology*, **3**(2), 190–194.

Kuroda M., Suda S., Sato M., Ayano H., Ohishi Y., Nishikawa H., Soda S. and Ike M. (2019). Biosynthesis of bismuth selenide nanoparticles using chalcogen-metabolizing bacteria. *Applied Microbiology and Biotechnology*, **103**(21–22), 8853–8861.

Liqiang J., Yichun Q., Baiqi W., Shudan L., Baojiang J., Libin Y., Wei F., Honggang F. and Jiazhong S. (2006). Review of photoluminescence performance of nano-sized semiconductor materials and its relationships with photocatalytic activity. *Solar Energy Materials and Solar Cells*, **90**(12), 1773–1787.

Lloyd J. R., Byrne J. M. and Coker V. S. (2011). Biotechnological synthesis of functional nanomaterials. *Current Opinion in Biotechnology*, **22**(4), 509–515.

Luo Q. Y., Lin Y., Li Y., Xiong L. H., Cui R., Xie Z. X. and Pang D. W. (2014). Nanomechanical analysis of yeast cells in CdSe quantum dot biosynthesis. *Small*, **10**(4), 699–704.

Mal J., Nancharaiah Y. V., Van Hullebusch E. D. and Lens P. N. L. (2016). Metal chalcogenide quantum dots : biotechnological synthesis and applications. *RSC Advances*, **6**(47), 41477–41495.

Mesbahi-nowrouzi M. and Mollania N. (2018). Purification of selenate reductase from *Alcaligenes sp.* CKCr-6A with the ability to biosynthesis of selenium nanoparticle: enzymatic behavior study in imidazolium based ionic liquids and organic solvent. *Journal of Molecular Liquids*, **249**, 1254–1262.

Mosallam F. M., El-Sayyad G. S., Fathy R. M. and El-Batal A. I. (2018). Biomolecules-mediated synthesis of selenium nanoparticles using *Aspergillus oryzae* fermented Lupin extract and gamma radiation for hindering the growth of some multidrug-resistant bacteria and pathogenic fungi. *Microbial Pathogenesis*, **122**, 108–116.

Nancharaiah Y. V. and Lens P. N. (2015a). Selenium biomineralization for biotechnological applications. *Trends in biotechnology*, **33**(6), 323–330.

Nancharaiah Y. V. and Lens P. N. (2015b). Ecology and biotechnology of selenium-respiring bacteria. *Microbiology and Molecular Biology Reviews*, **79**(1), 61–80.

Nguyen V. K., Park Y., Yu J. and Lee T. (2016). Bioresource technology microbial selenite reduction with organic carbon and electrode as sole electron donor by a bacterium isolated from domestic wastewater. *Bioresource Technology*, **212**, 182–189.

Pandey S. and Bodas D. (2020). High-quality quantum dots for multiplexed bioimaging: a critical review. *Advances in Colloid and Interface Science*, **278**, 102137.

Pantidos N. and Horsfall L. E. (2014). Biological synthesis of metallic nanoparticles by bacteria, fungi and plants. *Journal of Nanomedicine & Nanotechnology*, **5**(5), 1.

Park S. H., Choi J. Y., Lee Y. H., Park J. T. and Song H. (2015). Formation of metal selenide and metal–selenium nanoparticles using distinct reactivity between selenium and noble metals. *Chemistry–An Asian Journal*, **10**(7), 1452–1456.

Presentato A., Piacenza E., Anikovskiy M., Cappelletti M., Zannoni D. and Turner R. J. (2018). Biosynthesis of selenium-nanoparticles and -nanorods as a product of selenite bioconversion by the aerobic bacterium *Rhodococcus aetherivorans*. *New Biotechnology*, **41**, 1–8.

Qi S., Yang S., Yue L., Wang J., Liang X. and Xin B. (2019). Extracellular biosynthesis of $Cu_{2-x}Se$ nanocrystallites with photocatalytic activity. *Materials Research Bulletin*, **111**, 126–132.

Radhika R. S. R. and Gayathri S. (2015). Extracellular biosynthesis of Selenium nanoparticles using some species of *Lactobacillus*. *Indian Journal of Geo-Marine Sciences*, **43**(5), 766–775.

Ramya S., Shanmugasundaram T. and Balagurunathan R. (2020). Actinobacterial enzyme mediated synthesis of selenium nanoparticles for antibacterial, mosquito larvicidal and anthelminthic applications. *Particulate Science and Technology*, **38**(1), 63–72.

Rosenfeld C. E., Kenyon J. A., James B. R. and Santelli C. M. (2017). Selenium (IV, VI) reduction and tolerance by fungi in an oxic environment. *Geobiology*, **15**(3), 441–452.

Rotruck J. T., Pope A. L., Ganther H. E., Swanson A. B., Hafeman D. G. and Hoekstra W. (1973). Selenium: biochemical role as a component of glutathione peroxidase. *Science*, **179**(4073), 588–590.

Roy L., Bera D. and Adak S. (2018). Mycosynthesized nanoparticles: role in food processing industries. *Fungal Nanobionics: Principles and Applications*, 287–316.

Saikia K., Deb P. and Kalita E. (2013). Facile synthesis of highly luminescent ZnSe(S) alloyed quantum dot for biomedical imaging. *Current Applied Physics*, **13**(5), 925–930.

Shamberger R. J. (1981). Selenium in the environment. *Science of the Total Environment*, **17** (1), 59–74.

Shao M., Zhang R., Wang C., Hu B., Pang D. and Xie Z. (2018). Living cell synthesis of CdSe quantum dots : Manipulation based on the transformation mechanism of intracellular. *Nano Research*, **11**(5), 2498–2511.

Singh P., Kim Y., Zhang D. and Yang D. (2016). Biological synthesis of nanoparticles from plants and microorganisms. *Trends in Biotechnology*, **34**(7), 588–599.

Singh A., Gautam P. K., Verma A., Singh V., Shivapriya P. M., Shivalkar S., Sahoo A. K. and Samanta S. K. (2020). Green synthesis of metallic nanoparticles as effective alternatives to treat antibiotics resistant bacterial infections: a review. *Biotechnology Reports*, **25**, e00427.

Srivastava N. and Mukhopadhyay M. (2015a). Green synthesis and structural characterization of selenium nanoparticles and assessment of their antimicrobial property. *Bioprocess and Biosystems Engineering*, **38**(9), 1723–1730.

Srivastava N. and Mukhopadhyay M. (2015b). Biosynthesis and structural characterization of selenium nanoparticles using *Gliocladium roseum*. *Journal of Cluster Science*, **26**, 1473–1482.

Subedi G., Taylor J., Hatam I. and Baldwin S. A. (2017). Chemosphere Simultaneous selenate reduction and denitrification by a consortium of enriched mine site bacteria. *Chemosphere*, **183**, 536–545.

Suresh S. (2013). Semiconductor nanomaterials, methods and applications: a review. *Nanoscience and Nanotechnology*, **3**(3), 62–74.

Suresh A. K. (2014). Extracellular bio-production and characterization of small monodispersed CdSe quantum dot nanocrystallites. *Spectrochimica Acta Part A: Molecular and Biomolecular Spectroscopy*, **130**, 344–349.

Terra F. S., Mahmoud G. M., Nasr M. and Okr M. E. (2010). Preparation and structural properties of PbSe nanomaterial. *Surface and Interface Analysis*, **42**(6–7), 1239–1243.

Tian L., Zhou N., Liu X., Liu J., Zhang X., Huang H., Zhu T., Li L., Huang Q., Li W., Liu Y. and Yu H. (2017). A sustainable biogenic route to synthesize quantum dots with tunable fluorescence properties for live cell imaging. *Biochemical Engineering Journal*, **124**, 130–137.

Vogel M., Fischer S., Maffert A., Hübner R., Scheinost A. C., Franzen C. and Steudtner R. (2018). Biotransformation and detoxification of selenite by microbial biogenesis of selenium-sulfur nanoparticles. *Journal of Hazardous Materials*, **344**, 749–757.

Wang Y., Shu X., Hou J., Lu W., Zhao W., Huang S. and Wu L. (2018). Selenium nanoparticle synthesized by *Proteus mirabilis* YC801: an efficacious pathway for selenite biotransformation and detoxification. *International Journal of Molecular Sciences*, **19**(12), 3809.

Wang D., Xia X., Wu S., Zheng S. and Wang G. (2019). The essentialness of glutathione reductase GorA for biosynthesis of Se(0)-nanoparticles and GSH for CdSe quantum dot formation in *Pseudomonas stutzeri* TS44. *Journal of Hazardous Materials*, **366**, 301–310.

Wu S. M., Su Y., Liang R. R., Ai X. X., Qian J., Wang C., Chen J. Q. and Yan Z. Y. (2015). Crucial factors in biosynthesis of fluorescent CdSe quantum dots in *Saccharomyces cerevisiae*. *RSC Advances*, **5**(96), 79184–79191.

Xie Y., Su H., Li B. and Qian Y. (2000). A direct solvothermal route to nanocrystalline selenides at low temperature. *Materials Research Bulletin*, **35**(3), 459–464.

Xu J., Hu R., Wang Q., Wang P. and Bao H. (2019). Extracellular biosynthesis of biocompatible CdSe quantum dots. *IET Nanobiotechnology*, **13**(9), 962–966.

Yan Z., Qian J., Gu Y., Su Y., Ai X. and Wu S. (2014). Green biosynthesis of biocompatible CdSe quantum dots in living *Escherichia coli* cells. *Materials Research Express*, **1**(1), 015401.

Yamaguchi T., Tsuruda Y., Furukawa T., Negishi L., Imura Y., Sakuda S., Yoshimura E. and Suzuki M. (2016). Synthesis of CdSe quantum dots using *Fusarium oxysporum*. *Materials*, **9**(10), 855.

Yue L., Wang J., Qi S. and Xin B. (2016). Extracellular synthesis of cuprous selenide (Cu_2Se) nanospheres by a biological-chemical coupling reduction process in an anaerobic microbial system. *Biotechnology Progress*, **32**(5), 1264–1270.

Zhang Z., Chen J., Yang Q., Lan K., Yan Z. and Chen J. (2018). Eco-friendly intracellular microalgae synthesis of fluorescent CdSe QDs as a sensitive nanoprobe for determination of imatinib. *Sensors and Actuators B: Chemical*, **263**, 625–633.

Zhang H., Zhou H., Bai J., Li Y., Yang J., Ma Q. and Qu Y. (2019). Biosynthesis of selenium nanoparticles mediated by fungus *Mariannaea sp.* HJ and their characterization. *Colloids and Surfaces A*, **571**, 9–16.

Zhao P., Xu Q., Tao J., Jin Z., Pan Y. and Yu C. (2018). Near infrared quantum dots in biomedical applications : current status and future perspective. *Wiley Interdisciplinary Reviews: Nanomedicine and Nanobiotechnology*, **10**(3), 1483.

Zhou J., Zeng Q., Lv D., Sun L., Niu L., Fu W., Liu F., Shen Z., Jin C. and Liu Z. (2015). Controlled synthesis of high-quality monolayered α-In_2Se_3 via Physical Vapor Deposition. *Nano Letters*, **15**(10), 6400–6405.

Zonaro E., Lampis S., Turner R. J., Qazi S. J. S., Vallini G. and Nadeau J. (2015). Biogenic selenium and tellurium nanoparticles synthesized by environmental microbial isolates efficaciously inhibit bacterial planktonic cultures and biofilms. *Frontiers in Microbiology*, **6**, 584.

Chapter 12

Biosynthesis of selenium nanomaterials by anaerobic bacteria for environmental technologies

Piet N. L. Lens

12.1 INTRODUCTION

12.1.1 Nanotechnology for sustainable development

The world is facing formidable challenges in meeting the rising demands for fuels, potable water and consumer products. The health and welfare of people are closely connected to the availability of adequate, safe and affordable energy, water and food supplies (Lens, 2021). Also cost effective recovery methods of chemicals and resources from waste and wastewater play an increasing role in sustaining human civilization on Earth (Helland & Kastenholz, 2008).

Nanotechnology, comprising technologies working at the scale of typically 100 nm and below, is an enabling technology for a wide variety of traditional scientific disciplines (Fleischer & Grunwald, 2008). This has led to high expectations that nanotechnology will help improve peoples' standard of living, in the short-term by significantly improving existing processes and products, and in the long-term by providing revolutionary and life-changing advances, from cancer treatment and lightweight materials to renewable energy production and (waste)water purification (Lens *et al.*, 2012; Uskokovic, 2007). However, the novel properties that make nanotechnology so interesting have also raised many unanswered questions and concerns related to the impacts, negative and positive, nanotechnology may have on society and the environment (Helland *et al.*, 2006).

doi: 10.2166/9781789061055_0353

12.1.2 Nanotechnology and water treatment

Nanotechnology encompasses the creation and utilization of materials, devices and systems at the level of atoms and molecules, cutting across disciplines such as chemistry, physics, biology, engineering and material science (Masciangioli & Zhang, 2003). Nanomaterials often exhibit novel and significantly changed physical, chemical and biological properties. These are the result of their structure, larger surface area per unit volume and quantum effects that occur at the nanoscale. For environmental technology, the potential impact areas comprise three categories: treatment and remediation, sensing and detection, and prevention (Theron *et al.*, 2008).

Within the category of treatment and remediation, nanotechnology has the potential to contribute to long-term availability and viability of water resources of high quality (Bottero *et al.*, 2006), through the use of advanced filtration materials (nanostructured and nanoreactive membranes), bioactive nanoparticles for water disinfection (silver or titanium dioxide (TiO_2) nanoparticles) and engineered nanoparticles and nanomaterials that remove pollutants (e.g., carbon nanotubes sorbing heavy metals or organics; nanoparticles with enhanced catalytic properties that degrade organic matter, including nanoscale semiconductor photocatalysts, single-enzyme and zero-valent iron nanoparticles) (Zhang, 2003). Figure 12.1 illustrates different application areas of selenium (and other chalcogen) based nanomaterials for environmental technology and resource recovery.

12.1.3 Nanobiomanufacturing of nanomaterials

As the future of materials science is closely linked to nanotechnology, there is a need to improve the manufacturing of nanomaterials in terms of its environmental and economic impact. Nanoproducts can be prepared by different methods, such as colloidal aqueous and micellar solution synthesis methods, using ultrasonic waves, microwaves or gamma-irradiation (Yong *et al.*, 2003). In most cases, particles prepared by these methods have some problems including poor reproducibility, control of particle size, distribution and shape (Mandal *et al.*, 2005). Some reactions require high temperature, and/or high pressure for initiating the reaction, and/or inert atmosphere protection, and/or using toxic matters such as H_2S, toxic templates and stabilizers, and metallic precursors (Gericke & Pinches, 2006; Yong *et al.*, 2003). In addition, nanomaterial production has to be competitive, and the high costs associated with many existing processes limit the development and transfer of new nanotechnologies into the marketplace.

Thus, there is an increasing interest in the development of environmentally clean synthetic procedures for the formation of nanoproducts. An environmentally acceptable solvent system as well as eco-friendly reducing and capping agents

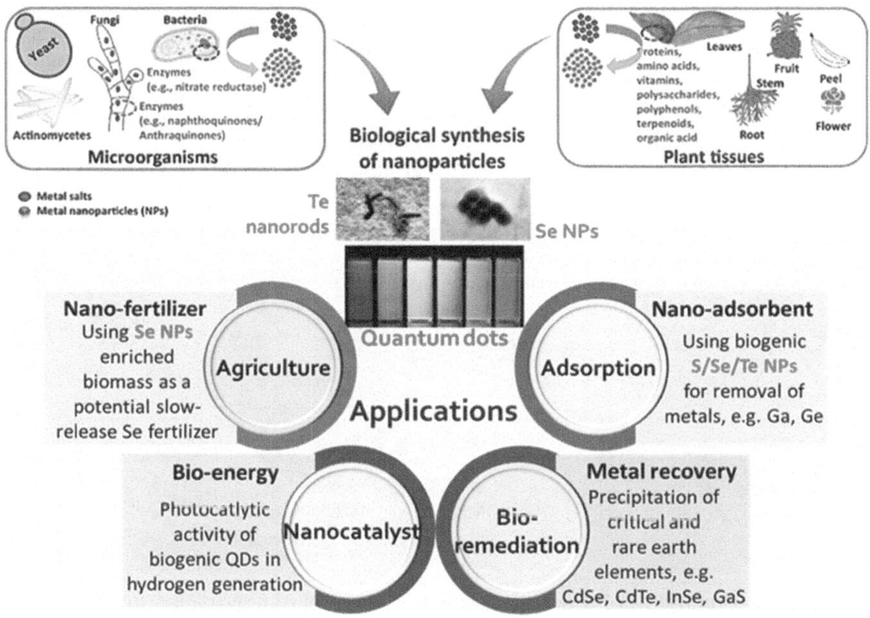

Figure 12.1 Biosynthesis and applications of chalcogen nanoparticles. Agricultural aspects are further detailed in Chapters 8 and 9, some nanocatalytic properties of selenium nanoparticles are detailed in Chapter 11.

are essential elements for a green nanoparticle synthesis. A biological approach to materials synthesis satisfies these criteria and is of benefit to the materials manufacturing industry. Using the capability of bacterial cells to manufacture useful bionanominerals offers a revolutionary method of materials synthesis that eliminates toxic organic solvents, minimizes expensive high-temperature processing and can involve the use of industrial waste as the starting material. Figure 12.2 illustrates the concept of the biomanufacturing of inorganic nanoparticles from metal and metalloid contaminated wastewaters using the anaerobic bacteria present in the granular sludge of an upflow anaerobic sludge bed (UASB) reactor, a reactor type commonly used for wastewater treatment (Jacob *et al.*, 2016; Mal *et al.*, 2017a).

12.2 THE BIOGEOCHEMICAL SELENIUM CYCLE

12.2.1 The element selenium and environmental pollution by selenium

Selenium (Se) belongs to the periodic table group 16 (together with oxygen (O), sulfur (S), and tellurium (Te)), which are called chalcogens. Se is called

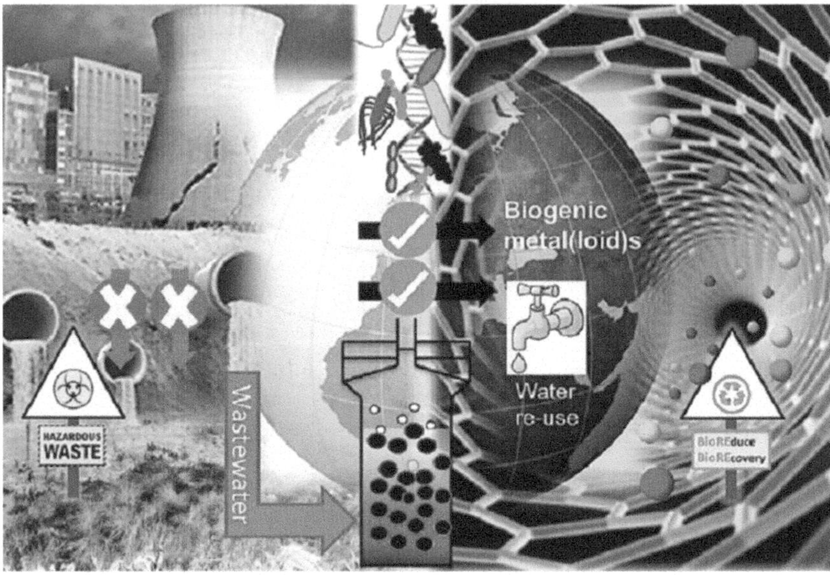

Figure 12.2 Illustration of the production of functional nanoparticles from selenium and/or metal contaminated wastewater using anaerobic bacteria present in the commonly used upflow anaerobic sludge bed (UASB) wastewater treatment reactors.

an 'essential toxin', as it is required for certain cell processes and enzymes, but becomes toxic at greater doses (Stolz *et al.*, 2006). Se is a key trace element found in representative species from all three domains of life: *Bacteria*, *Archaea* and *Eukaryota* (Oremland *et al.*, 2004; Sharma & Singh, 1984).

Generally, selenium occurs at trace levels in the environment. However, in a few areas, selenate (SeO_4^{2-}) or selenite (SeO_3^{2-}) concentrations are appreciable and selenium becomes toxic to bacteria or the animals drinking these waters (Ohlendorf, 2002; van Fleet-Stalder *et al.*, 2000). Anthropogenic activities interfere with the global selenium cycle influencing it crucially. It has been estimated that 35–40% of the total selenium emissions to the atmosphere are due to anthropogenic activities (Wen & Carignan, 2007), including foremost the combustion of coal and oil, non-ferrous metal melting and utilization of agricultural products (Lenz & Lens, 2009). Agricultural drainage waters, oil refining wastewaters and coal combustion residues contaminate the lotic, lentic and marine environment. Commonly, selenium oxyanion contamination occurs concomitantly with sulfate in different waste streams (Table 12.1). Selenium is introduced in the terrestrial compartment mainly as fertilizer or by mining activities.

Table 12.1 overview of sulfate and selenium concentrations in selenium contaminated waste streams.

Waste stream	SeO_4^{2-} [mg L^{-1}]	Se [μg L^{-1}]	Reference
Acid mine drainage	1880	16	España *et al.* (2006)
Acid mine drainage	1146	492	Simmons *et al.* (2002)
Acid seeps (Moreno shale, USA)	12,500	420	Presser (1994)
Flue gas desulfurization purge waters	815	536	Cantafio *et al.* (2001)

12.2.2 Microbial Se metabolism

Although bioreduction of selenium oxyanions (SeO_4^{2-} and SeO_3^{2-}) to Se^0 or selenide (Se^{2-}) by anaerobic microorganisms is an important part of the biogeochemical selenium cycle, it has thus far been rather poorly studied compared to the carbon, nitrogen, phosphorus and sulfur cycles. The selenium cycle is driven by prokaryotes (Chasteen & Bentley, 2003; Dungan & Frankenberger, 1998), which readily metabolize selenium for a range of metabolic functions including assimilation, methylation, detoxification and anaerobic respiration (Oremland *et al.*, 2004).

Microbes reduce Se(IV) or Se(VI) for a number of different reasons, including detoxification and energy conservation (see Chapter 3). For example, in *Bacillus selenitireducens*, the Se(IV) (e.g., SeO_3^{2-}) reduction mechanism involves energy conservation by oxidation of lactate coupled to growth via respiratory reduction of Se(IV) using Se-specific dissimilatory enzymes (Blum *et al.*, 1998). Fe(III) reducers, for example, *Shewanella oneidensis* (Klonowska *et al.*, 2005) and *Geobacter sulfurreducens* (Pearce *et al.*, 2008), can also reduce Se(IV) with c-type cytochromes implicated in electron transfer to the metalloid. Unlike *Shewanella* and *Geobacter* species, *Veillonella atypica* does not possess high concentrations of cytochromes and produces Se^0 nanospheres from Se(IV) via a hydrogenase-coupled reduction, mediated by ferredoxin.

12.2.3 Microbial synthesis of Se nanoparticles

Over the past three decades, the finer aspects of the selenium biogeochemical cycle have begun to emerge. The dissimilatory reduction of soluble SeO_4^{2-} via SeO_3^{2-} to insoluble elemental selenium (Se^0) has been shown to be a significant and rapid environmental process (Lloyd & Lovley, 2001). A particular feature of selenium is that Se^0 can be formed in a single reduction step (Figure 12.3), in contrast to elemental sulfur (S^0) which requires a two-step process (complete reduction to sulfide followed by partial oxidation to S^0). Environmental conditions (e.g., pH,

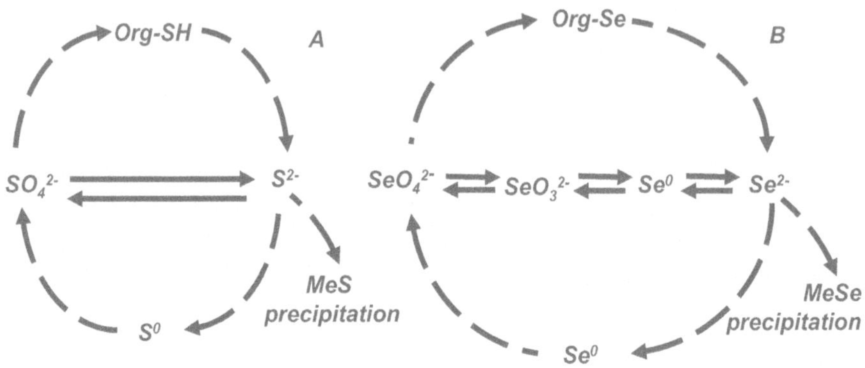

Figure 12.3 Comparison of the biological sulfur (a) and selenium (b) cycles.

temperature and salinity) and (waste)water geochemistry (electron donors and acceptors) influence the biochemical production and excretion of Se^0, as well as the physical and chemical properties of the Se^0 nanospheres. There is still much unknown regarding the factors that determine the bioreduction end product: Se^0, selenide or a mixture of oxidation states, and thus the possibility for metal selenide precipitation.

12.3 CHARACTERISTICS OF SELENIUM NANOPARTICLES

The discovery that selenium-respiring bacteria form Se^0 (Oremland *et al.*, 2004) and metal selenide (Pearce *et al.*, 2008) nanospheres offers an appealing technological potential. Indeed, these nanoparticles possess optical and semiconducting properties that are highly desired for nanotechnological applications (Figure 12.4).

12.3.1 Elemental (Se^0) nanoparticles

Se^0 particles can be formed intracellularly, extracellularly or on the cell surface. The Se^0 exospheres have a uniform diameter (0.2–0.3 mm). They eventually slough off the cell surface into the medium (Gonzalez-Gil *et al.*, 2016), from which they can be harvested and further processed. Lenz *et al.* (2008) showed that the Se^0 nanoparticle diameter and zeta potential is electron donor (acetate or hydrogen) dependent. Moreover, X-ray absorption near edge structure spectroscopy (XANES) showed that the nanospheres not only contain native Se^0, but also various selenide compounds (Figure 12.5).

Biogenic Se^0 nanospheres formed by *S. barnesii*, *B. selenitireducens* and *S. shriftii*, when grown with selenium oxyanions, were found to have unusual photo-optical and semiconducting physical properties with potential industrial

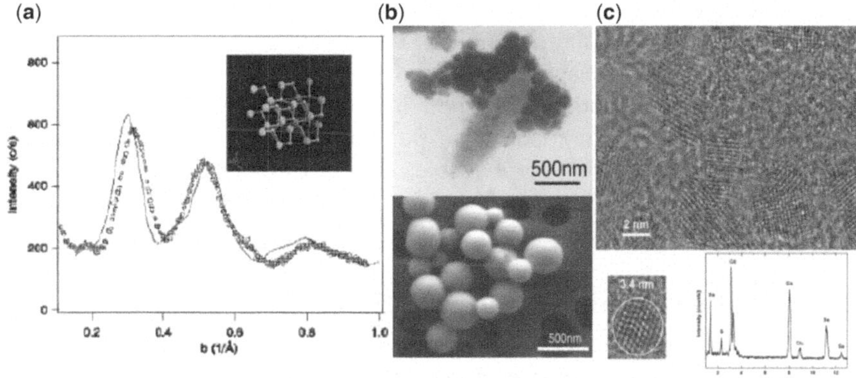

Figure 12.4 Examples of biosynthetic chalcogen nanomaterials with technological potential. (a) wide-angle X-ray scattering (WAXS) of *Schizosacchromyces pombe* CdS nanoparticles, proposed to be used in the fabrication of diodes (Kowshik *et al.*, 2002a, 2002b, 2003), (b) Transmission electron microscopy (TEM, top) and scanning electron microscopy (SEM, bottom) of Se^0 produced by *S. barnesii*, with potential applications in photocopiers and microelectronic circuits (van Fleet-Stalder *et al.*, 2000) and (c) TEM micrographs and energy-dispersive X-ray (EDX) spectra of size-fractionated 2-mercaptoethanol-stabilized CdSe quantum-dots, produced using biologically produced Se^{2-} by $SeO_3{}^{2-}$ fed *V. atypica* (Pearce *et al.*, 2008). *V. atypica* also excreted ZnSe and CdSe nanoparticles, but these were too large (30 nm) for quantum-dot type applications.

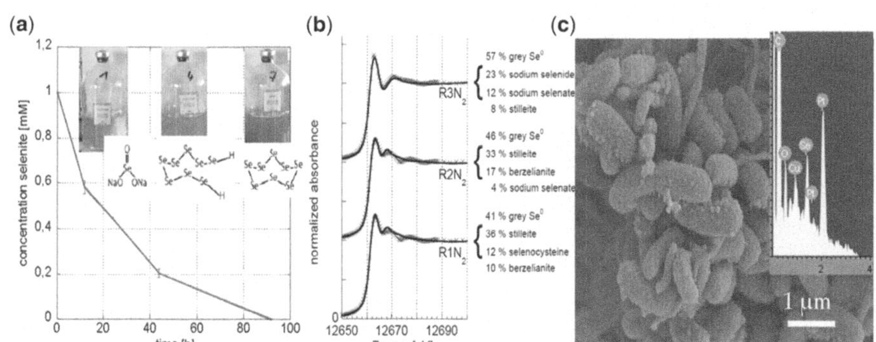

Figure 12.5 Microbiologically formed Se^0 exospheres obtained by the reduction of sodium selenite (1 mM) by a mixed culture of anaerobic microorganisms growing in a wastewater treatment bioreactor: color changes in the medium coupled to structural changes of the Se^0 (a), K-edge Fourier transforms of Se^0 precipitates (b), and scanning electron micrograph with energy dispersive X-ray analysis of Se exospheres (c) (from Lenz *et al.*, 2008).

applications in devices such as photocopiers and microelectronic circuits (Oremland *et al.*, 2004). Moreover, Dixit *et al.* (2021) described the magnetic properties of biogenic selenium nanomaterials produced by anaerobic granular sludge and elaborated on their potential application in magnetic refrigeration. Jain *et al.* (2015a, 2017) further showed biogenic Se^0 nanoparticles have a high sorption capacity for divalent cations, for example, Zn and Cu, and can thus be used as nanosorbents. Biogenic Se^0 can also be used as biofertilizer, either as soil amendment or as foliar application (Li *et al.*, 2021; see also chapter 9).

12.3.2 Metal selenide nanoparticles

A variety of regular spherical and non-spherical nanocolloids (including wires, tubes, disks, and more exotic structures like prisms and branched structures) of metal chalcogenides have been synthesized (Shanbhag *et al.*, 2007). These nanocolloids have been prepared from materials such as CdS, CdSe, CdTe, PbS, PbSe and ZnSe through a number of synthetic and biological routes (Cho *et al.*, 2005; Cozzoli *et al.*, 2005).

Chalcogen quantum-dots (QDs) are semiconducting crystals comprised of sulfur (S), selenium (Se) or tellurium (Te) in combination with transition metals, to form compounds such as zinc sulfide (ZnS), zinc selenide (ZnSe), cadmium selenide (CdSe), lead selenide (PbSe) and cadmium telluride (CdTe). QDs are prepared from a range of organic, inorganic and organometallic materials via a number of synthetic routes. As a result of the production process, QDs are coated with an organic capping agent. However, inorganic passivating agents can also be used to produce core/shell-type structures with shells formed of chalcogen compounds, for example, CdSe/CdS, CdSe/ZnS, ZnSe/ZnS, and a range of other materials, for example, composite spheres of silica and CdSe nanocrystals. As with the regular spherical QDs, the morphology of these nanocrystals can be varied, with non-spherical structures including wires, tubes, disks, and more exotic structures like prisms and branched structures.

There is a lot of excitement in the scientific community about the chemical, optical and electronic properties of such 'quantum-confined' structures, governed by their chemical composition, size and shape. Despite this excitement, there remain many unanswered questions about the mechanisms of formation of such complex nanoscale systems and the possibility of achieving these through advancements in bio-engineering to develop an industrial biotechnological process. However, initial results have highlighted the potential for using a novel biological approach in an environmentally friendly, aqueous-based synthesis to precipitate nanoscale, luminescent, semiconducting CdSe and ZnSe QD (Figure 12.6).

12.3.3 Toxicity of selenium nanoparticles

The implementation of selenium based nanoparticles (e.g., CdSe, CdTe, and PbSe) is, however, hindered by the presence of toxic elements, like Cd or Pb.

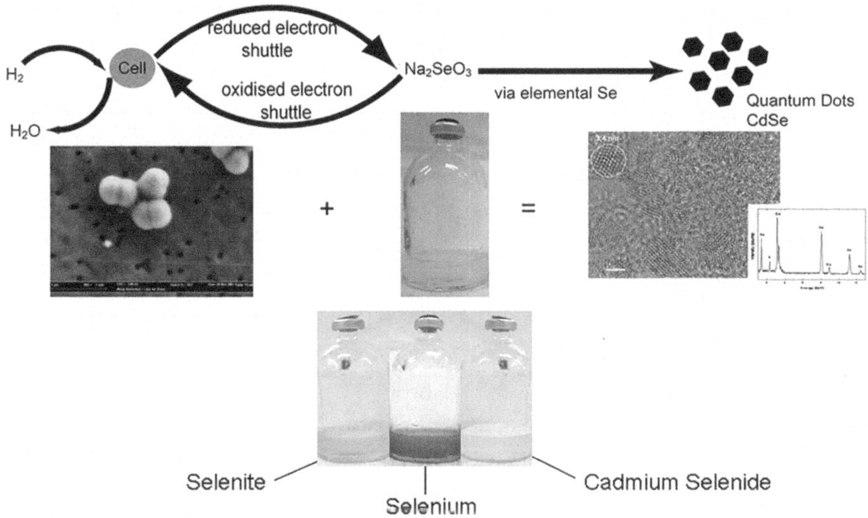

Figure 12.6 Microbial synthesis of chalcogenide-based nanoparticles via the reduction of sodium selenite using the metal-reducing bacterium *Veillonella atypica* (from Pearce *et al.*, 2008).

Technological applications thus also need to address the potential toxicity of this type of quantum-dots. This is of concern primarily because many quantum-dot core metals, such as zinc, cadmium, lead and selenium are toxic at relatively low concentrations (parts per million) and thus of considerable human health and environmental concern (Hardman, 2006; Chapter 1). Little information is available on routes of exposure to quantum-dots, their stability, aerosolization, and how they partition into different environmental compartments (e.g., soil, sediment, water, air or biota). Environmental exposure could result from, for example, leakages and spillage during manufacturing and transport (Zhang *et al.*, 2019). The toxicological properties of selenium as well as these metals in the form of nanoparticles may differ substantially from the bulk materials, but are nonetheless clearly of concern.

Hardman (2006) reviewed studies on the toxicity of quantum-dots. While some studies in cell culture systems (*in vitro*) have shown that quantum-dots are not toxic to cells, others have shown they can have toxic effects. Overall, these studies showed that toxicity depends on many factors, including the physiochemical properties (size, electrical charge, concentration, outer coating bioactivity and stability) of the quantum-dots and environmental conditions (Hardman, 2006). The main reasons for the cytotoxicity of CdSe/CdTe quantum-dots are desorption of Cd (i.e., QD core degradation), free radical formation, and interaction with intracellular components or bioavailability (uptake) of QDs. Biogenic nano-Se^0 synthesized by anaerobic granular sludge was 10-fold less toxic than chemically synthesized nano-Se^0 (Mal *et al.*, 2017c). This indicates that the presence of extracellular polymeric substances (EPS) increases the physiochemical stability of biogenic nano-Se^0 and prevents their dissolution. The presence of EPS on the surface of biogenic nano-Se^0 plays a major role in lowering the bioavailability (uptake) and toxicity of Se^0 nanoparticles. Still, detailed studies are required on the toxicity of biogenic CdSe/CdTe nanoparticles and the role EPS plays in it. It is also important to focus on biological synthesis of 'Cd-free' QDs (e.g., CuSe/CuTe) in the near future (Xu *et al.*, 2016; Supreet & Singh, 2020).

12.4 NANOSORBENTS

Biogenic Se^0 nanospheres have adsorptive properties with considerable technological potential. For instance, Johnson *et al.* (2008) reported that unstabilized nano-selenium in dry powder as well as in impregnated cloth was successful for the *in situ*, real time suppression of mercury (Hg) vapor escape following the fracture of compact fluorescent lamps. Nano-selenium was surprisingly much more effective in capturing Hg than sulfur (micro-sulfur and sulfur nanotubes), metals (micro- and nano-particles of zinc, nickel, copper and silver) and carbon (black carbon and activated carbon) nanoparticles. This unknown feature of Se^0 was attributed to the extremely high affinity of Se for mercury and the strong Hg/Se binding (Ralston *et al.*, 2007). Again, the efficiency of biogenic Se^0 needs to be demonstrated, but it is highly likely that similar Hg removal efficiencies can be obtained.

12.4.1 Metal adsorption onto biogenic nano-Se^0

The adsorption of Zn, as a model divalent heavy metal, onto biogenic nano-Se^0 produced by anaerobic granular sludge (Jain *et al.*, 2015a) and fungal *Phanerochaete chrysosporium* pellets (Espinosa-Ortiz *et al.*, 2016) has been investigated. The adsorption of Zn onto biogenic nano-Se^0 at acidic pH values follow a ligand-like (type II) adsorption mechanism (Jain *et al.*, 2015a). The adsorption of Zn onto biogenic nano-Se^0 followed a two-step process at near-neutral pH. X-ray photoelectron spectroscopy (XPS) suggested the precipitation of one of the Zn species which could have been ZnO, ZnSe, Zn $(OH)_2$, or $ZnCO_3$. Preliminary extended X-ray absorption of fine structure (EXAFS) data analysis suggests the absence of ZnSe, thus discarding the occurrence of disproportionation of elemental selenium. The ζ-potential of biogenic nano-Se^0 became less negative at higher loading of Zn, resulting in lower colloidal stability, thus leading to higher retention of biogenic nano-Se^0 on the filter when compared to biogenic nano-Se^0 after filtration but without adsorption (Jain *et al.*, 2015a).

The selective adsorption of heavy metals onto biogenic nano-Se^0 was explored by Jain *et al.* (2016a, b). It was found that the metals to biogenic nano-Se^0 ratio (v:v) and pH can be manipulated to optimize the selective adsorption of Cu. At the metal to biogenic nano-Se^0 ratio of 1:1 (v:v) and theoretical pH of 4.3, Cu was found to adsorb 4.7 times more onto biogenic nano-Se^0 than the total sum of Cd and Zn adsorbed, when an equimolar mixture of Cu, Cd and Zn was used in the adsorption experiments. The selective preference of Cu onto biogenic nano-Se^0 depends on the intrinsic properties of Cu: smaller ionic radius, higher electronegativity, higher ratio of ionization potential to ionic radius and higher first stability constant of metal hydroxo species and acetate complexes (Jain *et al.*, 2016b). The selective preference of Cu also depends on the presence of functional groups such as hydroxyl and carboxyl on the surface of biogenic nano-Se^0. Indeed, Fourier-transform infrared spectroscopy (FT-IR) analysis of biogenic nano-Se^0 loaded with heavy metals confirmed the interaction of hydroxyl and carboxyl groups on the surface of biogenic nano-Se^0 with heavy metals (Jain *et al.*, 2016b).

12.4.2 Role of EPS in the fate of selenium in the environment and bioreactors

Figure 12.7 demonstrates the role of the extracellular polymeric substances (EPS) in the formation and characteristics of biogenic nano-Se^0. EPS is mainly composed of polysaccharides, proteins, humic substances, lipids and nucleic acids (D'Abzac *et al.*, 2010; Sheng *et al.*, 2010). Proteins and polysaccharides are the major components of EPS. Generally, EPS is known to retard or prevent the dispersion of nanomaterials such as silver nanoparticles (Kang *et al.*, 2014; Tourney & Ngwenya, 2014). In contrast, Jain *et al.* (2015b) showed that the EPS provides colloidal stability to the biogenic nano-Se^0 due to less negative ζ-potential values (Buchs *et al.*, 2013; Dhanjal & Cameotra, 2010; Jain *et al.*, 2015a). Bare elemental selenium has been reported to have a ζ-potential of −10 mV as compared to −30 mV observed for biogenic nano-Se^0 (Dhanjal & Cameotra, 2010; Jain *et al.*, 2015a). This colloidal stability of the biogenic nano-Se^0 is the reason for their presence in bioreactor effluents (Lenz *et al.*, 2008) as well as the high mobility of biogenic nano-Se^0 in the environment (Buchs *et al.*, 2013).

The presence of EPS on the surface of biogenic nano-Se^0 determines the mechanism of interaction of the biogenic nano-Se^0 with heavy metals. The capture of mercury from the vapor phase by elemental selenium is due to the precipitation of mercury selenide on the surface of the elemental selenium (Fellowes *et al.*, 2011; Johnson *et al.*, 2008). However, the interaction of Zn with biogenic nano-Se^0 does not lead to formation of ZnSe as observed by Jain *et al.* (2015a). The interaction of heavy metals with biogenic nano-Se^0 is essentially an interaction of heavy metals with the EPS layer present on the surface of the biogenic nano-Se^0. Indeed, EPS is known to interact with heavy metals (D'Abzac *et al.*, 2010). The presence of amine and carboxylate groups on the surface can lead to the adsorption of heavy metals by ligand-like (Type II) adsorption, as

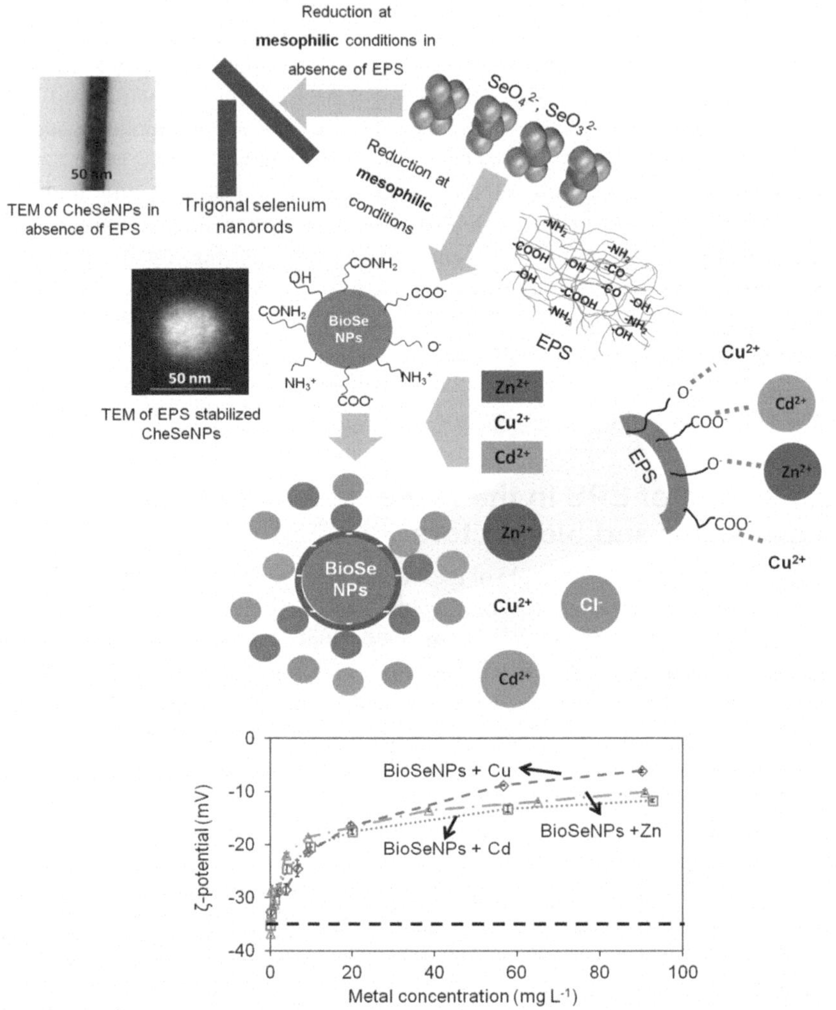

Figure 12.7 Summary of the role of EPS in determining the properties of the biogenic nano-Se^0 (Jain, 2014).

observed during the adsorption of Cu on cellulose modified with poly(glycidyl methacrylate) and polyethyleneimine (Navarro *et al.*, 2001). The presence of these surface groups on biogenic nano-Se^0 was confirmed by XPS and FT-IR analyses (Jain *et al.*, 2015a, b). Thus, the presence of such a layer of EPS on biogenic nano-Se^0 is responsible for ligand-like (Type II) adsorption of Zn onto biogenic nano-Se^0 at acidic pH (Jain *et al.*, 2015a). The presence of hydroxyl and carboxyl groups on the surface of biogenic nano-Se^0 can be attributed to the presence of EPS. The FT-IR data confirm that these groups interact with the heavy metals. The presence of these groups and the higher first stability constant of metal hydroxo and metal acetate complexes for Cu lead to a higher preference of biogenic nano-Se^0 towards Cu (Jain *et al.*, 2015a; Sitko *et al.*, 2013). Thus, the presence of EPS on the surface of biogenic nano-Se^0 is further affecting the affinity of the biogenic nano-Se^0 towards different heavy metals.

12.5 PHOTONIC NANOCRYSTALS AND PHOTOCATALYSTS

12.5.1 Photonic colloidal nanocrystals – quantum-dots

In recent years, nanoscience and nanotechnology have brought a great revolution in different areas (Grim *et al.*, 2015). In particular, the synthesis of transition metal nanoparticles has been of great relevance for their use in areas such as biomedicine (see Chapter 10), antimicrobial properties or catalytic applications for chemical synthesis (see Chapter 11). A particular class of these materials are colloidal quantum-dots (QDs), solution-processed semiconductor nanocrystals of 2–10 nanometers (10–50 atoms) in size. QDs have unique properties because, at this size, they behave differently to their bulk equivalents and exhibit unprecedented tunability, enabling completely new applications in science and technology (see Chapter 11). They are particularly suited to initiate a future generation of photonic devices (Figure 12.4).

Most of the research interest regarding semiconducting QDs has been focused on their electrical conductivity and unique optical (Figure 12.8) properties, which can be greatly altered and controlled by an external stimulus (e.g., voltage or photon flux), making them critical components of many different kinds of electrical circuits and optical applications. The properties of quantum structures are strongly dependent on their size and shape, and the quality of their internal crystal structure (e.g., structural defects) can serve as trapping sites for electrons or holes.

12.5.2 Biomanufacturing of quantum-dots using bacteria during wastewater treatment

Strategies in QD synthesis focus on the development of methodologies that allow the creation of these nanomaterials in a simple, efficient and sustainable way (Sakimoto *et al.*, 2017). For this reason, researchers have started to apply biological entities as an elegant approach for the direct synthesis of metal chalcogen nanoparticles. Microorganisms have the ability to induce the formation of QDs, as outlined in Section 12.3.2, sometimes even controlling the size and structural shape, and avoiding aggregation problems (Xu *et al.*, 2016). Figure 12.6 illustrated this microbial synthesis route, based on the precipitation of the divalent metal and the microbially formed selenide (Figure 12.3(b)).

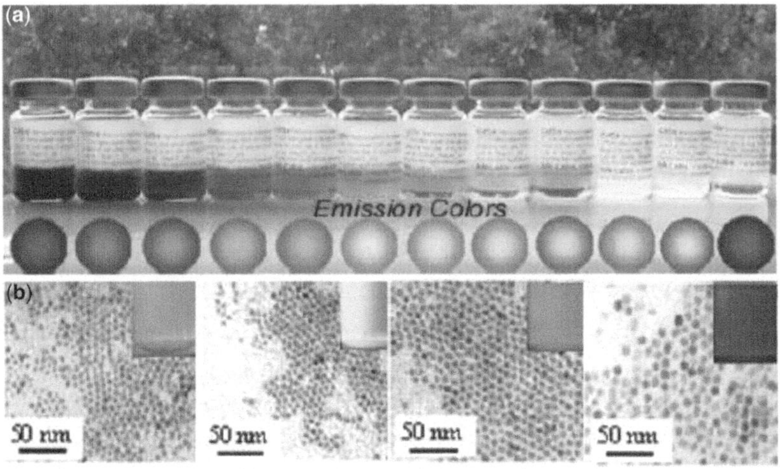

Figure 12.8 Commercially available quantum-dots, prepared via chemical synthesis. (a) 12 Wavelength CdSe Kit of quantum-dots with different emission wavelengths between 480 and 640 nm and (b) TEM micrographs with corresponding emission of quantum-dot aliquots, marketed by Nanomaterials & Nanofabrication Laboratories (Image from www.nn-labs.com).

Metal ions such as cadmium or zinc present in the selenium-rich wastewaters influence the microbial reduction and the fate of biogenic Se(0) nanoparticles (Mal *et al.*, 2016a). When metal ions are available, they can either form metal-selenide nanoparticles after reacting with selenide (HSe^-) generated during the microbial reduction of Se-oxyanions or they form metal–Se(0) complexes due to adsorption of metals onto Se(0) nanoparticles (Jain *et al.*, 2016b; Mal *et al.*, 2016b). Mal *et al.* (2017a) showed that microbial reduction of selenite (Se(IV)) in the presence of Cd(II) by anaerobic granular sludge results in the formation and deposition of cadmium selenide (CdSe) nanoparticles by the reaction of Cd (II) with biogenic HSe^-, but also Se(0) nanoparticles and Se(0)-Cd complexes can be formed. In further research, Wadgaonkar *et al.* (2018) showed that mixed microbial cultures present in anaerobic wastewater treatment reactors can also form selenium (CdSe, ZnSe and PbSe) as well as tellurium (CdTe and ZnTe) based nanoparticles, which are capped by a layer of extracellular polymers (Figure 12.9).

Little is known about the mechanism of metal selenide nanoparticle formation and their recovery from a bioreactor. The principles of nucleation, crystal growth and agglomeration also apply to metal selenide formation and some of the processes described for metal sulfide formation can be adopted to metal selenide formation. For the former, the focus is more on the formation of large, well-settling crystals of metal sulfides of base metals (e.g., ZnS, CuS, NiS, CoS, CdS and PbS) for efficient bioseparation (Esposito *et al.*, 2006). Metals are

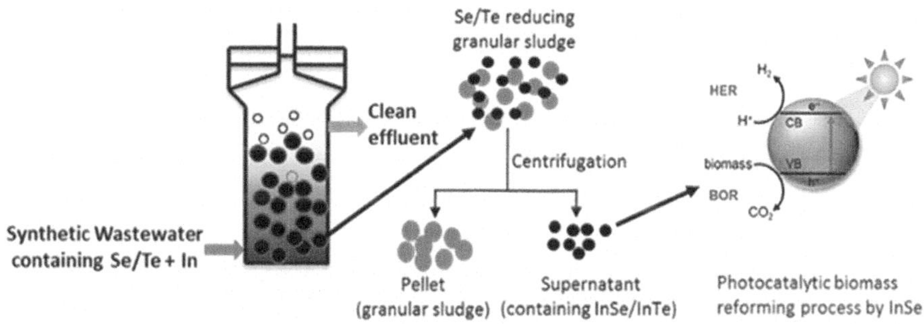

Figure 12.9 Schematic representation of microbial synthesis of chalcogenide-based nanoparticles using anaerobic granular sludge in an upflow anaerobic granular sludge bed (UASB) bioreactor. Thus, nanoparticles of ZnS (Jacob *et al.*, 2020), elemental selenium (Se^0; Lenz *et al.*, 2008), CdSe (Mal *et al.*, 2016b), elemental tellurium (Mal *et al.*, 2017b) and CdTe (Wadgaonkar *et al.*, 2018) have been synthesized. Mal *et al.* (2020) showed the nanoparticles are mainly present in the loosely bound fraction of the EPS, and can thus be harvested by centrifugation of the sludge. Most reactors were operated at mesophilic (30°C) temperatures, with a few studies investigating psychrophilic (Zeng *et al.*, 2019) and thermophilic (Dessì *et al.*, 2016) conditions. BOR: biomass oxidation reaction; HER: hydrogen evolution reaction.

typically separated or concentrated as sulfides using Na_2S, NaHS or H_2S (Veeken *et al.*, 2003). Biogenic sulfide produced by sulfate reducing bacteria (SRB) is an interesting alternative sulfide source, especially when besides metals also sulfate is present in the wastewater, for example, wastewaters from metal refineries and acid mine drainage (White & Gadd, 1999). Metal sulfides have very low solubility products, and nucleation thus occurs much faster than crystal growth, so that nanoparticles and fines of many metals, for example, CdS, ZnS and PbS, are also formed during bioprecipitation (Klaus-Joerger *et al.*, 2001). Similarly, Jacob *et al.* (2020) showed ZnS nanoparticles can be formed using sulfate reduction by anaerobic granular sludge. The CdSe nanoparticles are present in the loosely bound fraction of the EPS of anaerobic granules, and can thus be removed from the granular sludge by centrifugation (Mal *et al.*, 2020). In some cases, however, metal sulfide formation can result in a complete encrustation of the sulfate reducing cell (Jiang *et al.*, 2014) and it can be preferred to produce the nanoparticles in a separate reactor, downstream of the sulfide or selenide producing bioreactor. Thus, König *et al.* (2006) developed dynamic modeling and feedback control of the ZnS precipitation process. A dynamic mass balance model for ZnS precipitation was developed and an adequate strategy for controlling the sulfide (using a pS-electrode) and pH-level in a continuous flow stirred tank reactor (CFSTR) was designed in order to create appropriate conditions for precipitation, independent of the conditions in the sulfide producing bioreactor. A similar approach could be developed for the production of metal selenide nanoparticles, using biologically produced selenide in an upfront bioreactor.

12.5.3 Photocatalytic processes using chalcogen nanoparticles

In recent years, metal sulfide and selenide nanoparticles, particularly ZnS and CdS, have been intensively studied as active photocatalysts in virtue of their unique catalytic functions (Qin *et al.*, 2011). ZnS is a promising photocatalyst for degradation of organic pollutants (Jacob *et al.*, 2020), photoreduction of CO_2 (Meng *et al.*, 2017) and H_2 production (Kuehnel & Reisner, 2018) because of the rapid generation of electron–hole pairs by photoexcitation and highly negative reduction potentials of excited electrons (Chen *et al.*, 2013; Qin *et al.*, 2011). However, the bandgap of ZnS is 3.66 eV (Yoneyama, 1997), which is too large for visible light response. Doping with metal ions and combining with various narrow bandgap semiconductors have been applied to make ZnS have visible light activity. Ni, Cu and Pb-doped ZnS showed visible-light photocatalytic activities for H_2 production (Chen *et al.*, 2013).

Although many inorganic semiconductors have been used for photocatalytic CO_2 conversion, the performance of most of these materials is still rather sluggish due to the low conduction band level and/or poor activation ability of CO_2 molecules (Tu *et al.*, 2014). A promising strategy is to couple the inorganic semiconductor with an organic metal complex as a photosensitizer/electrocatalyst to achieve a more efficient reaction rate (Meng *et al.*, 2017; Sekizawa *et al.*, 2013). The drawback of these hybrid photocatalysts is that they generally need to be operated in an organic medium (e.g., acetonitrile or alcohol) in the presence of organic sacrificial agents (e.g., triethanolamine or alcohol). Therefore, a reaction system that can achieve efficient and stable CO_2 conversion is urgently needed. Chalcogen based nanoparticles, in particular metal selenides, are good candidates for this.

12.5.4 Nanobiohybrids

12.5.4.1 Microorganism – quantum-dot nanohybrids

Nanobiohybrids are a nanoscale combination of engineered nanomaterials with bioactive substances (Palomo, 2019). Various kinds of components can be utilized in preparation of nanobiohybrids. Among them, naturally obtained materials, for both the nano and bio part, are gaining interest because of their eco-friendliness and bio-compatibility (Chinnaiyan *et al.*, 2019). In order to maximize applicability of nanobiohybrids, it is important to develop various nanomaterials with tailor-made physico-chemical properties as well as to control the interactions between the nanomaterials and bio-component of the hybrids.

An important issue is the selection of the biological entity of the nanohybrid. One of the main strategies described in the literature for this green synthesis approach is based on the use of microorganisms (Jacob *et al.*, 2016). Prokaryotic or eukaryotic microorganisms have been employed for the preparation of nanoparticles of different metals (Au, Ag, Cd, Pt, Zn and Fe_3O_4) under moderate pressures and temperatures. The advantage is that microorganisms secrete large quantities of enzymes, which participate in the enzymatic reduction of metal ions (Nancharaiah & Lens, 2015). The localization and morphology of the nanoparticles depend on the microbial species because of the different enzymes and proteins that can be involved in the biomineralization, generating different sizes of nanoparticles.

One of the candidates for the nano-part of bacterial nanohybrids are quantum-dots (Sakimoto *et al.*, 2017). Quantum-dots are semiconductor nanocrystals with diameters less than 100 nm, and are composed of heavy metals (Cd, Zn, …) and chalcogens (S, Se or Te) (Xu *et al.*, 2016; see section 12.5.1). Because of their very specific light spectra, they are particularly suited to initiate a future generation of photonic devices. The discovery of the quantum confinement effect, now just over 30 years ago, and the first organic 'hot-injection' quantum-dot synthesis developed in 1993 have led to a significant advance in the production of high-quality size-tunable, monodisperse nanocrystals with high photoluminescence quantum efficiency (Grim *et al.*, 2015). This new material class has received substantial attention ever since, with an increasing range of applications, including photoactive compounds in nanobiohybrids.

12.5.4.2 Photosensitized nanohybrids

Quantum-dot bacteria nanobiohybrids combine the best of two worlds: the light-harvesting capabilities of semiconductors with the catalytic power of biology. Several quantum-dot based nanobiohybrids have been constructed, either with enzymes or microbial cells as the biological part of the hybrid. Electron transfer from illuminated quantum-dots to enzymes has been shown to support enzyme activity in hybrids of CdS with nitrogenase (Brown *et al.*, 2016a), CdSe with $NADP^+$ reductase (Brown *et al.*, 2016b) and CdTe with hydrogenase (Brown *et al.*, 2010). Sakimoto *et al.* (2016) showed the self-photosensitization of the non-photosynthetic bacterium *Moorella thermoacetica* by CdS bioprecipitation, and the photosynthetic production of acetate from CO_2 by this nanobiohybrid (Figure 12.10). Ding *et al.* (2019) further showed that CdS/ZnS core-shell quantum-dots penetrate *Azotobacter vinelandii* and *Cupriavidus necator* cells, where they bind to, respectively, the histi-tagged MoFe nitrogenase and FeS clusters of hydrogenases and quinones (Figure 12.11). Upon illumination, these nanobiohybrids demonstrated conversion of high yields of target products (ethylene and polyhydroxybutyrate) from CO_2, comparable or even exceeding (>150%) native production levels (glucose as substrate).

Recently, nanobiohybrids using other inorganic light harvesters have been developed. Guo *et al.* (2018) evidenced that yeast (*Saccharomyces cerevisiae*) cells functionalized with indium phosphide (InP) quantum-dots harvested

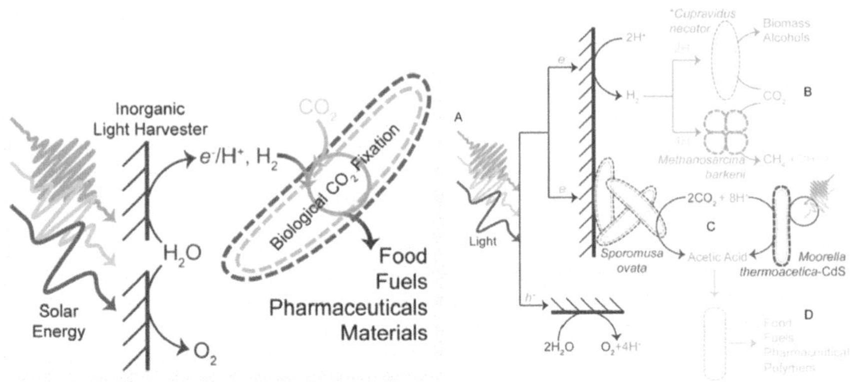

Figure 12.10 *Left*: Schematic presentation of the photosensitization of biohybrid systems, which combines the strengths of inorganic materials and biological catalysts by exploiting semiconductor broadband light absorption to capture solar energy and subsequently transform it into valuable CO_2-derived chemicals using the metabolic pathways in living organisms. *Right*: Overview of different nano bio-hybrid architectures. Utilizing electrons derived from a semiconductor light harvester (a), photosynthetic biohybrid systems channel reducing equivalents to generate H_2 (b) to feed CO_2 reducing microorganisms. These electrons may also go directly to the bacterium (c) to generate reduced CO_2 products, such as acetic acid. This acetic acid may then be fed to genetically engineered organisms (d) to upgrade to a wide range of products (from Sakimoto *et al.*, 2017, and references therein).

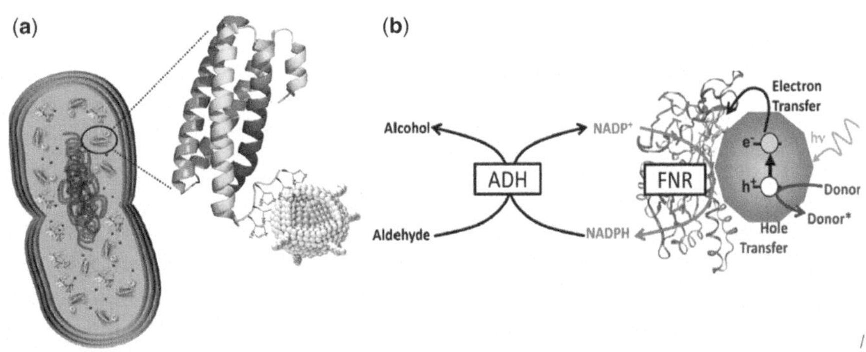

Figure 12.11 Schematic presentation of (a) the binding of a Cd/ZnS core shell quantum-dot to intracellular enzymes via histi-tag bonding upon incubating *Azotobacter vinelandii* cells in medium containing the quantum-dots (after Ding *et al.*, 2019) and (b) electron transfer in biohybrids of CdSe quantum-dots and ferredoxin NADP$^+$ reductase for photocatalytic regeneration of NADPH (After Brown *et al.*, 2016a, b). ADH: alcohol dehydrogenase; NADP$^+$: nicotinamide adenine dinucleotide phosphate; FNR: ferredoxin NADP$^+$-reductase.

photogenerated electrons from the illuminated nanoparticles and used them for the cytosolic regeneration of redox cofactors. Xu *et al.* (2019) reported on bioplastic production with light energy by coupling *Ralstonia eutropha* with the photocatalyst graphitic carbon nitride. Also intracellular gold nanoclusters can be used to photosensitize bacteria for solar fuel production (Zhang *et al.*, 2018).

12.5.4.3 Photobioreactors for photosensitized nanobiohybrids

Light based production of fuels and fine chemicals using quantum-dot bacteria nanobiohybrids can make use of the vast experience of growing microalgae. Alternatively, solar fuels and biocommodities can be produced by growing the photosensitized nanobiohybrids in cell suspension bioreactors (Lens *et al.*, 2003). Several commonly used reactor systems for microalgae growth, including raceway ponds (Karya *et al.*, 2013), flat plate reactors (Rada-Ariza *et al.*, 2019), photo-active granules (Bing *et al.*, 2018) and algal biofilms (Moreno Osorio *et al.*, 2019) can be used. Microalgal culture systems have low biomass concentrations and problems with biomass/liquid separation, which constitute the main drawbacks in the scale-up and industrial application of microalgal processes (Gupta *et al.*, 2016). Microalgal biofilm systems could provide an alternative approach to resolve the microalgae-based biomass production challenges (Moreno Osorio *et al.*, 2019), and can thus also be explored for photosynthetic biohybrid systems. Biofilm photobioreactors can be grouped into three categories: permanently submerged, intermittently submerged and perfused systems (Berner *et al.*, 2015). However, photobioreactors that are used for algae axenic biomass cultivation do not always show equal results in wastewater treatment (Han *et al.*, 2017). There is thus a need to determine the efficiency and performance of photobioreactors when using quantum-dot bacteria nanobiohybrids for H_2 production and CO_2 derived fuels.

REFERENCES

Berner F., Heimann K. and Sheehan M. (2015). Microalgal biofilms for biomass production. *Journal of Applied Phycology*, **27**, 1793–1804.

Bing Z., Lens P. N. L., Shi W., Zhang R., Zhang Z, Guo Y., Bao X. and Cui F. (2018). The attachment potential and *N*-acyl-homoserine lactone-based quorum sensing in aerobic granular sludge and algal-bacterial granular sludge. *Applied Microbiology and Biotechnology*, **102**, 5343–5353.

Blum J. S., Bindi A. B., Buzzelli J., Stolz J. F. and Oremland R. S. (1998). *Bacillus arsenicoselenatis* sp. nov. and *Bacillus selenitireducens* sp. nov.: two haloalkaliphiles from Mono Lake, California that respire oxyanions of selenium and arsenic. *Archives Microbiology*, **171**, 19–30.

Bottero J. Y., Rose J. and Wiesner M. R. (2006). Nanotechnologies, tools for sustainability in a new wave of water treatment processes. *Integrated Environmental Assessment and Management*, **2**, 391–395.

Brown K. A., Dayal S., Ai X., Rumbles G. and King P. W. (2010). Controlled assembly of hydrogenase-CdTe nanocrystal hybrids for solar hydrogen production. *Journal American Chemical Society*, **132**, 9672–9680.

Brown K. A., Wilker M. B., Boehm M., Hamby H., Dukovic G. and King P. W. (2016a). Photocatalytic regeneration of nicotinamide cofactors by quantum dot–enzyme biohybrid complexes. *ACS Catalysis*, **6**, 2201–2204.

Brown K. A., Harris D. F., Wilker M. B., Rasmussen A., Khadka N., Hamby H., Keable S., Dukovic G., Peters J. W., Seefeldt L. C. and King P. W. (2016b). Light-driven dinitrogen reduction catalyzed by a CdS:nitrogenase MoFe protein biohybrid. *Science*, **352**, 448–450.

Buchs B., Evangelou M. W.-H., Winkel L. and Lenz M. (2013). Colloidal properties of nanoparticular biogenic selenium govern environmental fate and bioremediation effectiveness. *Environmental Science and Technology*, **47**, 2401–2407.

Cantafio A. W., Hagen K. D., Lewis G. E., Bledsoe T. L., Nunan K. M. and Macy J. M. (2001). Pilot-scale selenium bioremediation of san joaquin drainage water with *Thauera selenatis*. *Applied and Environmental Microbiology*, **62**, 3298–3303.

Chasteen T. G. and Bentley R. (2003). Biomethylation of selenium and tellurium: microorganisms and plants. *Chemical Reviews*, **103**, 1–25.

Chen J., Xin F., Qin S. and Yin X. (2013). Photocatalytically reducing CO_2 to methyl formate in methanol over ZnS and Ni-doped ZnS photocatalysts. *Chemical Engineering Journal*, **230**, 506–512.

Chinnaiyan S. K., Deivasigamani K. and Gadela V. R. (2019). Combined synergetic potential of metformin loaded pectin-chitosan biohybrids nanoparticle for NIDDM. *International Journal of Biological Macromolecules*, **125**, 278–289.

Cho K.-S., Talapin D. V., Gaschler W. and Murray C. B. (2005). Designing PbSe nanowires and nanorings through oriented attachment of nanoparticles. *Journal of the American Chemical Society*, **127**, 7140–7147.

Cozzoli P., Manna L., Curri M., Kudera S., Giannini C., Striccoli M. and Agostiano A. (2005). Shape and phase control of colloidal ZnSe nanocrystals. *Chemistry of Materials*, **17**, 1296–1306.

D'Abzac P., Bordas F., Van Hullebusch E., Lens P. N. L. and Guibaud G. (2010). Extraction of extracellular polymeric substances (EPS) from anaerobic granular sludges: comparison of chemical and physical extraction protocols. *Applied Microbiology and Biotechnology*, **85**, 1589–1599.

Dessì P., Jain R., Singh S., Seder-Colomina M., van Hullebusch E. D., Rene E. R., Ahammad S. Z., Carucci A. and Lens P. N. L. (2016). Effect of temperature on selenium removal from wastewater by UASB reactors. *Water Research*, **94**, 146–154.

Dhanjal S. and Cameotra S. S. (2010). Aerobic biogenesis of selenium nanospheres by Bacillus cereus isolated from coalmine soil. *Microbial Cell Factories*, **9**, 52.

Ding Y., Bertram J. R., Eckert C., Bommareddy R. R., Patel R., Conradie A., Bryan S. and Nagpal P. (2019). Nanorg microbial factories: light-driven renewable biochemical synthesis using quantum dot-bacteria nanobiohybrids. *Journal American Chemical Society*, **141**, 10272–10282.

Dixit R., Gupta A., Jordan N., Zhou S., Schild D., Weiss S., Guillon E., Jain R. and Lens P. (2021). Magnetic properties of biogenic selenium nanomaterials. *Environmental Science and Pollution Research*. https://doi.org/10.1007/s11356-020-11683-2

Dungan R. S. and Frankenberger W. T. (1998). Reduction of selenite to elemental selenium by *Enterobacter cloacae* SLD1a-1. *Journal of Environmental Quality*, **27**, 1301–1306.

España J. S., Pamo E. L., Pastor E. S., Andrés J. R. and RubÍ J. A. M. (2006). The impact of acid mine drainage on the water quality of the odiel river (Huelva, Spain): evolution of recipitate mineralogy and aqueous geochemistry along the Concepción-Tintillo segment. *Water Air Soil Pollution*, **173**, 21–49.

Espinosa-Ortiz E. J., Shakya M., Jain R., Rene E. R., van Hullebusch E. D. and Lens P. N. L. (2016). Sorption of zinc onto elemental selenium nanoparticles immobilized in *Phanerochaete chrysosporium* pellets. *Environmental Science and Pollution Research*, **23**(21), 21619–21630.

Esposito G., Veeken A., Weijma J. and Lens P. N. L. (2006). Effect of the use of biogenic sulfide on ZnS precipitation under different process conditions. *Separation and Purification Technology*, **51**, 31–39.

Fellowes J. W., Pattrick R. A. D., Green D. I., Dent A., Lloyd J. R. and Pearce C. I. (2011). Use of biogenic and abiotic elemental selenium nanospheres to sequester elemental mercury released from mercury contaminated museum specimens. *Journal of Hazardous Materials*, **189**, 660–669.

Fleischer T. and Grunwald A. (2008). Making nanotechnology developments sustainable. A role for technology assessment? *Journal of Cleaner Production*, **16**, 889–898.

Gericke M. and Pinches A. (2006). Biological synthesis of metal nanoparticles. *Hydrometallurgy*, **83**, 132–140.

Gonzalez-Gil G., Lens P. N. L. and Saikaly P. E. (2016). Selenite reduction by anaerobic microbial aggregates: microbial community structure, and proteins associated to the produced selenium spheres. *Frontiers in Microbiology*, **7**, 571.

Grim J. Q., Manna L. and Moreels I. (2015). A sustainable future for photonic colloidal nanocrystals. *Chemical Society Reviews*, **44**, 5897–5914.

Guo J., Suástegui M., Sakimoto K. K., Moody V. M., Xiao G., Nocera D. G. and Joshi N. S. (2018). Light-driven fine chemical production in yeast biohybrids. *Science*, **362**, 813–816.

Gupta P. L., Choi H. J. and Lee S. M. (2016). Enhanced nutrient removal from municipal wastewater assisted by mixotrophic microalgal cultivation using glycerol. *Environmental Science and Pollution Research*, **23**, 10114–10123.

Han T., Lu H. F., Ma S. S., Zhang Y. H., Liu Z. D. and Duan N. (2017). Progress in microalgae cultivation photobioreactors and applications in wastewater treatment: a review. *International Journal Agrictural and Biological Engineering*, **10**, 1–29.

Hardman R. (2006). A toxicologic review of quantum dots: toxicity depends on physiochemical and environmental factors. *Environmental Health Perspectives*, **114**, 165–172.

Helland A. and Kastenholz H. (2008). Development of nanotechnology in light of sustainability. *Journal of Cleaner Production*, **16**, 885–888.

Helland A., Kastenholz H., Arnfalk P., Tidell A. and Deppert K. (2006). Nanoparticulate materials and regulatory policy in Europe: an analysis of stakeholder perspectives. *Journal of Nanoparticle Research*, **8**, 709–719.

Jacob M. J., Lens P. N. L. and Mohan R. B. (2016). Microbial synthesis of chalcogenide semiconductor nanoparticles: a review. *Microbial Biotechnology*, **9**, 11–21.

Jacob J. M., Sinharoy A. and Lens P. N. L. (2020). Photocatalytic degradation of Congo Red by zinc sulfide quantum dots produced by anaerobic granular sludge. *Environmental Technology*. doi: 10.1080/09593330.2020.1856940.

Jain R. (2014). Biogenic Nanoparticles of Elemental Selenium: Synthesis, Characterization and Relevance in Wastewater Treatment. PhD thesis, UNESCO-IHE, Delft, The Netherlands, p. 259.

Jain R., Jordan N., Schild D., van Hullebusch E. D., Weiss S., Franzen C., Farges F., Hübner R. and Lens P. N. L. (2015a). Adsorption of zinc by biogenic elemental selenium nanoparticles. *Chemical Engineering Journal*, **260**, 855–863.

Jain R., Jordan N., Weiss S., Foerstendorf H., Heim K., Kacker R., Hübner R., Kramer H., van Hullebusch E. D. and Lens P. N. L. (2015b). Extracellular polymeric substances govern the surface charge of biogenic elemental selenium nanoparticles. *Environmental Science & Technology*, **49**, 1713–1720.

Jain R., Dominic D., Jordan N., Rene E. R., Weiss S., van Hullebusch E. D., Hübner R. and Lens P. N. L. (2016a). Higher Cd adsorption on biogenic elemental selenium nanoparticles. *Environmental Chemistry Letters*, **14**, 381–386.

Jain R., Dominic D., Jordan N., Rene E. R., Weiss S., van Hullebusch E. D., Hübner R. and Lens P. N. L. (2016b). Preferential adsorption of Cu in a multi-metal mixture onto biogenic elemental selenium nanoparticles. *Chemical Engineering Journal*, **284**, 917–925.

Jain R., Jordan N., Tsushima S., Hübner R., Weiss S. and Lens P. N. L. (2017). Shape change of biogenic elemental selenium nanomaterials from nanospheres to nanorods decreases their colloidal stability. *Environmental Science: Nano*, **4**(5), 1054–1063.

Jiang N., Ying G.-L., Liu S.-Y., Shen L., Hu J., Dai L.-J., Yang X.-Y., Tian G. and Su B.-L. (2014). Amino acid-based biohybrids for nano-shellization of individual desulfurizing bacteria. *Chemical Communications*, **50**, 15407s–115410.

Johnson N. C., Manchester S., Sarin L., Gao Y., Kulaots I. and Hurt R. H. (2008). Mercury vapor release from broken compact fluorescent lamps and in situ capture by new nanomaterial sorbents. *Environmental Science and Technology*, **42**, 5772–5778.

Kang F., Alvarez P. J. and Zhu D. (2014). Microbial extracellular polymeric substances reduce Ag^+ to silver nanoparticles and antagonize bactericidal activity. *Environmental Science and Technology*, **48**, 316–322.

Karya N., Van der Steen N. P. and Lens P. N. L. (2013). Photo-oxygenation to support nitrification in an algal–bacterial consortium treating artificial wastewater. *Bioresource Technology*, **134**, 244–250.

Klaus-Joerger T., Joerger R., Olsson E. and Granqvist C-G. (2001). Bacteria as workers in the living factory: metal-accumulating bacteria and their potential for materials science. *Trends in Biotechnology*, **19**, 15–20.

Klonowska A., Heulin T. and Vermeglio A. (2005). Selenite and tellurite reduction by *Shewanella oneidensis*. *Applied and Environmental Microbiology*, **71**, 5607–5609.

König J., Keesman K. J., Veeken A. and Lens P. N. L. (2006). Dynamic modelling and process control of ZnS precipitation. *Separation Science Technology*, **41**, 1025–1042.

Kowshik M., Deshmukh N., Vogel W., Urban J., Kulkarni S. K. and Paknikar K. M. (2002a). Microbial synthesis of semiconductor CdS nanoparticles, their characterization and their use in the fabrication of an ideal diode. *Biotechnology and Bioengineering*, **78**, 583–588.

Kowshik M., Vogel W., Urban J., Kulkarni S. K. and Paknikar K. M. (2002b). Microbial synthesis of semiconductor PbS nanocrystallites. *Advanced Materials*, **14**, 815–818.

Kowshik M., Ashtaputre S., Kharrazi S., Vogel W., Urban J., Kulkarni S. K. and Paknikar K. M. (2003). Extracellular synthesis of silver nanoparticles by a silver-tolerant yeast strain MKY3. *Nanotechnology*, **14**, 95–100.

Kuehnel M. F. and Reisner E. (2018). Solar hydrogen generation from lignocellulose. *Angewandte Chemie*, **57**, 3290–3296.

Lens P. N. L. (2021). RESB: 20 years of environmental science and bio/technology for sustainable development. *Reviews in Environmental Science and Bio/Technology*, **20**(1), 1–3.

Lens P., Gastesi R. and Lettinga G. (2003). Use of a cell suspension bioreactor for biological flue gas desulfurisation. *Biodegradation*, **14**, 229–240.

Lens P. N. L., Virkutyte J., Jegatheesan V., Kim S.-H. and Al-Abed S. (2012). Nanotechnology for Water and Wastewater Treatment. IWA Publishing, London, UK, p. 435.

Lenz M. and Lens P. N. L. (2009). The essential toxin: the changing perception of selenium in environmental sciences. *Science of the Total Environment*, **407**(12), 3620–3633.

Lenz M., Hommes G., Corvini Ph. F. X., van Hullebusch E. D. and Lens P. N. L. (2008). Removal of selenate in sulfate reducing and methanogenic upflow anaerobic sludge bed reactors. *Water Research*, **42**(8–9), 2184–2194.

Li J., Ferrer I., Lens P. N. L. and Du Laing G. (2021). Evaluation of selenium-enriched microalgae produced on domestic wastewater as biostimulant and biofertilizer for growth of selenium-enriched crops. *Journal of Applied Phycology* (In press).

Lloyd J. R. and Lovley D. R. (2001). Microbial detoxification of metals and radionuclides. *Current Opinion in Biotechnology*, **12**, 248–253.

Mal J., Nancharaiah Y. V., van Hullebusch E. D. and Lens P. N. L. (2016a). Effect of heavy metal co-contaminants on selenite bioreduction by anaerobic granular sludge. *Bioresource Technology*, **206**, 1–8.

Mal J., Nancharaiah Y. V., van Hullebusch E. and Lens P. N. L. (2016b). Metal Chalcogenide quantum dots: biotechnological synthesis and applications. *RSC Advances*, **6**, 41477–41495.

Mal J., Nancharaiah Y. V., Bera S., Maheswari N., van Hullebusch E. D. and Lens P. N. L. (2017a). Biosynthesis of CdSe nanoparticles by anaerobic granular sludge. *Environmental Science: Nano*, **4**, 824–833.

Mal J., Nancharaiah Y. V., Maheswari N., van Hullebusch E. D. and Lens P. N. L. (2017b). Continuous removal and recovery of tellurium in an upflow anaerobic granular sludge bed reactor. *Journal of Hazardous Materials*, **327**, 79–88.

Mal J., Veneman W., Nancharaiah Y. V., van Hullebusch E. D., Peijnenburg W., Vijver M. G. and Lens P. N. L. (2017c). A comparison of fate and toxicity of selenite, biogenically and chemically synthesized selenium nanoparticles to the Zebrafish (*Danio rerio*) embryogenesis. *Nanotoxicology*, **11**(1), 87–97.

Mal J., Nancharaiah Y. V., Bourven I., Simon S., van Hullebusch E. D., Guibaud G. and Lens P. N. L. (2020). Cadmium selenide formation influences the production and characteristics of extracellular polymeric substances of anaerobic granular sludge.

Applied Biochemistry and Biotechnology, **193**, 965–980. doi: 10.1007/s12010-020-03464-x.

Mandal S., Phadtare S. and Sastry M. (2005). Interfacing biology with nanoparticles. *Current Applied Physics*, **5**, 1127–1218.

Masciangioli T. and Zhang W.-X. (2003). Environmental technologies at the nanoscale. *Environmental Science and Technology*, **37**, 102–108.

Meng X., Yu Q., Liu G., Shi L., Zhao G., Liu H., Li P., Chang K., Kako T. and Ye J. (2017). Efficient photocatalytic CO_2 reduction in all-inorganic aqueous environment: cooperation between reaction medium and Cd(II) modified colloidal ZnS. *Nano Energy*, **34**, 524–532.

Moreno Osorio J. H., Pinto G., Pollio A., Frunzo L., Lens P. N. L. and Esposito G. (2019). Start-up of a nutrient removal system using *Scenedesmus vacuolatus* and *Chlorella vulgaris* biofilms. *Bioresources and Bioprocess*, **6**, Article number: 27. doi.org/10.1186/s40643-019-0259-3

Nancharaiah Y. V. and Lens P. N. L. (2015). Selenium biomineralization for biotechnological applications. *Trends in Biotechnology*, **33**, 323–330.

Navarro R. R., Tatsumi K., Sumi K. and Matsumura M. (2001). Role of anions on heavy metal sorption of a cellulose modified with poly(glycidyl methacrylate) and polyethyleneimine. *Water Research*, **35**, 2724–2730.

Ohlendorf H. M. (2002). The birds of Kesterson Reservoir: a historical perspective. *Aquatic Toxicology*, **57**, 1–10.

Oremland R. S., Herbel M. J., Switzer Blum J., Langley S. and Beveridge T. J. (2004). Structural and spectral features of selenium nanospheres formed by Se-respiring bacteria. *Applied and Environmental Microbiology*, **70**, 52–60.

Palomo J. M. (2019). Nanobiohybrids: a new concept for metal nanoparticles synthesis. *Chemical Communication*, **55**, 9583–9589.

Pearce C. I., Coker V. S., Charnock J. M., Pattrick R. A. D., Mosselmans J. F. W., Law N., Beveridge T. J. and Lloyd J. R. (2008). Microbial manufacture of chalcogenide-based nanoparticles via the reduction of selenite using *Veillonella atypica*: an *in situ* EXAFS study. *Nanotechnology*, **19**, 1–13.

Presser T. (1994). The Kesterson effect. *Environmental Management*, **18**, 437–454.

Qin L., Guo B. D., Yu J. G., Ran J. R., Zhang B. H., Yan H. J. and Gong J. R. (2011). Highly efficient visible-light-driven photocatalytic hydrogen production of CdS-cluster decorated graphene nanosheets. *Journal American Chemical Society*, **133**, 10878–10884.

Rada-Ariza A. M., Fredy D., Lopez-Vazquez C. M., Van der Steen N. P. and Lens P. N. L. (2019). Ammonium removal mechanisms in a microalgal-bacterial sequencing batch photobioreactor at different solids retention times. *Algal Research*, **39**, 101468.

Ralston N. V. C., Blackwell J. L., III and Raymond L. J. (2007). Importance of molar ratios in selenium-dependent protection against methylmercury toxicity. *Biological Trace Element Research*, **119**, 255–268.

Sakimoto K. K., Wong A. B. and Yang P. (2016). Self-photosensitization of nonphotosynthetic bacteria for solar-to-chemical production. *Science*, **351**, 74–77.

Sakimoto K. K., Kornienko N. and Yang P. (2017). Cyborgian material design for solar fuel production: the emerging photosynthetic biohybrid systems. *Accounts of Chemical Research*, **50**, 476–481.

Sekizawa K., Maeda K., Domen K., Koike K. and Ishitani O. (2013). Artificial Z-scheme constructed with a supramolecular metal complex and semiconductor for the photocatalytic reduction of CO_2. *Journal American Chemical Society*, **135**, 4596–4599.

Shanbhag S., Tang Z. and Kotov N. A. (2007). Self-organization of Te nanorods into V-shaped assemblies: A brownian dynamics study and experimental insights. *ACS Nano*, **1**, 126–132.

Sharma S. and Singh R. (1984). Selenium in soil, plant, and animal systems. *Critical Reviews Environmental Control*, **13**, 23–50.

Sheng G.-P., Yu H.-Q. and Li X.-Y. (2010). Extracellular polymeric substances (EPS) of microbial aggregates in biological wastewater treatment systems: a review. *Biotechnology Advances*, **28**, 882–894.

Simmons J., Ziemkiewicz P. and Black C. (2002). Use of steel slag leach beds for the treatment of acid mine drainage. *Mine Water and the Environment*, **21**, 91–99.

Sitko R., Turek E., Zawisza B., Malicka E., Talik E., Heimann J., Gagor A., Feist B. and Wrzalik R. (2013). Adsorption of divalent metal ions from aqueous solutions using graphene oxide. *Dalton Transactions*, **42**, 5682–5689.

Stolz J. F., Basu P., Santini J. M and Oremland R. S. (2006). Arsenic and selenium in microbial metabolism. *Annual Reviews Microbiology*, **60**, 107–230.

Supreet and Singh G. (2020). Recent advances on cadmium free quantum dots-liquid crystal nanocomposites. *Applied Materials Today*, **21**, 100840.

Theron J., Walker J. A. and Cloete T. E. (2008). Nanotechnology and water treatment: applications and emerging opportunities. *Critical Reviews in Microbiology*, **34**, 43–69.

Tourney J. and Ngwenya B. T. (2014). The role of bacterial extracellular polymeric substances in geomicrobiology. *Chemical Geology*, **386**, 115–132.

Tu W., Zhou Y. and Zou Z. (2014). Photocatalytic conversion of CO_2 into renewable hydrocarbon fuels: state-of-the-art accomplishment, challenges, and prospects. *Advanced Materials*, **26**, 4607–4626.

Uskokovic V. (2007). Nanotechnology: what we don't know. *Technology in Society*, **29**, 43–61.

van Fleet-Stalder V., Chasteen T. G., Pickering I. J., George G. N. and Prince R. C. (2000). Fate of selenate and selenite metabolized by *Rhodobacter sphaeroides*. *Applied and Environmental Microbiology*, **66**, 4849–4853.

Veeken A. H. M., de Vries S., van der Mark A. and Rulkens W. H. (2003). Selective precipitation of heavy metals as controlled by a sulfide-selective electrode. *Separation Science and Technology*, **38**, 1–19.

Wadgaonkar S. L., Mal J., Nancharaiah Y. V., Maheshwari N., Esposito G. and Lens P. N. L. (2018). Formation of Se(0), Te(0) and Se(0)-Te(0) nanostructures during simultaneous bioreduction of selenite and tellurite in a UASB reactor. *Applied Microbiology and Biotechnology*, **102**, 2899–2911.

Wen H. and Carignan J. (2007). Reviews on atmospheric selenium: emissions, speciation and fate. *Atmospheric Environment*, **41**, 7151–7165.

White C. and Gadd G. M. (1999). Copper accumulation by sulfate-reducing biofilms. *FEMS Microbiology Letters*, **183**, 313–318.

Xu G., Zeng S., Zhang B., Swihart M. T., Yong K.-T. and Prasad P. N. (2016). New generation cadmium-free quantum dots for biophotonics and nanomedicine. *Chemical Reviews*, **116**, 12234–12327.

Xu M., Tremblay P.-L., Jiang L. and Zhang T. (2019). Stimulating bioplastic production with light energy by coupling *Ralstonia eutropha* with the photocatalyst graphitic carbon nitride. *Green Chemistry*, **21**, 2392–2400.

Yoneyama H. (1997). Photoreduction of carbon dioxide on quantized semiconductor nanoparticles in solution. *Catalysis Today*, **39**, 169–175.

Yong P., Rowson N. A., Farr J. P. G., Harris I. R. and Macaskie L. E. (2003). A novel electrobiotechnology for the recovery of precious metals from spent automotive catalysts. *Environmental Technology*, **24**, 289–297.

Zeng T., Rene E. R., Hu Q. and Lens P. N. L. (2019). Continuous biological removal of selenate in the presence of cadmium and zinc in UASB reactors at psychrophilic and mesophilic conditions. *Biochemical Engineering Journal*, **141**:102–111.

Zhang W.-X. (2003). Nanoscale iron particles for environmental remediation: an overview. *Journal of Nanoparticle Research*, **5**, 323–332.

Zhang H., Liu H., Tian Z., Lu D., Yu Y., Cestellos-Blanco S., Sakimoto K. K. and Yang P. (2018). Bacteria photosensitized by intracellular gold nanoclusters for solar fuel production. *Nature Nanotechnology*, **13**, 900–905.

Zhang W., Ke S., Sun C., Xu X., Chen J. and Yao L. (2019). Fate and toxicity of silver nanoparticles in freshwater from laboratory to realistic environments: a review. *Environmental Science and Pollution Research*, **26**, 7390–7404.

Index

IWA Publishing's authorised EU representative for General Product Safety Regulations is Diane D'Arras, 15 rue Duret, 75116 Paris, France, e-mail: safety@iwap.co.uk.

Printed and bound by CPI Group (UK) Ltd, Croydon, CR0 4YY
12/05/2026
02108863-0004